MODERN ELECTRONICS MATHEMATICS

MODERN ELECTRONICS MATHEMATICS

Richard L. Sullivan

DELMAR PUBLISHERS INC.

COVER PHOTO: Image by Chuck O'Rear; West Light

Delmar Staff
Administrative Editor: Mark W. Huth
Project Editor: Jonathan Plant
Production Editor: Eleanor Isenhart

For information, address Delmar Publishers Inc.,
2 Computer Drive West, Box 15-015,
Albany, New York 12212

Printed in the United States of America
Published simultaneously in Canada
by Nelson Canada,
A Division of International Thomson Limited

10 9 8 7 6 5 4 3 2 1

Library of Congress Cataloging in Publication Data

Sullivan, Richard L.
Modern electronics mathematics.

"Certain portions of this work are from Modern electronics math by Jerrold R. Clifford and Martin Clifford, c1976, Tab Books"—P. iv.
Includes index.
1. Electronics—Mathematics. I. Clifford, Jerrold R. Modern electronics math. II. Title. TK7835.S85 1985 510'.24'6213
84-15564
ISBN 0-8273-2088-4
ISBN 0-8273-2089-2 (instructors guide)

CONTENTS

Preface *ix*

CHAPTER 1 Basic Mathematics for Electronics *1*

UNIT 1–1 INTRODUCTION TO ELECTRONIC CALCULATORS AND BASIC MATHEMATICS *1*
UNIT 1–2 RATIO AND PROPORTION *16*
UNIT 1–3 FRACTIONS, DIRECT AND INVERSE RELATIONSHIPS, AND PERCENTS *24*
UNIT 1–4 TOLERANCES AND SIGNIFICANT FIGURES *38*
UNIT 1–5 PROBLEM REVIEW *43*

CHAPTER 2 The Decimal Number System *47*

UNIT 2–1 DECIMAL SYSTEM FUNDAMENTALS *47*
UNIT 2–2 DECIMAL AND FRACTION CONVERSIONS *55*
UNIT 2–3 ADDITION, SUBTRACTION, MULTIPLICATION, AND DIVISION OF DECIMALS *57*
UNIT 2–4 EXPONENTIAL NOTATION *60*
UNIT 2–5 CONVERSIONS AND ELECTRONIC UNITS OF MEASUREMENT *72*
UNIT 2–6 POWERS AND ROOTS IN ELECTRONICS *80*
UNIT 2–7 PROBLEM REVIEW *87*

CHAPTER 3 Mathematics for Direct Current Circuits *91*

UNIT 3–1 OHM'S LAW AND THE POWER LAW IN SERIES CIRCUITS *91*
UNIT 3–2 PARALLEL AND SERIES-PARALLEL CIRCUITS *103*
UNIT 3–3 KIRCHHOFF'S LAWS *127*

UNIT 3–4 CIRCUIT ANALYSIS THEOREMS *136*
UNIT 3–5 CAPACITORS AND INDUCTORS IN DIRECT CURRENT CIRCUITS *148*
UNIT 3–6 INTRODUCTION TO GRAPHS *161*
UNIT 3–7 PROBLEM REVIEW *171*

CHAPTER 4 Algebra for Electronics *179*

UNIT 4–1 THE LANGUAGE OF ALGEBRA *179*
UNIT 4–2 EXPONENTS AND RADICALS *184*
UNIT 4–3 ADDITION AND SUBTRACTION OF ALGEBRAIC EXPRESSIONS *191*
UNIT 4–4 MULTIPLICATION AND DIVISION OF ALGEBRAIC EXPRESSIONS *194*
UNIT 4–5 SPECIAL PRODUCTS AND FACTORING *199*
UNIT 4–6 ADDITION AND SUBTRACTION OF ALGEBRAIC FRACTIONS *207*
UNIT 4–7 MULTIPLICATION AND DIVISION OF ALGEBRAIC FRACTIONS *210*
UNIT 4–8 LOGARITHMS *212*
UNIT 4–9 PROBLEM REVIEW *233*

CHAPTER 5 Linear and Quadratic Equations *239*

UNIT 5–1 SIMPLE EQUATIONS AND FORMULAS *239*
UNIT 5–2 SIMULTANEOUS LINEAR EQUATIONS *247*
UNIT 5–3 KIRCHHOFF'S LAWS AND SIMULTANEOUS EQUATIONS *259*
UNIT 5–4 QUADRATIC EQUATIONS *270*
UNIT 5–5 GRAPHING LINEAR EQUATIONS *279*
UNIT 5–6 GRAPHING QUADRATIC EQUATIONS *294*
UNIT 5–7 PROBLEM REVIEW *298*

CHAPTER 6 Trigonometry for Electronics *305*

UNIT 6–1 THE LANGUAGE OF TRIGONOMETRY *305*
UNIT 6–2 THE TRIGONOMETRIC FUNCTIONS *318*
UNIT 6–3 RIGHT TRIANGLES *332*
UNIT 6–4 IMAGINARY AND COMPLEX NUMBERS *337*
UNIT 6–5 PROBLEM REVIEW *350*

CHAPTER 7 Trigonometry for Alternating Current Circuits in Series *353*

UNIT 7–1 THE SINE WAVE *353*
UNIT 7–2 VECTORS AND PHASORS *369*
UNIT 7–3 ALTERNATING CURRENT CIRCUITS IN SERIES *383*
UNIT 7–4 PROBLEM REVIEW *423*

CHAPTER 8 Trigonometry for Alternating Current Circuits in Parallel *429*

UNIT 8–1 ALTERNATING CURRENT CIRCUITS IN PARALLEL *429*
UNIT 8–2 PROBLEM REVIEW *445*

CHAPTER 9 Computer Mathematics: Binary Number System *449*

UNIT 9–1 THE BINARY NUMBER SYSTEM *449*
UNIT 9–2 BINARY ADDITION AND SUBTRACTION *460*
UNIT 9–3 BINARY MULTIPLICATION AND DIVISION *467*
UNIT 9–4 PROBLEM REVIEW *471*

CHAPTER 10 Computer Mathematics: Octal and Hexadecimal Numbering Systems *473*

UNIT 10–1 THE OCTAL NUMBER SYSTEM *473*
UNIT 10–2 THE HEXADECIMAL NUMBER SYSTEM *477*
UNIT 10–3 BINARY-OCTAL-HEXADECIMAL CONVERSIONS *484*
UNIT 10–4 PROBLEM REVIEW *491*

CHAPTER 11 Computer Mathematics: Boolean Algebra *495*

UNIT 11–1 INTRODUCTION TO LOGIC GATES *495*
UNIT 11–2 BOOLEAN EXPRESSIONS *506*
UNIT 11–3 BOOLEAN ALGEBRA FUNDAMENTALS *518*
UNIT 11–4 PROBLEM REVIEW *529*

APPENDIX 532

TABLE 1 ELECTRONICS ABBREVIATIONS 532
TABLE 2 CONVERSION FACTORS 533
TABLE 3 RESISTOR COLOR CODE 534
TABLE 4 POWERS OF TEN 534
TABLE 5 THE GREEK ALPHABET 534
TABLE 6 LOGARITHMS 535
TABLE 7 TRIGONOMETRIC FUNCTIONS 536

Glossary 540

Answers to Odd-Numbered Problems 549

Index 566

PREFACE

Modern Electronics Mathematics is designed to provide a strong mathematics background for those pursuing the study of electronics. The areas of mathematics most needed by the modern electronics technician are presented. By studying the various mathematical topics relating to electronics, the technician will recognize the relationship between mathematics and electronics. Because the concepts, theorems, and laws of electronics are founded on mathematical principles, a true understanding of electronics can be obtained only through an understanding of mathematics.

The content of this text parallels the electronics content typically covered in a direct current course, an alternating current course, and an introductory solid state course.

Organization

Modern Electronics Mathematics is divided into chapters and units. Each unit contains an exercise based on the content presented within that unit. The final unit in each chapter contains review problems taken from all the units covered in that chapter.

A brief description of the content of each chapter follows:

Chapter 1 introduces electronic calculators. In addition, this chapter contains a review of basic mathematics and includes topics such as ratio, proportion, fractions, percents, tolerances, and significant figures.

Chapter 2 covers the decimal number system, exponential notation, common units of measurement used in electronics, powers, and roots.

Chapter 3 focuses on Ohm's Law and the power law in series and parallel circuits. It also presents Kirchhoff's Laws, circuit analysis theorems, capacitors and inductors in direct current circuits, and an introduction to graphing.

Chapter 4 covers exponents, radicals, operations with algebraic expressions, factoring expressions, and logarithms.

Chapter 5 includes formulas as well as linear, quadratic, and simultaneous equations and the graphing of linear and quadratic equations.

Chapter 6 focuses on the trigonometric functions, right triangles, and imaginary and complex numbers.

Chapter 7 includes the sine wave, phasors, and computing values for quantities in alternating current circuits in series.

Chapter 8 includes computing values for quantities in alternating current circuits in parallel.

Chapter 9 introduces the binary number system including addition, subtraction, multiplication, and division of binary numbers.

Chapter 10 introduces octal numbers, hexadecimal numbers, and binary-octal-hexadecimal conversions.

Chapter 11 introduces logic gates, Boolean expressions, and Boolean algebra fundamentals.

Special Features

Modern Electronics Mathematics contains a number of special features to assist the reader in acquiring the necessary mathematical skills to be successful in the field of electronics.

- **Objectives**—Each unit contains objectives. The objectives alert the reader to the important concepts and procedures contained within the unit and indicate the desired behavior upon completion of the unit.
- **Calculator Tips**—Throughout the text the reader will find calculator tips demonstrating how to solve complex electronics problems using an electronic calculator.
- **Review Problems**—Problems are found at the end of each unit and at the conclusion of the chapter. These problems afford the reader an opportunity to apply the content contained within the unit or chapter before moving on to new material.
- **Examples**—Numerous examples are found throughout the text. Examples of procedures ensure an understanding of the content.
- **Illustrations**—Figures and tables are found throughout the text to supplement the text materials.
- **Key Equations**—A list of key equations is found inside the front and back covers with corresponding text pages. These key equations appear in color in the text when first introduced.
- **Appendix**—The appendix contains tables of electronics abbreviations, conversion factors, resistor color codes, powers of ten, the Greek alphabet, logarithms, and trigonometric functions.
- **Glossary**—A glossary has been included to provide a quick reference to unfamiliar terms.
- **Answers to the Odd-Numbered Problems**—These answers provide a quick check for the odd-numbered problems.

To the Reader

Modern Electronics Mathematics was written for you. The following hints are presented to help you master the content contained between these covers. Please read each item carefully and good luck in your study of electronics mathematics.

1. Read carefully! Be sure that you understand the content and/or problem.
2. When applicable, be sure that all similar values in a given problem are expressed in the same units of measure.
3. When a schematic is not given, it may be helpful to draw and label the circuit.
4. Look carefully at all subscripts. For example, current number one is written I_1 while the voltage across resistor number four is written E_{R_4}.
5. Be sure to use the appendix until you have gained a thorough knowledge of the numerous abbreviations and terms.
6. Many problems in this text require an electronic calculator. Prior to selecting an electronic calculator, it may be helpful to read Unit 1–1.
7. Both English and metric units of measure are used in this text. Refer to the appendix to answer any questions on units of measure. Metric spacing of numbers is used throughout the text. For example, 16 800 volts or 0.025 65 ampere.
8. The answers to odd-numbered problems are given at the end of the text. Check these answers as you work through each unit.
9. The direction of current flow in this text is from negative to positive, or electron flow.
10. In order for you to master mathematics for electronics, it will be necessary for you to practice, practice, practice!! Take advantage of the numerous practice problems to ensure that you totally understand the content.

Acknowledgments

I would like to thank Jerrold R. Clifford and Martin Clifford for the use of certain portions of *Modern Electronics Math* and also Joseph De Guilmo for his permission to use materials from *Electricity/Electronics: Principles and Applications.* I would also like to thank the reviewers who have helped to develop this text. These reviewers include Ray Chandos of Saddleback Community College; Fred Fisher of North Seattle Community College; Joel W. Griffin formerly of TSTI—Amarillo; Charles H. Hollins of El Camino College; David Nelson of Western Wisconsin Technical Institute; and John Tyler of National Education Center—Dallas.

I am very grateful to my best friend Jerry, and to H. Pierce, and G. Marx whose humor kept me going during the preparation of Modern Electronics Mathematics. Finally, I must thank my wife, Emily, and our sons Bryan and Shawn for their support, patience, and encouragement.

CHAPTER 1

BASIC MATHEMATICS FOR ELECTRONICS

Welcome to the world of modern electronics! You are entering an exciting, rapidly changing field that involves a variety of mathematical concepts and operations. This textbook will help prepare you for a successful career in electronics.

UNIT 1·1 INTRODUCTION TO ELECTRONIC CALCULATORS AND BASIC MATHEMATICS

Objectives:

After studying this unit, you should be able to

- select an appropriate electronic calculator.
- solve basic mathematical operations using an electronic calculator.

Selecting a Calculator

In the past, technicians and engineers used slide rules to perform calculations involving vacuum tubes. In 1948, the announcement was made that Bell Laboratories had developed a new device called a *transistor*. The ensuing decades ushered in a myriad of solid-state devices, discrete and integrated circuits, and microprocessors. The modern electronics technician now performs mathematical computations with computers and electronic calculators.

Selecting an appropriate calculator to help in your study of electronics is extremely important. There are many types and models of calculators on the market. Which one is right for you? Which functions and features will be the most useful? The following guidelines will be helpful as you select a calculator.

- Size is an important factor to consider. The student wishing to carry a calculator to class may shop for one that will fit in a pocket. An engineer may want a larger, desk-top model. When your calculator is not in use it should be stored in an appropriate case.
- Check the calculator's power requirements and power options. Some units may operate on nonrechargeable batteries for thousands of hours of operation. Rechargeable battery packs are also available. A power-saving feature is the automatic turn-off function which removes the display after a short period of nonuse. Similarly, after a few minutes of nonuse the calculator will automatically shut off.
- The two most common types of display readouts are light emitting diodes (LED) and liquid crystal displays (LCD). Try both types in a variety of lights to see which you prefer.
- Some of the first generation electronic calculators could not follow the rules of algebra. Check the calculator to ensure that it operates with an algebraic method of entry.
- Most of the available electronic calculators are capable of storing numbers in a memory. In some cases, these numbers are cleared when the power is off. Other models have a *constant memory* which will retain these numbers when the power is off. This constant memory is extremely useful and should be seriously considered when selecting a calculator.
- Calculators are often *programmable.* This mode of operation allows the user to design a specific program within the memory of the calculator. The owner's manual will contain directions for programming.
- Every calculator has the capability of performing specific operations or functions. These are executed by depressing the appropriate *keys.* Which functions should you look for? Table 1–1 lists the functions required in this text. Also identified are the typical key symbols for each function. Each of the functions will be explained within this text. It should be pointed out that most scientific calculators are designed to perform these functions and many more.
- Be sure to try a few different models before making your final selection. Also, ask other students, faculty, and colleagues for their recommendations. Once you have purchased an electronic calculator it is very important that you thoroughly study the owner's manual.
- The final considerations are cost and warranty. The total cost of the calculator and all accessories should be within your budget. In addition, read all available information to ensure that you understand the duration and coverage of the warranty. Finally, be sure that you understand your responsibilities in terms of sales receipts, original packaging, and repair procedures.

Take care of your calculator. As you work through this text the two of you will become inseparable!

TABLE 1–1

Typical Key Symbols and Functions for a Scientific Calculator

Key(s)	Function(s)
[+], [−], [×], [÷], [=]	Basic arithmetic
[+/−]	Change sign
[π]	Pi
[(], [)]	Parentheses
[EE]	Scientific notation
[Eng]	Engineering notation
[STO], [RCL], [EXC]	Memory or memories
[x^2], [$\sqrt{x}$]	Square and square root
[y^x]	*y* to the *x* power
[1/x]	Reciprocal
[%]	Percent
[log], [ln x]	Logarithm and natural logarithm
[DRG]	Degrees, radians, and graduations
[sin], [cos], [tan]	Trigonometric functions
[P↔R]	Polar and rectangular conversions
[INV]	Inverse function

Basic Mathematical Operations

The scope of this text assumes previous knowledge in the area of basic mathematics. However, the following tips are provided to afford you an opportunity to become familiar with your calculator and to practice using the keys to solve simple mathematics problems. As these tips are general in nature, it may be necessary to refer to your owner's manual for specific sequences.

Calculator Tip

Most electronic calculators will automatically add numbers as they are entered. The sequence to add 6, 17, 42, 9, and 20 would be as follows:

6 [+] 17 [+] 42 [+] 9 [+] 20 [=] 94

When positive and negative numbers appear in the problem, it is necessary to inform the calculator that a number is negative or that a number is to be subtracted. Indicating that a number is negative may be accomplished through use of the change sign key [+/−]. To indicate that a number is to be subtracted, the subtract key [−] is used. The change sign key is used *after* a number is entered and will change the sign each time the key is depressed. The subtract key is used *prior* to entering a number and indicates that the next number entered is to be subtracted.

The sequence to add −32, 15, 72, and −104 may be completed as follows:

32 [+/−] [+] 15 [+] 72 [+] 104 [+/−] [=] −49

Calculator Tip

In order to subtract signed numbers, it is necessary to inform the calculator which numbers are negative, and do so in the proper sequence. While this sequence may vary with different calculators, some typical examples might include:

Simplify: $-37 - (-14) - 12$

Solution: [−] 37 [−] [(] 14 [+/−] [)] [−] 12 [=] −35

Simplify: $-(-19) + (-46) - (27)$

Solution: [−] [(] 19 [+/−] [)] [+] [(] 46 [+/−] [)] [−] [(] 27 [)] [=] −54

Calculator Tip

The multiply key [×] instructs your calculator to multiply the displayed value by the next quantity to be entered. The sequence to multiply 14 and 37 would be as follows:

14 [×] 37 = 518

The sequence to multiply 8, 11, and 123 would be as follows:

8 [×] 11 [×] (88 now displayed) 123 [=] 10 824

In order to multiply signed numbers it is necessary to inform the calculator which numbers are negative, and do so in the proper sequence. While this sequence may vary with different calculators, a typical example might be the multiplication of −9 and 7.

[−] 9 [×] 7 [=] −63, or

9 [+/−] [×] 7 [=] −63

The sequence to multiply −12 and −18 would be as follows:

[−] 12 [×] 18 [+/−] [=] 216, or

12 [+/−] [×] 18 [+/−] [=] 216

Calculator Tip

In order to complete division problems, it is necessary to use the divide key [÷]. Also, it is necessary to inform the calculator which numbers are signed numbers, and do so in the proper sequence. While this sequence may vary with different calculators, a typical example might be the division of 63 by −9.

63 [÷] 9 [+/−] [=] −7

The sequence to divide −273 by −21 would be as follows:

273 [+/−] [÷] 21 [+/−] [=] 13

Combined Mathematical Operations

The four basic functions, addition, subtraction, multiplication, and division, are used to solve many problems in electronics. Even when problems involve more advanced forms of mathematics, some portion will include these four operations, either alone or in conjunction with advanced mathematical functions. The solution to problems in electronics not only requires expertise but also an ability to understand and confidently use the more elementary forms covered in this chapter.

When more than one of the four arithmetic functions appear in a problem, a specific sequence of operations must be followed. Operations within grouping signs (parentheses and brackets) are completed first. In terms of the four functions, multiplication and division operations are performed first, in order from left to right. Operations involving addition and subtraction are then performed, in order from left to right.

Example: Simplify $12 \div (3 \times 2) + 5 - 2 \times 2$.

Solution:

$$12 \div (3 \times 2) + 5 - 2 \times 2$$
$$12 \div 6 + 5 - 4 =$$
$$2 + 5 - 4 =$$
$$= 3$$

Example: Simplify $\dfrac{83 - (-7 \times 2) + 3 \times 6}{-32 + 15(16 \div 8) + 7}$.

Solution: Simplify the upper expression.

$$83 - (-7 \times 2) + 3 \times 6 =$$
$$83 - (-14) + 18 =$$
$$83 + 14 + 18 = 115$$

Simplify the lower expression.

$$-32 + 15(16 \div 8) + 7 =$$
$$-32 + 15(2) + 7 =$$
$$-32 + 30 + 7 = 5$$

Divide the expressions.

$$115 \div 5 = 23$$

Calculator Tip

The majority of modern electronic calculators are programmed to perform mathematics functions in the correct sequence. To check your calculator enter 8 [+] 2 [×] 7 [=]. Based upon convention, the result is 22 since the calculator correctly multiplies prior to performing the addition. In a problem such as $48 \div 6 \times 4$, convention dictates that the division occurs first. This results in $48 \div 6$ or 8. The 8 is then multiplied by the 4 to produce 32. Your calculator should give the same result.

In order to complete complex problems it may be necessary to utilize the memory function of your calculator. While some calculators are capable of storing many different numbers, the following example will assume the capability of storing only one number at a time.

Simplify: $-18(32 - 7)(14 \div -2) + \dfrac{3(-7 \times 2)}{-2}$

Solution: [−] 18 [×] [(] 32 [−] 7 [)] [×]

[(] 14 [÷] 2 [+/−] [)]

[=] [STO] 3 [×] [(] 7 [+/−] [×] 2 [)]

[=] [÷] 2 [+/−] [+] [RCL] [=] 3 171

EXERCISE 1·1

The following problems are provided for both the review and practice of basic mathematics. Note that a variety of electronic symbols, circuits, and formulas are used. Previous knowledge regarding these will not be necessary to solve the addition, subtraction, multiplication, and division problems. The purpose for introducing these at this point is to help develop an acquaintance with terms that will be with you throughout your career. In the event of questions concerning symbols, units of measurement, abbreviations, or terms, it may be helpful to refer to the appropriate appendices in this text.

1. Add 4 V, 37 V, 19 V, and 24 V.
2. Add 3 A, −6 A, 7 A, and −8 A.
3. Five resistors are connected in series with the total resistance (R_t) being the sum of the individual resistances. Find R_t if the values of the individual resistors are 3 900 Ω, 560 Ω, 47 Ω, 1 800 Ω, and 10 000 Ω.
4. Add −137, −62, 483, and −209.
5. Four coils are connected in series with no mutual inductance. What is the total inductance (L_t) if L_1 = 200 mH, L_2 = 150 mH, L_3 = 1 500 mH, and L_4 = 15 mH? ($L_t = L_1 + L_2 + L_3 + L_4$)
6. Add the following frequencies: 175 823 Hz, 682 987 Hz, 19 872 Hz, and 450 000 Hz.

Use Figure 1–1 for problems 7–10.

7. Calculate total resistance (R_t).
8. Compute the output voltage of the power supply (E_s).
9. What would be the value of R_t if R_1 were increased to 5 700 Ω?
10. With R_1 = 5 700 Ω, the approximate voltages in the circuit are E_1 = 37 V, E_2 = 3 V, and E_3 = 8 V. What is the value of E_s?

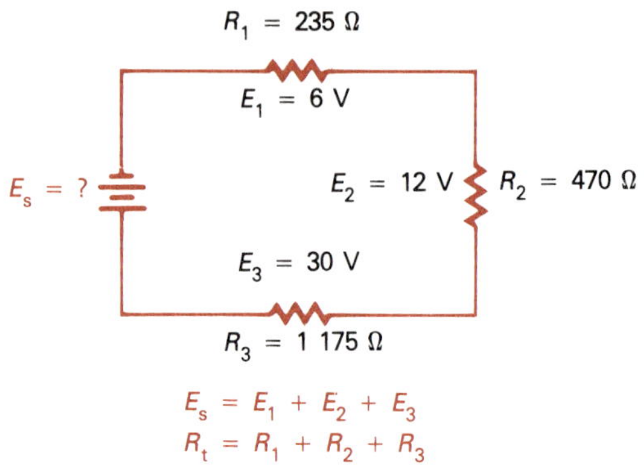

FIGURE 1–1

11. A power supply contains five printed circuit boards. Separately the boards contain 72, 104, 89, 89, and 114 components. What is the total number of components?
12. Kirchhoff's Current Law states that the sum of currents into a point (or node) equals the current out of the point. What is the current out of the node in Figure 1–2?

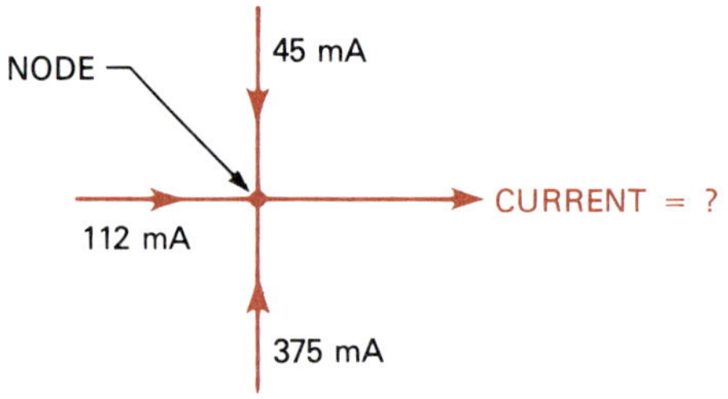

FIGURE 1–2

13. A delivery van travels 18 kilometres, 27 kilometres, 6 kilometres, and 32 kilometres. How far has the van traveled?
14. Find total capacitance (C_t) in Figure 1–3.

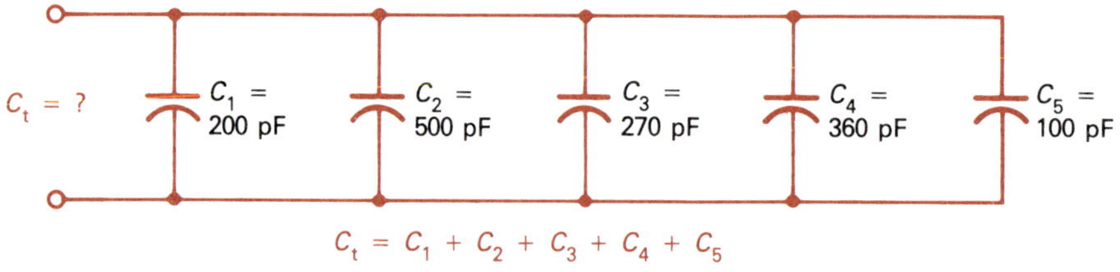

FIGURE 1–3

15. The peak-to-peak voltage of an alternating waveform is found by adding the positive and negative peak voltages. Find the peak-to-peak voltage in Figure 1–4.

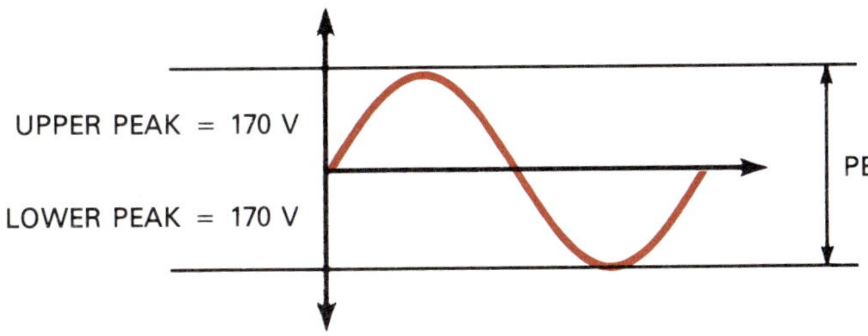

FIGURE 1–4

16. Multiply these numbers.
    a. 12 × 15
    b. (163)(−13)
    c. (35 V)(6)
    d. −18 × −42
    e. (−4 A)3
    f. 20 W × 131
    g. (−16)(37)(−4)
    h. 41(8 A)(11)
    i. −623 × 238 × 29
    j. (−42)(−42)(−42)(−42)
17. Multiply these numbers.
    a. 143 × 8
    b. (82)(17 V)
    c. 238(−21)
    d. (60 Hz)(110)
    e. (−55)(3)(−46)
    f. (11 mH)111
    g. −75 × 15 × 35
    h. (61)(−19)(2)(−19)
18. A parts delivery van travels at 85 kilometres per hour. How far has it traveled after four hours?
19. The current gain of a common emitter transistor amplifier is considered a constant. To find the collector current (I_c) multiply the current gain (β) times the base current (I_b). Find the collector current in microamperes (μA) in Figure 1–5.

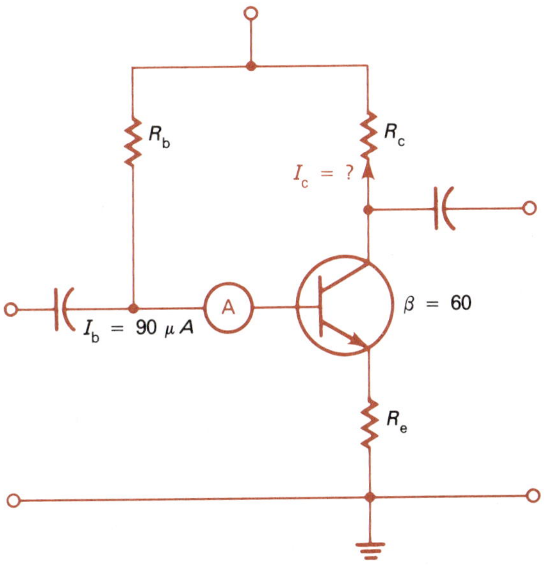

FIGURE 1–5

20. Find the collector current (I_c) when the current gain (β) is 75 and the base current (I_b) is 120 μA.
21. Ohm's Law states that voltage (E) in volts equals current (I) in amperes times resistance (R) in ohms. What is the voltage drop across a 22-ohm resistor if it is conducting 3 amperes of current?
22. Find the voltage drop (E) in volts across a 5-ohm resistor (R) if it is conducting 25 amperes of current (I).
23. Power, the rate of doing work, is measured in watts. Power (P) in watts is equal to the voltage (E) in volts across a device times the current (I) in amperes through the device. Find the power dissipated by the load in Figure 1–6.

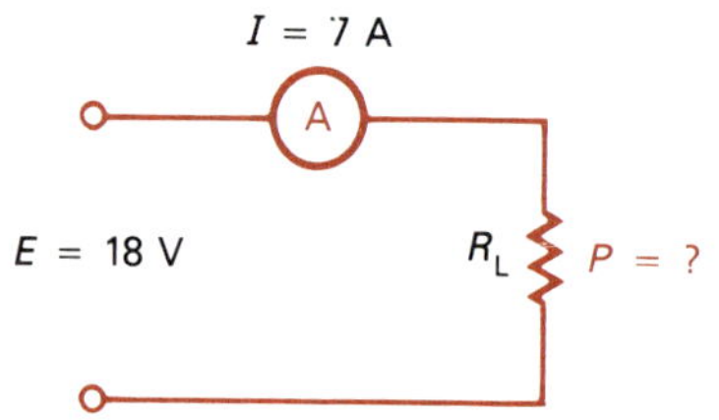

FIGURE 1–6

24. Find the power in watts dissipated by a load if it has a voltage drop of 66 volts and a current of 3 amperes passing through it.
25. A 12-volt battery supplies 35 amperes of current to a starter motor. What is the power in watts dissipated by the motor?
26. A certain type of wire has a resistance of 3 ohms per 1 000 feet. Find the total resistance of 27 000 feet of this wire.
27. An oscillator generates a signal at a frequency determined by its components. An oscillator's output is 455 000 cycles each second (hertz). How many cycles does it produce in 12 seconds?
28. The output voltage of an amplifier is found by multiplying the input voltage by the amplifier's voltage gain. Find the output voltage in millivolts (mV) in Figure 1–7.

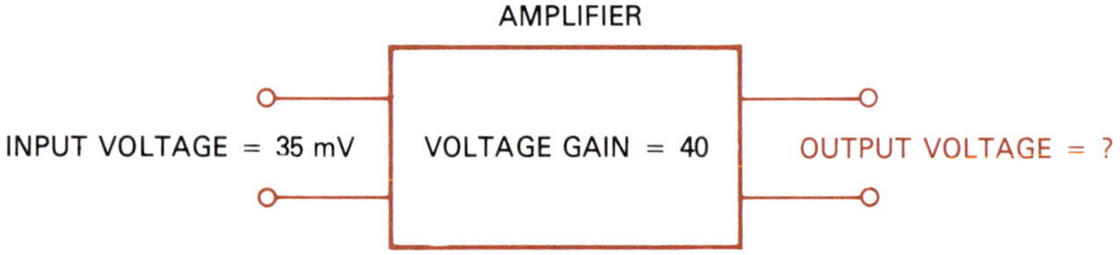

FIGURE 1–7

29. A 10 to 1 probe is attached to a voltmeter to read larger than normal voltages. The actual voltage will be 10 times that shown on the meter. The meter reads 123 volts. What is the actual voltage being measured?
30. A technician repairs an average of 33 units per week. How many could be repaired during a 12-week period?

31. Simplify these expressions.
    a. 42 V − (−18 V)
    b. −367 − (−283)
    c. (38 A) − (24 A)
    d. 21 − (−74) − (13)
    e. 427 V − 186 V − (−328 V)
    f. −62 − 50 − (−180) − 7
    g. 75 mA − (58 mA) − 75 mA
    h. −(−98 mV) − (−426 mV)
32. Simplify these expressions.
    a. 69 − (14) − (55)
    b. 1 028 V − 328 V − (−230 V)
    c. −(−107 mA) − (−81 mA)
    d. 19 A − 16 A − 21 A − (−32 A)
    e. −187X − 293X − (−371X)
    f. −287 mV − (33 mV) − (−321 mV)
    g. 62 − (−17) − (−39) − 438
    h. −(−143 V) − 89 V − (−37 V)
33. The power source in Figure 1–8 supplies 180 amperes of current to two cabinets of equipment. How many amperes of current does cabinet B conduct?

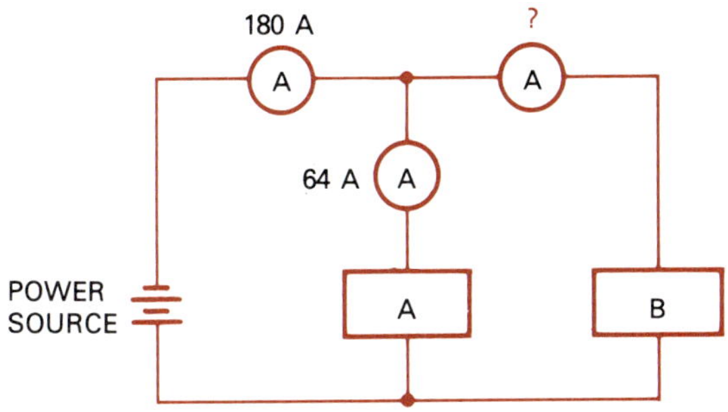

FIGURE 1–8

34. A technician has 887 integrated circuits in stock. During a four-week period 32, 18, 54, and 38 integrated circuits were used or sold. How many remain in stock?
35. An electronic mixer can produce the difference between two input frequencies. The input signals are 850 000 hertz and 1 305 000 hertz. Find the difference frequency.

Use the series circuit in Figure 1–9 for problems 36–38.

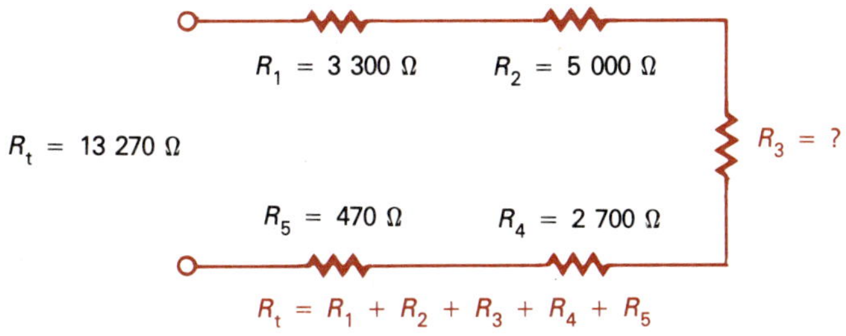

FIGURE 1–9

36. Find the value of R_3.
37. If R_4 were to be replaced by a short (zero ohms), what would be the value of R_t?
38. The total resistance (R_t) is increased to 15 000 ohms. Find the value of R_3.
39. An antenna must be 35 metres high. A 10-metre section and a 12-metre section are used. How long should the third section be to complete the antenna?
40. The peak-to-peak voltage of the square wave in Figure 1–10 is 42 volts. Find the value of the upper portion of the wave.

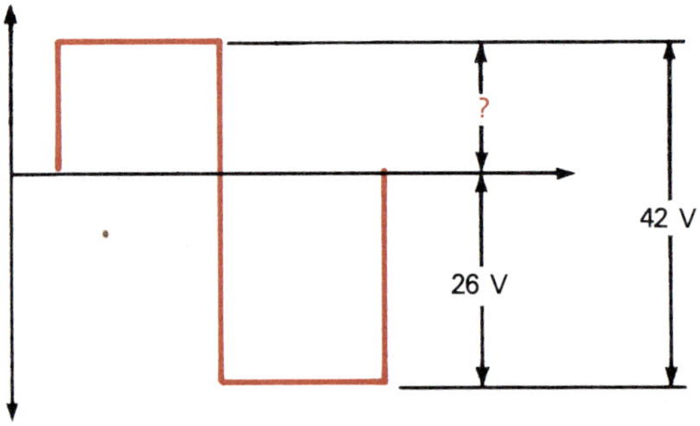

FIGURE 1–10

41. A cable television company has 125 000 feet of wire to connect the houses of new customers. During a four-week period 8 729 feet, 11 283 feet, 10 075 feet, and 9 830 feet of wire are installed. How many feet of wire remain?
42. A power supply is designed to produce 400 volts. It is set to 293 volts. How many more volts can it produce?

Use the series circuit in Figure 1–11 for problems 43–44.

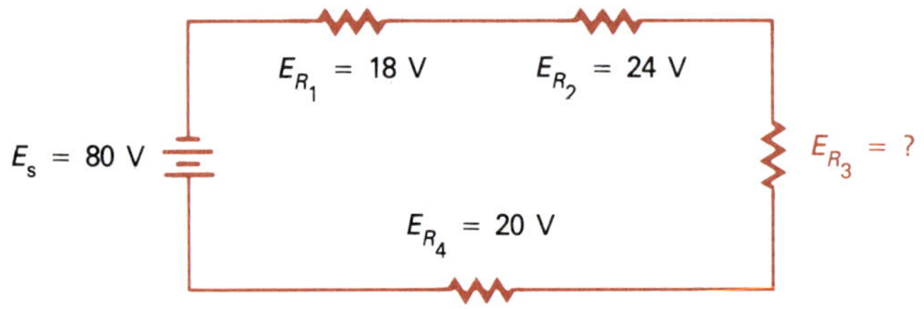

FIGURE 1–11

43. Kirchhoff's Voltage Law states that the sum of the voltages around a closed loop equals the source voltage (E_s). Find the value of E_{R_3}.
44. The resistors are replaced, making $E_{R_1} = 28$ V, $E_{R_2} = 6$ V, and $E_{R_3} = 11$ V. Find the value of E_{R_4}.
45. Divide these quantities.

a. 87 W ÷ 3

b. $\frac{-1\,680}{21}$

c. 1 818 g ÷ 18

d. 552 A ÷ 2

e. $\frac{56\,763}{-159}$

f. 576 V ÷ 48

g. 3 287 Ω ÷ 19
h. 10 368 ÷ 81
i. 258 324 ÷ 627
j. −1 431 ÷ −9

46. Divide these quantities.
a. 168 hours ÷ 7
b. $\frac{68\ 000}{32}$
c. 1 222 mA ÷ 94
d. 725 289 ÷ −577
e. 7 663 ÷ 79
f. 42 000 ÷ −16
g. 744 cm ÷ 31 cm
h. 576 μA ÷ 64 μA
i. 100 000 ÷ 6 250
j. 26 230 Hz ÷ 215

47. Current gain (β) of a common emitter circuit equals the output current (I_c) divided by the input current (I_b). Use Figure 1–12 to find β.

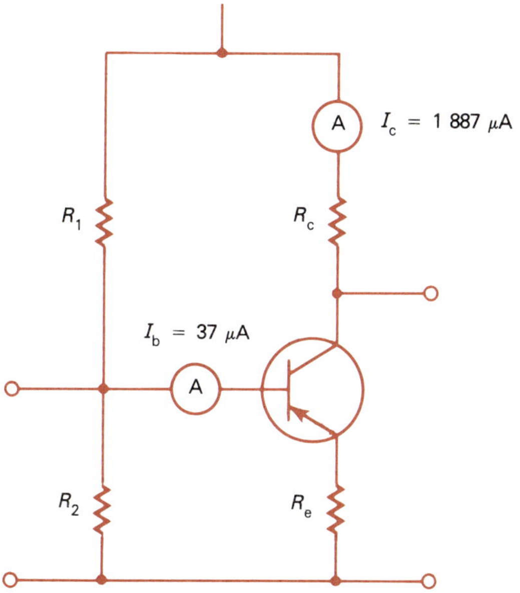

FIGURE 1–12

48. A company manufacturers 4 250 power supplies. The total income is $165 750. Find the price of each supply.
49. When batteries are connected in series the total voltage is the sum of the individual voltages. How many 24-volt batteries are needed to obtain 312 volts?
50. The voltage gain of an amplifier is the output voltage divided by the input voltage. The input is 45 millivolts and the output is 2 025 millivolts. Find the gain.
51. To assemble a stereo kit requires 42 hours. Working 3 hours each day, how many days does it take to assemble the stereo?

The following symbols and equations are used in problems 52–58.

E = voltage in volts
I = current in amperes
R = resistance in ohms
P = power in watts

$I = E \div R$ $\quad R = E \div I$ $\quad I = P \div E$ $\quad E = P \div I$

52. The resistor in Figure 1–13 conducts 4 amperes of current. Find the resistance in ohms.

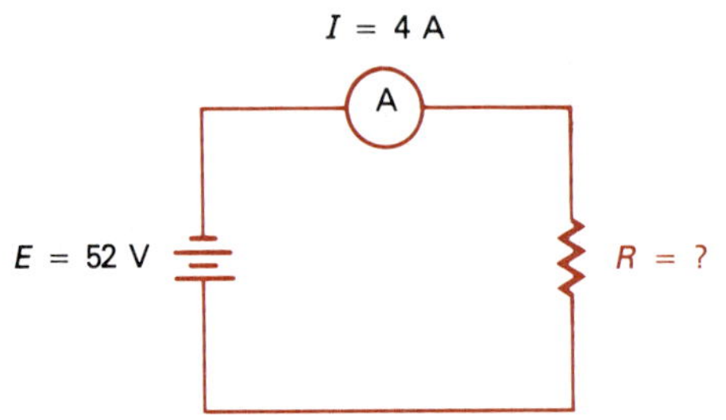

FIGURE 1–13

53. A 15-ohm load is connected to 180 volts. Find the current in amperes.
54. A 2-ohm resistor is dissipating 72 watts of power. Find the current in amperes if it is connected to 12 volts.
55. An 80-watt load is conducting 4 amperes of current. Find the applied voltage if the load resistance is 5 ohms.
56. A motor, designed to operate at 12 volts, dissipates 60 watts of power. Find the current this motor conducts in amperes.
57. The resistor in Figure 1–14 is connected to 48 volts. Find the current in amperes.

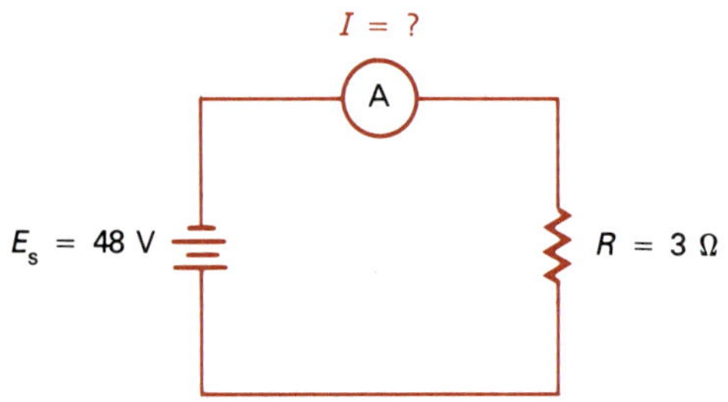

FIGURE 1–14

58. A load dissipates 100 watts of power. Find the applied voltage if the load current is 1 ampere.
59. Simplify these problems.
    a. $-19(35 - 47) - 64 \div 8$
    b. $185 \div 5 + 7 - 2 \times 6 + 3$
    c. $-25 - (-7 \times 2) + (13 + 27) \div (-4)$
    d. $[92 \div (12 - 8)]13 - 27$
    e. $[37 - (-3)(16)] \div (35 \div 7)$
    f. $[45(18 + 2 \times 3)][169 - (-37 \times 2)]$
    g. $-110 - (-13 - 27) - [-82(-10)]$
    h. $\dfrac{568 - 25(-3) + 2}{11 - (-32)}$

i. $\left[\dfrac{-248 - (-1\ 078 \times 2)}{-825 \div (-5) + 936 \div 3}\right]25$

j. $(18 - 25 \times 2)\left(\dfrac{397 - (-13) + 102}{(-2)(2)(-2)(-2)(-2)}\right)$

60. Simplify these problems.
    a. $-73 - (-19) + 27(2 - 11) + 14$
    b. $4\ 083 + (299 - 3\ 078) + 14 \times 10$
    c. $37 - (-37) - 37(37 + 37)$
    d. $[-142 \div (13 - 15)][19(-19)]$
    e. $(208 \div 4)16 - 10$
    f. $\dfrac{82 + 17}{6 + (-9)} \div [8 - (-3)]$
    g. $429 \div (12 \div 4) - (-47 \times 6)$
    h. $\left[\dfrac{75 \times 9 - 327}{-42 - (-7)} \times 35\right] - (-99 \div 3)$
    i. $(97 \times 6)(13 - 61) - (-31\ 084)$
    j. $[-628(-284 + 316)] \div (37 \times 4 + 9)$

The following symbols and equations are used in problems 61–66.

E = voltage in volts　　R = resistance in ohms
I = current in amperes　　P = power in watts

$$E = I \times R \qquad P = I \times E$$
$$I = E \div R \qquad I = P \div E$$
$$R = E \div I \qquad E = P \div I$$

61. The current passing through a resistor is 2 amperes. Find the resistance in ohms if the applied voltage is 48 volts.
62. A 7-ohm load is connected to a 35-volt source and dissipates 175 watts of power. Find the current.
63. The load in Figure 1–15 dissipates 36 watts of power. Calculate the load resistance and the source voltage.

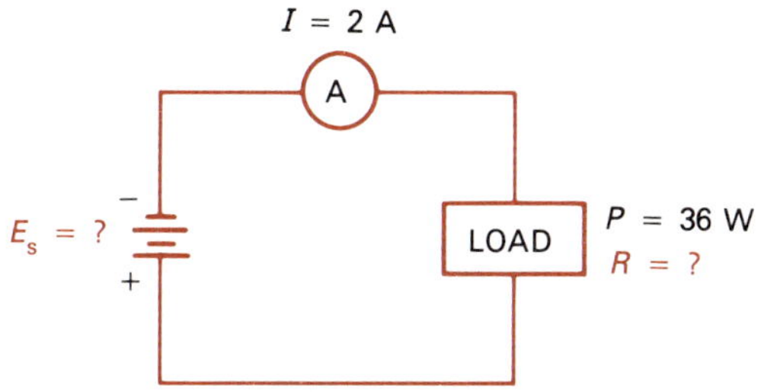

FIGURE 1–15

64. A 22-ohm resistor has 1 ampere of current passing through it. Find the voltage drop across the resistor.

65. A load is connected to 20 volts and dissipates 100 watts of power. Calculate the amount of current and load resistance.
66. The resistance in Figure 1–16 conducts 15 amperes of current. Find both the value of this resistance and the power it dissipates.

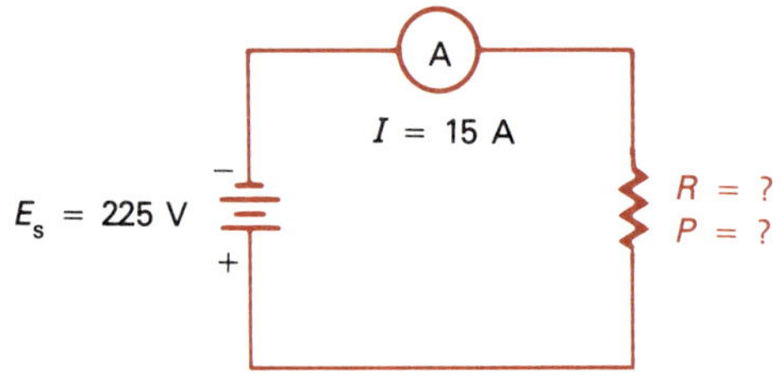

FIGURE 1–16

UNIT 1·2 RATIO AND PROPORTION

Objectives:

After studying this unit, you should be able to

- set up and simplify ratios.
- set up proportions and solve for the missing values.
- discuss electronic applications of ratio and proportion.

Ratio

Problems in electronics involve number relationships. In a network consisting of R_1 and R_2 in series, both resistors are related to the total resistance. If R_1 is 200 ohms and R_2 is 800 ohms, then the overall resistance is 1 000 ohms. However, R_2 has four times the value of R_1. We can express this relationship in one form of division known as a *ratio*. The ratio of R_1 to R_2 is

$$\frac{R_1}{R_2} = \frac{200\ \Omega}{800\ \Omega} = \frac{1}{4}$$

The ratio of these two resistors, or any other pair of values, can be expressed in four ways:

$$1 \text{ to } 4;\ 1{:}4;\ 1/4;\ \text{or } \frac{1}{4}$$

These are all forms of division, and that is just what a ratio is, a division. The numerator of the ratio is referred to as the *antecedent* (or we could call it the dividend) while the denominator is known as the *consequent* (or we could call it the divisor). In the following example, 8 is the numerator and 4 is the denominator.

Example: What is the ratio of 8 to 4?

Solution: $\frac{8}{4} = \frac{2}{1}$

The ratio of 8 to 4 is 2:1.

As indicated in the previous example, ratios are reduced to their lowest possible terms. This is accomplished by dividing both the numerator and denominator by a number such that the two quotients are whole numbers. This is repeated until the ratio is an expression of two numbers that share no common divisor. At this point the ratio has been reduced to lowest possible terms.

Example What is the ratio of 70 volts to 20 volts?

Solution: $\frac{70 \text{ volts}}{20 \text{ volts}} = \frac{70 \div 2}{20 \div 2} = \frac{35}{10} = \frac{35 \div 5}{10 \div 5} = \frac{7}{2}$ or 7:2

Example: What is the ratio of 65 to 195?

Solution: $\frac{65}{195} = \frac{65 \div 5}{195 \div 5} = \frac{13}{39} = \frac{13 \div 13}{39 \div 13} = \frac{1}{3}$ or 1:3

Proportion

A *proportion* is a statement of equality between two ratios. For example,

$$\frac{8}{4} = \frac{2}{1}$$

is a proportion since the two ratios are equal. This proportion could also be written as 8:4 = 2:1. In any proportion, the first and last terms are known as the *extremes;* the second and third terms are called the *means,* as shown in Figure 1–17. This proportion is typically read as 8 is to 10 as 4 is to 5.

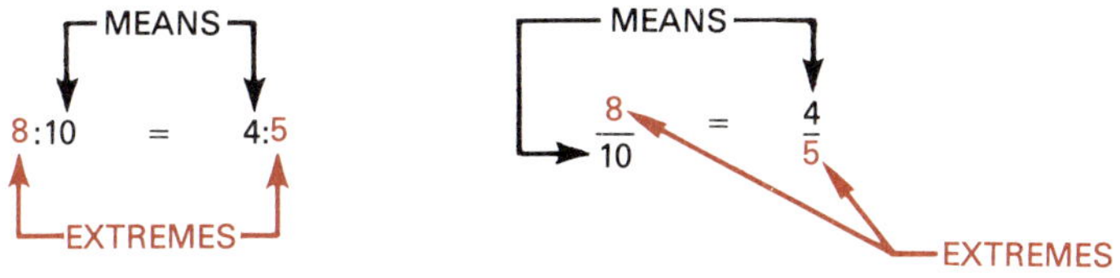

FIGURE 1–17 The product of the means is equal to the product of the extremes.

Rules of a Proportion

A proportion involves four quantities and generally one of the quantities is an unknown value. There are three rules that can be used for solving proportion problems. **The first rule is that the product of the means is equal to the product of the extremes.**

Example: Given the proportion 6:9 = 2:3, show that the product of the means is equal to the product of the extremes.

Solution: The extremes are 6 and 3, and $6 \times 3 = 18$.
The means are 9 and 2, and $9 \times 2 = 18$.

The second rule states that in any proportion, the product of the means divided by either extreme supplies the other extreme.

Example: Given the proportion 15:25 = 3:5, show that the product of the means divided by either extreme supplies the other extreme.

Solution:

$15:25 = 3:5$	(Original proportion)
$25 \times 3 = 75$	(Product of the means)
$75 \div 15 = 5$	(Division of product produces extreme)
$75 \div 5 = 15$	(Division of product produces extreme)

The third rule, a variation of the second, states that in any proportion, the product of the extremes divided by either mean supplies the other mean.

Example: Given the proportion 8:12 = 2:3, show that the product of the extremes divided by either mean supplies the other mean.

Solution:

$8:12 = 2:3$	(Original proportion)
$8 \times 3 = 24$	(Product of extremes)
$24 \div 12 = 2$	(Division of product produces mean)
$24 \div 2 = 12$	(Division of product produces mean)

Solving for an Unknown Value

In the proportions supplied so far, all the values were given. In electronic problems, sometimes one of the terms in the proportion is unknown. The proportional formula used in electronics is the $a:b = c:d$ arrangement. In this formula, a divided by b is equal to c divided by d.

$$\frac{a}{b} = \frac{c}{d}$$

The formula can be rearranged for manipulation by *cross multiplying* the means and extremes. This results in:

$$a \times d = b \times c \qquad \text{(Product of means = Product of extremes)}$$

In electronic problems, one value, a or b or c or d, may be *unknown* while the values of the others are either given or can somehow be derived or calculated. In the $a \times d = b \times c$ type of formula, any letter can be transposed to the left side of the equals sign, with all the other letters on the right side, by using division. As an example, assume a is to be the unknown.

$$a \times d = b \times c$$

As a first step, divide *both* sides by d.

$$\frac{a \times d}{d} = \frac{b \times c}{d}$$

On the left side we have d divided by d, equal to 1. The setup then becomes:

$$a = \frac{b \times c}{d}$$

The other letters can also be transposed in the same manner. Solving formulas for missing values will be covered in depth in a later chapter.

Example: What is d in the proportion $4:9 = 24:d$?

Solution: Write the proportion. $\frac{4}{9} = \frac{24}{d}$

Cross multiply. $4d = (9)(24)$

Solve for d. $d = \frac{9 \times 24}{4}$

$d = 54$

Check by substitution. $\frac{4}{9} = \frac{24}{54}$

$\frac{4}{9} = \frac{4}{9}$

The Shunt Law

When resistors are wired in parallel, the same voltage drop must appear across each, as indicated in Figure 1–18. The two resistors are coded as R_1 and

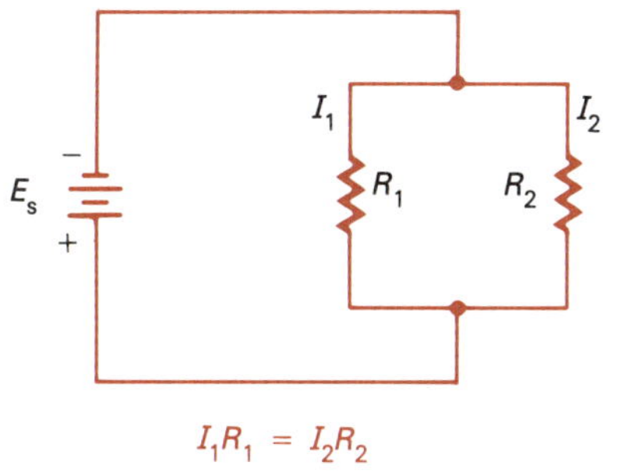

FIGURE 1–18 The voltage drops across resistors in parallel are equal.

R_2. Unlike a series circuit, the currents through the two resistors can be of different values, and so the current through R_1 can be identified as I_1 and the current through R_2 as I_2. Since voltage is equal to the product of current and resistance:

$$I_1 \times R_1 = I_2 \times R_2$$

This is a four-part formula. Through suitable transpositions, the original formula arrangement can be made to appear as four different formulas.

$$I_1 = \frac{I_2 \times R_2}{R_1} \quad (1\text{–}1)$$

$$R_1 = \frac{I_2 \times R_2}{I_1} \quad (1\text{–}2)$$

$$I_2 = \frac{I_1 \times R_1}{R_2} \quad (1\text{–}3)$$

$$R_2 = \frac{I_1 \times R_1}{I_2} \quad (1\text{–}4)$$

Example: In a dc circuit, a pair of parallel resistors are connected across a voltage source, as shown in Figure 1–19. Resistor R_1 is 18 ohms, R_2 is 54 ohms. Current I_2 is 2 A. What is the value of I_1?

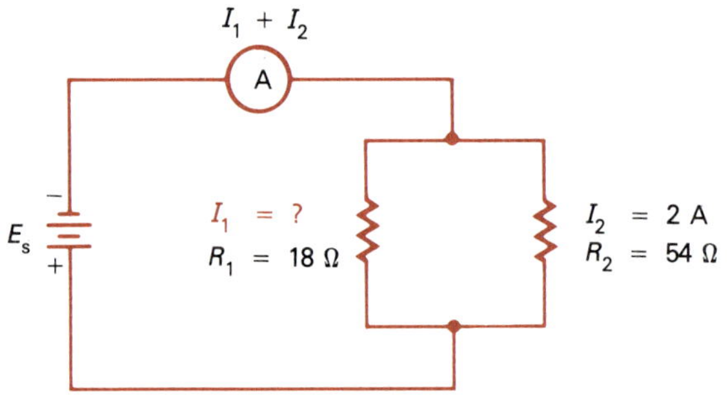

FIGURE 1–19

Solution: It will not be necessary to remember the four formulas involving resistances and their currents. All that is necessary to know is that $I_1R_1 = I_2R_2$ and the required formula can be derived. In this case it is formula 1–1.

$$I_1 = \frac{I_2 \times R_2}{R_1} = \frac{2\text{ A} \times 54\ \Omega}{18\ \Omega} = \frac{108\text{ V}}{18\ \Omega} = 6\text{ A}$$

The problem can be easily checked. Since $I_1R_1 = I_2R_2$, then:

$(6\text{ A})(18\ \Omega) = (2\text{ A})(54\ \Omega)$

$108\text{ V} = 108\text{ V}$

Transformer Turns Ratio

Another example of a four-part formula involves the currents flowing through the primary and the secondary windings of a transformer, or the voltages that appear across the windings. The turns ratio of a transformer is the relationship of the primary turns to the secondary turns. Expressed as a formula:

$$T_r = \frac{N_p}{N_s}$$

In this formula, T_r is the turns ratio and is simply a number. N_p is the number of primary turns, and N_s is the number of secondary turns. Both primary and secondary voltages and the primary and secondary currents are directly dependent on the turns ratio. In the case of voltage:

$$T_r = \frac{E_p}{E_s} = \frac{N_p}{N_s} \tag{1–5}$$

and in the case of current

$$T_r = \frac{I_s}{I_p} = \frac{N_p}{N_s} \tag{1–6}$$

From these basic formulas we can obtain a variety of interesting and useful variations of either the three-part or the four-part formula type. Since:

$$T_r = \frac{E_p}{E_s}$$

we can have formulas for E_s, E_p, or T_r as the unknown. The same rule is true for

$$T_r = \frac{I_s}{I_p}$$

Memorization of such a long list of formulas is unnecessary since it is easier to work with the original formula and then to develop the required formula as needed.

Example: A certain power transformer has a 6 to 1 turns ratio and a current of 90 mA flowing in the secondary winding. What is the value of the primary current?

Solution: In problems of this type, it is unnecessary to convert to basic units. The turns ratio is given as 6 to 1 or 6:1. This indicates that N_p is 6 and N_s is 1. This does *not* mean the primary has 6 turns and the secondary 1. It does mean that the total number of primary turns divided by the total number of secondary turns will yield a quotient of 6. If the primary winding has 6 000 turns then the secondary winding will have 1 000 turns, since $\frac{6\,000}{1\,000} = 6$.

$$T_r = \frac{I_s}{I_p} \qquad \text{(Original formula)}$$

$$\frac{T_r}{1} = \frac{I_s}{I_p} \qquad \text{(Expressed as a proportion)}$$

$$\frac{6}{1} = \frac{90 \text{ mA}}{I_p}$$ (Substituting values)

$$6I_p = 90 \text{ mA}$$ (Cross multiplication)

$$I_p = \frac{90 \text{ mA}}{6}$$

$$I_p = 15 \text{ mA}$$

EXERCISE 1·2

1. Simplify the following ratios.
   a. 12:144
   b. 72:18
   c. 6:99
   d. 500:5
   e. 84:132
   f. 729:6 561
   g. 32 V:160 V
   h. 98 W:343 W
   i. 990 A:45 A
   j. 46 800 Hz:1 170 Hz
2. Simplify the following ratios.
   a. 16:12
   b. 132:48
   c. 39 Ω:381 Ω
   d. 348 V:87 V
   e. 105 miles:195 miles
   f. 1 096:137
   g. 90 W:135 W
   h. 112 A:21 A
   i. 512:1 280
   j. 774 m:430 m
3. Find the missing quantity in each of these proportions.
   a. $3:4 = 12:d$
   b. $24 \text{ V}:60 \text{ V} = c:500 \text{ V}$
   c. $a:3 = 16:4$
   d. $56\ \Omega:14\ \Omega = c:7\ \Omega$
   e. $64 \text{ mA}:b = 8 \text{ mA}:1 \text{ mA}$
   f. $868:14 = 62:d$
   g. $18:b = 110:385$
   h. $10\ \Omega:4\ \Omega = c:6\ \Omega$
   i. $960:8 = 1\,920:d$
   j. $a:32 = 168:7$
4. Find the missing quantity in each of these proportions.
   a. $a:112 \text{ V} = 40 \text{ V}:16 \text{ V}$
   b. $35\ \Omega:b = 165\ \Omega:825\ \Omega$
   c. $198 \text{ A}:24 \text{ A} = c:4 \text{ A}$
   d. $76:494 = 10:d$
   e. $a:280 = 480:12$
   f. $6 \text{ W}:b = 3 \text{ W}:81 \text{ W}$
   g. $243:27 = c:11$
   h. $84:132 = 924:d$
   i. $a:65 = 1\,458:81$
   j. $66 \text{ V}:b = 198 \text{ V}:33 \text{ V}$
5. Use the diagram in Figure 1–20 to find the ratio of the meter resistance (R_m) to the shunt resistance (R_s).

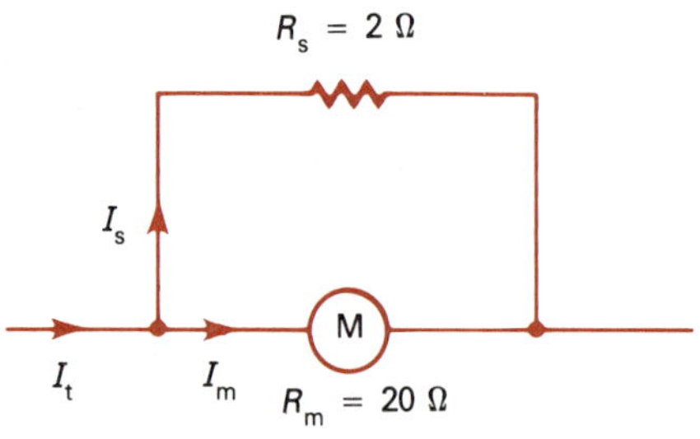

FIGURE 1–20

6. The input impedance of an amplifier is 60 kilohms. The output

impedance is 5 kilohms. Find the ratio of input impedance to output impedance.

Use the transformer diagram and the formulas in Figure 1–21 for problems 7–9.

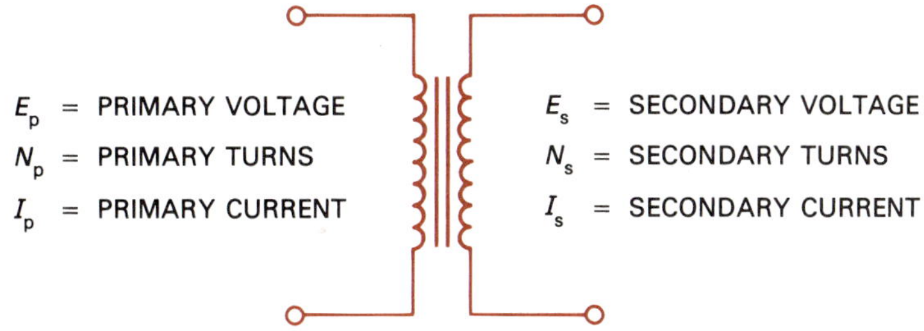

TURNS RATIO: $T_r = \frac{N_p}{N_s}$

$$\frac{N_p}{N_s} = \frac{E_p}{E_s}$$

$$\frac{N_p}{N_s} = \frac{I_s}{I_p}$$

FIGURE 1–21

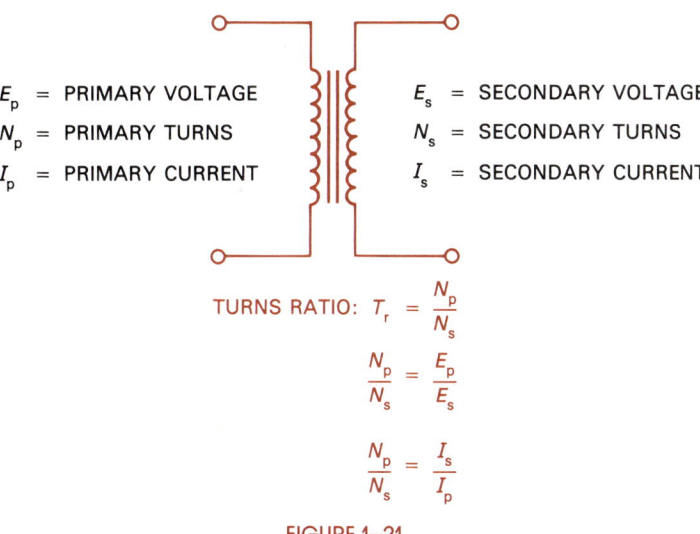

7. A transformer has a 400-turn primary and a 50-turn secondary. Find the secondary voltage if 120 volts are applied.
8. A transformer supplies a secondary current of 1 ampere. The primary turns are 150 and the secondary turns are 600. Find the primary current.
9. A 1 200-turn primary has an applied voltage of 60 V. Find the number of turns in the secondary if the secondary voltage is 24 V.

Use the Wheatstone bridge and balanced circuit formula in Figure 1–22 for problems 10–12.

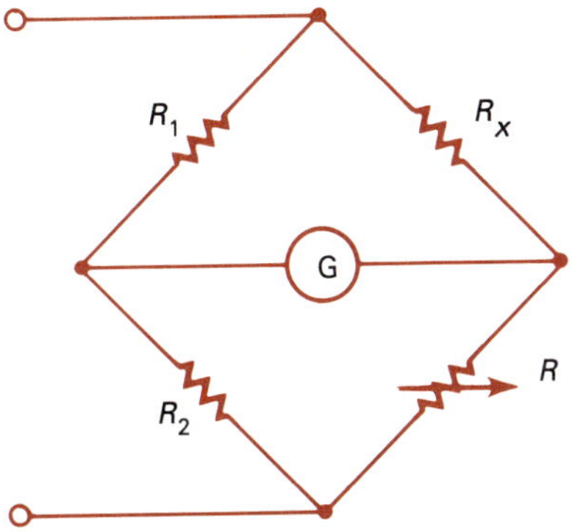

$$\frac{R_1}{R_2} = \frac{R_x}{R}$$

FIGURE 1–22

10. $R_1 = 10\ k\Omega$, $R_2 = 40\ k\Omega$, and $R = 32\ k\Omega$. Find the value of R_x.
11. $R_1 = R_2 = 25\ k\Omega$, and $R = 33\ k\Omega$. Find the value of R_x.
12. The ratio of R_1 to R_2 is 5. Find the value of R_x if $R = 20\ k\Omega$.

UNIT 1·3 FRACTIONS, DIRECT AND INVERSE RELATIONSHIPS, AND PERCENTS

Objectives:

After studying this unit, you should be able to

- simplify fractions and determine the lowest common denominator of two or more fractions.
- perform the four arithmetic functions with common fractions, improper fractions, and mixed numbers.
- discuss direct and inverse relationships.
- calculate reciprocals and determine reciprocals through use of a calculator.
- calculate percentages and use a calculator to determine percentages.
- perform conversions among fractions, percents, and decimals.

Proper and Improper Fractions

A fraction is another form or name for division. There are various kinds of common fractions, one of which is the *proper fraction,* a form of division in which the result has a value smaller than one. Number arrangements such as

$$\tfrac{3}{4},\ \tfrac{1}{2},\ \text{or}\ \tfrac{7}{8}$$

are proper fractions. The upper number is the *numerator;* the lower, the *denominator.* A fraction is an indication that division is to be performed.

$$\tfrac{1}{2} = 1 \div 2 = 0.5$$

This latter expression is a *decimal* fraction and will be discussed in Chapter 2.

Another type of common fraction is the *improper fraction* in which the numerator is equal to or greater than the denominator; for example

$$\tfrac{8}{5}$$

is an improper fraction and has a value equal to or greater than one. When the division called for by an improper fraction is completed, the result is a whole number or a whole number plus a fraction.

Example: Simplify $\tfrac{8}{5}$.

Solution: $\frac{8}{5} = 8 \div 5 = 1 + \frac{3}{5} = 1\frac{3}{5}$

In the example $\frac{8}{5}$ is considered an improper fraction. When expressed as $1\frac{3}{5}$, the correct term is *mixed number,* which indicates a whole number and a fraction. This mixed number would be read as "one and three-fifths."

Rules for Fractions

If the numerator and denominator of a fraction are multiplied by the same amount, the value of the fraction remains unchanged.

$$\frac{3}{4} \times \frac{2}{2} = \frac{6}{8}$$

The value of $\frac{6}{8}$ is the same as $\frac{3}{4}$. Note that $\frac{2}{2} = 1$. The original ratio remains the same since it has been multiplied by a value equal to 1. Multiplying by a value equal to 1 will not change the original value.

Adding Fractions

When fractions are added, only the numerators are involved in the addition process. The denominator remains unchanged. **Fractions with identical denominators are added by adding the numerators and placing the result over the common denominator.**

$$\frac{1}{4} + \frac{2}{4} = \frac{3}{4}$$

Fractions with different denominators are added after the fractions are expressed as equivalent fractions with identical denominators.

Example: Add $\frac{1}{4}$ and $\frac{1}{2}$.

Solution:
$$\begin{array}{rcl} \frac{1}{4} & = & \frac{1}{4} \\ +\frac{1}{2} & = & \frac{2}{4} \\ \hline & & \frac{3}{4} \end{array}$$

A fraction can be reduced by dividing the numerator and denominator by the same amount. While any pair of numbers can be used to expand the fraction by multiplication, there is a definite limit to number availability for division. In some instances, division may not be possible. A number such as $\frac{4}{8}$, for example, can be reduced by dividing both numerator and denominator by 2 or by 4. If 2 is used as the divisor,

$$\frac{4 \div 2}{8 \div 2} = \frac{2}{4}$$

If 4 is used as the divisor,

$$\frac{4 \div 4}{8 \div 4} = \frac{1}{2}$$

A fraction such as $\frac{1}{2}$ cannot be reduced any more since there is no number except 1 that will evenly divide both the numerator and the denominator. The numerator and denominator of a fraction are sometimes called the *terms* of the fraction. When a fraction cannot be reduced any more, it is in *lowest terms*.

Example: Reduce $\frac{12}{16}$ to lowest terms.

Solution: $\frac{12 \div 2}{16 \div 2} = \frac{6}{8} = \frac{6 \div 2}{8 \div 2} = \frac{3}{4}$

Lowest Common Denominator

When fractions and mixed numbers are to be used in addition or subtraction problems they must have common denominators. One quick method to find a common denominator is to compute the product of the denominators. To convert each fraction to this *common* denominator simply divide the product by each denominator and multiply the result by the numerator.

Example: Express the following fractions as equivalent fractions with the same denominator.
$\frac{3}{4}$, $\frac{2}{3}$, and $\frac{1}{2}$

Solution: The common denominator is $4 \times 3 \times 2 = 24$.
$\frac{3}{4} = \frac{18}{24}$, $\frac{2}{3} = \frac{16}{24}$, and $\frac{1}{2} = \frac{12}{24}$

While a combination of fractions may have many common denominators, the most useful of these will be the *lowest common denominator* or *LCD*. The LCD is the lowest number that is a multiple of each of the denominators. In order to determine the LCD it will be necessary to break each denominator into its lowest multiples. Follow this example carefully.

Example: Find the LCD for $\frac{5}{18}$, $\frac{37}{60}$, and $\frac{17}{35}$.

Solution: Factor each denominator into it lowest multiples:

$$18 = 3 \times 3 \times 2$$
$$60 = 2 \times 2 \times 3 \times 5$$
$$35 = 5 \times 7$$

For each multiple determine which denominator contains the factor the largest number of times. The multiple is used that number of times when finding the LCD.
The maximum number of times 3 appears is twice.
The maximum number of times 2 appears is twice.
The maximum number of times 5 appears is once.
The maximum number of times 7 appears is once.
The LCD is $3 \times 3 \times 2 \times 2 \times 5 \times 7 = 1\ 260$.

Adding and Subtracting Fractions and Mixed Numbers

As mentioned previously, in order to add fractions the denominators must be the same. This is also true in subtracting fractions.

Example: Add the following.
$12\frac{1}{4} + 6\frac{3}{5} + 10\frac{7}{16}$

Solution:

$$\begin{array}{rcl} 12\frac{1}{4} & = & 12\frac{20}{80} \\ 6\frac{3}{5} & = & 6\frac{48}{80} \\ +10\frac{7}{16} & = & 10\frac{35}{80} \\ \hline & & 28\frac{103}{80} = 29\frac{23}{80} \end{array}$$

Example: Subtract $5\frac{17}{36}$ from $9\frac{3}{48}$.

Solution:

$$\begin{array}{rcccl} 9\frac{3}{48} & = & 9\frac{9}{144} & = & 8\frac{153}{144} \\ -5\frac{17}{36} & = & 5\frac{68}{144} & = & 5\frac{68}{144} \\ \hline & & & & 3\frac{85}{144} \end{array}$$

Converting Mixed Numbers to Improper Fractions

When performing the four arithmetic functions with combinations of whole numbers, fractions, and mixed numbers it may be necessary that each term have a common denominator, preferably the LCD.

Example: Express $3\frac{1}{2}$ as an improper fraction with a denominator of 6.

Solution: Change $3\frac{1}{2}$ to an improper fraction by multiplying the denominator by the whole number, adding the numerator, and placing the result over the denominator.

$$3\frac{1}{2} = \frac{(3 \times 2) + 1}{2} = \frac{7}{2} = \frac{21}{6}$$

Example: Express 12 as an improper fraction with a denominator of 6.

Solution: $12 = \frac{12}{1} = \frac{72}{6}$

Example: Add these mixed numbers using improper fractions.

$4\frac{7}{8} + 3\frac{3}{4}$

Solution: $4\frac{7}{8} + 3\frac{3}{4} = \frac{39}{8} + \frac{15}{4} = \frac{39}{8} + \frac{30}{8} = \frac{69}{8} = 8\frac{5}{8}$

Example: Subtract these mixed numbers using improper fractions.

$9\frac{3}{48} - 5\frac{17}{36}$

Solution: $9\frac{3}{48} - 5\frac{17}{36} = \frac{435}{48} - \frac{197}{36} = \frac{1\,305}{144} - \frac{788}{144} = \frac{517}{144} = 3\frac{85}{144}$

Multiplying Fractions and Mixed Numbers

To multiply fractions or mixed numbers follow these steps:

1. **Change the mixed numbers to improper fractions.**
2. **Multiply the numerators.**
3. **Multiply the denominators.**
4. **Reduce the product to lowest terms.**

Example: Multiply these fractions.

a. $\frac{3}{5} \times \frac{2}{3}$

b. $2\frac{1}{2} \times \frac{1}{2}$

c. $5 \times 6\frac{5}{8}$

d. $14\frac{1}{3} \times 9\frac{15}{16}$

Solution:
a. $\frac{3}{5} \times \frac{2}{3} = \frac{6}{15} = \frac{2}{5}$
b. $2\frac{1}{2} \times \frac{1}{2} = \frac{5}{2} \times \frac{1}{2} = \frac{5}{4} = 1\frac{1}{4}$
c. $5 \times 6\frac{5}{8} = \frac{5}{1} \times \frac{53}{8} = \frac{265}{181} = 33\frac{1}{8}$
d. $14\frac{1}{3} \times 9\frac{15}{16} = \frac{43}{3} \times \frac{159}{16} = \frac{6\,837}{48} = 142\frac{21}{48} = 142\frac{7}{16}$

Dividing Fractions and Mixed Numbers

To divide fractions or mixed numbers, change the mixed numbers to improper fractions, invert the divisor, and multiply.

Example: Divide these fractions.
a. $\frac{5}{8} \div \frac{1}{4}$
b. $6\frac{2}{3} \div \frac{2}{3}$
c. $10 \div 4\frac{5}{6}$
d. $46\frac{7}{32} \div 3\frac{5}{8}$

Solution:
a. $\frac{5}{8} \div \frac{1}{4} = \frac{5}{8} \times \frac{4}{1} = \frac{20}{8} = 2\frac{1}{2}$
b. $6\frac{2}{3} \div \frac{2}{3} = \frac{20}{3} \times \frac{3}{2} = \frac{60}{6} = 10$
c. $10 \div 4\frac{5}{6} = \frac{10}{1} \times \frac{6}{29} = \frac{60}{29} = 2\frac{2}{29}$
d. $46\frac{7}{32} \div 3\frac{5}{8} = \frac{1\,479}{32} \times \frac{8}{29} = \frac{11\,832}{928} = 12\frac{3}{4}$

Direct and Inverse Relationships

In a simple formula such as $a = b$, the effect of the statement is that the value of a is equal to that of b. The value of the letter a can be divided or multiplied provided the same action is performed on the letter b. Thus, if $a = b$, then

$$5a = 5b \quad \text{or} \quad \frac{a}{5} = \frac{b}{5}$$

Another implication of the $a = b$ formula is that if the value of a increases, so will the value of b, and in the same amount. Conversely, if a decreases, so must the value of b, and again by the same amount. A relationship of this kind is known as a *direct proportion.* A direct proportion can also exist in formulas of the $a = b \times c$ type. If the value of the product $b \times c$ increases, then the value of a will also increase.

In another type of formula,

$$a = \frac{1}{b}$$

the value of a is *inversely* proportional to the value of b. If the value of b decreases, the value of a must increase. Conversely, if b is raised in value, then the value of a must go down. This inverse proportionality is also called a *reciprocal relationship.*

Example: Given $a = \frac{1}{b}$, find a if $b = 5$, and then if $b = 8$.

Solution: $a_1 = \frac{1}{5} = 0.2$

$a_2 = \frac{1}{8} = 0.125$

In this inverse relationship, as b increases in value, a decreases in value.

Resistance and Conductance

The electrical opposition to the flow of current through a conductor is known as resistance. A similar concept is to consider the ease or facility with which a current passes through a conductor. This ease of current flow is called *conductance* and its symbol is the letter G. Resistance (R) and conductance (G) are *reciprocals* because they describe ideas which are completely opposite. Resistance is expressed in ohms; conductance is expressed in siemens. In terms of a formula:

$$G = \frac{1}{R} \quad \text{or} \quad R = \frac{1}{G} \tag{1–7}$$

Example: A resistor has a value of 10 ohms. What is its equivalent conductance in siemens?

Solution: $G = \frac{1}{R} = \frac{1}{10\ \Omega} = 0.1\ \text{S}$

Electronics formulas involve both direct and inverse relationships. Formulas such as $E = I \times R$ or $P = E \times I$ are representative of the direct type. Formulas such as

$$\lambda = \frac{1}{f} \qquad \text{where } \lambda = \text{wavelength in metres}, \quad f = \text{frequency in hertz}$$

and

$$T = \frac{1}{f} \qquad \text{where } T = \text{period of a wave in seconds}, \quad f = \text{frequency in hertz}$$

are examples of the inverse type. These examples are one type of relationship or the other, that is, direct or inverse. However, it is possible for a single formula to contain both relationships. One form of Ohm's Law, for example,

$$I = \frac{E}{R}$$

shows us that the current I is directly proportional to E and inversely proportional to R. What this means is that if the voltage increases and the resistance remains the same, the current I will also increase. However, if R increases with a constant voltage, the current will decrease.

Example: A 100-volt power supply is connected to a 10-ohm resistor. Calculate the amount of current flow for this circuit, (a) if R increases to 20 ohms, (b) if E_s increases to 200 volts and R remains at 10 ohms.

Solution: $I_1 = \frac{E_s}{R} = \frac{100\text{ V}}{10\ \Omega} = 10\text{ A}$

a. $I_2 = \frac{E_s}{R} = \frac{100\text{ V}}{20\ \Omega} = 5\text{ A}$ (Inverse relationship)

b. $I_3 = \frac{E_s}{R} = \frac{200\text{ V}}{10\ \Omega} = 20\text{ A}$ (Direct relationship)

Formula Relationships

At the start of this chapter, a number of formulas were supplied that were all related in the sense that they were all solved by simple addition. The formulas for resistors in series, $R_t = R_1 + R_2 \ldots$, and capacitors in parallel, $C_t = C_1 + C_2 \ldots$, are examples of this relationship. While this kind of electronic kinship is rather obvious, there are other formula types that are also related. These include the directly proportional, inversely proportional, and combined proportionality formulas.

Another type of formula similarity is based on the way components behave. Thus, resistors in parallel and capacitors in series use the same basic formula arrangement. For resistors in parallel:

$$R_t = \frac{R_1 \times R_2}{R_1 + R_2}$$

and for capacitors in series:

$$C_t = \frac{C_1 \times C_2}{C_1 + C_2}$$

The only difference in these formulas is in the codes used to identify the components; otherwise they are identical. Both of these formulas are fractions and will be discussed in another chapter.

Calculator Tip

There are many electronics formulas which require that the reciprocal of a number be computed. This involves dividing 1 by the number. The reciprocal key [1/x] instructs the calculator to divide 1 by the displayed number.

An excellent example of the use of this function is the computation of the total resistance of three parallel resistors. Given values of 9 ohms, 36 ohms, and 36 ohms, find the total resistance. The formula is as follows:

$$R_t = \frac{1}{\dfrac{1}{9\ \Omega} + \dfrac{1}{36\ \Omega} + \dfrac{1}{36\ \Omega}}$$

The computational sequence is as follows:

9 [1/x] [+] 36 [1/x] [+] 36 [1/x] [=] [1/x] [=] 6 Ω

Proportionality

The same formula can be composed of direct and inverse components. In the formula

$$I = \frac{E}{R}$$

the right side of the formula is a fraction, with a single-value numerator and a single-value denominator. There are some formulas, though, with more than one value used for the numerator or denominator. Consider a type such as:

$$X_c = \frac{1}{6.28 \times f \times C}$$

This is the formula for the reactance of a capacitor, coded as X_c. On the right side of the equals sign, 6.28 is a constant, while f is the frequency in hertz and C is the capacitance in farads. What this formula tells us, in effect, is that if we want to decrease the reactance of a capacitor, there are three ways of doing so. We can increase the frequency, we can increase the capacitance, or we can increase the frequency and the capacitance at the same time. But the formula tells us more than just that. On the right side of the formula we see the factors that affect capacitive reactance. Except in the case of tuning capacitors, capacitance usually remains fixed. Frequency, however, is quite often a variable quantity. But if the frequency changes, so will the capacitive reactance. If frequency increases, capacitive reactance decreases. Conversely, when frequency decreases, capacitive reactance increases. In solving a formula of this kind, when the values of f and C are supplied, it is important to remember that the answer is valid for this one condition only. When a capacitor, as an example, is used as a coupling unit between two circuits, its reactance will vary, and quite possibly over a wide range, depending on the applied frequency.

Effects of Temperature

Because formulas are involved with mathematics and because mathematics is often regarded as a precise science, formulas are often regarded as immutable or unchanging. Within limits, they are, but the trouble is that the people using the formulas frequently do not take into consideration that external factors can influence the behavior of components. In $E = I \times R$, for example, the formula does not indicate that R, representing resistance, can change for a number of reasons. The resistor may have a tolerance of 10 percent, meaning that its value can be anywhere up to 10 percent greater or 10 percent smaller than the value indicated by the resistor color code. When the resistor is wired into the circuit, the very act of assembly, the use of a hot soldering iron, can change the value of the resistor. If the resistor is mounted near a heat-emitting component, the value of the resistor can change again. And if the resistor is measured by a test instrument, the instrument may also have a certain amount of error. But E still equals $I \times R$, for the formula does not take any of these possibilities into consideration. The R in the formula, then, must be *the actual value of* R. If R should vary because of temperature changes, then I will also change. If $E = I \times R$ and R rises because of some reason, then an increase in resistance will cause the current to decrease.

A component has a positive temperature coefficient if its resistance increases with a rise in temperature. It has a negative temperature coefficient if the resistance varies inversely with temperature, decreasing when temperature rises and increasing when temperature falls. Some components are designed to have zero temperature coefficients and are not affected by changes in temperature.

While there are a number of different temperature scales, the Fahrenheit and Celsius scales, shown in Figure 1–23, are most commonly used in work in electronics. With the Fahrenheit scale, the freezing point of water is 32°; with the Celsius it is 0°. With the Fahrenheit scale, the boiling point of water is 212°; with the Celsius it is 100°. The advantage of the Celsius is that it has 100 divisions between the freezing and boiling points of water, while the Fahrenheit has 180°. The Celsius scale, then, is closer to our existing number system, the decimal system.

The resistance of a component, such as wire, is often specified at some particular temperature because wire has a positive temperature coefficient. Since the temperature may be given either in degrees Fahrenheit or Celsius, it is convenient to be able to convert from one scale to the other. To convert degrees Celsius to degrees Fahrenheit:

$$°F = (°C \times \tfrac{9}{5}) + 32° \qquad (1–8)$$

This is a formula that involves basic arithmetic: multiplication, division, and addition.

Example: A roll of wire has a resistance of 18 ohms when measured at 25°C. What is the equivalent temperature in degrees Fahrenheit?

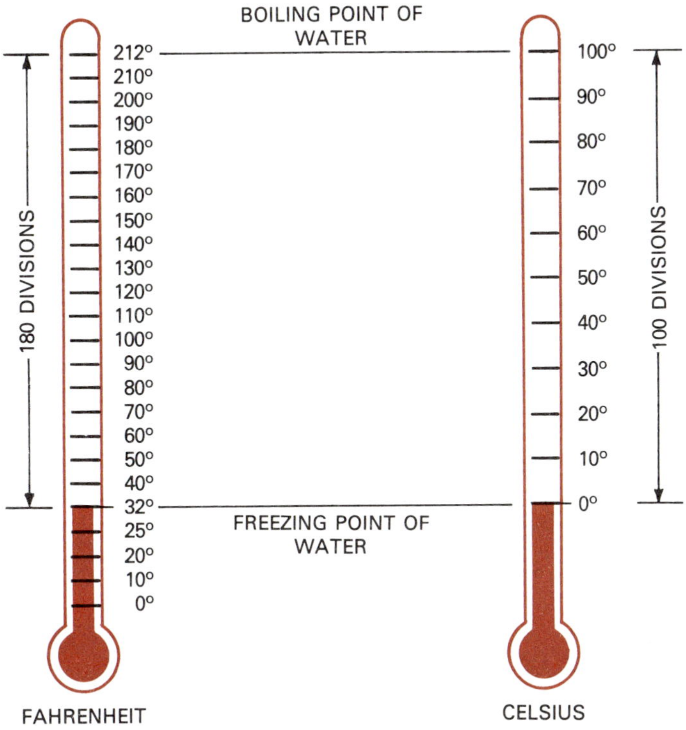

$$°F = \left(°C \times \frac{9}{5}\right) + 32°$$

$$°C = (°F - 32°)\,\frac{5}{9}$$

FIGURE 1–23 Relationship between degrees Fahrenheit and degrees Celsius

Solution:
$$\begin{aligned} °F &= (°C \times \tfrac{9}{5}) + 32° \\ &= (25° \times \tfrac{9}{5}) + 32° \\ &= 45° + 32° \\ &= 77° \end{aligned}$$

Sometimes, however, the temperature is supplied in Fahrenheit but the requirements of the job may call for the temperature in terms of degrees Celsius. The formula for making the conversion is:

$$°C = (°F - 32°) \times \tfrac{5}{9} \qquad \textbf{(1–9)}$$

Example: During a life test, the case of an iron-core transformer was found to have a temperature of 104° Fahrenheit. What was the temperature in degrees Celsius?

Solution:
$$\begin{aligned} °C &= (°F - 32°) \times \tfrac{5}{9} \\ &= (104° - 32°) \times \tfrac{5}{9} \\ &= 72° \times \tfrac{5}{9} \\ &= 40° \end{aligned}$$

Percent

Percent begins with a special kind of fraction, one which has 100 in the denominator. Percent means the number of parts per 100, and so a fraction such as $\frac{63}{100}$ is one which contains 63 possible parts out of a maximum of 100.

When the division indicated by the fraction is completed, the result is a *decimal.* The result of dividing 63 by 100, for example, is 0.63. A fraction such as $\frac{85}{100}$ is equal to 0.85. Note that two digits follow the decimal point. A fraction such as $\frac{7}{100}$ is equal to 0.07. Again, two digits follow the decimal point. No actual division is needed since the numerator is used directly and all that is required for the conversion of the fraction to a decimal is the correct placement of the decimal point. To convert a fraction having 100 in the denominator to a decimal just count off two decimal places in the numerator. For converting other fractions to decimals, the actual division may be required.

A decimal can be converted to a percent by moving the decimal point two places to the right. Thus, 0.85 is the same as 85 percent. The number 0.09 is equal to 9 percent, and 0.37 is 37 percent. Percents are often used in connection with electronics problems and are also frequently related to a manufacturer's specifications of a component.

Conversion of a Fraction to Percent

To change a fraction to percent, perform the division indicated by the fraction. This will result in a decimal. Move the decimal point two places to the right to obtain the percent.

Example: Convert $\frac{3}{4}$ to a percent.

Solution: $\frac{3}{4} = 3 \div 4 = 0.75$
Move the decimal point two places to the right.
$0.75 = 75\%$

Conversion of Percent to Decimal or Fraction

The conversion of percent to decimal is the reverse of the process just described. **To change a percent to a decimal, remove the percent symbol and move the decimal point two places to the left.**

Example: Convert 48% to a decimal.

Solution: $48\% = 0.48$

To change the percent to a fraction, remove the percent symbol and divide by 100. Reduce the fraction to lowest terms.

Example: Convert 48% to a fraction.

Solution: $48\% = \frac{48}{100} = \frac{12}{25}$

Although it might appear that 100 percent is a maximum number, it is possible to have larger values. To change any percent to a number, follow the same rule—move the decimal point two places to the left. Thus, $125\% = 1.25$.

Finding a Percentage

To find the percentage, convert the percent to its equivalent decimal value and then multiply by the number.

Example: What is 37% of 85?

Solution: 37% = 0.37
$0.37 \times 85 = 31.45$

Finding a Percent

To determine the relationship one number has to another in terms of a percent, divide. The number following the word "of" is the divisor.

Example: 17 is what percent of 85?

Solution: $\frac{17}{85} = 17 \div 85 = 0.2 = 20\%$

Percent seems to imply fractions in which the numerator is smaller than the denominator. This is often the case, but not always.

Example: 175 is what percent of 25?

Solution: $\frac{175}{25} = 175 \div 25 = 7 = 700\%$

Calculator Tip

Some examples of typical sequences to solve percent or percentage problems include:

a. Find 85% of 500: 500 [×] 85 [%] [=] 425

b. What is 360 + 10%?: 360 [+] 10 [%] [=] 396

c. What is 260 − 5%?: 260 [−] 5 [%] [=] 247

d. Convert 68/100 to a percent: 68 [÷] 100 [=] .68 or 68%

Finding the Base Number

In the problem "17 is what percent of 85?" this latter number, 85, is sometimes referred to as the base number. In some problems, the base number may not be known. **To find the base, divide the given number by the percent.**

Example: 38 is 12% of what number?

Solution: Change 12% to a decimal. 12% = 0.12

Divide. $\frac{38}{0.12} = 316.66$

Efficiency

No electrical device, or any other kind of device for that matter, is 100 percent efficient. Some engines have efficiencies of 5 percent to 15 percent. Transformers can have efficiencies of 90 percent, or possibly more. With a power transformer, for example, its efficiency is a measure of how much power supplied to the primary winding reaches the secondary winding for delivery to a load. If a transformer has an efficiency of 90 percent, for every 10 watts delivered to the primary, 9 watts will be available through the secondary. The 1 watt difference will be lost in the form of heat, or a form of leakage flux, that is, in the form of energy lost in the core or windings. Whatever the reasons, less power is available from the secondary than is supplied to the primary. In terms of a formula:

$$\textbf{Output} = \textbf{Input} \times \textbf{Efficiency} \qquad \textbf{(1–10)}$$

Example: A transformer connected to the 110-V ac line has a primary current of 1 A and an efficiency of 85%. How much power is delivered to the secondary?

Solution: The power supplied to the primary by the power line can be calculated from the formula

$P = E \times I$

$= 110 \text{ V} \times 1 \text{ A}$

$= 110 \text{ W}$ (primary power)

The output can be calculated from the formula

output = input × efficiency

output = 110 W × 0.85 = 93.5 W

EXERCISE 1·3

1. Reduce these fractions to their lowest terms. Express improper fractions as mixed numbers.
   a. $\frac{18}{48}$ b. $\frac{35}{75}$ c. $\frac{108}{80}$ d. $\frac{544}{153}$ e. $\frac{84}{192}$
2. Reduce these fractions to their lowest terms. Express improper fractions as mixed numbers.
   a. $\frac{35}{56}$ b. $\frac{140}{64}$ c. $\frac{222}{384}$ d. $\frac{414}{276}$ e. $\frac{495}{528}$
3. Find the LCD for the following.
   a. $\frac{1}{2}, \frac{7}{8}, \frac{3}{5}$
   b. $\frac{2}{3}, \frac{3}{4}, \frac{5}{6}, \frac{7}{9}$
   c. $1\frac{5}{8}, 3\frac{13}{16}, \frac{11}{15}, \frac{7}{10}$
4. Find the LCD for the following.
   a. $\frac{9}{16}, \frac{17}{25}, \frac{2}{3}$
   b. $\frac{1}{8}, \frac{7}{6}, \frac{20}{21}$
   c. $4\frac{7}{9}, 2\frac{1}{6}, \frac{19}{27}$
5. Perform the indicated operations.
   a. $6\frac{7}{8} + \frac{13}{16} - \frac{3}{4}$ c. $(6\frac{13}{16} + 1\frac{1}{2}) \times 1\frac{1}{2}$
   b. $3\frac{1}{2} \times 2\frac{1}{4}$ d. $12\frac{1}{2} \div 2\frac{1}{2}$

e. $10\frac{19}{32}(2\frac{3}{4} - 1\frac{1}{8})$
f. $(7\frac{1}{8} + 12\frac{1}{64}) \div 4\frac{3}{8}$
g. $104\frac{3}{5} - 82\frac{1}{2} \times \frac{3}{4}$
h. $(65\frac{19}{25} + 5\frac{2}{3}) \div 4\frac{4}{5}$

6. Perform the indicated operations.
   a. $-11\frac{5}{8} - (-32\frac{7}{16})$
   b. $6\frac{8}{9} \times \frac{3}{8} + \frac{3}{8}$
   c. $(108\frac{2}{3} - 69\frac{7}{12}) \times \frac{5}{6}$
   d. $47\frac{9}{64}(100 \div 2\frac{1}{2})$
   e. $(-3\frac{31}{32})(7\frac{4}{9}) + 16\frac{5}{9}$
   f. $13\frac{4}{5} \times 6\frac{3}{4} - 104\frac{1}{8}$
   g. $1\frac{3}{8}(4\frac{3}{4} \div \frac{1}{4})$
   h. $-5\frac{3}{8}(1\frac{1}{2} \div \frac{1}{16})$
7. Find the total resistance (R_t) in the parallel circuit in Figure 1–24.

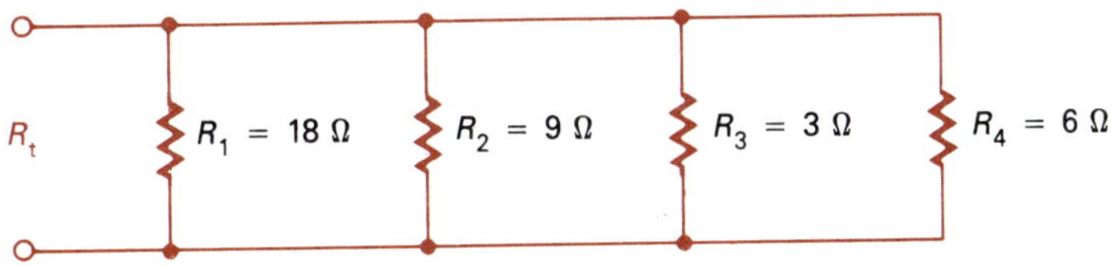

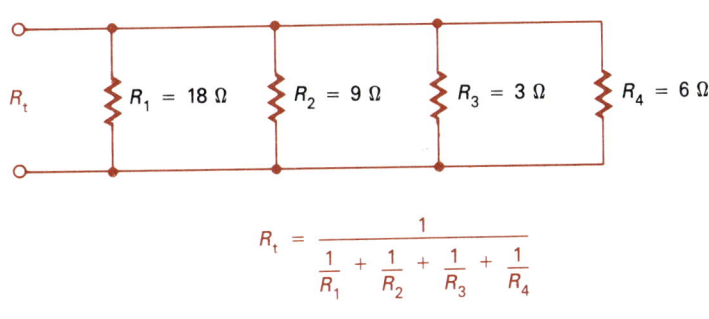

$$R_t = \frac{1}{\frac{1}{R_1} + \frac{1}{R_2} + \frac{1}{R_3} + \frac{1}{R_4}}$$

FIGURE 1–24

8. A $10\frac{1}{2}$-ohm resistor is placed in parallel with an $8\frac{3}{4}$-ohm resistor. Find total resistance. Note:

$$R_t = \frac{R_1 \times R_2}{R_1 + R_2}$$

9. Convert 35°C to Fahrenheit.
10. Convert 68°F to Celsius.
11. Convert 176°F to Celsius.
12. Convert 65°C to Fahrenheit.

Use this equation for problems 13–14.

$$x = \frac{y}{z + w}$$

13. As y increases and the values of z and w remain the same, will x increase or decrease?
14. As w decreases in value, with z and y remaining the same, will x increase or decrease?
15. Find 47% of 183.
16. What is 375 + 15%?
17. What is 1 084 − 108%?
18. Find 380% of 16.
19. What is −275 + 19%?
20. What is 75 − (62% × 93)?

UNIT 1·4 TOLERANCES AND SIGNIFICANT FIGURES

Objectives:

After studying this unit, you should be able to

- use a calculator to determine tolerance values.
- identify significant figures.
- round numbers to a specified number of places.

Tolerances

Because electronics is based on mathematics, there is a connotation that electronics, like mathematics, is a precise science. Perhaps from a formula point of view it is, but certainly not in practice. All electronic equipment is constructed using components that have a reasonable approximation to the values calculated by precise formulas. Parts such as resistors, capacitors, transistors, coils, and conductors used in electronic equipment do not have exact values but deviate from those values by certain percents known as *tolerance.* A resistor, for example, may have a tolerance of 5 percent or 10 percent. Since 5 percent is 0.05 and 10 percent is 0.10, a 1 000-ohm resistor with a tolerance of 5 percent could have an actual value of anywhere between 1 000 ohms plus 5 percent of 1 000 ohms to 1 000 ohms minus 5 percent of 1 000 ohms:

$$1\,000\ \Omega \times 0.05 = 50\ \Omega$$
$$1\,000\ \Omega + 50\ \Omega = 1\,050\ \Omega$$
$$1\,000\ \Omega - 50\ \Omega = 950\ \Omega$$

In a receiver or transmitter, or other electronic apparatus, then, a 1 000-ohm resistor with a 5 percent tolerance may actually have any value between 950 ohms and 1 050 ohms. This is a 100-ohm spread for a 1 000-ohm resistor. But this isn't all! Again, as mentioned earlier, Ohm's Law does not take the effects of temperature into consideration, and neither does tolerance. And so this resistor, now anywhere between 950 ohms and 1 050 ohms, might be 1 100 ohms under actual working conditions.

Example: A 5 100-ohm resistor has a tolerance of 10 percent. What should be the maximum and minimum possible values of this resistor when measured with a resistance bridge?

Solution: $10\% = 0.10$

$$5\,100\ \Omega \times 0.10 = 510\ \Omega$$
$$5\,100\ \Omega + 510\ \Omega = 5\,610\ \Omega$$
$$5\,100\ \Omega - 510\ \Omega = 4\,590\ \Omega$$

The possible range of resistance values is from 4 590 ohms to 5 610 ohms.

Tolerances can be plus-minus, plus only, or minus only, depending on manufacturer's specifications.

Calculator Tip

The memory function of the calculator may be useful in solving tolerance problems. Finding the maximum and minimum values of a 3 300-ohm resistor with a 5% tolerance can be computed without stopping to write down intermediate steps or answers. Try the following sequence:

Enter Number	*Key*	*Display After Key*
3 300	+	3 300
5	%	165
	=	3 465
	STO	3 465
3 300	−	3 300
5	%	165
	=	3 135 (minimum)
	EXC	3 465 (maximum)
	OR RCL	3 465 (maximum)

Example: For each of the following numbers use your calculator to find the high and low tolerance value.

a. 150 ± 10%
b. 47 000 ± 20%
c. 25 μF + 10%
d. 370 Ω ± 5%
e. 100 000 − 20%

Solution:

a. 150 + 10% = 150 + 15 = 165
150 − 10% = 150 − 15 = 135

b. 47 000 + 20% = 47 000 + 9 400 = 56 400
47 000 − 20% = 47 000 − 9 400 = 37 600

c. 25 μF + 10% = 25 μF + 2.5 μF = 27.5 μF
25 μF − 0% = 25 μF − 0 μF = 25 μF

d. 370 Ω + 5% = 370 Ω + 18.5 Ω = 388.5 Ω
370 Ω − 5% = 370 Ω − 18.5 Ω = 351.5 Ω

e. 100 000 + 0% = 100 000 + 0 = 100 000
100 000 − 20% = 100 000 − 20 000 = 80 000

Meaning of Significant Figures

While the parts used to construct electronic equipment can have all sorts of values within their tolerance ranges, and sometimes outside them as well, numbers are ordinarily regarded as being exact. The numbers used in expressing a value, such as the values of resistance, or capacitance, or inductance, etc. are known as *significant figures*. If, for example, you have a current of 12 amperes flowing through a wire, this is an accuracy having two significant figures and these are 1 followed by 2. Note that the current could be fractionally higher or lower than 12 amperes. All we know, though, is the accuracy of measurement to the nearest ampere. However, if you write 12.0 amperes, you become accurate to three significant figures. This means the measurement is now correct, not only to the nearest whole ampere, but to the nearest tenth of an ampere. However, the number is not absolutely accurate, for 12.0 amperes does not take into account that it might include variations of hundredths of an ampere. If you write 12.00 amperes, then the accuracy is correct to four significant figures.

The following rules may be used to determine the number of significant digits.

1. **All nonzero digits are significant.**
2. **Zeros used as place holders are *not* significant unless we know the number is exact.**
3. **Zeros not used as place holders are significant.**

Typically, some confusion arises concerning the number of significant figures when zeros are present. Assume for a moment that you count 100 fuses in a carton. You know that the number is *exact* because you counted each fuse. However, most of the numbers in electronics are *approximate*. A resistor on a schematic is marked 1 000 ohms. As this resistor has a tolerance, it is doubtful if the value is exactly 1 000 ohms. This number uses three zeros as place holders and, therefore, has one significant figure.

Suppose a resistor must be exactly 1 000 ohms. Then the schematic would indicate 1 000. ohms and the number would have four significant figures. Examine closely the following values and the number of significant figures for each.

Number	Significant figures
350	2
350.08	5
40 720	4
0.0078	2
0.604	3
100 000	1
100 000.	6
9.010	4

Rounding Off Numbers

A number can be made to represent a higher degree or order of accuracy by increasing the quantity of significant figures: 314.0 is more accurate than 314. However, from an economic point of view, unnecessary accuracy can be costly. It is more expensive to make resistors having a 1 percent tolerance than those having a 10 percent tolerance. To insist on 1 percent tolerance resistors for a low-cost radio would increase the cost of the receiver, but would not necessarily result in better performance. The solution to problems in electronics can often be simplified and made easier to handle by not insisting on extreme number accuracy. As an example, the number 3.141 6 is often used in electronics problems. This number, having five significant figures, can frequently be rounded off by reducing it to three significant figures, making it 3.14.

The rule for rounding off is simple. **If the digit to the right of the digit being rounded, has a value of less than 5, just drop it. However, if it has a value of 5 or greater, drop it, but make its adjacent left digit increase by 1.**

Example: Round 147.536 815 to five, four, three, two, one and no decimal places.

Solution:

five decimal places:	147.536 82
four decimal places:	147.536 8
three decimal places:	147.537
two decimal places:	147.54
one decimal place:	147.5
no decimal place:	148

In computations involving numerous operations it is possible that rounding off may lead to a result that is slightly different than the correct response. Throughout this text there will be problems that must be rounded off at two, three, or four decimal places. As you work these problems attempt to use the memory functions of your calculator to avoid writing down long numbers or, worse yet, rounding off a value that is to be used in subsequent operations. Because of this problem there will be many times when your answer will differ slightly from that found in the answer key. When your answer differs considerably, check your calculations without rounding off intermediate steps.

Absolute Error

The formulas used in electronics do not take practical electronics into consideration. In using Ohm's Law, for example, the computation you do may show you require a resistor having a value of 1 037 Ω. The closest actually available physical unit may be 1 000 Ω. The difference between these two numbers is known as the *absolute error* and in this case it is 1 037 Ω − 1 000 Ω = 37 Ω. Absolute error is the difference between the calculated value, based on the use of some formula, perhaps, and the actual value. The actual value is the true value. A 100-Ω resistor with a tolerance of 10 percent, in all probability, is not a 100-Ω resistor. Its actual value could be 103 Ω. If the calculation you have used calls for a resistance of 99 Ω, then in this case the absolute error is 103 Ω − 99 Ω = 4 Ω.

Relative Error

The word *ratio,* as explained earlier, means division. A fraction is a ratio. The ratio of absolute error to an exact value is called *relative error.* If your calculations call for a resistor having a value of 87 Ω, but the true value of the component is 90 Ω, then the absolute error is 90 Ω − 87 Ω = 3 Ω. If you now divide 3 Ω by the exact value, 90 Ω in this case, you will have 0.033 33. This can be converted to a percent by moving the decimal point two places to the right. The *relative error* in this example, then, is 3.333 percent.

Combined Operations

While addition, subtraction, multiplication, and division are the four basic mathematics operations, the solution to a problem in electronics may use just one of these, or all four, or may involve the repetitive use of any one or more. And, while the very human tendency when confronted by a problem is to look for the single formula that will solve it, it is often necessary to use a number of formulas before arriving at a point where a solution becomes possible.

Problems in electronics can frequently be solved in a number of different ways. While the same answer, or reasonably approximate facsimiles of the same answer, can be obtained no matter which route is taken toward a solution, there is no point in doing more mathematics than is necessary. And so, when an electronics problem seems capable of solution in a number of different ways, the practical approach is to make a quick, rough outline of the steps to be taken, and then to select the shortest and easiest route.

There is still one other factor to be considered. The greater the number of steps taken in the solution of an electronics problem, the greater is the chance for error.

Before starting a problem, also consider whether it is practical to round off numbers. This may not be so important if you are using an electronic calculator as an aid, but if you are using pencil and paper, you may find that rounding off supplies a reasonable answer with much less work and less chance for error.

Finally, if your electronics problems involve formulas, and they often do, write the formulas as a necessary part of the work in problem solution. This has two advantages. Writing formulas is a way of memorizing them and so each time you write a formula you impress it on your memory. The other important factor is that, in any problem, the letters of the formula must be replaced by numbers somewhere along the line. It is easier to substitute numbers for letters if the letters are right in front of you. And there's also less chance for error.

EXERCISE 1·4

In problems 1–10 find the maximum and minimum values based upon the given tolerance.

1. 550 Ω ± 10%
2. 15 V ± 20%
3. 2 200 Ω ± 5%
4. 38 V + 3%
5. 50 μF ± 15%
6. 110 V ± 1%
7. 455 000 Hz ± 2%
8. 1 800 Ω ± 10%
9. 27 000 Ω ± 1%
10. 27 000 Ω ± 10%

In problems 11–20 indicate the number of significant figures present in each approximate number.

11. 60
12. 120.03
13. 6 200
14. 0.009 05
15. 10 000.5
16. 0.08
17. 10.103
18. 100
19. 47 000 000
20. 500 702

In problems 21–25 round the value to the number of decimal places indicated.

21. 1.37 (1 decimal place)
22. 46.872 36 (3 decimal places)
23. 700.597 (2 decimal places)
24. 64 206.291 4 (3 decimal places)
25. 1 989.73 (no decimal places)

UNIT 1·5 PROBLEM REVIEW

1. Add. 370 Ω + 5 600 Ω + 3 900 Ω
2. Add. 47 000 Ω + 100 000 Ω + 84 000 Ω
3. Three series-connected resistors are marked as 100 ohms, 700 ohms, and 1 000 ohms. What is the total resistance?
4. Add −167 V, 283 V, 64 V, and −197 V.
5. Four resistors are connected in series. Find the total resistance if these values are 56 000 Ω, 39 000 Ω, 2 700 Ω, and 1 800 Ω.
6. Three coils are connected in series so that their magnetic fields do not interact. Find the total inductance if L_1 = 500 mH, L_2 = 700 mH, and L_3 = 30 mH.
7. Three batteries are connected in series-aiding. The first has an output of 12 volts, the second an output of 9 volts, the third is 6 volts. What is the total voltage?

8. What is the product of 913 and 37?
9. What is the product of 135 and 9 824?
10. Multiply these numbers.
    a. $-83(13)(6)$
    b. $157 \times 93 \times 10$
    c. $(-42\text{ V})(-133)$
    d. $(-5)(-2)(10)(-5)$
    e. $(783)(3\,991)(14)$
11. The base current (I_b) of a transistor is 75 microamperes. Find the collector current (I_c) if the gain (β) is 85. Note: $I_c = \beta I_b$
12. How much power is dissipated by a 4-ohm resistor when a current of 6 amperes flows through it?
13. Subtract 389 from 1 040.
14. Subtract -37 from 101.
15. Subtract -27 from -46.
16. A voltage of 24 V is connected in series-opposing with a voltage of 8 V. What is the resultant voltage?
17. An electronic mixer can produce the difference between two input frequencies. Find the difference frequency if the input frequencies are 1 495 000 hertz and 1 040 000 hertz.
18. Two coils, L_1 and L_2, are connected in series-aiding. Coil L_1 is 10 H, L_2 is 4 H, and the mutual inductance (L_M) is $\frac{1}{2}$ H. What is the total inductance (L_t)? Note: $L_t = L_1 + L_2 + 2L_M$
19. Two coils, each having an inductance of 5 H, are connected in series-opposing. The mutual inductance (L_M) is 1 H. What is the total inductance? Note: $L_t = L_1 + L_2 - 2L_M$
20. Divide these numbers.
    a. $-481\text{ V} \div 37$
    b. $414\ \mu\text{A} \div 46\ \mu\text{A}$
    c. $180 \div -12$
21. Divide these numbers.
    a. $-891 \div -33$
    b. $1\,919\text{ V} \div 101$
    c. $4\,158 \div -77$
22. What is the voltage drop across a 48-Ω resistor when 2 A flows through it?
23. How much current flows through a 22-Ω resistor connected across a 154-V power supply?
24. A current of 2 A flows through a resistor when the resistor is connected across 238 V. What is the value of the resistance?
25. A current of 3 A flows through a resistor when it is connected across 91 V. How much power is dissipated by the resistor?
26. A current of 4 A flows through a resistor that is using 260 W. What is the amount of voltage across the resistor?
27. How much current flows through a resistor dissipating 840 W when it is connected across a 120-V source?

28. Simplify these problems.
    a. $(-101 \times 11) + (-4\ 728)$
    b. $[-8(62 \times 3)] \div 31 + 387$
    c. $[(386 \times 4) - (-1\ 928)] \div [(491 - 46 \times 10)]$
29. Simplify these problems.
    a. $(1\ 060 \div 4)(11 - 32) + 785$
    b. $[(75 \times 75) \div 5 + 11] \times -42$
    c. $-18 \div [(69 \div 3)(-16 + 2) + 304]$
30. What is the turns ratio of a transformer having 600 secondary turns and 200 primary turns?
31. A transformer having a 6:1 turns ratio has 120 V across its primary winding. What is the secondary voltage?
32. Find the missing quantity in these proportions.
    a. 78 V:13 V = c:147 V
    b. a:1 W = 841 W:29 W
    c. 172:49 192 = 11:d
33. A transformer conducts a primary current of 2 A. Find the secondary current if the primary voltage is 120 V and the secondary voltage is 8 V.
34. Simplify these problems.
    a. $9\frac{7}{8} + 3\frac{1}{2} \times 2\frac{1}{3}$
    b. $(6\frac{15}{32} - 2\frac{7}{16}) \div \frac{5}{6}$
    c. $-4\frac{1}{2}(7\frac{2}{5} + \frac{2}{3} - \frac{11}{20})$
    d. $108\frac{5}{6} \div 10\frac{3}{4}$
    e. $(8\frac{4}{7} \times 9\frac{2}{3} - 6) \div 8$
35. Simplify these problems.
    a. $16\frac{37}{64} - 7\frac{3}{4} + 4\frac{13}{16}$
    b. $1\frac{3}{8}(11\frac{1}{2} - 3\frac{9}{11})$
    c. $(5\frac{25}{64} \times 7) \div 2\frac{1}{2}$
36. Three resistors are connected in parallel. Find total resistance (R_t) if these values are 8 ohms, 12 ohms, and 24 ohms.
    Note:

$$R_t = \frac{1}{\frac{1}{R_1} + \frac{1}{R_2} + \frac{1}{R_3}}$$

37. A resistor has a value of 100 ohms. What is its conductance in siemens?
38. At the end of three hours of continuous operation, the temperature of the metal case of a power transformer was measured at 140°F. What is this temperature in degrees Celsius?
39. A transformer operating at 90 percent efficiency from a 220-V line has a primary current of 4 amperes. What is the amount of secondary power?
40. What are the possible minimum and maximum values of a 1 200-ohm resistor with a 5% tolerance?

41. Round off the following numbers to the place indicated.
    a. 733.2 (no decimal places)
    b. 3.816 (2 decimal places)
    c. 201 (tens place)
42. a. What is 43% of 497?
    b. What is 18% of 929?
43. Convert 60% to a fraction and then reduce it to its lowest terms.
44. 43 is 20% of what number?
45. 68 is 40% of what number?
46. A resistor is marked 69 000 Ω ± 5%. What are the maximum and minimum values of this resistor?
47. A 150-μF capacitor has a minimum tolerance of −5% and a maximum of +10%. What are the lower and upper limits?
48. Round 53.749 to one decimal place.
49. Round 149.513 to no decimal places.
50. Round 1 999.956 to two decimal places.

CHAPTER 2

THE DECIMAL NUMBER SYSTEM

The decimal number system, the most commonly used number system, is a system that uses 10 different symbols to represent numbers. Other number systems, such as the binary and the hexadecimal, can be part of a computer system, but in the final analysis, the symbols of all other number systems must ultimately be converted to decimal form. A computer may be used for billing purposes, but the end product, the bills, are in decimal form.

UNIT 2·1 DECIMAL SYSTEM FUNDAMENTALS

Objectives:

After studying this unit, you should be able to

- discuss the foundations of the decimal number system.
- discuss circulating and repeating decimals.
- perform conversions between English and metric units of measurement.

Radix

The number of symbols used by a system is known as its *radix* or *base*. The radix or base of the decimal system is 10 because it uses 10 symbols, 0, 1, 2, 3, 4, 5, 6, 7, 8, and 9. In the binary system, the base is 2. The two symbols used in the binary system are 0 and 1. Any number larger than 1 in the binary system consists of an arrangement of 0 and 1. This should not be surprising, since we follow a similar setup in the decimal system. In the decimal system, any number larger than 9 consists of some combination of the original 10 symbols. A number such as 367 consists of 3, 6, and 7, all of which are part of the basic symbols of the decimal system.

In the octal system, the radix is 8 and the symbols used are 0, 1, 2, 3, 4, 5, 6, and 7. In the trinary system, the radix is 3, and the symbols are 0, 1, and 2.

Subscripts

Using the same symbols for a variety of number systems does create a problem. We must know whether 110 is a number in the decimal system, the binary system, the trinary, the octal, or some other system. If 110 is in the decimal system, the value is one hundred ten. But if it is in the binary system, it has a decimal equivalent value of six.

To avoid confusion because of similarity of symbols, a radix subscript is sometimes used to identify the system. A number such as 110_{10} indicates definitely that the symbols are part of the decimal system. A number such as 110_2 means that the symbols belong to the binary. When no radix subscript is present, the number is understood to be to the base ten.

Radix subscripts may not always be used where the identity of the symbols is obvious. If you are working with a problem and are using decimal notation throughout, there would be no point in attaching a radix subscript to every number. In fact, it would take time and accomplish nothing. But if you have a problem in which number systems are intermixed, then it will be necessary to identify the numbers by a subscript to avoid misinterpretation.

Decimals

Just as it is possible to have a complete pie, or to have it sliced into segments, so too can we have whole numbers or parts of whole numbers. A number such as 1 is a whole number. It is a complete number and can be thought of as representing a complete unit of something. Because 1 is a whole number, every number that is a multiple of 1 is also a whole number. For example, 2 is a whole number since it consists of the sum of 1 + 1, and 3 is a whole number because it is made up of 1 + 1 + 1.

A portion or part of a whole number is known as a fraction:

$$\frac{3}{4}$$

The number in the denominator tells us into how many parts the unit is divided, while the number in the numerator tells us how many of these parts are being discussed. In the fraction shown, the number 1 has been divided into 4 parts and we are concerned with three of these four parts. A fraction is an instruction to perform division. When the division is completed, the result is a decimal. When 3 is divided by 4, the result is 0.75.

Decimal Point or Radix Point

In a number such as 0.75, the point placed between the 0 and the 7 is called a radix point, but because the number is in the decimal system, it is more commonly known as the decimal point. In the binary system, the radix point would then be called the binary point, in the trinary system, the trinary point, and so on. Numbers to the left of the decimal point are whole numbers; those to the right of it are decimals. If the decimal point is followed by a decimal, but there is no whole number, then the symbol zero is often used in the units column. A number such as .35 is preferably written as 0.35 since the 0 indicates a zero whole number condition and also calls attention to the existence of the decimal point.

Decimal Columnar Position

Just as the value of a whole number is dependent on its columnar position, so too is the value of a decimal determined by its columnar placement. In the case of a whole number, the lowest value is occupied by the units column, or the least significant digit (LSD) column. Whole numbers have increasingly larger values as they move column by column toward the left, until the most significant digit (MSD) or, left-most column is reached.

For decimals, however, the decimal having the greatest numerical value is the one immediately following the decimal point. Decimal place values are shown in Figure 2–1. The first column is the tenths column, the next the hun-

DECIMAL POINT	TENTHS	HUNDREDTHS	THOUSANDTHS	TEN THOUSANDTHS	HUNDRED THOUSANDTHS	MILLIONTHS	
•	1	6	3	5			$= 0.163\ 5 = \frac{1\ 635}{10\ 000}$
•	2	8					$= 0.28 = \frac{28}{100}$
•	1	9	5				$= 0.195 = \frac{195}{1\ 000}$
•	0	0	0	0	7		$= 0.000\ 07 = \frac{7}{100\ 000}$

FIGURE 2–1 Lines can be used to indicate columnar positioning, but are usually omitted. The first number following the decimal point, the number in the tenths column, is the high-order digit or MSD of the decimal. Note the use of zeros as placeholders in 0.000 07.

dredths, and so on. And, in the case of decimals, whether used in conjunction with whole numbers or not, the decimal point must always be included. When writing just a whole number, however, the decimal point is often omitted, unless there is some special reason for including it or emphasizing it.

As in the case of whole numbers, a decimal number may be considered as the addition of numbers occupying various decimal columns. A number such as 0.387 9, for example, may be regarded as:

$$
\begin{array}{r}
0.300\ 0 \\
0.080\ 0 \\
0.007\ 0 \\
+0.000\ 9 \\
\hline
0.387\ 9
\end{array}
$$

Use of the Number Zero

It is rather easy to make errors when using the number zero since zero is commonly (but mistakenly) thought of as a number that has no value, as contrasted with a number such as 1 or 4 or 7. The number zero can also be used improperly even when it does not alter the value of an existing number.

Consider a number such as 348, for example. It has three significant places, emphasized when the number is written as 348 with the decimal point following the LSD. This number is now regarded as accurate to the nearest unit. If the same number is written as 348.0, its numerical value does not change, but now the number is correct to four significant figures.

In the case of decimals, the inclusion of zeros following the LSD column, changes the number of significant figures but not the value. For example, 0.35 has the same value as 0.350 0 but not the same number of significant figures.

The number zero also has a few other interesting characteristics. It is always located at the same distance from a pair of oppositely signed numbers. For example, zero is halfway between +7 and −7, or +5 and −5, or +9 and −9. Zero is always the answer to a problem in addition when numbers of equal value, but opposite sign, are added:

$$(+7) + (-7) = 0$$
$$(-5) + (+5) = 0$$

Multiplication of any number by zero whether it is a whole number, decimal, or fraction, always results in zero:

$$6 \times 0 = 0$$
$$\tfrac{1}{5} \times 0 = 0$$
$$0.786\ 5 \times 0 = 0$$

Division by zero is undefined. The reason for this is that the result of any division operation involving division by zero is infinity. To understand why this is so, review this example very carefully.

Example: Divide 4 by (a) 2, (b) 1, (c) $\frac{1}{2}$, (d) $\frac{1}{4}$, and (e) $\frac{1}{100}$.

Solution:

a. $4 \div 2 = 2$
b. $4 \div 1 = 4$
c. $4 \div \frac{1}{2} = \frac{4}{1} \times \frac{2}{1} = 8$
d. $4 \div \frac{1}{4} = \frac{4}{1} \times \frac{4}{1} = 16$
e. $4 \div \frac{1}{100} = \frac{4}{1} \times \frac{100}{1} = 400$

Notice that as 4 is divided by smaller and smaller numbers, the resulting quotients increase in value. As the divisor approaches zero the quotients will rapidly increase towards infinity. As infinity is undefined, division by zero is also undefined.

Circulating and Repeating Decimals

As mentioned earlier, every decimal is the arithmetic result of the division called for by a fraction. The fraction $\frac{18}{19}$ is an instruction to perform division and when that instruction is followed the result or quotient of the fraction is 0.947 368 to six decimal places.

It would seem that a fraction and its decimal would have identical values, but this will not always be the case. A fraction such as $\frac{1}{2}$ can be converted to a decimal, 0.500 00, which is the exact equivalent. Consider a fraction such as $\frac{1}{3}$. Its decimal equivalent may be 0.3, or 0.33, or 0.333, or 0.333 3. A number such as 0.33 is larger than 0.3, and 0.333 is larger than 0.33. Yet each of these decimals may be used as the decimal equivalent of $\frac{1}{3}$. In this sense, then, $\frac{1}{3}$ can be regarded as a more exact number than 0.3. The number of places in the decimal depends upon how precise the answer needs to be.

In some instances, a fraction will produce a decimal which contains a group of digits used over and over again. For example:

$$\frac{15}{37} = 0.405\ 405\ 405\ .\ .\ .$$

If 405 is called a triad, or grouping of three numbers, then 0.405 405 405 contains three identical triads. This is a type of decimal known as a *circulating decimal.* A *repeating decimal* is one in which a single number is repeated. For example:

$$\frac{1}{3} = 0.333\ 3\ .\ .\ .$$

The decimal number 0.333 3 is an example of a repeating form of a circulating decimal.

The Metric and English Systems

While the decimal system conveniently uses 10 symbols, not all measurements are based on 10. In the English system, an ordinary ruler is 12 inches long, not 10 inches. The inch is subdivided into eighths and sixteenths, not into tenths. The yardstick is 36 inches long, not 30 or 40 inches (30 and 40 are multiples of 10).

The metric system, widely used in the solution of problems in electronics, uses 10 as a base. A metre stick, for example, is similar to a yardstick, except that all of its divisions and markings are based on 10. This makes it easier to use than a yardstick.

In the English system, all sorts of multiples must be used. Some equivalent units of English length measurements are listed.

$$
\begin{aligned}
1 \text{ foot (ft)} &= 12 \text{ inches (in)} \\
1 \text{ yard (yd)} &= 3 \text{ feet (ft)} \\
1 \text{ mile (mi)} &= 1\,760 \text{ yards (yd)} \\
1 \text{ mile (mi)} &= 5\,280 \text{ feet (ft)}
\end{aligned}
$$

In the metric system, however, all units of measurement are related by the number 10.

1 metre = 10 decimetres = 100 centimetres = 1 000 millimetres
1 kilometre = 10 hectometres = 100 dekametres = 1 000 metres

Conversions between the English and metric systems occur often in the electronics field. The majority of these conversions involve linear measure. Table 2–1 contains common conversions that will be useful in understanding the examples and problems presented in this unit.

Example: Convert the following measurements as indicated. Decimal values should be rounded off to three decimal places.

a. 2 miles to kilometres
b. 795 centimetres to metres
c. 2 yards 1 foot $3\frac{3}{4}$ inches to metres

Solution:

a. $\dfrac{2 \text{ mi}}{1} \times \dfrac{1.609 \text{ km}}{\text{mi}} = 3.218 \text{ km}$

b. $\dfrac{795 \text{ cm}}{1} \times \dfrac{1 \text{ m}}{100 \text{ cm}} = 7.95 \text{ m}$

c. $2 \text{ yd } 1 \text{ ft } 3\frac{3}{4} \text{ in} = 72 \text{ in} + 12 \text{ in} + 3.75 \text{ in} = 87.75 \text{ in}$

$$\frac{87.75 \text{ in}}{1} \times \frac{2.54 \text{ cm}}{\text{in}} \times \frac{1 \text{ m}}{100 \text{ cm}} = 2.229 \text{ m}$$

TABLE 2–1

Linear Equivalents

ENGLISH TO METRIC		METRIC TO ENGLISH	
English	Metric	Metric	English
1 inch	25.4 millimetres	1 millimetre	0.039 4 inches
1 inch	2.54 centimetres	1 centimetre	0.394 inches
1 foot	0.304 8 metres	1 metre	39.37 inches
1 yard	0.914 4 metres	1 metre	3.281 feet
1 mile	1.609 kilometres	1 kilometre	0.621 4 miles

Calculator Tip

Your calculator will be extremely useful in converting between systems of measurement. Note the sequence followed to convert 60 miles per hour to metres per second.

A. The first step is to set up the correct multiplication factors that will change the original units to the desired units.

$$\frac{60\ \cancel{mi}}{1\ \cancel{h}} \times \frac{1.609\ \cancel{km}}{1\ \cancel{mi}} \times \frac{1\ 000\ m}{1\ \cancel{km}} \times \frac{1\ \cancel{h}}{60\ \cancel{min}} \times \frac{1\ \cancel{min}}{60\ s}$$

Note that in the above process that all units "cancel" out leaving metres per second (m/s).

B. The sequence followed on the calculator would be as follows. Note that the reciprocal function is used to simplify the process.

60 [×] 1.609 [×] 1 000 [×] 60 [1/x] [×] 60 [1/x] [=] 26.817 or 26.817 m/s.

Problems involving conversions among different units of measurement are often simplified by setting up the expression to produce the desired units. This helps to ensure that the numbers involved in the conversions are subjected to the correct operations. The following examples demonstrate the method used in canceling units to obtain the correct response.

Example: Convert 52 metres per second to miles per hour. Decimal values in the final answer should be rounded off to three decimal places.

Solution: Setting up the value as a ratio often makes the conversion clearer in terms of units of measurement. In this case we have:

$$\frac{52\ \text{metres}}{1\ \text{second}}$$

To convert metres to miles it is first necessary to convert metres to kilometres. We need to multiply by a second ratio such that the original value remains unchanged, but that the resulting units are kilometres per second. As there are 1 000 metres in a kilometre, our conversion becomes:

$$\frac{52\ \cancel{\text{metres}}}{1\ \text{second}} \times \frac{1\ \text{kilometre}}{1\ 000\ \cancel{\text{metres}}} = \frac{52\ \text{kilometres}}{1\ 000\ \text{seconds}}$$

Since 1 kilometre = 0.621 4 miles, our next step becomes:

$$\frac{52\ \cancel{\text{kilometres}}}{1\ 000\ \text{seconds}} \times \frac{0.621\ 4\ \text{miles}}{1\ \cancel{\text{kilometre}}} = \frac{32.312\ 8\ \text{miles}}{1\ 000\ \text{seconds}}$$

Follow the conversion of 1 000 seconds to minutes very carefully:

$$\frac{32.312\ 8\ \text{miles}}{1\ 000\ \cancel{\text{seconds}}} \times \frac{60\ \cancel{\text{seconds}}}{1\ \cancel{\text{minute}}} \times \frac{60\ \cancel{\text{minutes}}}{1\ \text{hour}} = 116.326\ \text{miles/hour}$$

It may be helpful to review the conversion by examining the units of measurement only:

$$\frac{\cancel{\text{metres}}}{\cancel{\text{second}}} \times \frac{\cancel{\text{kilometres}}}{\cancel{\text{metre}}} \times \frac{\text{miles}}{\cancel{\text{kilometre}}} \times \frac{\cancel{\text{seconds}}}{\cancel{\text{minute}}} \times \frac{\cancel{\text{minutes}}}{\text{hour}} = \frac{\text{miles}}{\text{hour}}$$

Example: Convert 55 800 miles in 0.3 seconds to kilometres per second.

Solution:

$$\frac{55\ 800\ \cancel{\text{mi}}}{0.3\ \text{s}} \times \frac{1.609\ \text{km}}{1\ \cancel{\text{mi}}} = \frac{89\ 782.2\ \text{km}}{0.3\ \text{s}} = 299\ 274\ \text{km/s}$$

EXERCISE 2·1

Convert the following measurements as indicated. Decimal values may be rounded off to three decimal places.

1. 3.5 miles to inches
2. 1.2 kilometres to feet
3. 35.56 centimetres to inches
4. 9 feet to centimetres
5. $7\frac{1}{2}$ inches per second to centimetres per second
6. 100 metres to feet
7. $17\frac{1}{2}$ feet to yards
8. 482 centimetres to feet
9. 21 inches to millimetres
10. 100 centimetres to millimetres
11. 520 yards to millimetres
12. 62 metres per 0.5 seconds to miles per hour
13. 98 decimetres per 0.02 minutes to miles per hour
14. 35 miles per hour to inches per second
15. 147.2 metres per second to feet per hour
16. 12.29 miles to decimetres
17. 4 yards 2 feet $11\frac{3}{8}$ inches to metres
18. 18 480 feet per hour to miles per hour
19. 115 centimetres to feet
20. 8 metres per 0.8 seconds to kilometres per hour

UNIT 2·2 DECIMAL AND FRACTION CONVERSIONS

Objectives:

After studying this unit, you should be able to

- convert fractions to decimals.
- convert decimals to fractions.

Converting Fractions to Decimals

The conversion of a common fraction into a decimal is always a problem involving the division of the numerator by the denominator. The result or quotient is always a decimal number.

Example: Change the following to decimal numbers.

a. $\frac{3}{8}$
b. $\frac{13}{16}$
c. $2\frac{1}{4}$
d. $\frac{3}{22}$
e. $\frac{1}{13}$

Solution: a. $3 \div 8 = 0.375$
b. $13 \div 16 = 0.812\ 5$
c. $2\frac{1}{4} = \frac{9}{4} = 9 \div 4 = 2.25$
d. $3 \div 22 = 0.136\ 363\ 6\overline{36}$ (repeating decimal)
e. $1 \div 13 = 0.076\ 923$ (rounded to the millionths place)

Converting Decimals to Fractions or Mixed Numbers

The conversion of a decimal to a fraction involves division by 10 (or some multiple of 10) and then reducing the resulting fraction to lowest terms.

Example: Change the following to fractions or mixed numbers.

a. 0.8
b. 0.625
c. 5.687 5
d. 0.44

Solution: a. $0.8 = \frac{8}{10} = \frac{4}{5}$
b. $0.625 = \frac{625}{1\ 000} = \frac{125}{200} = \frac{5}{8}$
c. $5.687\ 5 = 5\frac{6\ 875}{10\ 000} = 5\frac{1\ 375}{2\ 000} = 5\frac{275}{400} = 5\frac{11}{16}$
d. $0.44 = \frac{44}{100} = \frac{22}{50} = \frac{11}{25}$

Table of Decimal Equivalents

As an aid in the conversion of decimals to fractions, it is helpful to have the use of a table of decimal equivalents, such as Table 2–2.

TABLE 2–2

Table of Decimal Equivalents

Fraction	Decimal	Fraction	Decimal	Fraction	Decimal	Fraction	Decimal
$\frac{1}{64}$	0.015 625	$\frac{17}{64}$	0.265 625	$\frac{33}{64}$	0.515 625	$\frac{25}{32}$	0.781 25
$\frac{1}{32}$	0.031 25	$\frac{9}{32}$	0.281 25	$\frac{17}{32}$	0.531 25	$\frac{51}{64}$	0.796 875
$\frac{3}{64}$	0.046 875	$\frac{19}{64}$	0.296 875	$\frac{35}{64}$	0.546 875	$\frac{13}{16}$	0.812 5
$\frac{1}{16}$	0.062 5	$\frac{5}{16}$	0.312 5	$\frac{9}{16}$	0.562 5	$\frac{53}{64}$	0.828 125
$\frac{5}{64}$	0.078 125	$\frac{21}{64}$	0.328 125	$\frac{37}{64}$	0.578 125	$\frac{27}{32}$	0.843 75
$\frac{3}{32}$	0.093 75	$\frac{11}{32}$	0.343 75	$\frac{19}{32}$	0.593 75	$\frac{55}{64}$	0.859 375
$\frac{7}{64}$	0.109 375	$\frac{23}{64}$	0.359 375	$\frac{5}{8}$	0.625	$\frac{7}{8}$	0.875
$\frac{1}{8}$	0.125 0	$\frac{3}{8}$	0.375	$\frac{41}{64}$	0.640 625	$\frac{57}{64}$	0.890 625
$\frac{9}{64}$	0.140 625	$\frac{25}{64}$	0.390 625	$\frac{21}{32}$	0.656 25	$\frac{29}{32}$	0.906 25
$\frac{5}{32}$	0.156 25	$\frac{13}{32}$	0.406 25	$\frac{43}{64}$	0.671 875	$\frac{59}{64}$	0.921 875
$\frac{11}{64}$	0.171 875	$\frac{27}{64}$	0.421 875	$\frac{11}{16}$	0.687 5	$\frac{15}{16}$	0.937 5
$\frac{3}{16}$	0.187 5	$\frac{7}{16}$	0.437 5	$\frac{45}{64}$	0.703 125	$\frac{61}{64}$	0.953 125
$\frac{13}{64}$	0.203 125	$\frac{29}{64}$	0.453 125	$\frac{23}{32}$	0.718 75	$\frac{31}{32}$	0.968 75
$\frac{7}{32}$	0.218 75	$\frac{15}{32}$	0.468 75	$\frac{47}{64}$	0.734 375	$\frac{63}{64}$	0.984 375
$\frac{15}{64}$	0.234 375	$\frac{31}{64}$	0.484 375	$\frac{3}{4}$	0.750	$\frac{64}{64}$	1.000 0
$\frac{1}{4}$	0.250	$\frac{1}{2}$	0.50	$\frac{49}{64}$	0.765 625		

EXERCISE 2·2

Convert the following fractions to decimal numbers. If necessary round to 5 decimal places.

1. $\frac{3}{32}$
2. $\frac{11}{19}$
3. $\frac{15}{16}$
4. $\frac{1}{3}$
5. $\frac{4}{5}$
6. $\frac{34}{85}$
7. $\frac{17}{255}$
8. $\frac{23}{64}$
9. $\frac{9}{11}$
10. $\frac{91}{26}$

Convert the following decimal numbers to fractions or mixed numbers. Express fractions in lowest terms.

11. 0.5
12. 0.437 5
13. 0.68
14. 3.74
15. 0.008
16. 0.031 25
17. 1.137 5
18. 0.016 8
19. 0.252
20. 0.405

UNIT 2·3 ADDITION, SUBTRACTION, MULTIPLICATION, AND DIVISION OF DECIMALS

Objectives:

After studying this unit, you should be able to

- add decimal fractions.
- subtract decimal fractions.
- multiply decimal fractions.
- divide decimal fractions.

Adding and Subtracting Decimals

The operations required for the addition and subtraction of decimal fractions are the same as those for whole numbers. The key in working these problems is to ensure that each of the decimal points is placed directly beneath the previous decimal point. Review these two examples very carefully.

Example: Add 0.107, 33.2, 9.062 8, and 62.91.

Solution:

```
  ① ①
     0.107
   3 3.2
     9.062 8
 +6  2.91
 ----------
 10  5.279 8
```

Note that once the decimal points are lined up that normal addition occurs, including carries.

Example: Subtract 107.296 from 165.82.

Solution:

```
   ⑤①  ⑦⑪①
  1 6̸ 5 . 8̸ 2̸ 0
 −1 0 7 . 2 9 6
 ---------------
    5 8 . 5 2 4
```

Note that the zero does not change the value of the minuend, and makes the borrowing process easier to follow.

Multiplying Decimals

Multiplication of decimals, like addition and subtraction, is the same as for whole numbers. The one difference in decimals, however, is that the number of decimal places in the product is the sum of the decimal places in each of the multipliers. The following examples will demonstrate this rule.

Example: Multiply 0.1 by 0.1.

Solution: $0.1 \times 0.1 = 0.01$

Note the total of two decimal places in the multipliers and two in the product.

Example: Multiply 4.178 by 0.75.

Solution:

$$\begin{array}{r} 4.178 \\ \times\ 0.75 \\ \hline 208\ 90 \\ 2\ 924\ 6 \\ \hline 3.133\ 50 \end{array}$$

Note that there is a total of five decimal places in the multipliers and five decimal places in the product.

Dividing Decimals

Division of decimals is again similar to the division of whole numbers. The first step is to move the decimal point in the divisor to the right until the divisor becomes a whole number. As this movement of the decimal point is essentially multiplication by a multiple of ten (10, 100, 1 000, etc.) then the decimal point in the dividend must also be moved the same number of places. Once the divisor is a whole number then the division may proceed. The following examples will demonstrate this process.

Example: Divide 3.2 by 0.125.

Solution:

$0.125\overline{)3.200} \rightarrow 125\overline{)3\ 200.0}$ (3 places, 3 places)

$$\begin{array}{r} 25.6 \\ 125\overline{)3\ 200.0} \\ 2\ 50 \\ \hline 700 \\ 625 \\ \hline 75\ 0 \\ 75\ 0 \\ \hline \end{array}$$

Example: Divide 4 154.358 by 1.02.

Solution:

$1.02\overline{)4\ 154.358} \rightarrow 102\overline{)415\ 435.8}$ (2 places, 2 places)

$$\begin{array}{r} 4\ 072.9 \\ 102\overline{)415\ 435.8} \\ 408 \\ \hline 7\ 43 \\ 7\ 14 \\ \hline 295 \\ 204 \\ \hline 91\ 8 \\ 91\ 8 \\ \hline \end{array}$$

EXERCISE 2·3

Simplify the following problems. Decimal fractions may be rounded off to the nearest thousandth.

1. 2.75 Hz + 3.16 Hz
2. 1.1 cm + 0.8 cm + 13.09 cm

3. 183.4 mA + 16.7 mA + 90.04 mA
4. 0.38 s + 1.008 s + 4.2 s + 11 s
5. 94.1 Ω + 113 Ω + 12.802 Ω + 8.62 Ω
6. 45.28 kW + 16.4 kW + 7.09 kW
7. 114 kHz + 82.05 kHz + 112.98 kHz + 1 058.273 kHz
8. 15.28 g + 11.711 g + 19.702 g + 4.005 g
9. 16.25 A + 8.32 A + 17.96 A + 7.47 A
10. 1 016.25 m + 983.1 m + 685.662 m
11. 9.753 kV − 6.837 kV
12. 14.2 V − 7.8 V
13. 586.08 Ω − 250 Ω
14. 68 kHz − 53.08 kHz
15. 45.758 s − 16.97 s
16. 5.137 W − 4.986 W
17. 42.7 μF − 9.9 μF
18. 100 s − 24.173 s
19. 14.86 A − 12.9 A
20. 56.287 6 − 27.678 2
21. 3.28 × 11.05
22. 1.01 s × 0.01
23. 0.12 A × 12
24. 16.8 × 1.4 × 2.6
25. 12.14 × 8 × 0.5
26. 38.5 V × 6.21
27. 12.9 × 0.008 × 482
28. 120 V × 0.707
29. 16.2 × 0.18 × 72.9
30. 3.125 × 2.5 × 64 × 0.5 × 4
31. 10 Ω ÷ 0.025
32. 145.08 cycles ÷ 16.12
33. 4.05 ÷ 0.18
34. 9.963 kΩ ÷ 0.9
35. 1 ÷ 0.062 5
36. 29.195 V ÷ 6.5
37. 72 Ω ÷ 0.09
38. 37.2 H ÷ 14.1
39. 39.96 ÷ 2.7
40. 11.1 V ÷ 7.5
41. (18.28 − 1.03)(16.2 + 9.8)
42. (12.7 ÷ 0.1) × 0.5
43. 4.85 − (2 × 0.8)
44. (4.2 ÷ 1.6) + 0.375
45. 42.08 − 16.5 + 13.74
46. 100 − (2.16 × 11.8 − 32)
47. 38.26 ÷ (6.214 + 9.09)
48. 0.18 + 1.4 × 2.6 − 0.75

49. 16 − 2.5 × 3.9 × 0.8
50. 22 ÷ 1.6 + 3.8 × 2

Use Figure 2–2 for problems 51–52.

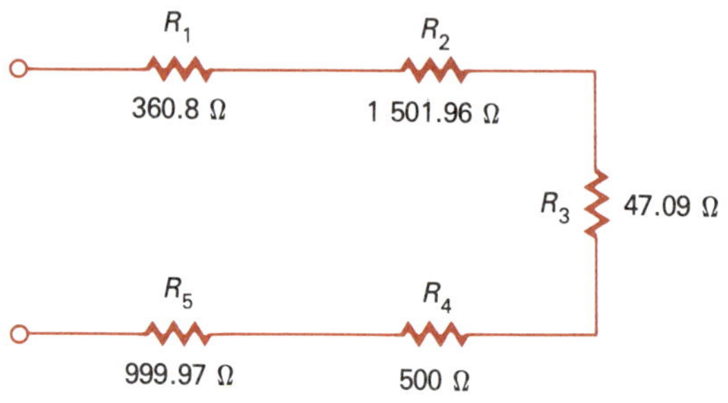

FIGURE 2–2

51. The values of resistors in series are added to form total resistance (R_t). Find total resistance in Figure 2–2.
52. Find R_t in Figure 2–2 if R_3 were to be shorted (zero ohms of resistance).
53. An alternating signal has a peak voltage (E_p) of 167 volts. Find E_{rms} if $E_{rms} = 0.707 \times E_p$.
54. The frequency (f), in hertz, of a signal can be found by calculating the reciprocal of the period (T) in seconds. If the period is 0.000 25 seconds, find the frequency in hertz.
55. The current gain of an amplifier is the output current divided by the input current. Find the gain if the input current is 0.000 8 A and the output is 0.072 A.

UNIT 2·4 EXPONENTIAL NOTATION

Objectives:

After studying this unit, you should be able to

- write numbers in exponential notation.
- multiply and divide exponential numbers.
- add and subtract exponential numbers.
- use a calculator to work with exponential numbers.

The Base and the Exponent

In a number such as 10^5, the digit 5 is the *exponent* while the number 10 is known as a *base.* The exponent indicates the number of times the base is to be multiplied by itself. Thus, $10^2 = 10 \times 10$ or 100. In this case, the exponent tells us that 10 is to be multiplied twice. So $10^3 = 10 \times 10 \times 10$, and $10^4 = 10 \times 10 \times 10 \times 10$.

Any number, or letter, or combination of numbers and letters can be a base: 5^3 is equivalent to $5 \times 5 \times 5$; 7^2 is 7 multiplied by itself, that is, 7×7

$= 7^2$; $(a + b)^2$ indicates $(a + b)$ is to be multiplied by $(a + b)$; $(3x)^2$ is the same as $3x$ multiplied by $3x$ and is equal to $9x^2$.

An exponent is applicable only to an immediately adjacent number or letter, or any group of combinations provided they are enclosed in parentheses. With a number such as 6^2, the exponent, 2, is immediately adjacent to the number 6 and so we have a clear indication that the number 6 is to be squared, that is, to be multiplied by itself. Similarly, in $(6x)^3$ the entire term inside the parentheses is to be cubed. The exponent is an instruction to do this: $(6x)(6x)(6x) = 216x^3$. However, if the term is written as $6x^3$, the exponent applies only to the letter x and not to the number 6. Compare: $(6x)^3$ is equivalent to 6^3 multiplied by x^3; a term such as $6x^3$ is equivalent to 6 multiplied by x^3.

Primes and Subscripts

In electronics problems it is often necessary to distinguish between similar quantities. A circuit might have various resistors which need to be identified. These can be written as $R1$, $R2$, $R3$. This does not mean $R \times 1$, but rather resistor number 1. The numbers can be written on the same line as the letter abbreviation for resistors (R), but are more often positioned a little lower, in which case they are known as *subscripts*. In R_1, the digit 1 is a subscript of R. The number does not enter into any calculations involving the resistor, but is used to distinguish this particular resistor from others, which could be identified as R_2 or R_3. Subscript notation is often preferable since it helps the quick recognition of the letter symbol. Thus $I1$ is sometimes confusing, but is less so when written as I_1. $R1$ is read as R one, while R_1 is read as R sub one.

Primes are used when two similar quantities are involved. The prime is a *superscript* and can be written as I'. Do not confuse this with I^1 in which the 1 is an exponent. The superscript in I' is called a *prime*. Sometimes more than one prime is marked as in I''. This is read as I double prime. The prime marks are useful when two currents flow in a circuit or two voltages are used. Prime marks, like subscripts, are not involved in any way in any arithmetic process using letters or numbers; they are simply identifiers.

Powers of 10

The number 10 followed by an exponent is a form of exponential notation called powers of 10. Here is a partial list of the powers of 10:

$10^0 = 1$	$10^5 = 100\ 000$
$10^1 = 10$	$10^6 = 1\ 000\ 000$
$10^2 = 100$	$10^7 = 10\ 000\ 000$
$10^3 = 1\ 000$	$10^8 = 100\ 000\ 000$
$10^4 = 10\ 000$	$10^9 = 1\ 000\ 000\ 000$

The exponent indicates the number of zeros following 1; $10^0 = 1$, that is, there are no zeros following one in this case. As each exponent increases in value by 1, the value on the right, though, has a value ten times greater than the preceding number; 100 is ten times greater than 10, 1 000 is ten times greater than 100, and so on.

Coefficients

A *coefficient* is a multiplier. With a number such as 200, we can regard 2 as a coefficient of 100; that is $2 \times 100 = 200$. Since 100 can also be written exponentially as 10^2, 200 can be written as 2×10^2. The 2 is a coefficient (multiplier) of 10^2.

A coefficient can be a whole number, a whole number plus a decimal, or simply a decimal. A number such as 265 can be written as:

$$265 \times 10^0$$
$$26.5 \times 10^1$$
$$2.65 \times 10^2$$
$$0.265 \times 10^3$$
$$0.026\,5 \times 10^4$$

Any number can be analyzed as consisting of coefficients and some powers of 10. For example:

$$312 = 300 + 10 + 2$$
$$312 = (3 \times 10^2) + (1 \times 10^1) + (2 \times 10^0)$$

Multiplying Exponential Numbers

When powers of 10 are involved in multiplication, the rule is to add the exponents and to bring the base along unchanged: $10^2 \times 10^1 = 10^3$. Note in this example that exponents 2 and 1 were added: $2 + 1 = 3$. The base 10 appears in the answer unchanged: $10^2 \times 10^1$ is the same as 100×10 since 10^2 equals 100 and 10^1 equals 10. And so $10^2 \times 10^1 = 10^3$ or $100 \times 10 = 1\,000$. In order to add exponents, the bases must be the same. Figure 2–3 shows another sample problem of this type.

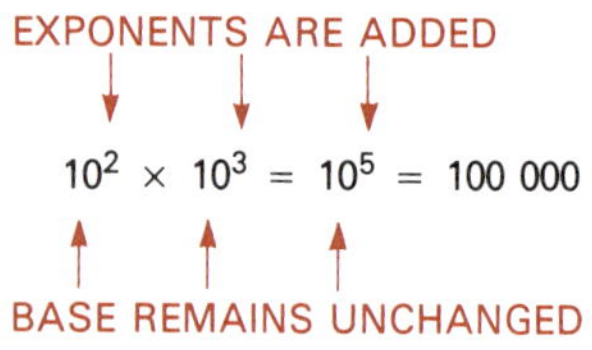

FIGURE 2–3 In multiplying exponential numbers, the exponents are added.

If the powers of 10 have coefficients, these are multiplied just as in the ordinary multiplication process. For example, 200 is 2×10^2 while 3 000 is 3×10^3. The multiplication of 200 by 3 000 is equal to $(2 \times 10^2)(3 \times 10^3) = 6 \times 10^5$ or 600 000. Figure 2–4 shows another sample problem of this type.

ADD THE EXPONENTS

$$(4 \times 10^3)(6 \times 10^2) = 24 \times 10^5 = 24 \times 100\ 000 = 2\ 400\ 000$$

MULTIPLY THE COEFFICIENTS

FIGURE 2–4 In multiplying exponential numbers, the coefficients are multiplied.

Calculator Tip

Your calculator is designed to work with exponents. Prior to the advent of calculators working with extremely large and small numbers was very time consuming. This is no longer the case.

Modern calculators are familiar with exponential notation and entering these numbers is simple. Here are some typical entries utilizing the exponent [EE] function.

Number	*To enter*
3×10^3	3 [EE] 3
0.25×10^6	.25 [EE] 6
857×10^{-8}	857 [EE] [+/−] 8

To multiply (4.8×10^3) by (10.6×10^4) follow this sequence:

4.8 [EE] 3 [×] 10.6 [EE] 4 [=] 5.088×10^8

To check our calculations:

$4.8 \times 10^3 = 4.8 \times 1\ 000 = 4\ 800$
$10.6 \times 10^4 = 10.6 \times 10\ 000 = 106\ 000$
$4\ 800 \times 106\ 000 = 508\ 800\ 000 = 5.088 \times 10^8$

Example: Multiply these values.

a. $(9.2 \times 10^2) \times (1.7 \times 10^3)$
b. $(0.06 \times 10^4) \times (37.2 \times 10^1)$
c. $(45 \times 10^0) \times (0.008 \times 10^2)$

Solution:

a. $(9.2 \times 10^2)(1.7 \times 10^3) = 15.64 \times 10^5 = 1.564 \times 10^6 = 1\ 564\ 000$
b. $(0.06 \times 10^4)(37.2 \times 10^1) = 2.232 \times 10^5 = 223\ 200$
c. $(45 \times 10^0)(0.008 \times 10^2) = 0.36 \times 10^2 = 36$

Example: One form of the power law states that power equals the current flow squared times the resistance. This is written as $P = I^2R$. Compute the power in Figure 2–5.

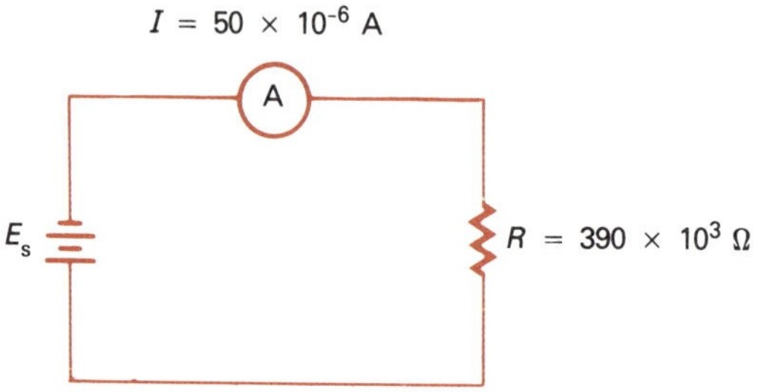

FIGURE 2–5

Solution:

$$\begin{aligned} P &= I^2R \\ &= (50 \times 10^{-6}\ \text{A})(50 \times 10^{-6}\ \text{A})(390 \times 10^3\ \Omega) \\ &= (50 \times 50 \times 390)(10^{-6} \times 10^{-6} \times 10^3)\ \text{W} \\ &= 975\,000 \times 19^{-9}\ \text{W} \\ &= 9.75 \times 10^{-4}\ \text{W} \end{aligned}$$

Calulator Tip

To solve the power problem just presented follow this sequence using the exponent [EE] and squared [x^2] keys.

50 [EE] [+/−] 6 [×] 50 [EE] [+/−] 6 [×] 390 [EE] 3

[=] 9.75×10^{-4},

or

50 [EE] [+/−] 6 [x^2] [×] 390 [EE] 3 [=] 9.75×10^{-4}

The reactance of a coil is a problem in multiplication because inductive reactance is directly proportional to the product of frequency and inductance, multiplied by a constant. The formula is:

$$X_L = 6.28 \times f \times L$$

where f is the frequency in hertz, and L is the inductance in henrys. The result, X_L, is the inductive reactance in ohms. The circuit in Figure 2–6 shows the practical application of exponents being used to compute inductive reactance. This type of calculation will be discussed in a later chapter.

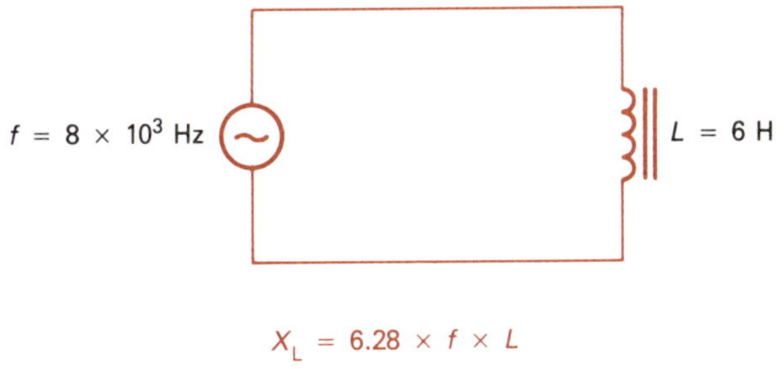

FIGURE 2–6 The use of exponents helps simplify the work involved in solving circuit problems.

Scientific Notation

As long as a problem in electronics involves only the use of relatively small numbers, solutions may not be difficult. But with large numbers, particularly when multiplication or division is involved, handling large numbers can be tiresome, and can easily lead to error. To overcome this difficulty, and to help in problem solving, a mathematical technique known as *scientific notation* is used to describe the product of a coefficient and some power of 10. Thus, to express 345 in scientific notation, we could write it as 3.45×10^2. A value of 65 000 in scientific notation is 6.5×10^4. Note that in both of these examples that the coefficients are between 1 and 10. When a number is expressed in scientific notation it is written as a value equal to, or greater than, 1 and less than 10 times a power of 10.

Expressing numbers in scientific notation was extremely important when values were computed by hand or with a slide rule. Electronic calculators do not require this format. This has resulted in a decrease in the use of scientific notation in recent years.

Example: Express each of the following numbers in scientific notation.

a. 1 470 000
b. 36.5
c. 0.000 85
d. 270×10^5

Solution:

a. $1\,470\,000 = 1.47 \times 10^6$
b. $36.5 = 3.65 \times 10^1$
c. $0.000\,85 = 8.5 \times 10^{-4}$
d. $270 \times 10^5 = 2.7 \times 10^7$

Calculator Tip

The exponent function of your calculator will automatically express numbers in scientific notation. The following sequence will express 0.000 365 in scientific notation.

0.000 365 [EE] [=] and 3.65×10^{-4} is displayed

Typically, once the calculator is instructed to operate in the scientific notation mode, all subsequent displays will also be expressed in scientific notation.

Dividing Exponential Numbers

When powers of 10 are divided, the exponent of the divisor is subtracted from the exponent of the dividend. The base, as in the case of multiplication, remains unchanged.

In a problem such as $10^8 \div 10^5$, all that is required is to subtract exponent 5 from exponent 8, thus: $10^8 \div 10^5 = 10^3$. Similarly, $10^2 \div 10^1 = 10^1 = 10$. As a check, convert 10^2 to 100 and 10^1 to 10. Then $100 \div 10 = 10$.

If the powers of 10 are accompanied by coefficients, these are treated as a simple problem in division, just as though they existed independently of the powers of 10. In the problem (12×10^7) divided by (4×10^3), the answer is 3×10^4. The 3 is obtained by dividing 12 by 4. The 10^4 is the subtraction of 10^3 from 10^7. See Figure 2–7 for another example.

$$\frac{18 \times 10^7}{3 \times 10^4} = 6 \times 10^3 = 6 \times 1\,000 = 6\,000$$

SUBTRACT THIS EXPONENT FROM THE EXPONENT IN THE NUMERATOR

DIVIDE THE COEFFICIENTS

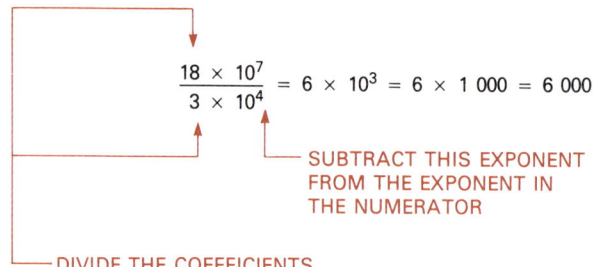

FIGURE 2–7 To divide exponential numbers, subtract the exponent in the denominator from the exponent in the numerator. The coefficients are handled as in ordinary division.

When two exponents are divided the base values must be the same, although the powers to which the bases are raised may differ. Given equal bases the smaller exponent is subtracted from the larger exponent. Review the following examples.

Example: Simplify $8^5 \div 8^3$.

Solution: $\frac{8^5}{8^3} = \frac{\cancel{8} \times \cancel{8} \times \cancel{8} \times 8 \times 8}{\cancel{8} \times \cancel{8} \times \cancel{8}} = \frac{8 \times 8}{1} = 8^2 = 64$

therefore, $8^5 \div 8^3 = 8^{5-3} = 8^2$

Example: Simplify $4^2 \div 4^4$.

Solution: $\frac{4^2}{4^4} = \frac{\cancel{4} \times \cancel{4}}{\cancel{4} \times \cancel{4} \times 4 \times 4} = \frac{1}{4 \times 4} = \frac{1}{16}$

Example: The formula to calculate frequency (f) in a circuit when inductive reactance (X_L) and inductance (L) are known is given by $f = \frac{X_L}{2\pi L}$. Find frequency if $X_L = 3.8 \times 10^6\ \Omega$ and $L = 120 \times 10^{-3}$ H.

Solution: $f = \frac{X_L}{2\pi L} = \frac{3.8 \times 10^6}{2 \times \pi \times 120 \times 10^{-3}} = \frac{3.8}{2 \times \pi \times 120} \times \frac{10^6}{10^{-3}}$

$= 0.005 \times 10^{6-(-3)} = 0.005 \times 10^9 = 5 \times 10^6$ Hz

A more extensive discussion of the multiplication and division of exponents will be presented in the chapter on algebra.

Combined Multiplication and Division

It is possible to have combined multiplication and division using powers of 10, just as it is possible to use these arithmetic functions with ordinary numbers. In a problem such as:

$$\frac{(15 \times 10^4)(20 \times 10^6)}{5 \times 10^2}$$

various initial steps can be taken. The coefficients in the numerator can be multiplied. Or, the coefficient in the denominator can be divided into either of the coefficients in the numerator. This latter step is preferred since its effect is to reduce the size of the problem. If 15 in the numerator is divided by 5 in the denominator, the result will be:

$$\frac{(3 \times 10^4)(20 \times 10^6)}{10^2}$$

The 10^2 in the denominator can divide either 10^4 or 10^6. Dividing 10^4 by 10^2 results in:

$$(3 \times 10^2)(20 \times 10^6)$$

Since $3 \times 20 = 60$, and $10^2 \times 10^6 = 10^8$, the answer is:

$$60 \times 10^8$$

The answer can be carried still another step by considering 60 as 6×10^1. We would then have:

$$6 \times 10^1 \times 10^8 = 6 \times 10^9$$

The form 6×10^9 has exactly the same value as 60×10^8. Which form would be used in the final answer would depend on the job requirements.

Adding and Subtracting Exponential Numbers

In order to add exponential numbers, the bases and the powers must be the same. In other words, coefficients of powers of 10 can be added or subtracted only if the exponents of the powers of 10 are the same.

Example: Add 23×10^2 and 14×10^2.

Solution:

$$\begin{array}{rcr} 23 \times 10^2 & = & 2\,300 \\ +14 \times 10^2 & = & 1\,400 \\ \hline 37 \times 10^2 & = & 3\,700 \end{array}$$

To add 230×10^3 and 15×10^5 one power of ten is changed. This will not present any difficulty as it is easy to shift the significant figures by changing the exponents. It is important to realize that either, or both, of the numbers could be changed. The following examples will demonstrate three methods of adding these numbers.

Example: Add 230×10^3 and 15×10^5 using the numbers expressed as 10 to the fifth power.

Solution: $230 \times 10^3 = 2.3 \times 10^5$
$15 \times 10^5 + 2.3 \times 10^5 = 17.3 \times 10^5 = 1\,730\,000$

Example: Add 230×10^3 and 15×10^5 using the numbers expressed as 10 to the fourth power.

Solution: $230 \times 10^3 = 23 \times 10^4$ and $15 \times 10^5 = 150 \times 10^4$
$23 \times 10^4 + 150 \times 10^4 = 173 \times 10^4 = 1\,730\,000$

Example: Add 230×10^3 and 15×10^5 using the numbers expressed as 10 to the third power.

Solution: $15 \times 10^5 = 1\,500 \times 10^3$
$230 \times 10^3 + 1\,500 \times 10^3 = 1\,730 \times 10^3 = 1\,730\,000$

Exponential numbers can also be involved in subtraction provided all the powers of ten are identical. It is possible to write $(26 \times 10^3) - (12 \times 10^3)$ and do the work required by subtracting one coefficient from the other.

$$\begin{array}{rcr} 26 \times 10^3 & = & 26\,000 \\ -12 \times 10^3 & = & -12\,000 \\ \hline 14 \times 10^3 & = & 14\,000 \end{array}$$

Just as in addition, when numbers with different powers of ten appear in the problem, change one or both so that all powers of ten are the same.

Example: Subtract 15×10^2 from 19×10^3 and express the result in scientific notation.

Solution: $19 \times 10^3 = 190 \times 10^2$

$$\begin{array}{r} 190 \times 10^2 \\ -\ \ 15 \times 10^2 \\ \hline 175 \times 10^2 \end{array} = 1.75 \times 10^4$$

You should understand that modern electronic calculators will automatically add and subtract exponential numbers with different powers of ten. However, it is important that the electronics technician also understands the basic rules that govern these operations.

Negative Powers of 10

Since powers of 10 can represent whole numbers, it is reasonable to expect that they will share the behavior of whole numbers. The reciprocal of a whole number is 1 divided by that number. The reciprocal of 10 is

$$\frac{1}{10}$$

Similarly, the reciprocal of a power of 10 is 1 divided by that power of 10. The reciprocal of 10^4 is

$$\frac{1}{10^4}$$

Another way of writing the reciprocal of a power of 10 is to put a *minus sign in front of the exponent.* Thus, the reciprocal of 10^3 is 10^{-3}. The reciprocal of 10^6 is 10^{-6}. We can set up a table of values for *negative* powers of 10 in a manner similar to that used for positive powers of 10.

$$10^{-1} = \frac{1}{10} = 0.1$$

$$10^{-2} = \frac{1}{10^2} = 0.01$$

$$10^{-3} = \frac{1}{10^3} = 0.001$$

$$10^{-4} = \frac{1}{10^4} = 0.000\ 1$$

$$10^{-5} = \frac{1}{10^5} = 0.000\ 01$$

$$10^{-6} = \frac{1}{10^6} = 0.000\ 001$$

To demonstrate why you should be able to evaluate values with negative exponents, closely follow this example.

Example: Simplify $10^4 \div 10^5$

Solution: The rule states that the exponent of the denominator is subtracted from that of the numerator. This results in 10^{4-5}, or 10^{-1}.

Since $\dfrac{10^4}{10^5} = \dfrac{\cancel{10} \times \cancel{10} \times \cancel{10} \times \cancel{10}}{\cancel{10} \times \cancel{10} \times \cancel{10} \times \cancel{10} \times 10} = \dfrac{1}{10}$

$10^{-1} = \dfrac{1}{10} = 0.1$

Multiplication of Positive and Negative Powers of 10

The rule for the multiplication of powers of 10 (exponents are to be added) applies even if the powers of 10 are negative. In a problem such as: $10^{-2} \times 10^{-2}$, the result is 10^{-4} since $(-2) + (-2) = -4$. If the negative powers of 10 are accompanied by coefficients, these are multiplied in the usual way. For example, $(-3 \times 10^{-2}) \times (4 \times 10^{-5}) = -12 \times 10^{-7}$. The multiplication of numbers having like signs results in a positive number and the multiplication of numbers having unlike signs yields a negative number.

Example: Multiply these exponential numbers.

a. $(6 \times 10^{-2})(-4 \times 10^{-3})$
b. $(-5 \times 10^{-1})(-3 \times 10^{-7})$
c. $(12 \times 10^{-9})(8 \times 10^{-7})$

Solution:

a. $(6 \times 10^{-2})(-4 \times 10^{-3}) = -24 \times 10^{-5}$
b. $(-5 \times 10^{-1})(-3 \times 10^{-7}) = 15 \times 10^{-8}$
c. $(12 \times 10^{-9})(8 \times 10^{-7}) = 96 \times 10^{-16}$

The rule for the addition of exponents is also applicable if the exponents have *different* signs: $(10^{-2})(10^4) = 10^2$ and $(10^8)(10^{-3}) = 10^5$.

Calculator Tip

Try solving the following problem without stopping to write down intermediate steps.

$$\frac{(83.2 \times 10^5)(1.4 \times 10^{-3}) + (19.8 \times 10^3)}{(37.9 \times 10^{-2}) - (6 \times 10^1)(2.5 \times 10^2)}$$

One sequence is as follows:

83.2 [EE] 5 [×] 1.4 [EE] [+/−] 3 [+] 19.8 [EE] 3

[=] [STO] 37.9 [EE] [+/−] 2 [−] 6 [EE] 1 [×] 2.5 [EE] 2

[=] [1/x] [×] [RCL] [=] −2.096 6

EXERCISE 2·4

Multiply these numbers and express the answers in scientific notation. Round answers to the nearest thousandth.

1. $(8.5 \times 10^{4})(0.03 \times 10^{-2})$
2. $(46 \times 10^{10})(1.2 \times 10^{0})(11.05 \times 10^{-19})$
3. $(658 \times 10^{-1})(39 \times 10^{4})(0.08 \times 10^{-2})$
4. $(144 \times 10^{3})(12 \times 10^{-3})$
5. $(1\ 086 \times 10^{-5})(429 \times 10^{4})$
6. $(13.7 \times 10^{-2})(0.68 \times 10^{4})(118 \times 10^{3})$
7. $(1.25 \times 10^{19})(5.9 \times 10^{23})(7.2 \times 10^{-14})$
8. $(19 \times 10^{4})(19 \times 10^{4})$

Divide these numbers and express the answers in scientific notation. If needed, round the answer to the nearest thousandth.

9. $62 \times 10^{3} \div 145 \times 10^{2}$
10. $0.8 \times 10^{6} \div 8 \times 10^{-5}$
11. $37.28 \times 10^{-3} \div 10.5 \times 10^{4}$
12. $3.14 \times 10^{-2} \div 22.8 \times 10^{-5}$
13. $1.25 \times 10^{-4} \div 50 \times 10^{3}$
14. $14.87 \times 10^{2} \div 14.87 \times 10^{-2}$
15. $0.9 \times 10^{-5} \div 1.5 \times 10^{-6}$
16. $112 \times 10^{4} \div 62 \times 10^{10}$

Perform the indicated operations. Express the answers in scientific notation. Round answers to the nearest thousandth.

17. $18 \times 10^{4} + 11 \times 10^{3} - 120 \times 10^{2}$
18. $1\ 600 \times 10^{-1} - 185 + 16 \times 10^{2}$
19. $62.9 \times 10^{-5} + 0.009 \times 10^{-1} - 1.529 \times 10^{-3}$
20. $94 \times 10^{3} - 10 \times 10^{4} - 32.8 \times 10^{5} + 42 \times 10^{5}$
21. $(4\tfrac{3}{4})^{2} - 1\tfrac{7}{8} \times 10^{2} + \tfrac{1}{5} \times 10^{3}$
22. $825 - 10 \times 10^{3} + 7.1 \times 10^{2}$
23. $1.5 \times 10^{-5} - 9.2 \times 10^{-6} + 125 \times 10^{-7}$
24. $4\ 200 \times 10^{-2} - 18 \times 10^{3}$

25. Simplify this problem:

$$\frac{(18.7 \times 10^{3})(28 \times 10^{-5}) - (4 \times 10^{6})(3.1 \times 10^{-3})}{(89 \times 10^{-2})(11.6 \times 10^{5})(100)}$$

26. Simplify this problem:

$$\frac{(-25 \times 10^{-3})(62 \times 10^{4})}{-118 \times 10^{-2}} + \frac{(19.2 \times 10^{-1}) + (8.9 \times 10^{3})}{(32 \times 10^{4})(8.5 \times 10^{-4})}$$

UNIT 2·5 CONVERSIONS AND ELECTRONIC UNITS OF MEASUREMENT

Objectives:

After studying this unit, you should be able to

- discuss the basic units of measurement used in electronics.
- convert units of measurement.
- utilize a calculator to work with units of measurement.
- add and subtract units of measurement.

Electronic Symbols

The field of electronics is made up of many complex units and numbers. In order to communicate correctly it is important that each person in the field use the same language. A set of standard units and symbols is used. Table 2–3 lists the units and symbols most commonly used in electronics.

TABLE 2–3

Quantities and Units of Measurement

Quantity	Unit	Example
Capacitance (C)	farad (F)	C_1 = 0.000 01 F
Capacitive reactance (X_C)	ohm (Ω)	X_C = 45 Ω
Current (I)	ampere (A)	I_1 = 0.5 A
frequency (f)	hertz (Hz)	f = 60 Hz
Impedance (Z)	ohm (Ω)	Z = 50 Ω
Inductance (L)	henry (H)	L_2 = 0.15 H
Inductive reactance (X_L)	ohm (Ω)	X_L = 37 Ω
Power (P)	watt (W)	P = 100 W
Reactance (X)	ohm (Ω)	X = 12 Ω
Resistance (R)	ohm (Ω)	R_2 = 15 Ω
Time (t)	second (s)	t = 0.8 s
Voltage (E or V)	volt (V)	E_s = 12 V

There is a need to make numbers like 68 000 000 ohms or 0.000 052 ampere easier to write. To simplify standard units only specific powers of ten are used. The powers used most are 10^6, 10^3, 10^{-3}, 10^{-6}, and 10^{-12}. Table 2–4 lists the prefixes used with electronic units of measure.

TABLE 2–4

Prefixes Used for Electronics

Prefix	Symbol	Numerical Value	Exponential Value
tera	*T*	1 000 000 000 000	10^{12}
giga	*G*	1 000 000 000	10^9
mega	*M*	1 000 000	10^6
kilo	*K*	1 000	10^3
hecto	*H*	100	10^2
deka	*dk*	10	10^1 or 10
deci	*d*	0.1	10^{-1}
centi	*c*	0.01	10^{-2}
milli	*m*	0.001	10^{-3}
micro	μ	0.000 001	10^{-6}
nano	*n*	0.000 000 001	10^{-9}
pico	*p*	0.000 000 000 001	10^{-12}

- **MEGA:** When mega is used as a prefix it stands for one million or 10^6. For example, 5 megavolts means 5×10^6 volts or 5 000 000 volts. The value 15 megohms means 15×10^6 ohms or 15 000 000 ohms. The letter that indicates mega is M. Therefore 5 megavolts is written 5 MV and 15 megohms is written 15 MΩ.
- **KILO:** When kilo is used as a prefix it stands for 10^3 or 1 000. For example, 5.6 kilohms means 5.6×10^3 ohms or 5 600 ohms. The value 0.85 kilohertz means 0.85×10^3 hertz or 850 hertz. The letter used to indicate kilo is k. Therefore 5.6 kilohms is 5.6 kΩ and 0.85 kilohertz is 0.85 kHz.
- **MILLI:** When milli is used as a prefix it stands for 0.001 or 10^{-3}. For example, 26 milliamperes means 26×10^{-3} ampere or 0.026 ampere. The value 8 milliwatts means 8×10^{-3} watts or 0.008 watts. The letter that indicates milli is m. Therefore 26 milliamperes is 26 mA and 8 milliwatts is 8 mW.
- **MICRO:** When micro is used as a prefix it stands for 10^{-6}, 0.000 001, or one millionth. For example, 75 microhenries means 75×10^{-6} henries or 0.000 075 henries. The value 359 microseconds means 359×10^{-6} second or 0.000 359 second. The letter that indicates micro is μ. This is the greek letter mu. Therefore 75 microhenries is 75 μH and 359 microseconds is 359 μs.
- **PICO:** When pico is used as a prefix it stands for 10^{-12}. For example, 150 picofarad means 150×10^{-12} farad or 0.000 000 000 15 farad. The letter that indicates pico is p. Therefore 150 picofarads is 150 pF.

Adding and Subtracting Electronic Units

There are many times when these units of measurement must be added and subtracted. When units are added or subtracted they must be in the same units. To add 470 ohms and 1.2 kilohms one value or the other must be expressed as the other unit. This would mean 470 ohms plus 1 200 ohms or 0.47 kilohms plus 1.2 kilohms. The answer of course would be 1 670 ohms or 1.67 kilohms.

Choosing the Prefix

There are many units a given answer or value could be expressed in. Assume you have a voltage of 0.25 volt. This could also be written as 250 millivolts, 250 000 microvolts, or 0.000 25 kilovolt. As a general rule, if there are more than three numbers preceding or following a decimal point then a change of units should be made. For a voltage of 0.25 volt, the 250 millivolts would be the most acceptable. Look at these other examples.

$$4\ 700\ \text{V} = 4.7\ \text{kV}$$
$$0.182\ 6\ \text{A} = 182.6\ \text{mA}$$
$$1\ 600\ 000\ \Omega = 1.6\ \text{M}\Omega$$

Conversion of Current

It is often necessary to convert values of resistance, current, inductance, capacitance, and voltage to basic units before using them in formulas. Use of powers of 10 not only helps in making these conversions, but also in the ultimate solutions of the problems in which they are used.

The basic unit of current is the ampere (A). The most used submultiples are the milliampere (mA) and the microampere (μA).

$$1\ \text{A} = 1\ 000\ \text{mA} = 10^3\ \text{mA}$$
$$1\ \text{A} = 1\ 000\ 000\ \mu\text{A} = 10^6\ \mu\text{A}$$
$$1\ \text{mA} = 0.001\ \text{A} = \frac{1}{1\ 000\ \text{A}} = \frac{1}{10^3\ \text{A}} = 10^{-3}\ \text{A}$$
$$1\ \mu\text{A} = 0.000\ 001\ \text{A} = \frac{1}{1\ 000\ 000\ \text{A}} = \frac{1}{10^6\ \text{A}} = 10^{-6}\ \text{A}$$

To assist in visualizing the relationships between these units examine this diagram.

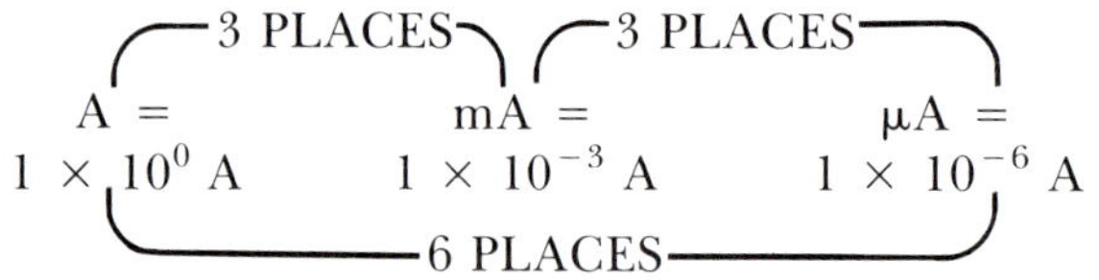

Sometimes it is also necessary to convert microamperes to milliamperes or milliamperes to microamperes

$$1\ \text{mA} = 1\ 000\ \mu\text{A} = 10^3\ \mu\text{A}$$

$$1\ \mu\text{A} = 0.001\ \text{mA} = \frac{1}{1\ 000\ \text{mA}} = \frac{1}{10^3\ \text{mA}} = 10^{-3}\ \text{mA}$$

Example: What is the voltage drop across a 2 000-ohm resistor when a current of 800 μA flows through it?

Solution: $R = 2\ 000\ \Omega = 2 \times 10^3\ \Omega$

$I = 800\ \mu\text{A} = 800 \times 10^{-6}\ \text{A} = 8 \times 10^{-4}\ \text{A}$

The formula is $E = I \times R$

$$= 2 \times 10^3\ \Omega \times 8 \times 10^{-4}\ \text{A}$$
$$= 16 \times 10^{-1}\ \text{V}$$
$$= 1.6 \times 10^0\ \text{V}$$
$$= 1.6\ \text{V}$$

Example: Express 45 mA (a) as amperes (b) as microamperes.

Solution: a. $45\ \text{mA} = 45 \times 10^{-3}\text{A} = 0.045\ \text{A}$

b. $45\ \text{mA} = 45 \times 10^3\ \mu\text{A} = 45\ 000\ \mu\text{A}$

Example: Convert 2.2 A (a) to milliamperes (b) to microamperes.

Solution: a. $2.2\ \text{A} = 2.2 \times 10^3\ \text{mA} = 2\ 200\ \text{mA}$

b. $2.2\ \text{A} = 2.2 \times 10^6\ \mu\text{A} = 2\ 200\ 000\ \mu\text{A}$

Example: Convert 800 μA (a) to milliamperes (b) to amperes.

Solution: a. $800\ \mu\text{A} = 800 \times 10^{-3}\ \text{mA} = 0.8\ \text{mA}$

b. $800\ \mu\text{A} = 800 \times 10^{-6}\ \text{A} = 0.000\ 8\ \text{A}$

The advantage of using powers of 10 in these conversions is that it avoids the use of zeros, with the consequent possibility of dropping or overlooking one of them. It is easier to write 25 μA than to write 0.000 025 A. Making conversions is easier by remembering that the ampere is a thousand times as large as a milliampere and a million times as large as a microampere.

Conversions of Voltage

The basic unit will not always be the largest unit used. The basic unit can be largest, as in the case of current, but this does not establish a precedent. In voltage, for example, the volt (V) is the basic unit, but a multiple, the kilovolt (kV), is a thousand times as large, while another multiple, the megavolt (MV), is a million times greater. The volt also has submultiples: the millivolt (mV) or thousandth of a volt, and the microvolt or millionth of a volt (μV). For voltage submultiples:

$$1\ \text{V} = 1\ 000\ \text{mV} = 10^3\ \text{mV}$$
$$1\ \text{V} = 1\ 000\ 000\ \mu\text{V} = 10^6\ \mu\text{V}$$

$$1\ \text{mV} = 0.001\ \text{V} = \frac{1}{1\ 000\ \text{V}} = \frac{1}{10^3\ \text{V}} = 10^{-3}\ \text{V}$$

$$1\ \mu\text{V} = 0.000\ 001\ \text{V} = \frac{1}{1\ 000\ 000\ \text{V}} = \frac{1}{10^6\ \text{V}} = 10^{-6}\ \text{V}$$

For voltage multiples:

$$1\ \text{V} = 0.001\ \text{kV} = \frac{1}{1\ 000\ \text{kV}} = \frac{1}{10^3\ \text{kV}} = 10^{-3}\ \text{kV}$$

$$1\ \text{V} = 0.000\ 001\ \text{MV} = \frac{1}{1\ 000\ 000\ \text{MV}} = \frac{1}{10^6\ \text{MV}} = 10^{-6}\ \text{MV}$$

$$1\ \text{kV} = 1\ 000\ \text{V} = 10^3\ \text{V}$$

$$1\ \text{MV} = 1\ 000\ 000\ \text{V} = 10^6\ \text{V}$$

To assist in visualizing these relationships examine this diagram.

3 PLACES	3 PLACES	3 PLACES	3 PLACES	
MV =	kV =	V =	mV =	μV =
1×10^6 V	1×10^3 V	1×10^0 V	1×10^{-3} V	1×10^{-6} V

The conversion factor between each submultiple or multiple is 1 000, used either as a dividing or multiplying factor. A kilovolt is 1 000 volts. A volt is 1 000 millivolts. However, to convert from millivolts to kilovolts is a double step. There is a 1 000 factor in going from millivolts to volts and another 1 000 factor in moving from volts to kilovolts. But $1\ 000 \times 1\ 000 = 1\ 000\ 000$ or 10^6. And so, in going from millivolts to kilovolts the factor becomes 10^6.

Example: The voltage across a resistor is 125 kV. What is the voltage in terms of millivolts?

Solution: First, convert the multiple to the basic unit and then convert the basic unit to the submultiple.
$125\ \text{kV} = 125 \times 10^3\ \text{V} = 1.25 \times 10^5\ \text{V}$
$1.25 \times 10^5\ \text{V} = 1.25 \times 10^5 \times 10^3\ \text{mV} = 1.25 \times 10^8\ \text{mV} =$
$125\ 000\ 000\ \text{mV}$

Conversions of Resistance

The basic unit of resistance is the ohm. There are submultiples such as the milliohm and the microhm but these are laboratory curiosities. More practical units are multiples such as the kilohm (kΩ), and the megohm (MΩ).

$$1\ \Omega = 0.001\ \text{k}\Omega = \frac{1}{1\ 000\ \text{k}\Omega} = \frac{1}{10^3\ \text{k}\Omega} = 10^{-3}\ \text{k}\Omega$$

$$1\ \Omega = 0.000\ 001\ \text{M}\Omega = \frac{1}{1\ 000\ 000\ \text{M}\Omega} = \frac{1}{10^6\ \text{M}\Omega} = 10^{-6}\ \text{M}\Omega$$

$$1\ \text{k}\Omega = 1\ 000\ \Omega = 10^3\ \Omega$$

$$1\ \text{M}\Omega = 1\ 000\ 000\ \Omega = 10^6\ \Omega$$

$$1\ \text{k}\Omega = 0.001\ \text{M}\Omega = \frac{1}{1\ 000\ \text{M}\Omega} = \frac{1}{10^3\ \text{M}\Omega} = 10^{-3}\ \text{M}\Omega$$

$$1\ \text{M}\Omega = 1\ 000\ \text{k}\Omega = 10^3\ \text{k}\Omega$$

Conversions of Capacitance, Inductance and Frequency

There are other basic units such as the farad for capacitance, the henry for inductance and the hertz for frequency. Conversions of these basic units follow the same procedure as those outlined for current, voltage, and resistance. The same relationships exist between the basic units, the multiples and the submultiples.

Calculator Tip

When working electronics problems involving different units of measurement it is only necessary to remember the exponents associated with each prefix. The calculator will compute the answer and typically display the answer in scientific notation.

A good example is the calculation of capacitive reactance (X_c) in a circuit. This topic will be addressed in a later chapter, but is also helpful at this point. Given this formula and values the following sequence will produce the answer.

$$X_c = \frac{1}{2\pi fC} \quad \text{where } X_c = \text{capacitive reactance in ohms}$$

π = the constant pi

f = frequency in hertz

C = capacitance in farads

If $f = 5$ kHz and $C = 1\ \mu$F then:

$$X_c = \frac{1}{2\pi(5 \text{ kHz})(1\ \mu\text{F})}, \text{ and}$$

2 [×] π [×] 5 [EE] 3 [×] 1 [EE] [+/−] 6 [=] [1/x] 3.183 1 × 10^1 = 31.831 Ω

Decimal Point Conversions

While conversions are easily handled using powers of 10, it is also possible to do so by moving the decimal point to the left or right as required. As indicated earlier, conversions involve division or multiplication by 1 000 or 1 000 000. Moving a decimal point three places to the right is equivalent to multiplication by 1 000, while moving it six places in the same direction is the same as multiplication by 1 000 000. Conversely, moving the decimal point to the left three places is equivalent to division by 1 000 and moving it six places to the left is the same as division by 1 000 000.

To convert megohms to ohms, multiply megohms by 1 000 000 or move the decimal point six places to the right.

$$2\ \text{M}\Omega = 2\ 000\ 000\ \Omega$$

To convert picofarads to microfarads, divide by 1 000 000 or move the decimal point six places to the left.

$$400\ \text{pF} = 0.000\ 4\ \mu\text{F}$$

The zeros following the digit 4 are dropped since they do not contribute to the value of the number.

Table 2–5 indicates how to move the decimal point when multiplying and dividing by the powers of 10.

TABLE 2–5

Decimal Point Conversions		
Number	To multiply by the number, move the decimal point to the right the number of places indicated.	To divide by the number, move the decimal point to the left the number of places indicated.
10	1	1
100	2	2
1 000	3	3
10 000	4	4
100 000	5	5
1 000 000	6	6

Which is the better technique? Use powers of 10 or the movement of the decimal point when making conversions? That depends on what you want to do with the results. If the purpose of the conversion is simply to change from one form to another such as from a basic unit to a multiple or submultiple, then either decimal point conversion or powers of 10 can be used. But if the results of the conversion are to be plugged into a formula, then it is advisable to use powers of 10 since these lend themselves more easily to the consolidation and simplification of the mathematics to be done.

EXERCISE 2·5

Express each of the following values in the units indicated.

1. 380 mA = _______ A
2. 455 kHz = _______ Hz
3. 280 μA = _______ mA
4. 0.08 kΩ = _______ Ω
5. 9 800 mV = _______ kV
6. 0.01 μF = _______ pF
7. 49 mH = _______ H
8. 380 Ω = _______ kΩ
9. $62\frac{1}{2}$A = _______ mA
10. 100 000 Ω = _______ MΩ
11. 75 kW = _______ W
12. 2 500 000 μA = _______ A
13. 4 500 ms = _______ s
14. 0.5 MΩ = _______ Ω
15. 590 MHz = _______ kHz
16. 0.25 A = _______ μA
17. 0.002 μF = _______ pF
18. 85 000 Hz = _______ MHz
19. 0.47 MΩ = _______ Ω
20. $\frac{1}{20}$ s = _______ μs

Complete these problems and express the answers in the units indicated.

21. 12.5 V + 820 mV = _______ V
22. 0.05 A − 1 500 μA = _______ mA
23. 0.5 MW − 37 000 W = _______ kW
24. 16 × 0.02 W = _______ mW
25. 4.5 MHz ÷ 3 = _______ kHz
26. 0.002 μF + 900 pF = _______ pF
27. 350 μA × 225 = _______ A
28. 150 Ω + 1.7 kΩ = _______ Ω
29. 0.8 MHz ÷ 25 = _______ Hz
30. $7\frac{1}{8}$ × 0.8 mA = _______ μA
31. Use the transistor circuit in Figure 2–8 to calculate the collector current (I_c) in milliamperes. Note that I_c is found by multiplying the base current (I_b) times the gain (β) of the transistor.

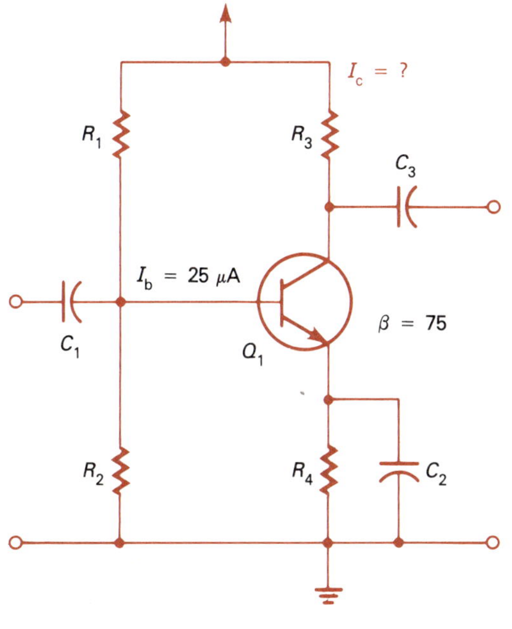

FIGURE 2–8

32. An electronic mixer can subtract two frequencies. The inputs are 0.8 MHz and 345 kHz. Find the difference, in kilohertz, between the frequencies.
33. Find the frequency (f), in kilohertz, of a signal with a period (T) of 40 μs. Note: $f = \frac{1}{T}$
34. The peak voltage of an ac signal is 2.8 V. Find the rms voltage in millivolts. Note: $E_{rms} = E_{peak} \times 0.707$
35. Total resistance (R_t) in a series circuit is the sum of the individual resistors. Find R_t in kilohms in Figure 2–9.

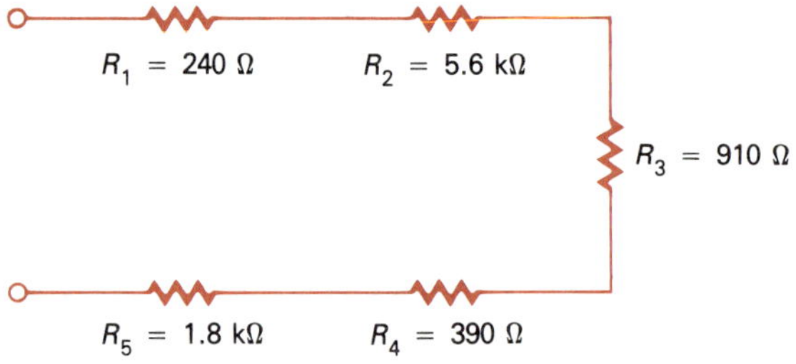

FIGURE 2–9

UNIT 2·6 POWERS AND ROOTS IN ELECTRONICS

Objectives:

After studying this unit, you should be able to

- raise numbers to powers.
- extract the roots of numbers.
- use a calculator to compute powers and roots.

Powers of Numbers

In the system of powers of 10, the multiplication of 10^1 by 10^1 results in 10^2. While this multiplication follows the laws of powers of 10—that is, the exponents are added—the same technique can be used for numbers other than 10. Thus, $6^1 \times 6^1 = 6^2$. But 6^1 is the same as 6, and so $6^1 \times 6^1 = 6 \times 6 = 36$. The exponent, then, indicates the number of times the base is to be multiplied by itself.

The multiplication of a number by itself also has a geometric implication. A square is a geometric figure all of whose sides have the same dimension. If one side of a square is 6, then so are all the other sides. The area of the square is obtained by multiplying 6 by itself, a method sometimes called *squaring*. Thus, 6 squared is the same as 6×6, or 6^2.

Squared numbers appear in various electronics formulas. In the basic formula for power, $P = E \times I$. According to Ohm's Law, $E = I \times R$. We then have:

$$P = E \times I$$
$$E = I \times R$$

Since E and $I \times R$ are identical (as shown by the equals sign), $I \times R$ can be substituted for E in the power formula.

$$P = E \times I$$
$$P = I \times R \times I$$

On the right side, any of the letters can be transposed without affecting the formula,

$$P = I \times I \times R$$

but I is the same as I^1, hence $I \times I = I^1 \times I^1 = I^2$
Consequently:

$$P = I^2R \text{ or } P = I^2 \times R \text{ or } P = (I^2)(R).$$

Still another power formula can be derived in this way; for, according to Ohm's Law, $I = \frac{E}{R}$. The power law is $P = E \times I$. Since $I = \frac{E}{R}$ we can substitute $\frac{E}{R}$ for the letter I in the power law formula:

$$P = E \times \frac{E}{R} = \frac{E}{1} \times \frac{E}{R} = \frac{E^2}{R}$$

Thus, instead of just one power formula, we have *three:*

$$P = E \times I,\ P = I^2R, \text{ and } P = \frac{E^2}{R}$$

In each of these formulas, P is the power in watts, I is the current in amperes, and R is the resistance in ohms. When multiples or submultiples are used, they must first be converted to basic units before plugging them into the formulas. The conversions to basic units can be handled by moving the decimal point or by using powers of 10.

Example: As shown in Figure 2–10, a current of 30 mA flows through a 400-ohm wirewound resistor. How much power is being used by the resistor?

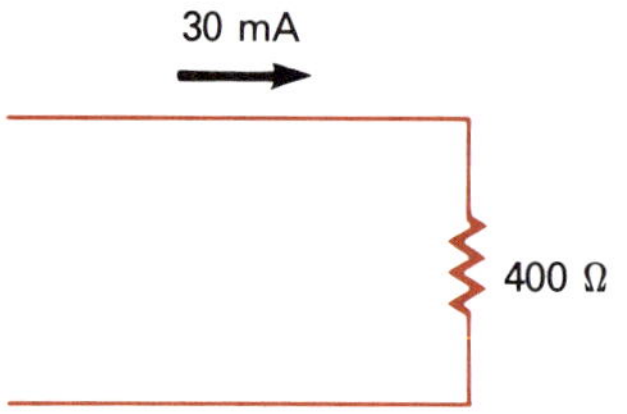

FIGURE 2–10

Solution:
$$P = I^2R$$
$$= (30 \times 10^{-3}\ \text{A})(30 \times 10^{-3}\ \text{A})(400\ \Omega)$$
$$= 3.6 \times 10^{-1}\ \text{W}$$
$$= 0.36\ \text{W}$$
$$= 360\ \text{mW}$$

Calculator Tip

By utilizing the square key [x^2] it is possible to square the number displayed. The sequences to square 3, −12.2, and 4.5×10^2 would be as follows:

a. 3^2: 3 [x^2] and 9 is displayed

b. −12.2: 12.2 [+/−] [x^2] and 148.84 is displayed

c. $(4.5 \times 10^2)^2$: 4.5 [EE] 2 [x^2] and 2.025×10^5 is displayed

Example: If a voltage of 500 mV is measured across a 300-ohm resistor, how much power is used?

Solution:
$$P = \frac{E^2}{R}$$
$$= \frac{(500 \times 10^{-3}\ \text{V})^2}{300\ \Omega}$$
$$= \frac{2.5 \times 10^{-1}}{300}$$
$$= 8.33 \times 10^{-4}\ \text{W}$$

$= 833\ \mu W$
$= 0.833\ mW$

The square of a number is the number multiplied by itself. The square of 9 is 9^2 or $9 \times 9 = 81$. In some instances, the number must be multiplied by itself more than twice. A number such as 5^3 is an instruction to *cube* the number, that is to multiply the digit 5, by itself, three times: $5^3 = 5 \times 5 \times 5 = 125$.

In addition to squaring (x^2) and cubing (x^3) a number may be multiplied by itself any number of times specified by the exponent. The expression 6^4 indicates that the *base* 6 is raised to the fourth *power*. In other words, multiplied by itself 4 times. This results in $6 \times 6 \times 6 \times 6$ or 1 296.

Example: Raise these bases to the powers indicated.
(a) -8^3 (b) 3^6 (c) $(\frac{3}{4})^3$ (d) 4.2^4

Solution: a. $-8^3 = -8 \times -8 \times -8 = -512$
b. $3^6 = 3 \times 3 \times 3 \times 3 \times 3 \times 3 = 729$
c. $(\frac{3}{4})^3 = \frac{3}{4} \times \frac{3}{4} \times \frac{3}{4} = \frac{27}{64} = 0.421\ 875$
d. $4.2^4 = 4.2 \times 4.2 \times 4.2 \times 4.2 = 311.169\ 6$

Calculator Tip

With most electronic calculators it is possible to raise displayed numbers to various powers. One common procedure is to utilize the "y to the x power" [y^x] function. The sequence to raise the base 2 to the 8th power $(2)^8$ would be as follows:

2 [y^x] 8 [=] 256

The sequence to raise the base 6.7 to the 3rd power $(6.7)^3$ would be:

6.7 [y^x] 3 [=] 300.763

Roots of Numbers

Squaring a number is an easy process since only basic multiplication is involved. The reverse process, finding the *square root* of a number, is more complicated only in the sense that it requires more work. Finding the square root of a number is a search for not one, but a pair of identical numbers which can be multiplied to supply the given number. The square root of 9 is 3 since 3 multiplied by itself is 9.

Certain square roots are obvious. To find the square root of 100, all we need to know is which number multiplied by itself will result in 100: $10 \times 10 = 100$ and so 10 is the square root of 100. The square root of most numbers may not be quite so apparent.

The square root sign ($\sqrt{\ }$) closely resembles the sign used for long division, since finding the square root of a number is a division process. To be technically correct, the square root sign is called the *radical sign* and is really $\sqrt[2]{\ }$ but the *index* 2 is often omitted. The horizontal bar joining the square root sign is called a *vinculum*. Its function is to make sure that all the terms beneath it are part of the square root. The square root of any number is also that number raised to the 1/2 power. And so we have, as an example:

$$\sqrt{625} = \sqrt[2]{625} = 625^{1/2} = 25$$

The cube root of a number asks the question, "What base, when multiplied by itself three times, equals the original number?" In fact, any root asks basically the same question. The fourth root of a number is a base such that when multiplied by itself four times it equals the number.

Calculator Tip

Roots are obtainable on some calculators through the square root key $\boxed{\sqrt{x}}$ or the "*y* to the *x* power" key $\boxed{y^x}$ used in conjunction with the inverse function $\boxed{\text{INV}}$. The following examples will demonstrate the use of these functions.

$$\sqrt[2]{345.96} = 345.96 \ \boxed{\sqrt{x}} \text{ and 18.6 is displayed}$$

$$\sqrt[3]{2\,197} = 2\,197 \ \boxed{\text{INV}} \ \boxed{y^x} \ 3 \ \boxed{=} \ 13$$

$$\sqrt[12]{4.2 \times 10^6} = 4.2 \ \boxed{\text{EE}} \ 6 \ \boxed{\text{INV}} \ \boxed{y^x} \ 12 \ \boxed{=} \ 3.564$$

Example: Find (a) the cube root of 64, (b) the fifth root of 243, (c) the 15th root of 32 768, and (d) the fourth root of 19.448 1

Solution:

a. $\sqrt[3]{64} = 64^{1/3} = 4$

b. $\sqrt[5]{243} = 243^{1/5} = 3$

c. $\sqrt[15]{32\,768} = 32\,768^{1/15} = 2$

d. $\sqrt[4]{19.448\,1} = 19.448\,1^{1/4} = 2.1$

Prior to the advent of electronic calculators the root of a number was determined by examining a root table, calculated through a process similar to long division, or logarithms were used. These processes are no longer used extensively as most calculators will perform the required calculations.

Calculator Tip

Roots may be identified through use of the root sign or a fractional exponent. For example:

$$\sqrt[4]{81} = 81^{1/4} = 81^{0.25}$$

These would be read as the fourth root of 81 or 81 to the one-fourth power.

Your scientific calculator is capable of extracting the root of a number using the y^x function or through fractional exponents. The following sequences produce the fifth root of 16 807.

16 807 [INV] [y^x] 5 [=] 7

or, 16 807 [y^x] 0.2 [=] 7

or, 16 807 [y^x] 5 [1/x] [=] 7

Example: Determine the roots of the following values.

a. $\sqrt[3]{10\,648}$
b. $(1\,024)^{1/5}$
c. $\sqrt[7]{187.65}$
d. $\sqrt[24]{1}$

Solution: a. $\sqrt[3]{10\,648} = 22$
b. $\sqrt[5]{1\,024} = 4$
c. $\sqrt[7]{187.65} = 2.112\,341\,3$
d. $\sqrt[24]{1} = 1$

Signs of Square Roots

The product of numbers having like signs is always a number with a plus sign.

$$(+2) \times (+2) = +4 \text{ and } (-2) \times (-2) = +4$$

The square root of a plus number, then, consists of two roots, one which is plus, the other minus. The square root of 16 is +4 and also −4, since 4 × 4

= 16 and -4×-4 also equals 16. For practical problems in electronics, though, the required root is usually considered to be plus, unless there is some specific reason for a negative root.

Square Root Problems

Even though an electronics formula may contain a square root sign, it is often convenient to use powers of 10 to handle much of the mathematics prior to finding the square root. A typical problem in electronics would be finding the frequency of a tuned circuit when the values of inductance and capacitance are known.

$$f_r = \frac{1}{6.28\sqrt{L \times C}}$$

where f_r is the resonant frequency in hertz, L the inductance in henrys, and C the capacitance in farads. When multiples or submultiples are used, these must first be converted to basic units prior to use in the formula.

Example: In the circuit shown in Figure 2–11, the inductance is 400 μH and the capacitance is 300 pF. What is the resonant frequency of the circuit?

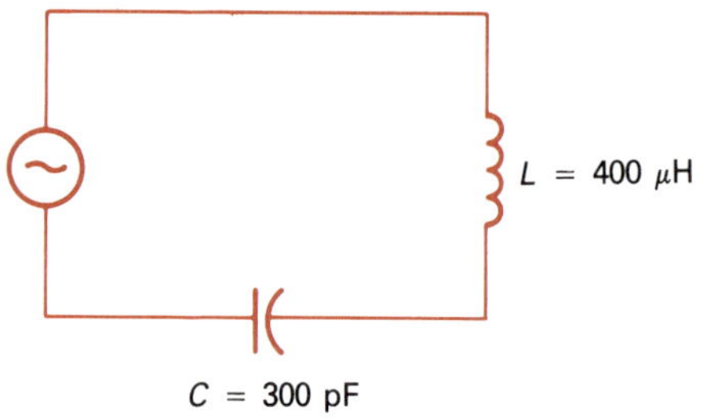

FIGURE 2–11 Finding the resonant frequency of this circuit requires the use of square root.

Solution:

$$\begin{aligned}
f_r &= \frac{1}{6.28\sqrt{L \times C}} \\
&= \frac{1}{6.28\sqrt{(400 \times 10^{-6}\ \text{H})(300 \times 10^{-12}\ \text{F})}} \\
&= \frac{1}{6.28\sqrt{1.2 \times 10^{-13}}} \\
&= \frac{1}{(6.28)(3.464 \times 10^{-7})} \\
&= \frac{1}{2.175 \times 10^{-6}} \\
&= 4.597 \times 10^{5}\ \text{Hz} \\
&= 459.7\ \text{kHz}
\end{aligned}$$

Calculator Tip

In order to demonstrate the usefulness of your calculator, use the following sequence to solve the resonant frequency problem previously described.

$$f_r = \frac{1}{6.28\sqrt{400\ \mu H \times 300\ pF}}$$

400 [EE] [+/−] 6 [×] 300 [EE] [+/−] [12] [=] [$\sqrt{x}$] [×] 6.28 [=] [1/x] and $4.596\,7 \times 10^5$ is displayed.

Using standard units this would be written 459.67 kHz.

EXERCISE 2·6

Perform the indicated operations. If needed, round the answer to the nearest thousandth.

1. 10^3
2. 1.8^4
3. $\sqrt{256}$
4. $(\frac{1}{8})^2$
5. $\sqrt{\frac{9}{64}}$
6. $\sqrt[4]{50\,625}$
7. $(3\frac{3}{8})^2$
8. 14^5
9. $\sqrt[6]{3.2}$
10. 0.05^2
11. $(1.1)^3$
12. $\sqrt[3]{333}$
13. $(1.01)^2$
14. 4^6
15. 6^4
16. $\sqrt[9]{9\,540}$
17. 11^5
18. 3.2^6
19. $\sqrt[3]{2\frac{1}{2}}$
20. $\sqrt{108.213}$
21. 25^8
22. 50^4
23. $8^{1/8}$
24. $\sqrt[10]{10}$
25. $(3.72)^4$
26. $(27.2)^{1/5}$
27. $\sqrt[3]{15\frac{7}{8}}$
28. $(83.7)^7$
29. $\sqrt[5]{187\,029}$
30. $\sqrt[16]{987\,628}$
31. Given $f = 2.5$ kHz and $L = 800$ mH. Calculate the inductive reactance (X_L) in ohms. Note: $X_L = 2\pi fL$
32. Given $f = 25$ kHz and $C = 0.001\ \mu$F. Calculate the capacitive reactance (X_c) in ohms. Note: $X_c = \dfrac{1}{2\pi fC}$

UNIT 2·7 PROBLEM REVIEW

In problems 1–5 convert the values as indicated.

1. $4\frac{1}{2}$ yards to metres
2. 370 millimetres to inches
3. 128 km to miles

4. 65 inches to metres
5. 1.8 metres to millimetres

In problems 6–10 convert the common fractions and mixed numbers to decimal fractions.

6. $\frac{27}{32}$
7. $4\frac{3}{5}$
8. $\frac{9}{17}$
9. $\frac{8}{25}$
10. $\frac{19}{20}$

In problems 11–15 convert the decimal fractions to common fractions or mixed numbers.

11. 0.15
12. 2.148
13. 0.35
14. 1.48
15. 0.078 125

Simplify the expressions in problems 16–30.

16. 8.08 A + 11.295 A − 16.2 A + 0.93 A
17. −32.19 W + 64.7 W + 3.2 W
18. 42.8 kHz × 2.5
19. 69.7 Ω ÷ 1.5
20. 2.5 mA × 186 + 10 mA
21. 0.25 × 100 V
22. 84 − 2.8 × 16 + 4
23. (32.8 × 1.4) ÷ 0.2
24. 148 + 2.9(16 ÷ 8) − 5
25. 4 500 Hz ÷ (28.8 ÷ 3.2)
26. (−0.05 × 6 280) ÷ 4.8
27. (−28.2 − 3)(14.8 ÷ 7.5)
28. 83.5 × 1.7 + 6.2 × 1.5
29. 2 × 3.14 × 68 000 × 0.15
30. −93.5(145 − 287) + 3 200
31. Radio signals travel at 186 000 miles per second. How far does a signal travel in 0.15 seconds?
32. A television signal travels 22 320 miles in 0.12 seconds. Find the speed in miles per second.

Simplify problems 33–45 and express the answers in scientific notation.

33. $6.5 \times 10^4 + 11.9 \times 10^3$
34. $(12.75 \times 10^4)(0.5 \times 10^{-4})$
35. $(2.9 \times 10^{-5}) \div (4.5 \times 10^{-7})$
36. $(17 \times 10^2) - (0.8 \times 10^3)1.5$
37. $6.28 \times 10^4 \div (3.3 \times 11.6)$
38. $1\ 800 \times 10^{-2} + 18$
39. $0.008 \times 10^6 \div (2.5 \times 10^{-2})$
40. $(35.8 \times 10^3)(1.8 \times 10^{-3})$

41. $\dfrac{(4.3 \times 10^5)(2.8 \times 10^{-4})}{11.5 \times 10^6}$

42. $\dfrac{(53 \times 10^3) + (0.1 \times 10^5)}{150 \times 10^{-2}} \div (6.5 \times 10^{-1})$

43. $\dfrac{(19.8 \times 10^6)(4.1 \times 10^5)}{4.51 \times 10^8}$

44. $\dfrac{(630 \times 10^3)(0.7 \times 10^{-2})}{387} \div 0.8 \times 10^6$

45. $\dfrac{(64 \times 10^2)(1.2 \times 10^3)}{(1.72 \times 10^6) + (2.12 \times 10^6)} \div \dfrac{(6.82 \times 10^{-6}) - (1.06 \times 10^{-6})}{(4.8 \times 10^{-2})(6 \times 10^{-5})}$

Complete problems 46–55 and express the answers in the units indicated.

46. 28 mA − 150 μA = ______ mA
47. 180 kHz ÷ 10 = ______ Hz
48. 50 V − 21 600 mV = ______ mV
49. 220 Ω + 3.9 kΩ = ______ kΩ
50. 0.8 MΩ − 120 kΩ = ______ kΩ
51. 3 V ÷ 1 200 = ______ μV
52. 0.01 s × $84\frac{1}{2}$ = ______ ms
53. 350 mA × 100 = ______ A
54. 350 mA ÷ 100 = ______ μA
55. 1.02 MHz + 62 kHz = ______ kHz

Perform the indicated operations in problems 56–66.

56. 21^5
57. $\sqrt[3]{47\,832.147}$
58. 0.04^3
59. $\sqrt[7]{5\,840}$
60. 2.7^4
61. $(6\frac{1}{2})^3$
62. 128^{12}
63. $\sqrt[12]{128}$
64. $\sqrt[4]{1\,128}$
65. $\sqrt{482\,600}$
66. 3.7^5

CHAPTER 3

MATHEMATICS FOR DIRECT CURRENT CIRCUITS

With an understanding of the four basic mathematics functions, decimals, powers of ten, units of measurement in electronics, and the ability to handle basic formula transpositions, it is now possible for you to work with a variety of electronics formulas, and, if required, to derive new ones.

UNIT 3·1 OHM'S LAW AND THE POWER LAW IN SERIES CIRCUITS

Objectives:

After studying this unit, you should be able to

- discuss the principles of series circuits.
- use Ohm's Law in series circuits.
- use the power law in series circuits.
- rearrange basic equations.

Ohm's Law

During the 19th century Georg Simon Ohm examined the relationships among voltage (E), current (I), and resistance (R). He found that if the voltage applied to a resistance was kept constant, and the resistance value was increased, that the current through the resistance decreased. Ohm also found that, for a constant resistance, when voltage was increased, current increased. These two findings can be expressed as follows:

1. Current is *inversely* proportional to resistance. At a constant voltage, as the resistance increases the current decreases and as the resistance decreases the current increases.
2. Current is *directly* proportional to voltage. At a constant resistance, as the voltage increases the current increases and as the voltage decreases the current decreases.

Expressing these relationships in a formula results in *Ohm's Law:*

$$I = \frac{E}{R} \tag{3–1}$$

where, I = current in amperes (A)
E = voltage in volts (V)
R = resistance in ohms (Ω)

By rearranging the terms in this formula two other forms of Ohm's Law can be derived:

$$E = I \times R \tag{3–2}$$

$$R = \frac{E}{I} \tag{3–3}$$

A helpful technique for remembering the three forms of Ohm's Law is the wheel shown in Figure 3–1. To find the formula for a specific unit simply cover the value of interest and the position of the two remaining values will indicate the operation to be performed.

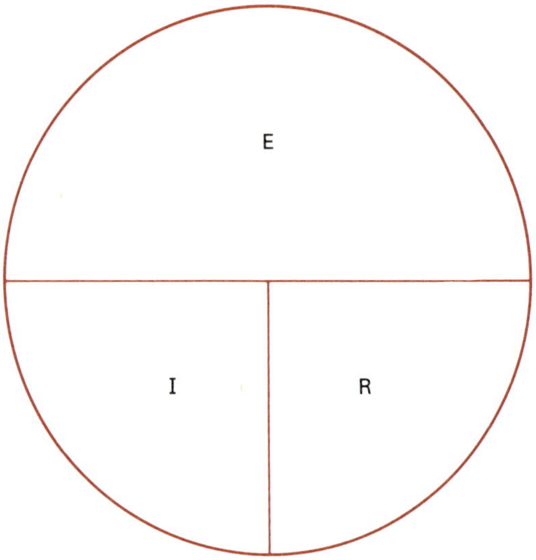

FIGURE 3–1 The Ohm's Law wheel

Example: Use the wheel in Figure 3–1 to find the Ohm's Law formulas for (a) E, (b) I, and (c) R.

Solution: a. In order to find the expression for voltage, cover the E and note that I and R are side-by-side, suggesting multiplication. The formula is as follows:

$$E = IR$$

b. Covering the I leaves E over R, suggesting division. The formula is as follows:

$$I = \frac{E}{R}$$

c. Covering the R leaves E over I, again suggesting division. The formula is as follows:

$$R = \frac{E}{I}$$

Example: A 50-Ω resistor is connected to 100 volts. What is the value of the current moving through the resistor?

Solution:

$$\begin{aligned} I &= \frac{E}{R} \\ &= \frac{100\ \text{V}}{50\ \Omega} \\ &= 2\ \text{A} \end{aligned}$$

Example: Use the circuit in Figure 3–2 to determine the applied voltage.

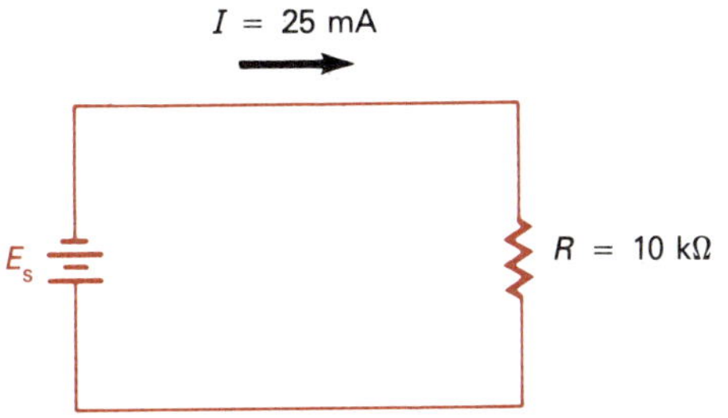

FIGURE 3–2 The current and resistance values in this series circuit can be used to determine the applied voltage.

Solution:

$$\begin{aligned} E_s &= I \times R \\ &= 25\ \text{mA} \times 10\ \text{k}\Omega \\ &= (25 \times 10^{-3}\ \text{A})(10 \times 10^{3}\ \Omega) \\ &= 250\ \text{V} \end{aligned}$$

Example: A current of 15 μA flows through a resistor when 6 V are applied. What is the value of the resistor?

Solution:

$$\begin{aligned} R &= \frac{E}{I} \\ &= \frac{6\text{V}}{15\ \mu\text{A}} \\ &= \frac{6\text{V}}{15 \times 10^{-6}\ \text{A}} \\ &= 4 \times 10^{5}\ \Omega \\ &= 400\ \text{k}\Omega \end{aligned}$$

Rearranging Equations

Ohm's Law describes the relationship between voltage, current, and resistance. This relationship is expressed in the form of an equation. An *equation* states that two factors or expressions are equal in value. In terms of Ohm's Law,

$$E = I \times R$$

states that the left side of the equation is equal (=) to the right side of the equation. This relationship is similar to a set of scales which are balanced. In order to keep the scales in balance any change on one side must be accompanied by an equal change on the other side.

There are many occasions when it will be necessary to rearrange an equation in order to *solve* for a specific value. This process will be presented in depth in a later chapter, but is introduced here to assist in solving Ohm's Law and power law problems.

Addition Axiom

If the same term is *added* to both sides of an equation the relationship expressed by that equation is unchanged.

Example: Given the following equation. Solve for R_t.

$$R_1 = R_t - R_2$$

Solution: Add R_2 to both sides.

$$R_1 = R_t - R_2$$
$$R_1 + R_2 = R_t - R_2 + R_2$$
$$R_1 + R_2 = R_t$$

Subtraction Axiom

If the same term is *subtracted* from both sides of an equation the relationship expressed by that equation is unchanged.

Example: Given the following equation. Solve for $\frac{1}{R_1}$.

$$\frac{1}{R_t} = \frac{1}{R_1} + \frac{1}{R_2} + \frac{1}{R_3}$$

Solution: As the other two terms are being added, it will be necessary to subtract them to shift both terms to the other side of the equation.

$$\frac{1}{R_t} = \frac{1}{R_1} + \frac{1}{R_2} + \frac{1}{R_3}$$
$$\frac{1}{R_t} - \frac{1}{R_2} - \frac{1}{R_3} = \frac{1}{R_1} + \frac{1}{R_2} + \frac{1}{R_3} - \frac{1}{R_2} - \frac{1}{R_3}$$
$$\frac{1}{R_t} - \frac{1}{R_2} - \frac{1}{R_3} = \frac{1}{R_1}$$

Division Axiom

If two expressions are equal, and are *divided* by the same term, the quotients are equal. Division by zero is not included, as this process is undefined.

Example: Given the following equation. Solve for I.

$$E = I \times R$$

Solution: As R is presently being multiplied by I it will be necessary to divide by R in order to move it to the other side of the equation.

$$E = I \times R$$

$$\frac{E}{R} = \frac{I \times R}{R}$$

$$\frac{E}{R} = I$$

Example: In the equation, $X_L = 2\pi fL$, solve for f.

Solution:

$$X_L = 2\pi fL \quad \text{(Initial expression)}$$

$$\frac{X_L}{2\pi L} = \frac{2\pi fL}{2\pi L} \quad \text{(Division by } 2\pi L\text{)}$$

$$f = \frac{X_L}{2\pi L} \quad \text{(Final expression)}$$

Multiplication Axiom

If two expressions are equal, and are *multiplied* by the same term, the products are equal.

Example: In the equation, $°F = \frac{9}{5}(°C) + 32°$, solve for $°C$.

Solution:

$$°F = \frac{9}{5}(°C) + 32° \quad \text{(Initial expression)}$$

$$°F - 32° = \frac{9}{5}(°C) \quad \text{(Subtraction of 32°)}$$

$$5(°F - 32°) = 9(°C) \quad \text{(Multiplication by 5)}$$

$$\frac{5(°F - 32°)}{9} = °C \quad \text{(Division by 9)}$$

$$°C = \frac{5}{9}(°F - 32°) \quad \text{(Final expression)}$$

Powers and Roots Axiom

If two expressions are equal, and are subjected to the same root or raised to the same power, the resulting expressions are equal.

Example: Given the equation $P = \frac{E^2}{R}$, solve for E.

Solution: As R is being divided into E^2 it will be necessary to multiply by R in order to move it to the other side of the equation.

$$P = \frac{E^2}{R}$$

$$P \times R = \frac{E^2}{R} \times \frac{R}{1}$$

$$P \times R = E^2$$

$$\sqrt{P \times R} = \sqrt{E^2}$$

$$E = \sqrt{P \times R}$$

Power Law

Power (P) is the rate of doing work. Power is also the rate at which energy is used. The basic formula for power in electronics is:

$$P = I \times E \tag{3–4}$$

where, I = current in amperes (A)
E = voltage in volts (V)
P = power in watts (W)

By rearranging the terms in this formula two other forms of the power law can be derived:

$$I = \frac{P}{E} \tag{3–5}$$

$$E = \frac{P}{I} \tag{3–6}$$

A helpful technique for remembering the three forms of the power law is the wheel shown in Figure 3–3. To find the formula for a specific unit simply cover the value of interest and the position of the two remaining values will indicate the operation to be performed.

Example: Use the wheel in Figure 3–3 to find the power law formulas for (a) P, (b) I, and (c) E.

Solution:
a. Covering the P leaves I times E.
b. Covering the I leaves P divided by E.
c. Covering the E leaves P divided by I.

In examining the power formula we see that power is directly proportional to current and voltage. When 1 V causes a current of 1 A to flow, the rate of energy use is 1 V × 1 A or 1 W. If either voltage or current were to increase, the rate of energy use would also increase.

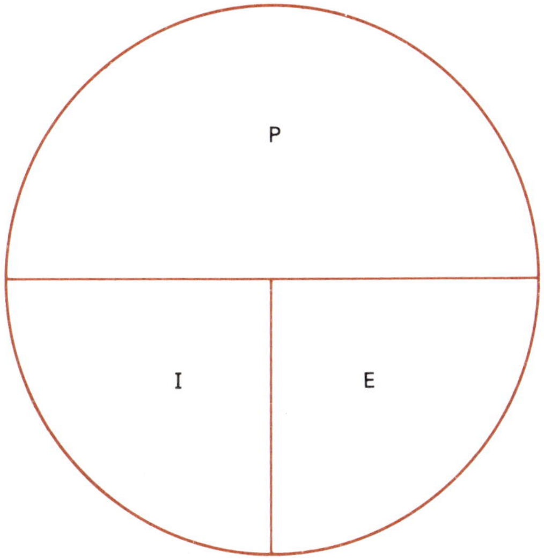

FIGURE 3–3 The power law wheel

Example: A resistor draws 2.5 amperes when connected to a 12-V source. Calculate the power delivered.

Solution:
$$P = I \times E$$
$$= 2.5\text{ A} \times 12\text{ V}$$
$$= 30\text{ W}$$

Example: Use the circuit in Figure 3–4 to determine the circuit current.

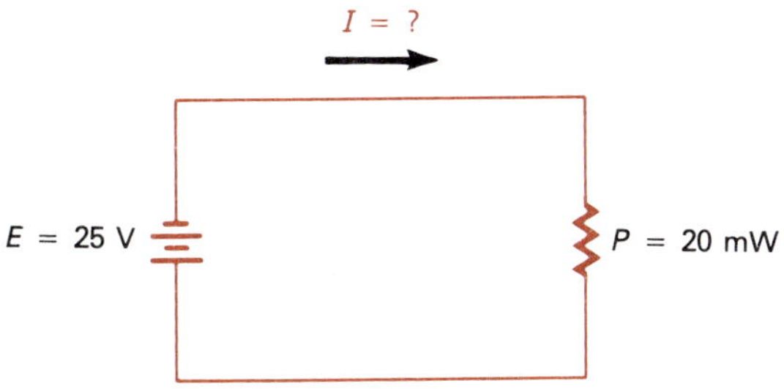

FIGURE 3–4 The voltage and power values in this series circuit can be used to determine circuit current.

Solution:
$$I = \frac{P}{E}$$
$$= \frac{20\text{ mW}}{25\text{ V}}$$
$$= \frac{20 \times 10^{-3}\text{ W}}{25\text{ V}}$$
$$= 0.000\,8\text{ A} = 0.8\text{ mA} = 800\ \mu\text{A}$$

Example: A 60-W incandescent lamp draws 500 mA. Calculate the applied voltage.

Solution:
$$E = \frac{P}{I} = \frac{60\ \text{W}}{500\ \text{mA}} = \frac{60\ \text{W}}{0.5\ \text{A}} = 120\ \text{V}$$

Combining Ohm's Law and the Power Law

In examining Ohm's Law and the power law it becomes obvious that current and voltage are common to both formulas. By rearranging these equations one can derive many other useful relationships.

$P = I \times E$ (Power Law)
$E = I \times R$ (Ohm's Law)

$P = I \times (I \times R)$ (Substitution of $I \times R$ for E)
$P = I \times I \times R$

A new formula for power in terms of current and resistance is derived.

$$P = I^2 \times R \qquad (3\text{–}7)$$

Example: A 390-Ω resistor conducts 15 mA of current. Calculate the power dissipation.

Solution:
$$P = I^2 \times R = (15\ \text{mA})^2(390\ \Omega) = (15 \times 10^{-3}\ \text{A})^2(390\ \Omega) = 87.75\ \text{mW}$$

A formula for power in terms of voltage and resistance is derived as follows.

$E = I \times R$ (Ohm's Law)
$I = \frac{E}{R}$ (Solve for I)

$P = I \times E$ (Power Law)
$P = \frac{E}{R} \times E$ (Substitution)

A new formula for power in terms of voltage and resistance is derived.

$$P = \frac{E^2}{R} \qquad (3\text{–}8)$$

A total of six new formulas can be derived by combining Ohm's Law and the power law. Two of these (equations 3–7 and 3–8) have been derived. The reader should derive the remaining four formulas.

$$I = \sqrt{\frac{P}{R}} \quad (3\text{–}9)$$

$$E = \sqrt{P \times R} \quad (3\text{–}10)$$

$$R = \frac{E^2}{P} \quad (3\text{–}11)$$

$$R = \frac{P}{I^2} \quad (3\text{–}12)$$

One common and useful method of presenting these twelve formulas is to use the wheel shown in Figure 3–5. Given any two values, the other two may be found.

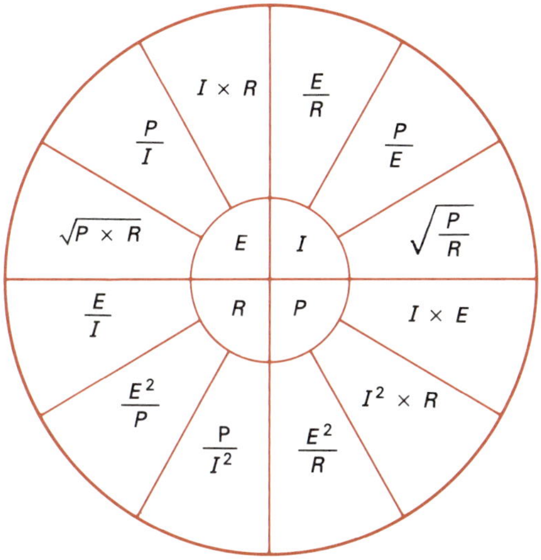

FIGURE 3–5 The Ohm's Law and power law formulas

Series Circuits

If the same current flows through two or more points, these points are said to be in series. The resistors in Figure 3–6 are connected in series as the 10 mA current must flow through each resistor. The current (I) in a series circuit is said to be constant. In other words, it is the same at each point in the circuit. Expressed as an equation.

$$I_t = I_{R_1} = I_{R_2} = I_{R_3} = \cdots = I_{R_n} \quad (3\text{–}13)$$

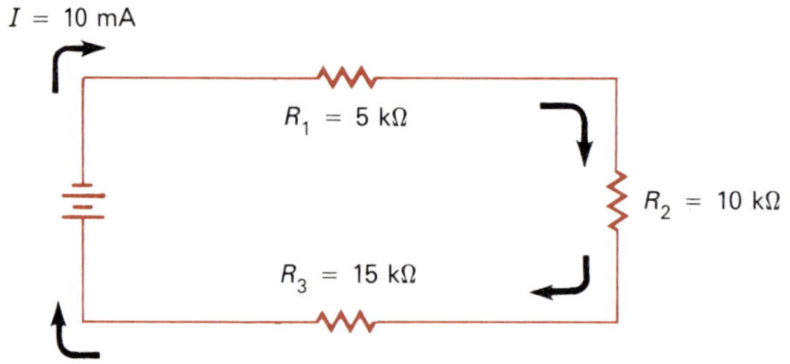

FIGURE 3–6 The current is the same at all points in a series circuit.

Note that in Figure 3–6 the current must flow through resistors of 5 kΩ, 10 kΩ, and 15 kΩ. The total resistance (R_t) is the sum of $R_1 + R_2 + R_3$, or 30 kΩ. The general formula for resistance in a series circuit is given by

$$R_t = R_1 + R_2 + \cdots + R_n \tag{3–14}$$

As the current flows through each resistor in Figure 3–6 there will be a change in potential, or voltage drop, across each resistor. These would be:

$$E_{R_1} = I \times R_1 = 10\text{ mA} \times 5\text{ k}\Omega = 50\text{ V}$$
$$E_{R_2} = I \times R_2 = 10\text{ mA} \times 10\text{ k}\Omega = 100\text{ V}$$
$$E_{R_3} = I \times R_3 = 10\text{ mA} \times 15\text{ k}\Omega = 150\text{ V}$$

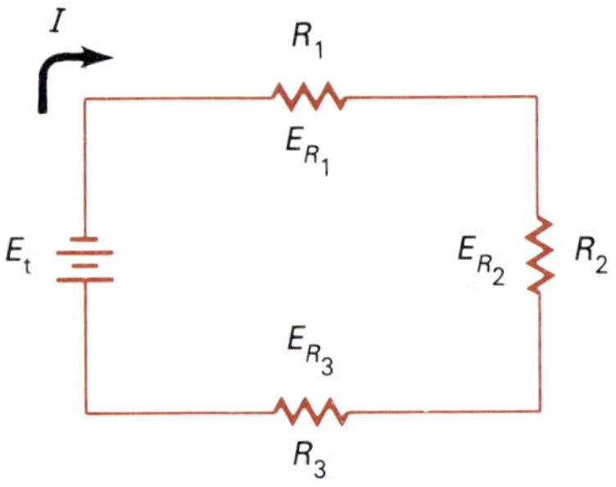

$$R_t = R_1 + R_2 + R_3$$

$$I = \frac{E_t}{R_t}$$

$$E_{R_1} = I \times R_1$$

$$E_{R_2} = I \times R_2$$

$$E_{R_3} = I \times R_3$$

$$P_t = I \times E_t$$

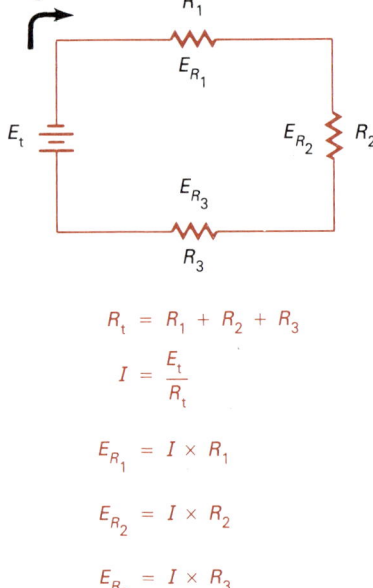

FIGURE 3–7 Formulas used to calculate circuit values in a series circuit

It appears that the total applied voltage (E_t) would be the sum of the individual voltage drops. The general formula for total voltage in a series circuit is given by

$$E_t = E_{R_1} + E_{R_2} + \cdots + E_{R_n} \tag{3–15}$$

It would also follow that total power (P_t) would be the product of circuit current and total voltage, or

$$P_t = I \times E_t \tag{3–16}$$

Figure 3–7 summarizes the formulas used to compute circuit values in a series circuit.

Example: Given the circuit in Figure 3–8 compute the following.

a. the circuit current (I)
b. the total voltage (E_t)
c. the voltage across resistor two (E_{R_2})
d. the power dissipated by resistor three (P_{R_3})
e. the total circuit power (P_t)

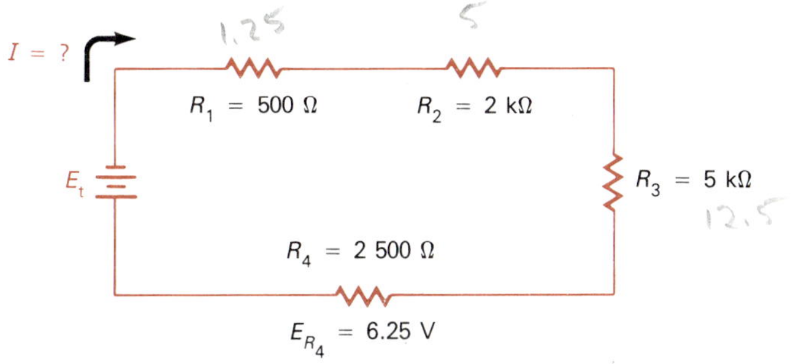

FIGURE 3–8 Use Ohm's Law and the power law to calculate the missing circuit values.

Solution:

a. $I = E_{R_4} \div R_4$
$= 6.25\text{ V} \div 2\,500\ \Omega$
$= 2.5\text{ mA}$

b. $R_t = 0.5\text{ k}\Omega + 2\text{ k}\Omega + 5\text{ k}\Omega + 2.5\text{ k}\Omega = 10\text{ k}\Omega$
$E_t = I \times R_t$
$= 2.5\text{ mA} \times 10\text{ k}\Omega$
$= 25\text{ V}$

c. $E_{R_2} = I \times R_2$
$= 2.5\text{ mA} \times 2\text{ k}\Omega$
$= 5\text{ V}$

d. $P_{R_3} = I^2 \times R_3$
$= (2.5\text{ mA})^2(5\text{ k}\Omega)$
$= 31.25\text{ mW}$

e. $P_t = I \times E_t$
$= 2.5 \text{ mA} \times 25 \text{ V}$
$= 62.5 \text{ mW}$

EXERCISE 3·1

1. A 470-Ω resistor is connected to 6 V. Find the current to the nearer tenth milliampere. 12.8 mA
2. A load is connected to 24 V and draws a current of 0.004 mA. Calculate the load resistance. 6 KΩ
3. An unmarked battery has a 6.8-kΩ load attached and delivers 1.6 mA of current. Find the output voltage of the battery with the load attached. 10.88 V
4. A load is connected to 12 V and draws a current of 75 mA. Find the power dissipated by the load. 900 KW
5. The 5-W zener diode in Figure 3–9 maintains a constant voltage across itself. Find the maximum current it can conduct without exceeding the wattage rating.

416.67 mA

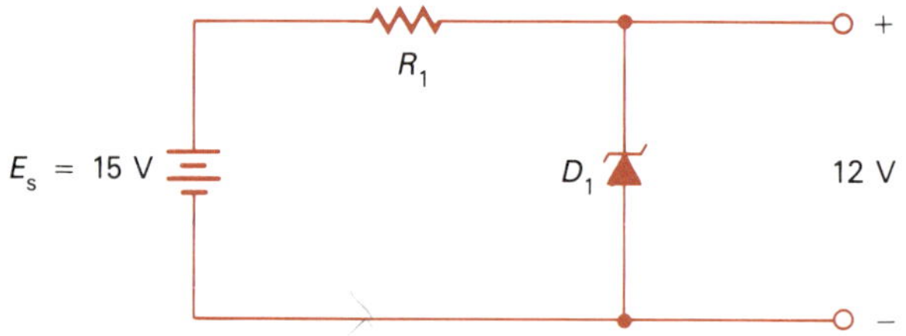

FIGURE 3–9

6. A 1.5-W load draws a current of 375 mA. What is the applied voltage? 4

Use the circuit in Figure 3–10 for problems 7–12.

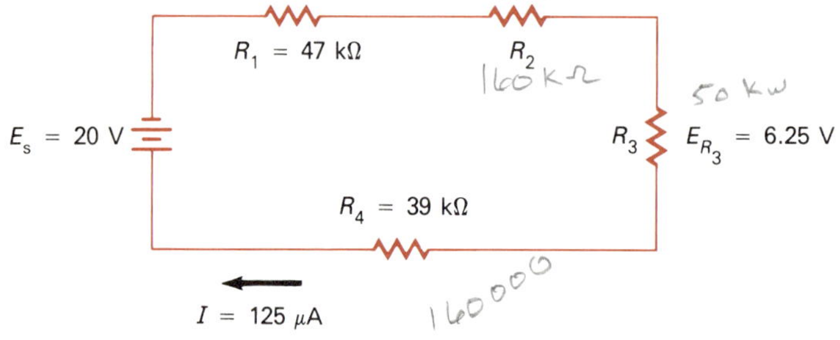

FIGURE 3–10

7. Calculate total resistance.
8. Find the voltage across R_4.
9. What is the value of R_2?

10. Find the value of R_3.
11. Calculate the power dissipated by R_1.
12. Calculate the total power dissipated in the circuit.
13. A resistor has a current of 20 mA passing through it. It is dissipating 80 mW of power. Compute the resistance.
14. A 2.7-kΩ load is dissipating 833 μW of power. Compute the value of the applied voltage.
15. A 1-Ω load dissipates 250 mW of power. Compute the current being drawn by the load.
16. Compute the power dissipated by a 2.2-kΩ load when it is connected to 6 V.
17. The current flowing from the emitter to collector of a transistor is 14.97 mA. Find the power being dissipated if the emitter to collector voltage is 8.08 V.
18. The current through a 56-kΩ resistor is 150 μA. Calculate the power dissipation.
19. Calculate the hot resistance of a 1-W bulb that conducts 100 mA of current.
20. A rheostat is set to 700 Ω and placed in series with a 6.2-kΩ resistor. Find the voltage drop across the rheostat if the resistor is dissipating 27.2 mW of power.

In problems 21–25 solve each equation for the indicated value.

21. $R = \dfrac{kL}{A}$, solve for L.
22. $X_c = \dfrac{1}{2\pi fC}$, solve for f.
23. $I_{ceo} = (\beta + 1)I_{cbo}$, solve for β.
24. $f_c = \dfrac{1}{\pi\sqrt{LC}}$, solve for L.
25. $f_L = f_r - \dfrac{BW}{2}$, solve for BW.

UNIT 3·2 PARALLEL AND SERIES-PARALLEL CIRCUITS

Objectives:

After studying this unit, you should be able to

- discuss the principles of parallel circuits.
- discuss the principles of series-parallel circuits.
- solve for circuit values in parallel and series-parallel circuits.
- calculate the required voltage, current, and resistance values in voltage divider circuits.
- calculate component values in resistive bridge circuits.

Parallel Circuits

Two or more electronic components are connected in *parallel* when both ends of each component are connected to common points. The circuits in Figure 3–11 show two resistors in parallel with a 12-V source.

Note that points A and C are the same electrical point, and are therefore common points. Points B and D are also common points. Points A and C are connected to the negative terminal of the voltage source and B and D to the positive terminal.

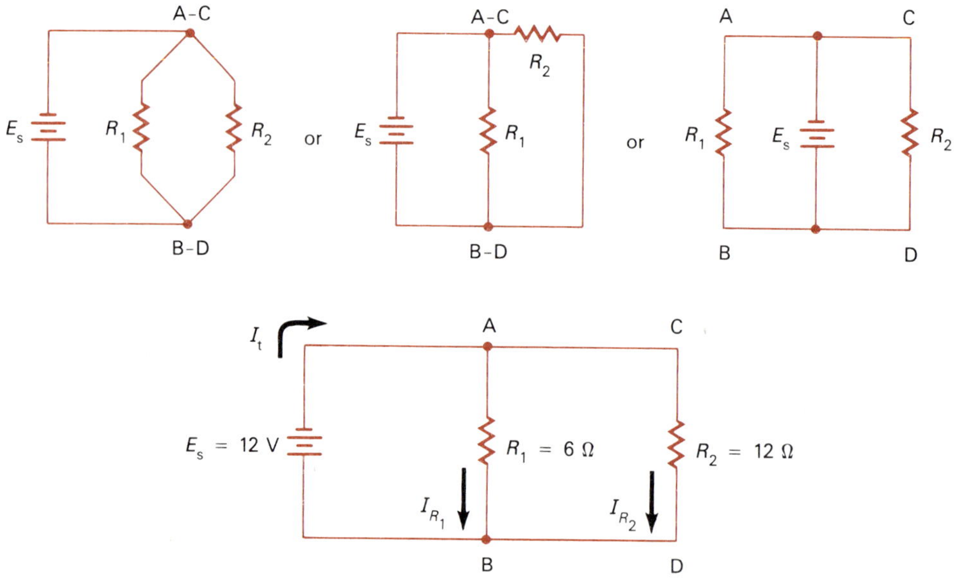

FIGURE 3–11 Resistors R_1 and R_2 are connected in parallel in all four of these circuits.

A voltmeter placed across the voltage source would indicate 12 V. As points A and B are common with the negative and positive terminals respectively the voltmeter would also indicate 12 V across R_1 and R_2. In a parallel circuit each component has the same voltage across its terminals. Expressed as an equation:

$$E_t = E_{R_1} = E_{R_2} = \cdots = E_{R_n} \tag{3–17}$$

In Figure 3–11 the voltage across R_1 and R_2 is 12 V. Using Ohm's Law the currents through these resistors are found by the following calculations:

$$I_{R_1} = \frac{E_s}{R_1} = \frac{12\ \text{V}}{6\ \Omega} = 2\ \text{A}$$

$$I_{R_2} = \frac{E_s}{R_2} = \frac{12\ \text{V}}{12\ \Omega} = 1\ \text{A}$$

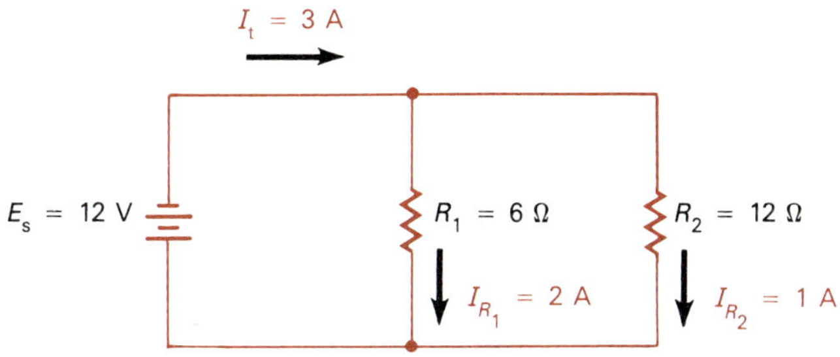

FIGURE 3–12 The total current delivered by the power supply splits and flows through each resistor.

These currents have been included in Figure 3–12. It is apparent that the total current delivered by the power supply must equal the sum of the *branch* currents.

$$I_t = I_{R_1} + I_{R_2} = 2\text{ A} + 1\text{ A} = 3\text{ A}$$

In terms of a general formula we have

$$\boldsymbol{I_t = I_{R_1} + I_{R_2} + \cdots + I_{R_n}} \qquad \textbf{(3–18)}$$

Example: Find the total current delivered by the power supply in Figure 3–13.

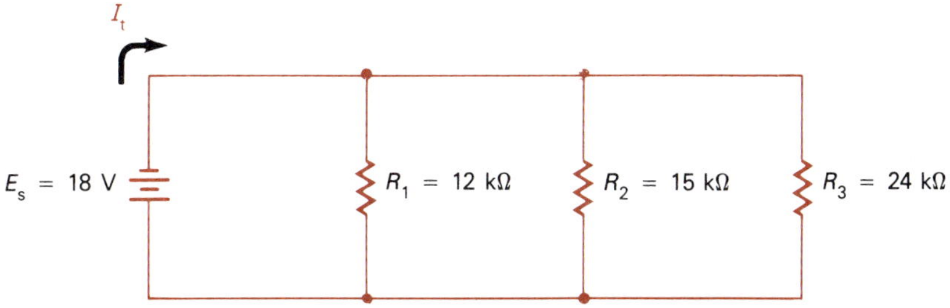

FIGURE 3–13

Solution:

$$I_{R_1} = \frac{18\text{ V}}{12\text{ k}\Omega} = 1.5\text{ mA}$$

$$I_{R_2} = \frac{18\text{ V}}{15\text{ k}\Omega} = 1.2\text{ mA}$$

$$I_{R_3} = \frac{18\text{ V}}{24\text{ k}\Omega} = 0.75\text{ mA}$$

$$\begin{aligned} I_t &= I_{R_1} + I_{R_2} + I_{R_3} \\ &= 1.5\text{ mA} + 1.2\text{ mA} + 0.75\text{ mA} \\ &= 3.45\text{ mA} \end{aligned}$$

Total Resistance in Parallel Circuits

Assume for a moment that in Figure 3–13 resistors R_2 and R_3 were removed. With only R_1 connected the total current would be 1.5 mA due to the 12 kΩ resistance. When R_2 is added the total current jumps to 1.5 mA plus 1.2 mA, or 2.7 mA. Since the voltage remains constant, and the current is increased, Ohm's Law indicates that the total resistance must have decreased. Total resistance in a parallel circuit is found by:

$$R_t = \frac{E}{I_t} \qquad (3\text{–}19)$$

The total, or equivalent, resistance in Figure 3–13 is

$$R_t = \frac{18\ \text{V}}{3.45\ \text{mA}} = 5\,217.4\ \Omega$$

Resistors R_1, R_2, and R_3 could be replaced by a 5 217.4-Ω resistor and the same total current would be delivered by the power supply. The power supply would *see* the same total resistance.

Example: Three resistors (680 Ω, 200 Ω, and 560 Ω) are placed in parallel and connected to a 48-V power supply. Calculate (a) the branch currents, (b) total current, and (c) the total resistance.

Solution: a. $I_{R_1} = \frac{48\ \text{V}}{680\ \Omega} = 70.6\ \text{mA}$

$I_{R_2} = \frac{48\ \text{V}}{200\ \Omega} = 240\ \text{mA}$

$I_{R_3} = \frac{48\ \text{V}}{560\ \Omega} = 85.7\ \text{mA}$

b. $I_t = 70.6\ \text{mA} + 240\ \text{mA} + 85.7\ \text{mA} = 396.3\ \text{mA}$

c. $R_t = \frac{48\ \text{V}}{396.3\ \text{mA}} = 121.12\ \Omega$

Note an important point concerning parallel resistance. The equivalent circuit resistance is always smaller than the smallest resistance value.

Total circuit resistance may be computed directly from the individual resistance values. Given the values in Figure 3–14 total circuit current is

$$I_t = I_1 + I_2$$

or in terms of voltage and resistance,

$$\frac{E}{R_t} = \frac{E}{R_1} + \frac{E}{R_2}$$

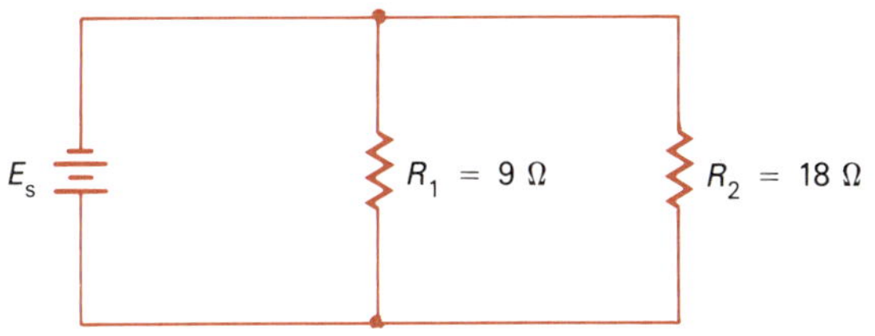

FIGURE 3–14

Dividing both sides by E gives one of the most common formulas for computing the equivalent resistance of two parallel resistors.

$$\frac{1}{R_t} = \frac{1}{R_1} + \frac{1}{R_2} \qquad (3\text{–}20)$$

Example: Compute the equivalent resistance in Figure 3–14.

Solution:

$$\frac{1}{R_t} = \frac{1}{R_1} + \frac{1}{R_2}$$
$$\frac{1}{R_t} = \frac{1}{9\ \Omega} + \frac{1}{18\ \Omega}$$
$$\frac{1}{R_t} = 0.111 + 0.055$$
$$\frac{1}{R_t} = 0.166$$
$$R_t = \frac{1}{0.166}$$
$$R_t = 6\ \Omega$$

Formula 3–20 can be rearranged to provide a second useful expression for two parallel resistances. Follow the development of this formula carefully.

$$\frac{1}{R_t} = \frac{1}{R_1} + \frac{1}{R_2} \qquad \text{(Initial expression)}$$
$$1 = R_t\left(\frac{1}{R_1} + \frac{1}{R_2}\right) \qquad \text{(Multiplication by } R_t\text{)}$$
$$R_t = \frac{1}{\frac{1}{R_1} + \frac{1}{R_2}} \qquad \left[\text{Division by } \left(\frac{1}{R_1} + \frac{1}{R_2}\right)\right]$$
$$R_t = \frac{1}{\frac{1}{R_1} + \frac{1}{R_2}} \times \frac{R_1R_2}{R_1R_2} \qquad \text{(Multiplication of the right side by 1)}$$

$$R_t = \frac{R_1R_2}{R_1R_2\left(\frac{1}{R_1} + \frac{1}{R_2}\right)} \quad \text{(Multiplication of numerators and denominators)}$$

$$R_t = \frac{R_1R_2}{\frac{R_1R_2}{R_1} + \frac{R_1R_2}{R_2}} \quad \text{(Multiplication to remove parentheses)}$$

The formula for the equivalent resistance of two parallel resistances may be expressed as follows:

$$R_t = \frac{R_1 \times R_2}{R_1 + R_2} \tag{3–21}$$

In terms of Figure 3–14,

$$\begin{aligned} R_t &= \frac{9\ \Omega \times 18\ \Omega}{9\ \Omega + 18\ \Omega} \\ &= \frac{162\ \Omega^2}{27\ \Omega} \\ &= 6\ \Omega \end{aligned}$$

For any number of resistors in parallel the basic formula is

$$\frac{1}{R_t} = \frac{1}{R_1} + \frac{1}{R_2} + \frac{1}{R_3} + \cdots + \frac{1}{R_n} \tag{3–22}$$

which, when rearranged, becomes

$$R_t = \frac{1}{\frac{1}{R_1} + \frac{1}{R_2} + \frac{1}{R_3} + \cdots + \frac{1}{R_n}} \tag{3–23}$$

For any number of equal value resistors the basic formula is

$$R_t = \frac{R}{n} \tag{3–24}$$

where, R = the value of 1 resistor
n = the number of parallel resistors

Example: Four resistors are connected in parallel to a 24-V power supply. Find the total current if each of the resistors has a value of 10 kΩ.

Solution: $R_t = \frac{10\ k\Omega}{4}$

$R_t = 2.5\ k\Omega$

$I_t = \frac{24\ V}{2.5\ k\Omega}$

$I_t = 9.6\ mA$

Calculator Tip

There will be many times in your study of electronics when it will be necessary to calculate the equivalent resistance of two parallel resistors. Formulas 3–20 and 3–21 can be used to calculate this resistance value. The following sequences demonstrate how your calculator can compute the equivalent resistance of 4.7 kΩ and 3.3 kΩ connected in parallel.

Using formula 3–20:

4.7 [EE] 3 [1/x] [+] 3.3 [EE] 3 [1/x] [=] [1/x] and 1.939 kΩ is displayed

Using formula 3–21:

4.7 [EE] 3 [×] 3.3 [EE] 3 [=] [STO] 4.7 [EE] 3 [+] 3.3 [EE] 3 [=] [1/x] [×] [RCL] [=] and 1.939 kΩ is displayed

Note that when all terms are in a common unit (such as kilohms) that it may not be necessary to enter the exponents, as the answer is assumed to be in kilohms.

Series-Parallel Circuits

In a series-parallel circuit, such as the one shown in Figure 3–15 the current leaving the voltage source is sometimes referred to as the *line current*. In this case, identified as I_{total} or I_t, it is the total circuit current and is not only the amount of current leaving the source, but returning to it as well. This is also the amount of current flowing through the first resistor, R_1. At the junction of the two parallel resistors, the current divides, forming, in this case, two *branch currents*. The branch currents are not necessarily equal but are inversely proportional to the values of resistance. The higher the value of R_2, the smaller the amount of I_{R_2}. Conversely, if R_2 decreases, the current flowing through R_2 increases. But no matter what is done to R_2 and R_3, the sum of their currents is always equal to the line current. That is:

$$I_t = I_{R_2} + I_{R_3}$$

The voltage across R_2 is equal to that across R_3. Since $E = I \times R$, $I_{R_2} \times R_2 = I_{R_3} \times R_3$.

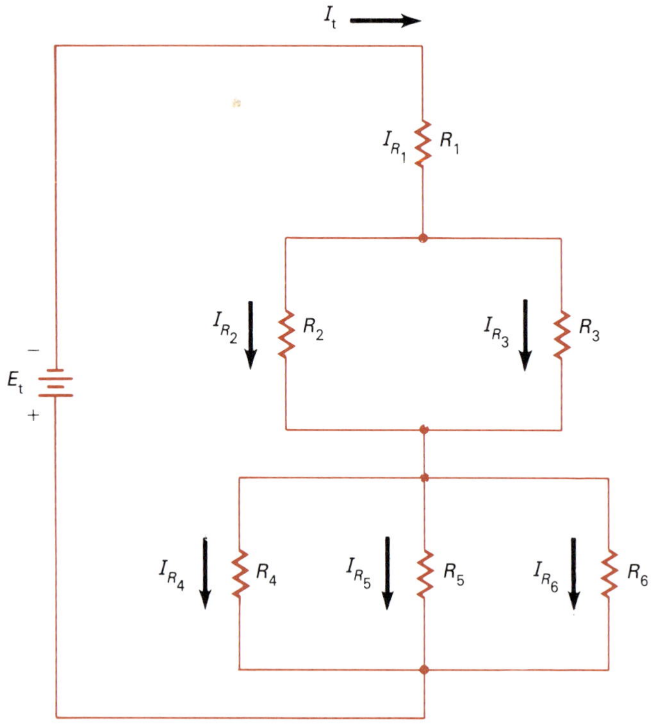

$$I_t = I_{R_1} = I_{R_2} + I_{R_3} = I_{R_4} + I_{R_5} + I_{R_6}$$

$$E_{R_2} = E_{R_3}$$

$$E_{R_4} = E_{R_5} = E_{R_6}$$

$$E_t = E_{R_1} + E_{R_2} + E_{R_4}$$

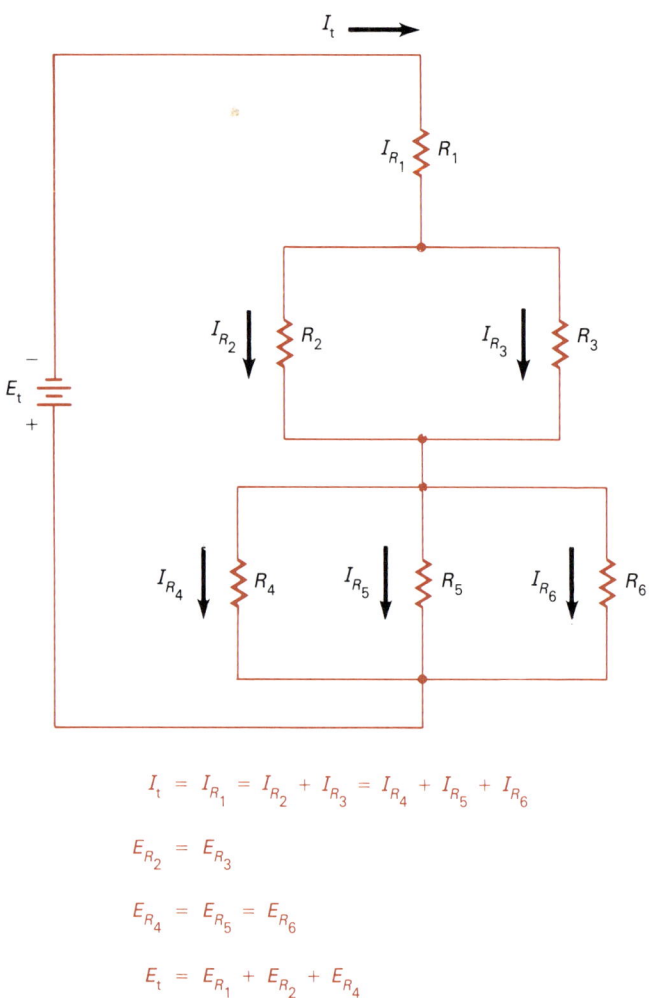

FIGURE 3–15 There are often several ways of solving circuit problems. The relationships supplied here indicate some of the methods that can be used.

The three remaining resistors, R_4, R_5, and R_6 are also in parallel. When the two branch currents, I_{R_2} and I_{R_3}, reach the junction of the three resistors, they change from two to three branch currents. Again, the current flowing through each of the resistors is inversely proportional to the value of resistance, in ohms. The voltage drop across R_4 is the same as that across R_5 and R_6. That is,

$$I_{R_4} \times R_4 = I_{R_5} \times R_5 = I_{R_6} \times R_6$$

The sum of the voltage drops across each of the networks, $E_{R_1} + E_{R_2} + E_{R_4}$, equals the source voltage.

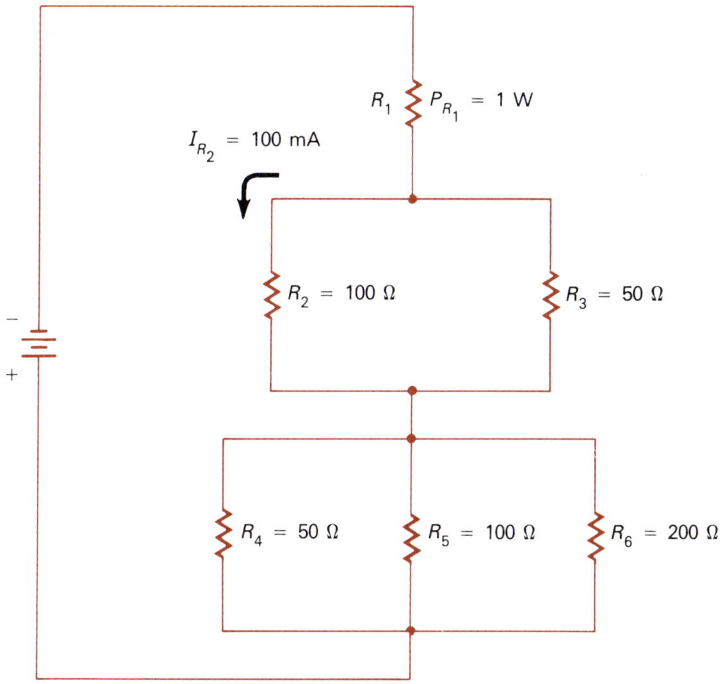

FIGURE 3–16 In this series-parallel circuit, only a small amount of data is supplied. The remaining information concerning the circuit must be derived.

The solution for problems involving circuits of this kind is easy if all the information concerning line and branch currents and resistor values is supplied, but this is often not the case. In Figure 3–16, the power dissipated by R_1 is given as 1 W. The resistor values are $R_2 = 100\ \Omega$, $R_3 = 50\ \Omega$, $R_4 = 50\ \Omega$, $R_5 = 100\ \Omega$, $R_6 = 200\ \Omega$. The current through R_2 is 100 mA. With this information, we are able to find:

- the total amount of circuit current.
- the current flowing through R_1.
- the voltage across R_1.
- the voltage across R_2.
- the voltage across R_3.
- the current through R_3.
- the power dissipated by R_2.
- the power dissipated by R_3.
- the current flowing through R_4, R_5, and R_6.

- the voltage across R_4, R_5, and R_6.
- the power dissipated by R_4, R_5, and R_6.
- the amount of source voltage, E_t.

Start the problem solution with R_2 and R_3, for most of the information is supplied by this part of the circuit. The current through R_2 is 100 mA or 100×10^{-3} A or 0.1 A. **The voltage across R_2** is

$$E = I \times R = 0.1 \text{ A} \times 100\ \Omega = 10 \text{ V}$$

The voltage across R_3 is equal to the voltage across R_2.

$$E_{R_3} = 10 \text{ V}$$

We can calculate **the current flowing through R_3.**

$$I = \frac{E}{R} = \frac{10 \text{ V}}{50\ \Omega} = 0.2 \text{ A}$$

This is an expected result; R_3 has half the value of resistance as R_2 so it should carry twice as much current.

We can now determine **the total amount of circuit current.** The circuit current is equal to the sum of the two branch currents.

$$0.1 \text{ A} + 0.2 \text{ A} = 0.3 \text{ A}$$

The current through R_1 is equal to the circuit current.

$$I_{R_1} = 0.3 \text{ A}$$

We can calculate **the voltage across R_1.**

$$E = \frac{P}{I} = \frac{1 \text{ W}}{0.3 \text{ A}} = 3.33 \text{ V}$$

We can calculate **the power dissipated by R_2.**

$$P = E \times I = 10 \text{ V} \times 0.1 \text{ A} = 1 \text{ W}$$

We can calculate **the power dissipated by R_3.**

$$P = E \times I = 10 \text{ V} \times 0.2 \text{ A} = 2 \text{ W}$$

Now consider the three parallel resistors, R_4, R_5, and R_6. To learn more about these three parallel resistors, we can combine the individual resistance values into an equivalent resistance.

$$R_{eq} = \frac{1}{\frac{1}{R_4} + \frac{1}{R_5} + \frac{1}{R_6}} = \frac{1}{\frac{1}{50\ \Omega} + \frac{1}{100\ \Omega} + \frac{1}{200\ \Omega}} = 28.57\ \Omega$$

For convenience, 28.57 Ω can be rounded off to 29 Ω, and this is the equivalent resistance of the three resistors, R_4, R_5, and R_6. The current flowing through this equivalent resistance, as calculated earlier, is 0.3 A. This puts us in the position of being able to calculate **the voltage across R_4, R_5, and R_6.**

$$E = I \times R = 0.3\ \text{A} \times 29\ \Omega = 8.7\ \text{V}$$

Now that we know the voltage we can easily calculate **the current flowing through R_4, R_5, and R_6.**

$$I = \frac{E}{R}$$

$$I_{R_4} = \frac{8.7\ \text{V}}{50\ \Omega} = 0.174\ \text{A}$$

$$I_{R_5} = \frac{8.7\ \text{V}}{100\ \Omega} = 0.087\ \text{A}$$

$$I_{R_6} = \frac{8.7\ \text{V}}{200\ \Omega} = 0.043\ 5\ \text{A}$$

This is an expected result; R_4 carries twice as much current as R_5 since the resistance value of R_4 is half that of R_5, and R_5 carries twice as much current as R_6 since the resistance value of R_5 is half that of R_6. **The power dissipated by R_4, R_5, and R_6** can be calculated.

$$P = E \times I$$

$$P_{R_4} = 8.7\ \text{V} \times 0.174\ \text{A} = 1.5\ \text{W}$$

$$P_{R_5} = 8.7\ \text{V} \times 0.087\ \text{A} = 0.8\ \text{W}$$

$$P_{R_6} = 8.7\ \text{V} \times 0.043\ 5 = 0.4\ \text{W}$$

The total applied voltage, the source voltage, is equal to the sum of all the voltage drops. The voltage across R_1 is 3.33 V; the voltage across R_2 and R_3 is 10 V; and the voltage across R_4, R_5, and R_6 is 8.7 V. **The amount of source voltage, E_t,** is approximately

$$3.33\ \text{V} + 10\ \text{V} + 8.7\ \text{V} = 22.03\ \text{V}$$

This answers all the required questions concerning the circuit but because of the large number of steps involved, it is possible to make a mistake. If an error is made during an early stage of the work, it can be carried along, with the result that a large number of steps will be incorrect. For these reasons, it is advisable to check such problems, preferably by following an alternative method of solution.

The check may not agree completely with the original answer to the problem. Rounding off and the use of numbers which may be carried out to more or fewer decimal places, conspire to produce dissimilarities in answers. Some judgment is required to decide if the check is sufficiently close in value to the solution to the problem.

Calculator Tip

The formula to determine R_1 in Figure 3–16 is given as follows:

$$R_1 = P_{R_1} \div I_t^2$$

but to find I_t

$$\begin{aligned} I_t &= I_{R_2} + I_{R_3} \\ &= 100 \text{ mA} + I_{R_3} \end{aligned}$$

but to find I_{R_3}

$$\begin{aligned} E_{R_3} &= I_{R_2} \times R_2 \\ &= 100 \text{ mA} \times 100\ \Omega \\ &= 10 \text{ V} \end{aligned}$$

so that I_{R_3} is

$$\begin{aligned} I_{R_3} &= E_{R_3} \div R_3 \\ &= 10 \text{ V} \div 50\ \Omega \\ &= 200 \text{ mA} \end{aligned}$$

and I_t becomes

$$\begin{aligned} I_t &= I_{R_2} + I_{R_3} \\ &= 100 \text{ mA} + 200 \text{ mA} \\ &= 300 \text{ mA} \\ &= 0.3 \text{ A} \end{aligned}$$

The value of R_1 therefore becomes

$$R_1 = 1\text{ W} \div (0.3\text{ A})^2$$
$$= 11.111\ \Omega$$

The formula for R_t is then given as

$$R_t = R_1 + \frac{R_2R_3}{R_2 + R_3} + \frac{1}{\dfrac{1}{R_4} + \dfrac{1}{R_5} + \dfrac{1}{R_6}}$$

with the complete sequence for finding R_t being:

11.111 [STO] 100 [×] 50 [=] [÷] [(] 100 [+] 50 [)] [=] [+] [RCL] [=] [STO] 50 [1/x] [+] 100 [1/x] [+] 200 [1/x] [=] [1/x] [+] [RCL] [=] 73.016 Ω

Conductance

Conductance is a measure of the ability of an electric circuit to pass current. The letter symbol for conductance is *G*. The metric unit of conductance is siemens. (The mho was the unit used for conductance in the English system of measurement.) The unit symbol for siemens is S.

Conductance is the reciprocal of resistance and is written as

$$G = \frac{1}{R} \tag{3–25}$$

Using the circuit shown in Figure 3–16, we can find the combined resistance of $R_4 + R_5 + R_6$ by converting each of the resistances into an equivalent conductance. Since we have three resistors:

$$G = \frac{1}{R_4} + \frac{1}{R_5} + \frac{1}{R_6} = \frac{1}{50\ \Omega} + \frac{1}{100\ \Omega} + \frac{1}{200\ \Omega} = 0.035\text{ S}$$

All we need to do now is to convert conductance back to resistance.

$$R = \frac{1}{G} = \frac{1}{0.035\text{ S}} = 28.57\ \Omega$$

This agrees with the value obtained earlier.

Voltage Dividers

Resistive voltage dividers have several functions. Connected across an electronic voltage source, resistive voltage dividers help discharge filter capacitors that form part of the power supply. They also help divide the total supply voltage so that each load receives its proper voltage share. These circuits are an excellent application of series-parallel circuits.

Figure 3–17 shows one possible arrangement of a voltage divider. The divider is shunted by three loads, each with its own voltage and current requirements. The problem is to find suitable resistance values for each of the divider resistors, R_1, R_2, R_3, and R_4. The total source voltage, as indicated, is 300 V. It is also assumed that the power supply can furnish the current demands of all the loads and the divider network.

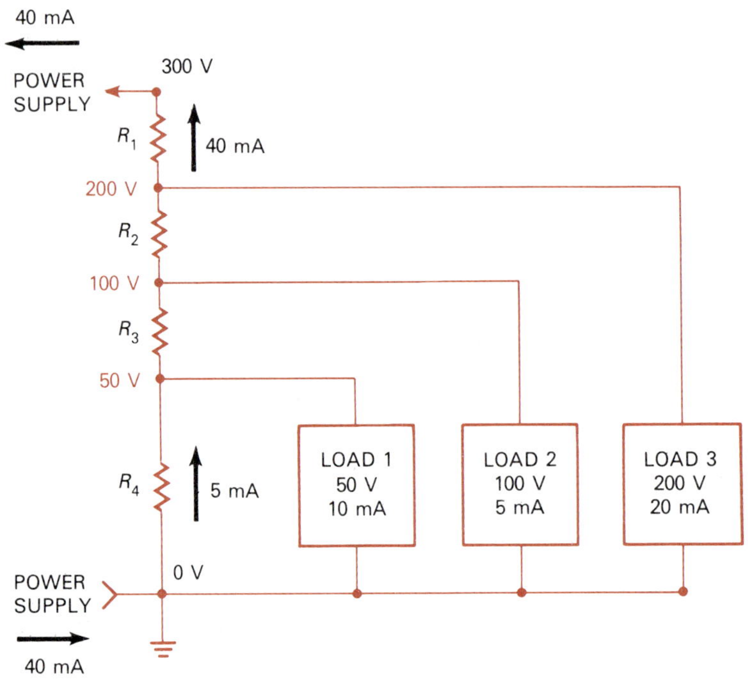

FIGURE 3–17 Shunt loads connected across a series voltage divider

The total current demand is easy to determine. This is equal to the sum of the three load currents added to the current flowing through R_4. It is not necessary to know the current quantities through R_1, R_2, and R_3, since this information can be obtained from an analysis of the diagram. The total current, then, is 5 mA flowing through R_4, plus 10 mA through load 1, plus 5 mA through load 2 and 20 mA through load 3. Simply adding, we have

$$5 \text{ mA} + 10 \text{ mA} + 5 \text{ mA} + 20 \text{ mA} = 40 \text{ mA}$$

As a first step, consider load 1. It has 50 V across it. But this load is in shunt with R_4, and so the same voltage must appear across it as well. The value of R_4 can be found by using Ohm's Law:

$$R = \frac{E}{I}$$

$$R_4 = \frac{50\ \text{V}}{5\ \text{mA}}$$

$$R_4 = 10\ \text{k}\Omega$$

The current through R_3 is obtained from two sources. It consists of 10 mA from load 1 and 5 mA flowing from R_4. The current through R_3, then, is

$$5\ \text{mA} + 10\ \text{mA} = 15\ \text{mA}$$

The voltage across R_3 is the difference between 100 V and 50 V. As indicated in the diagram, the voltage at the top of R_3 is 100 V. The bottom of R_3, however, is not connected to the ground, but to the top of R_4 which is a point 50 V above the ground. The voltage across R_3, then is

$$100\ \text{V} - 50\ \text{V} = 50\ \text{V}$$

The value of R_3 can now be calculated.

$$R = \frac{E}{I}$$

$$R_3 = \frac{50\ \text{V}}{15\ \text{mA}}$$

$$R_3 = 3.333\ \text{k}\Omega$$

The technique for finding the resistance value of R_2 follows the same procedure. The current through R_2 consists of the current it receives from R_3 plus the current it gets from load 2. This is

$$15\ \text{mA} + 5\ \text{mA} = 20\ \text{mA}$$

The voltage at the top of R_2 is 200 V. The voltage at its bottom end is 100 V. The voltage across R_2 is the difference in potential between these two voltages, or,

$$200\ \text{V} - 100\ \text{V} = 100\ \text{V}$$

Resistor R_2 has a value of

$$R = \frac{E}{I}$$

$$R_2 = \frac{100\text{ V}}{20\text{ mA}}$$

$$R_2 = 5\text{ k}\Omega$$

The final step is to determine the value of R_1. The current through R_1 is a combination of two currents: the current of 20 mA from load 3 and the current of 20 mA received from R_2. The current through R_1 is

$$20\text{ mA} + 20\text{ mA} = 40\text{ mA}$$

This figure is correct since the current flowing back into the power supply must be equal to the current taken from it.

The voltage across R_1 is 100 V, the difference between 300 V at the top of R_1 and 200 V appearing at the junction of R_1 and R_2. The value of R_1 is

$$R = \frac{E}{I}$$

$$R_1 = \frac{100\text{ V}}{40\text{ mA}}$$

$$R_1 = 2.5\text{ k}\Omega$$

The total resistance of the bleeder is

$$R_t = R_1 + R_2 + R_3 + R_4$$

$$R_t = 2.5\text{ k}\Omega + 5\text{ k}\Omega + 3.333\text{ k}\Omega + 10\text{ k}\Omega$$

$$R_t = 20.833\text{ k}\Omega$$

While problems of this kind may look complicated, they can often be solved by the application of one form of Ohm's Law, plus an analysis of the way the circuit behaves. As we have seen, Ohm's Law can be used to find the resistance of each load since load voltages and currents are already supplied.

The voltage divider circuit in Figure 3–18 contains a switch (S_1). With S_1 open, the voltage source will see $R_1 + R_2$ and supply a specific amount of current. When S_1 is closed, R_L will be parallel with R_2, the total circuit resistance will decrease, and the total circuit current will increase. Follow this example carefully.

Example: Calculate the voltage from point A to point B in Figure 3–18 with (a) S_1 open, and (b) S_1 closed.

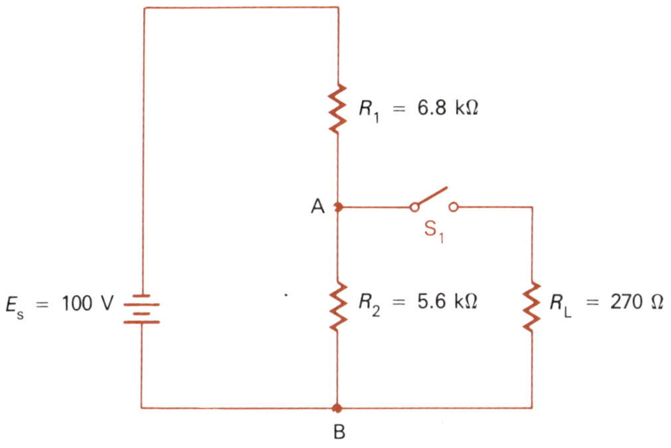

FIGURE 3–18

Solution: a. With S_1 open:

$$E_{A-B} = \left(\frac{R_2}{R_1 + R_2}\right)E_s$$
$$= \left(\frac{5.6\text{ k}\Omega}{6.8\text{ k}\Omega + 5.6\text{ k}\Omega}\right)(100\text{ V})$$
$$= 45.16\text{ V}$$

b. With S_1 closed:

$$R = \frac{R_2 \times R_L}{R_2 + R_L}$$
$$= \frac{5.6\text{ k}\Omega \times 0.27\text{ k}\Omega}{5.6\text{ k}\Omega + 0.27\text{ k}\Omega}$$
$$= 257.58\ \Omega$$
$$E_{A-B} = \left(\frac{R}{R_1 + R}\right)E_s$$
$$= \left(\frac{257.58\ \Omega}{6\,800\ \Omega + 257.58\ \Omega}\right)(100\text{ V})$$
$$= 3.65\text{ V}$$

Bridge Circuits

Current flows when there is a difference of potential between two points. In the battery circuit of Figure 3–19A, assume both batteries are identical in every respect. No current will flow through resistor R connected between the negative terminals of the two 10 V batteries because the voltage at point A is exactly equal to the voltage at point B.

In Figure 3–19B, the resistors and the batteries are of equal values in the left- and right-hand circuits. If a wire is connected to the precise midpoint center of R_1 and R_2, no current will flow from one circuit to the other because

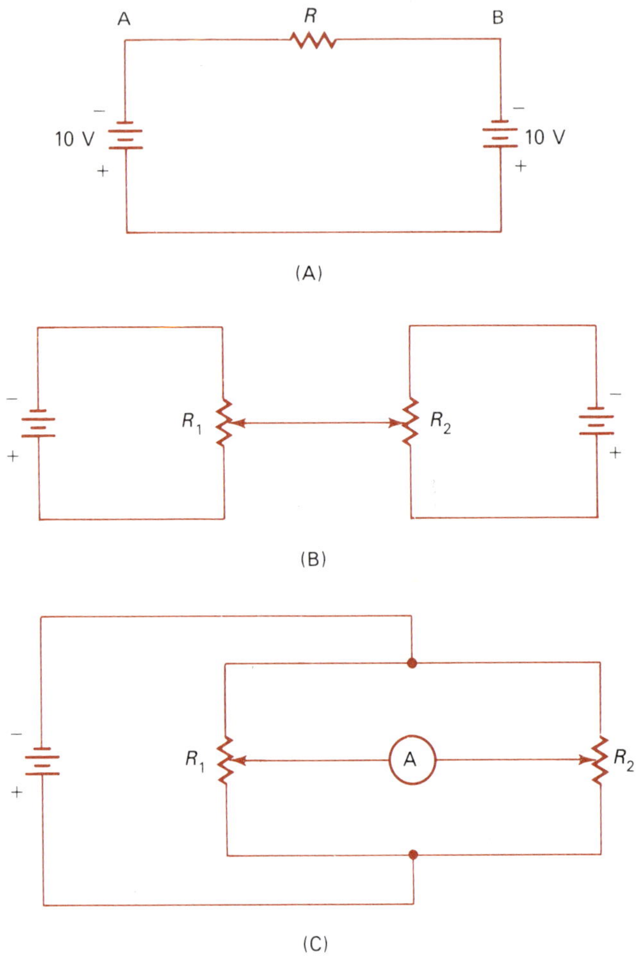

FIGURE 3–19 Steps in the development of a bridge circuit

the connecting points of the joining wire are at equipotential points. However, current will flow through the wire if either of the sliders are moved away from their center connecting points. The direction of current flow through the wire will depend on which end of the wire is made negative and which end is made positive.

A similar arrangement, but using just one voltage source, is illustrated in Figure 3–19C. The circle with the inscribed letter A indicates a current-reading meter, possibly a sensitive type such as a galvanometer. If the resistors have equal values, the *IR* drops across each will be the same. When the voltage at the contact point on R_1 is equal to the voltage at the contact point on R_2, no current will flow through the meter. Any number of positions of the slide arms will satisfy this requirement.

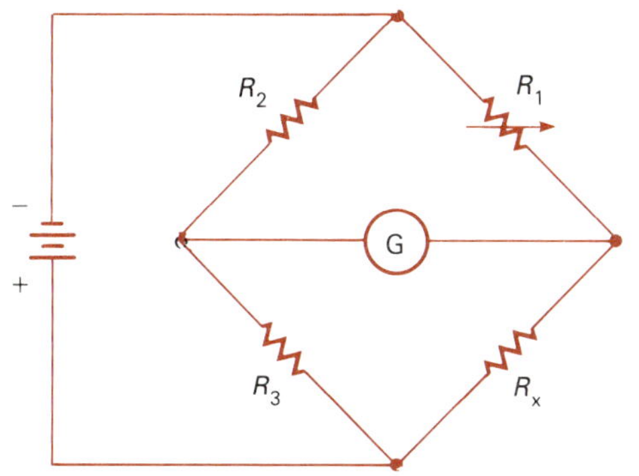

FIGURE 3–20 Wheatstone bridge

Figure 3–19C can be rearranged in the more familiar diamond pattern of the *Wheatstone bridge* shown in Figure 3–20. No current will flow through the meter when $R_1 = R_x$ and $R_2 = R_3$. When the bridge is *balanced*, that is, when the meter indicates zero:

$$R_x = \frac{\text{product of adjacent resistors}}{\text{opposite resistor}}$$

or

$$R_x = \frac{R_1 R_3}{R_2} \qquad (3\text{–}26)$$

In terms of a proportion:

$$\frac{R_1}{R_x} = \frac{R_2}{R_3} \qquad (3\text{–}27)$$

The Wheatstone bridge is used for making accurate measurements, in this case, of resistance. Here, R_x is the resistor whose value is unknown. Variable unit R_1 is adjusted until the galvanometer pointer is at zero. When this happens, the value of resistance can be read directly from a calibrated scale. The meter is center-zero reading type since the current can flow in either direction through the wires connected to the meter.

Example: An unknown resistor is connected to a Wheatstone bridge. The bridge arms have resistance values of $R_2 = 90\ \Omega$ and $R_3 = 150\ \Omega$. When the variable arm R_1 is adjusted for center-zero reading on the galvanometer, the resistance scale mounted on R_1 indicates $110\ \Omega$. What is the value of the unknown resistance?

Solution: $$R_x = \frac{R_1}{R_2} \times R_3 = \frac{110\ \Omega}{90\ \Omega} \times 150\ \Omega = 183.33\ \Omega$$

For a practical bridge, having R_2 not equal to R_3 makes it necessary to use the formula for calculating resistance. When using a Wheatstone bridge this can be avoided if bridge arms R_2 and R_3 have equal values. As an example, R_2 and R_3 could have resistances of 100 Ω each. Assume a resistor of unknown value is connected (R_x) and that the scale of R_1 indicates 195 Ω.

$$\begin{aligned} R_x &= \frac{R_1}{R_2} \times R_3 \\ &= \frac{195\ \Omega}{100\ \Omega} \times 100\ \Omega \\ &= 195\ \Omega \end{aligned}$$

This is exactly the value indicated by R_1. Thus we have proof that when R_2 and R_3 are made equal, the Wheatstone bridge can be made to indicate directly the unknown resistance value without the need for calculation.

Example: The values in Figure 3–20 are $R_1 = 5.6\ \text{k}\Omega$, $R_2 = 5\ \text{k}\Omega$ and $R_3 = 3.3\ \text{k}\Omega$. What value of R_x will balance the bridge?

Solution:

$$\begin{aligned} R_x &= \frac{R_1 R_3}{R_2} \\ &= \frac{(5.6\ \text{k}\Omega)(3.3\ \text{k}\Omega)}{5\ \text{k}\Omega} \\ &= 3\ 696\ \Omega \end{aligned}$$

Example: The value of the ratio of R_2 to R_3 in Figure 3–20 is 1.5. Find the value of R_x if $R_1 = 20\ \text{k}\Omega$.

Solution:

$$\begin{aligned} \frac{R_1}{R_x} &= \frac{R_2}{R_3} \\ \frac{20\ \text{k}\Omega}{R_x} &= \frac{1.5}{1} \\ (R_x)(1.5) &= (20\ \text{k}\Omega)(1) \\ R_x &= \frac{20\ \text{k}\Omega}{1.5} \\ R_x &= 13.33\ \text{k}\Omega \end{aligned}$$

EXERCISE 3·2

1. Find the equivalent resistance if a 40-kΩ resistor is placed in parallel with a 120-kΩ resistor.
2. Two resistors are placed in parallel and connected to a 12-V power supply. Calculate the total power dissipation if the values are 390 Ω and 220 Ω.
3. Two resistors are in parallel across a voltage source. Find the source voltage, if total current is 32 mA and the total circuit power is 0.208 W.

4. Four identical bulbs are wired in parallel across 18 V. Find the resistance of one lamp if the total current is 0.72 A.

Use Figure 3–21 for problems 5–10.

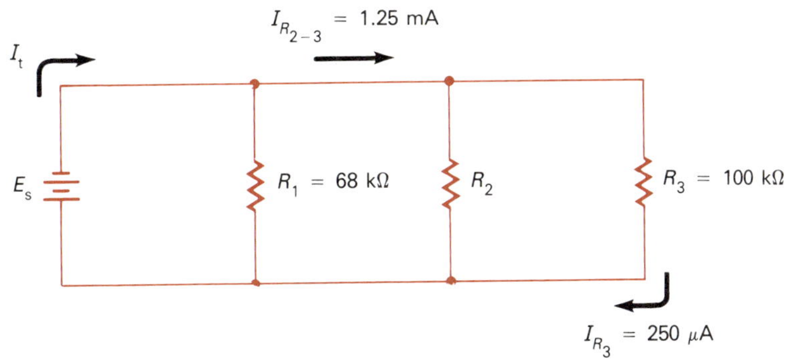

FIGURE 3–21

5. Calculate the output of the power supply.
6. Find the value of R_2.
7. What is the value of I_t?
8. Compute the power dissipation of R_1.
9. Find the total equivalent resistance of the circuit.
10. If R_2 were to open (infinite resistance), calculate the total current delivered by the power supply.
11. Calculate the equivalent resistance of 2.2 kΩ, 4 kΩ, and 8.2 kΩ in parallel.
12. A 5-kΩ resistor and another resistor are connected in parallel to a 50-V supply. The total current is 12 mA. Find the value of the second resistor.

Use Figure 3–22 for problems 13–20.

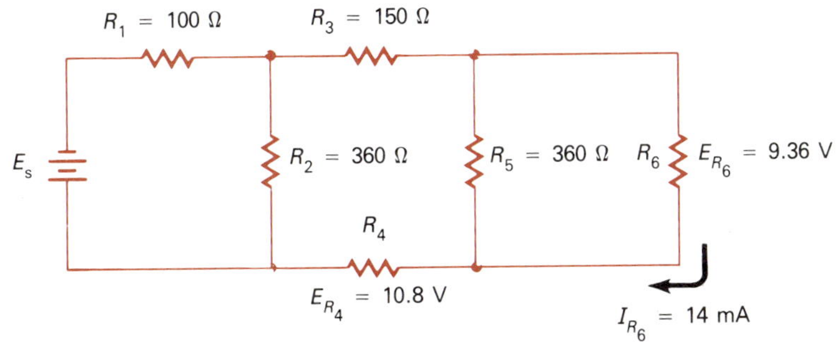

FIGURE 3–22

13. Determine the value of R_6.
14. What current flows through R_5?

15. Find the value of R_4.
16. Calculate the voltage across R_2.
17. Find the total current delivered by the power supply.
18. Determine the power dissipated by R_3.
19. What is the total equivalent resistance of the circuit?
20. Determine the total equivalent resistance if R_5 were to be replaced by a short (zero resistance).

Use Figure 3–23 for problems 21–25.

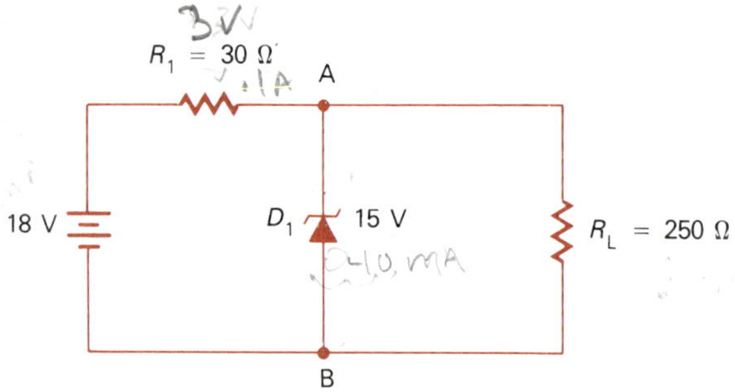

FIGURE 3–23 The zener diode (D_1) in this circuit maintains a constant 15 V between points A and B.

21. What is the voltage across R_1?
22. How much current is delivered by the power supply?
23. Find the current conducted by D_1.
24. Determine the power dissipated by the diode.
25. What is the power dissipation of R_L?

Use Figure 3–24 for problems 26–35.

26. What is the voltage from point B to the ground?
27. What is the voltage from point C to the ground?
28. What is the voltage from point D to the ground?
29. Calculate the voltage across a 120-kΩ load connected from point C to the ground.
30. With a 120-kΩ load from C to the ground, what voltage would be measured from point D to the ground?
31. A load is connected between B and the ground and conducts 1.2 mA. The voltage from D to the ground is 24 V. What is the load resistance?
32. A 39-kΩ resistor is connected between C and D. What current flows through the resistor?
33. A 5.6-kΩ load is connected from point A to the ground. If R_2 were to open, what voltage would appear across the load?
34. A 75-kΩ load is connected between B and the ground. If R_3 were to open, what voltage would be measured across the load?
35. The following loads are connected between the voltage divider points and the ground: A: 9.1 kΩ, B: 47 kΩ, C: 47 kΩ, D: 2.7 kΩ. Calculate the total current delivered by the power supply.

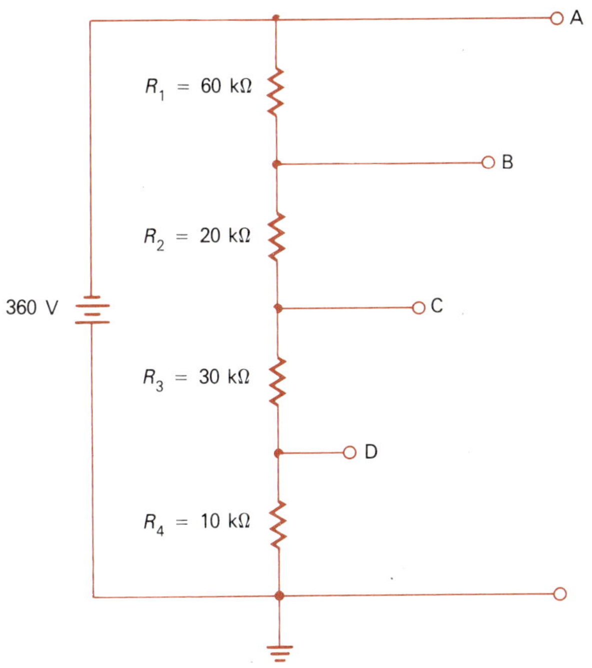

FIGURE 3–24

Use Figure 3–25 for problems 36–45.

36. Determine the value of R_4.
37. How much current flows through R_3?
38. Calculate the resistance of R_{L_2}.
39. What is the value of R_3?
40. How much power is dissipated by R_2?
41. Find the resistance of R_2.
42. Calculate the resistance of R_1.
43. What is the total current delivered by the power supply?
44. With S_1 open calculate the voltage across R_{L_3}. Note: resistors R_1 through R_4 are the same as previously calculated.
45. With S_1 open calculate the current through R_{L_1}. Note: resistors R_1 through R_4 are the same as previously calculated.

Use Figure 3–26 for problems 46–50.

46. Assuming that R_2 = 10 kΩ, R_3 = 40 kΩ, and R_1 = 30 kΩ, find the value of R_x.
47. Given R_2 = R_3 = 68 kΩ, and R_1 = 6.8 kΩ, what is the value of R_x?
48. The bridge has been balanced such that no current flows through the galvanometer. If the voltage across R_3 is 4.7 V, and the current through R_1 is 2.5 mA, find the value of R_x.

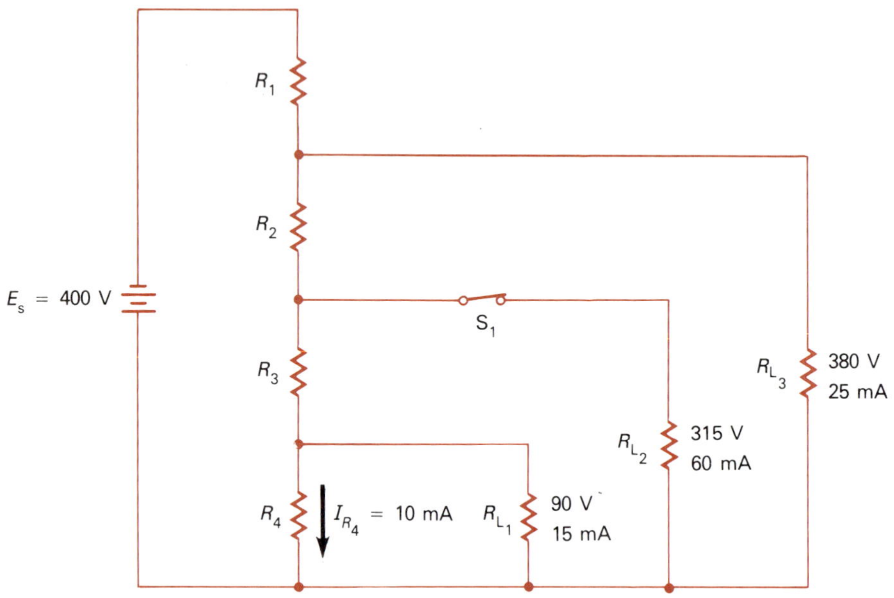

FIGURE 3–25

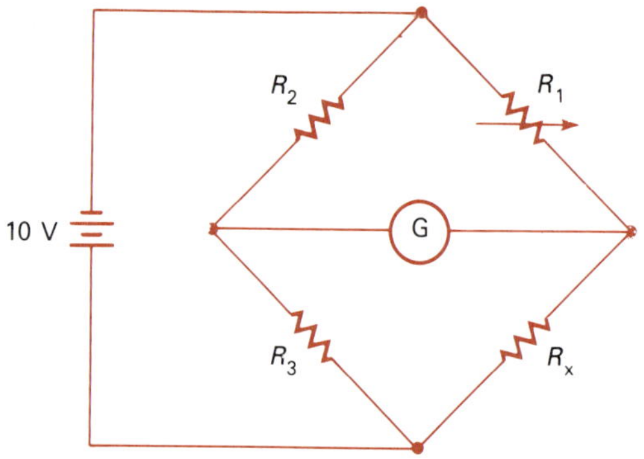

FIGURE 3–26

49. The value of the ratio of R_2 to R_3 is 0.2. Find the value of R_x if $R_1 = 48\ \text{k}\Omega$.
50. Assuming that $R_2 = 8.2\ \text{k}\Omega$, $R_3 = 5.6\ \text{k}\Omega$, and $R_1 = 22\ \text{k}\Omega$, find the value of R_x.

UNIT 3·3 KIRCHHOFF'S LAWS

Objectives:

After studying this unit, you should be able to

- use Kirchhoff's Voltage Law.
- use Kirchhoff's Current Law.
- use the voltage divider rule.
- use the current divider rule.

Kirchhoff's Voltage Law

In some circuits, especially if there is more than one voltage source, it may not be feasible to resort to Ohm's Law. Fortunately, a number of other circuit solution techniques are available. One of these methods is known as *Kirchhoff's Voltage Law.* This law states that **the algebraic sum of the voltages in a closed loop (or closed network) is equal to zero.** Figure 3–27 shows a battery shunted by three resistors, R_1, R_2, and R_3. If we assume that moving along a voltage drop or voltage source from minus to plus is an *increase,* then we can represent this increase by a plus (+) sign. In that case, moving along a voltage drop or voltage source from plus to minus (that is, in the opposite direction) would be a *decrease.* This can be indicated by a minus sign (−).

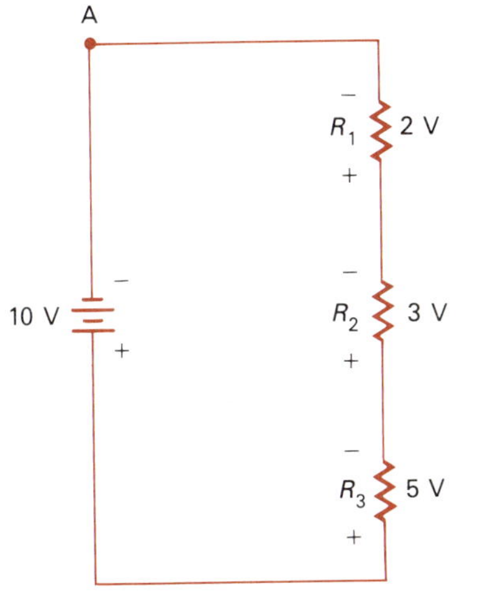

FIGURE 3–27 The algebraic sum of the voltage drops in a closed network is equal to zero.

Kirchhoff's Voltage Law may be expressed in an equation as

$$E_s + E_{R_1} + E_{R_2} + E_{R_3} + \cdots + E_{R_n} = 0 \qquad (3\text{–}28)$$

or,

$$\sum E_{source} = \sum E_{drops} \qquad (3\text{–}29)$$

which indicates that the algebraic sum (Σ) of the source voltages in a closed series circuit is equal to the sum of the voltage drops.

The words "algebraic sum" merely indicate that number signs must be considered when working with Kirchhoff's laws. The algebraic sum of $5 + (-5) = 0$.

In Figure 3–27 there is a 2-V drop across R_1, 3 V across R_2, and 5 V across R_3. The source is 10 V. To use Kirchhoff's Voltage Law, we can start at any point in the circuit, provided we finish at the same point. If we start at point A and move down through R_1, we will be moving from minus to plus. The increase is 2 V and since the travel is from minus to plus, it is better represented by +2. Continuing through R_2 we have +3 V and then through R_3 we have +5 V. This brings us to the bottom of the circuit and from there to the plus terminal of the battery. In moving through the battery we go from plus to minus, indicating a *decrease*, represented by a *minus* sign. In this example we have −10 V. With the trip to the battery, we are now back at point A and have made a complete circuit of this loop. Summing the results:

$$(+2) + (+3) + (+5) - 10 = 0$$

Example: Use Kirchhoff's Voltage Law to determine the voltage across R_2 in Figure 3–28.

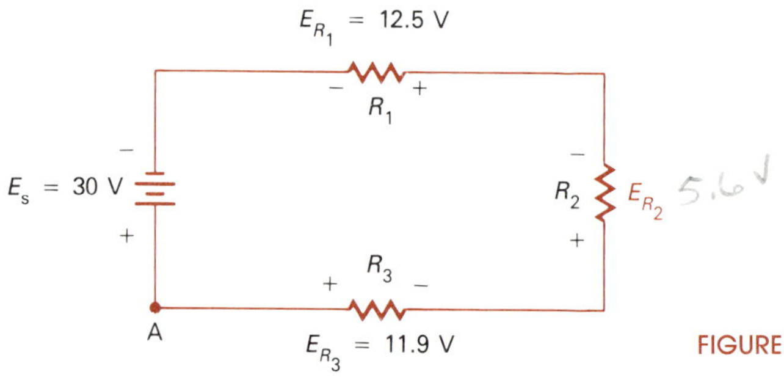

FIGURE 3–28

Solution:

$$E_s + E_{R_1} + E_{R_2} + E_{R_3} = 0$$
$$-30\text{ V} + 12.5\text{ V} + E_{R_2} + 11.9\text{ V} = 0$$
$$-5.6\text{ V} + E_{R_2} = 0$$
$$E_{R_2} = 5.6\text{ V}$$

Check: $E_s = E_{R_1} + E_{R_2} + E_{R_3}$

$30\text{ V} = 12.5\text{ V} + 5.6\text{ V} + 11.9\text{ V}$

$30\text{ V} = 30\text{ V}$

Example: Use Kirchhoff's Voltage Law to determine the circuit current in Figure 3–29A.

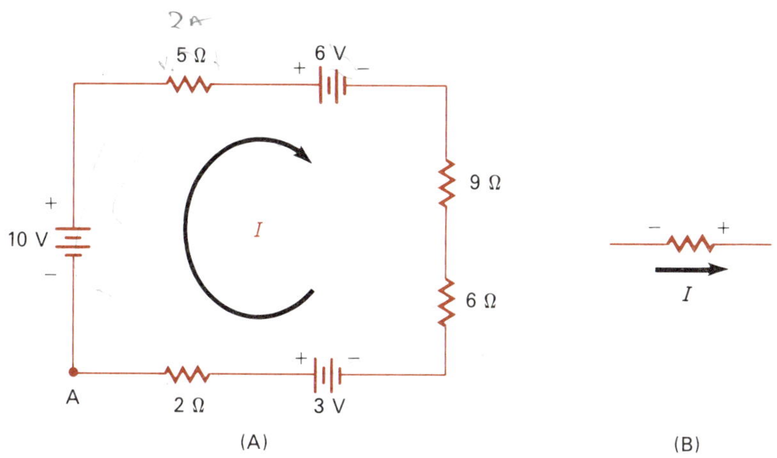

FIGURE 3–29 The *assumed* current direction produces the polarity as shown in B.

Solution: Beginning at point A and assuming clockwise current gives

$$10\text{ V} + I(5\ \Omega) - 6\text{ V} + I(9\ \Omega) + I(6\ \Omega) + 3\text{ V} + I(2\ \Omega) = 0$$

$$7\text{ V} + I(22\ \Omega) = 0$$

$$I(22\ \Omega) = -7\text{ V}$$

$$I = \frac{-7\text{ V}}{22\ \Omega}$$

$$I = -0.318\text{ A}$$

The negative sign indicates that current is actually flowing in the opposite direction selected.

Voltage Divider Rule

Another characteristic of a series circuit is expressed by the *voltage divider rule.* According to this rule, **the ratio between any two voltage drops in a series circuit is the same as the ratio of the two resistances across which these voltage drops occur.**

Example: Use the voltage divider rule to find the voltage drops across each resistor in Figure 3–30.

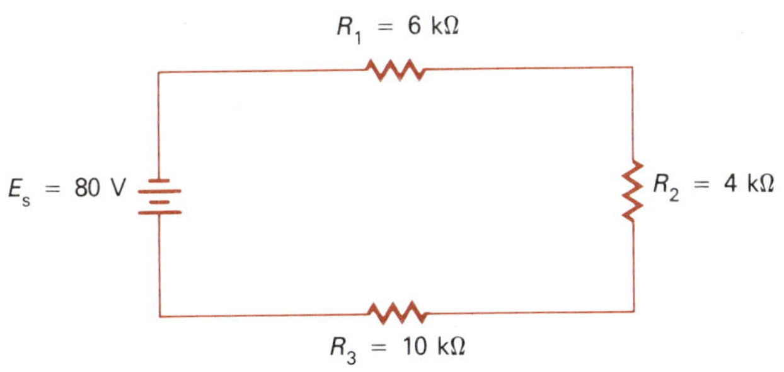

FIGURE 3–30

Solution:

$$R_t = R_1 + R_2 + R_3$$
$$= 6\ k\Omega + 4\ k\Omega + 10\ k\Omega$$
$$= 20\ k\Omega$$

$$E_{R_1} = \frac{R_1}{R_t}(E_s)$$
$$= \frac{6\ k\Omega}{20\ k\Omega}(80\ V)$$
$$= 24\ V$$

$$E_{R_2} = \frac{R_2}{R_t}(E_s)$$
$$= \frac{4\ k\Omega}{20\ k\Omega}(80\ V)$$
$$= 16\ V$$

$$E_{R_3} = \frac{R_3}{R_t}(E_s)$$
$$= \frac{10\ k\Omega}{20\ k\Omega}(80\ V)$$
$$= 40\ V$$

Check:

$$E_s = E_{R_1} + E_{R_2} + E_{R_3}$$
$$80\ V = 24\ V + 16\ V + 40\ V$$
$$80\ V = 80\ V$$

Example: Two resistors are in series with 12 V. One is a 500-Ω resistor with a 2.5-V drop across it. Use the voltage divider rule to determine the value of the second resistor.

Solution:

$$E_{R_2} = E_s - E_{R_1}$$
$$= 12\ V - 2.5\ V$$
$$= 9.5\ V$$

Setting up the following proportion:

$$\frac{E_{R_2}}{E_{R_1}} = \frac{R_2}{R_1}$$
$$\frac{9.5\ V}{2.5\ V} = \frac{R_2}{500\ \Omega}$$
$$R_2 = 1\ 900\ \Omega$$

Kirchhoff's Current Law

Kirchhoff's Current Law states that **the algebraic sum of the currents entering and leaving a node is zero.** Figure 3–31 shows three resistors meeting at a

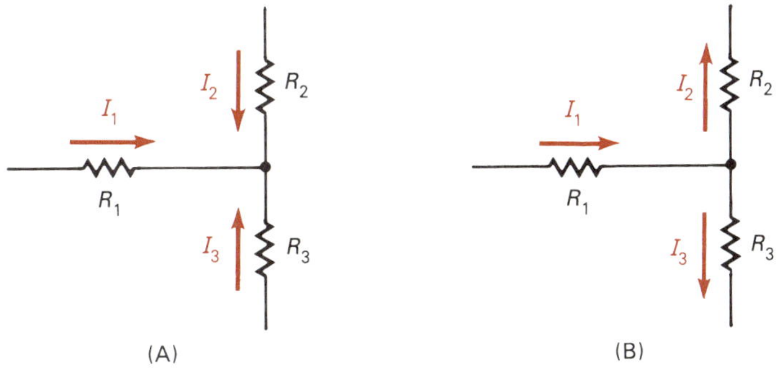

FIGURE 3–31 The algebraic sum of the currents flowing toward a junction is equal to zero ($I_1 + I_2 + I_3 = 0$). This is simply a mathematical concept and does not mean the currents suddenly disappear at the junction point. In drawing B, $I_1 = (-I_2) + (-I_3)$.

common node or junction with currents I_1, I_2, and I_3 flowing through the individual resistors as marked on the diagram. Stated in terms of an equation:

$$I_1 + I_2 + I_3 = 0$$

Since each of these currents flows in the same direction, toward their junction, and since each current is preceded by a plus sign (not stated in this case, but implied) a current flowing in the opposite direction could be represented by a minus sign.

The original equation, $I_1 + I_2 + I_3 = 0$, can be modified by transposing I_2 and I_3 to the other side of the equals symbol by changing their signs:

$$I_1 = (-I_2) + (-I_3)$$

All that has happened here is that the *direction* of flow of currents I_2 and I_3 has been changed. Instead of flowing toward the junction, they are now moving away from it, as seen in Figure 3–31B. This can be verbally interpreted as the algebraic sum of the currents flowing away from a junction must be equal to the algebraic sum of the currents flowing toward it.

This statement is not new, for in previous examples emphasis has been placed on the fact that the amount of current leaving a battery terminal must ultimately return to it.

Kirchhoff's Current Law may be expressed in an equation as

$$\sum I_{\text{entering}} = \sum I_{\text{leaving}} \qquad \textbf{(3–30)}$$

which indicates that the sum (Σ) of the currents entering a node equals the sum of the currents leaving a node.

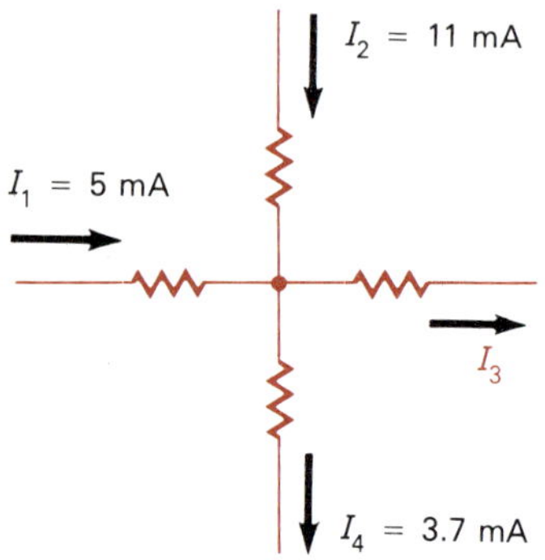

FIGURE 3–32

Example: Use Kirchhoff's Current Law to determine I_3 in Figure 3–32.

Solution:

$$I_1 + I_2 + I_3 + I_4 = 0$$
$$5\text{ mA} + 11\text{ mA} - I_3 - 3.7\text{ mA} = 0$$
$$16\text{ mA} - 3.7\text{ mA} = I_3$$
$$I_3 = 12.3\text{ mA}$$

Current Divider Rule

According to the *current divider rule,* **the ratio between any two parallel branch currents is the inverse of their resistance ratio, and the same as the ratio of their conductances.**

Refer to Figure 3–33.

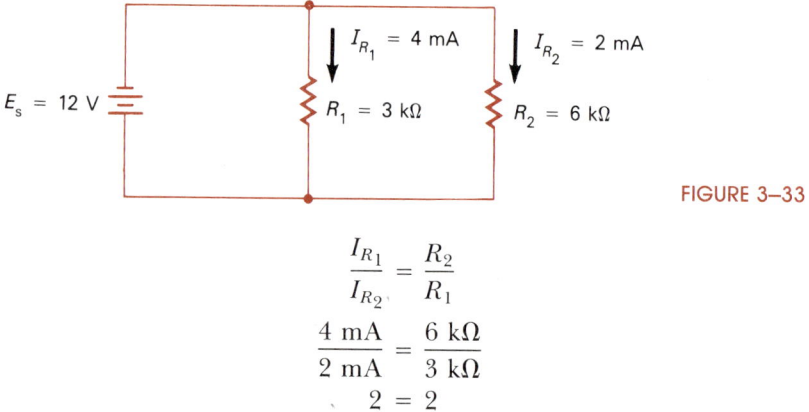

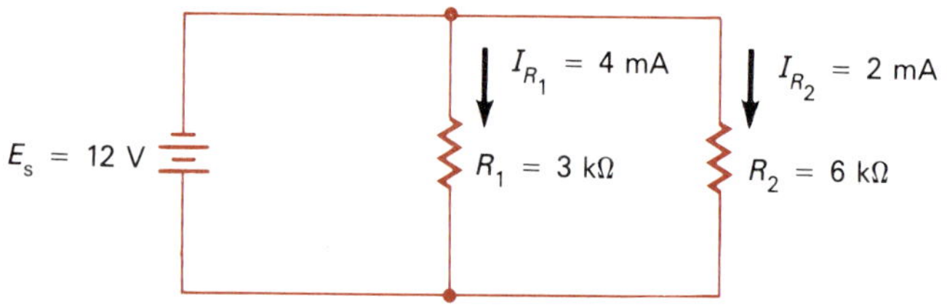

FIGURE 3–33

$$\frac{I_{R_1}}{I_{R_2}} = \frac{R_2}{R_1}$$
$$\frac{4\text{ mA}}{2\text{ mA}} = \frac{6\text{ k}\Omega}{3\text{ k}\Omega}$$
$$2 = 2$$

Example: Use the current divider rule to determine I_{R_1} in Figure 3–34.

Solution:

$$\frac{I_{R_1}}{I_{R_2}} = \frac{R_2}{R_1}$$
$$\frac{I_{R_1}}{0.82\text{ mA}} = \frac{1.8\text{ k}\Omega}{0.36\text{ k}\Omega}$$
$$I_{R_1} = 4.1\text{ mA}$$

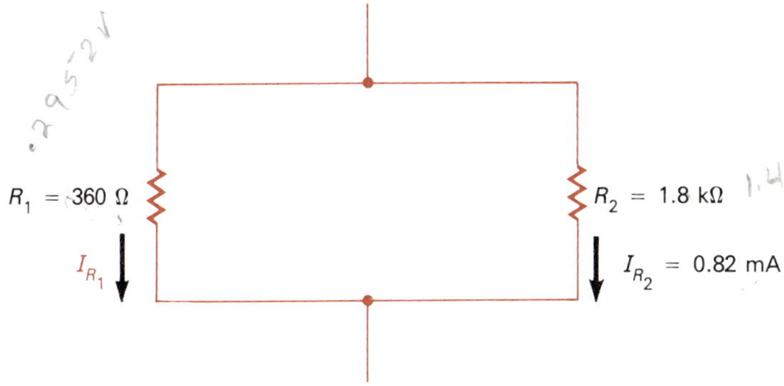

FIGURE 3–34

Applications of Kirchhoff's Laws

The use of Kirchhoff's Current and Voltage Laws is somewhat more involved than Ohm's Law and more steps may be required, but these laws are an excellent tool for solving circuit problems.

For example, Figure 3–35 shows a series network consisting of a number of series resistors and two voltage sources. When the direction of current flow through a resistor is known, the head end of the arrow is regarded as plus; the tail end as minus. Assume a direction of current flow. As shown in Figure 3–35, it is taken as clockwise. Starting at point A, the first resistor has a value of 1 Ω. Since $E = I \times R$, the voltage drop across this resistor is equal to $I(1\ \Omega)$. The next resistor is 2 Ω. The voltage across it is $I(2\ \Omega)$. Continuing through the circuit, the next voltage source is a 92-V battery, but in moving through this voltage from plus to minus we have a voltage decrease (-92) shown by the minus sign in front of 92. The remaining voltage drops are $I(3\ \Omega)$, $I(4\ \Omega)$, $I(5\ \Omega)$, and $I(6\ \Omega)$. Finally, in the trip through this loop we reach

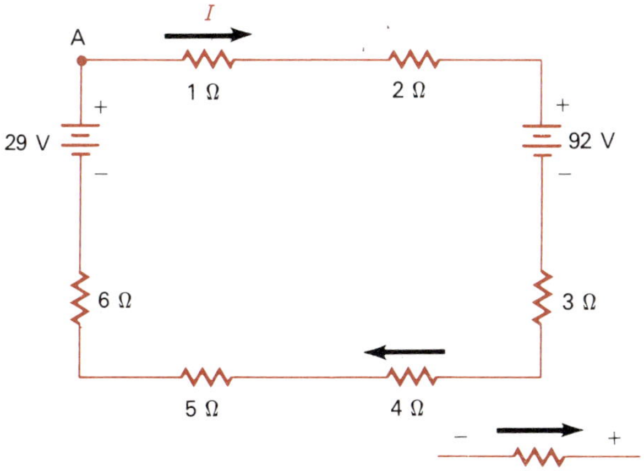

FIGURE 3–35 Example of problem solved by using Kirchhoff's Voltage Law

the 29-V battery. In going through the battery we move from minus to plus, representing a voltage increase.

The complete trip through this loop, starting at point A and returning to it can be written as:

$$I(1\ \Omega) + I(2\ \Omega) - 92\ \text{V} + I(3\ \Omega) + I(4\ \Omega) + I(5\ \Omega) + I(6\ \Omega) + 29\ \text{V} = 0$$
$$I(21\ \Omega) - 63\ \text{V} = 0$$
$$I(21\ \Omega) = 63\ \text{V}$$
$$I = \frac{63\ \text{V}}{21\ \Omega}$$
$$I = 3\ \text{A}$$

A problem of the type illustrated in Figure 3–35 can be checked by substituting the actual value of current for I. In this case I has a value of 3 A. When making this voltage substitution using $E = IR$, we have

$$3\ \text{V} + 6\ \text{V} - 92\ \text{V} + 9\ \text{V} + 12\ \text{V} + 15\ \text{V} + 18\ \text{V} + 29\ \text{V} = 0$$
$$92\ \text{V} - 92\ \text{V} = 0$$
$$0 = 0$$

If the current had been assumed to flow counterclockwise instead of clockwise as shown in the illustration, the only effect would be that the answer for the value of I would have been a negative value. That is: $I = -3$ A. The presence of the minus sign is an indication to change the direction of the current arrow.

EXERCISE 3·3

1. Use Kirchhoff's Voltage Law to determine the voltage across R_3 in Figure 3–36.

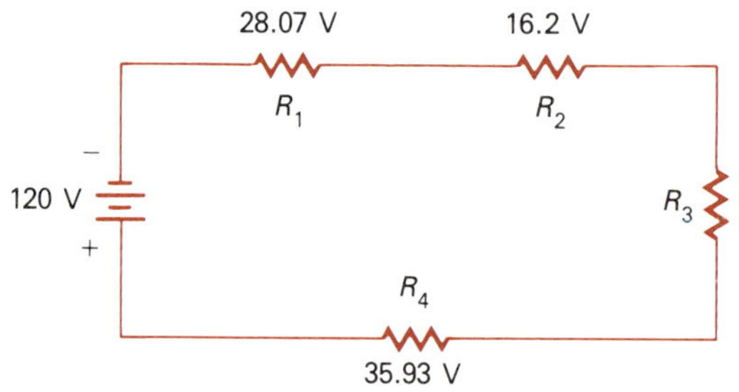

FIGURE 3–36

2. Four resistors are connected in series and attached to a 48-V power supply. Find the voltage across each if their values are 2.7 kΩ, 3.9 kΩ, 1.5 kΩ, and 2 kΩ.
3. If the 2.7-kΩ resistor in problem 2 was to short, what would the voltage drop across the 3.9-kΩ resistor change to?
4. A 3.6-kΩ resistance and a 9-kΩ resistance are in series with a power supply. Determine the source voltage if the voltage across the 3.6-kΩ resistance is 7.5 V.

5. Two resistors are in series and connected to 15 V. If the first resistor is twice the second, find the voltage across each.

Use Figure 3–37 for problems 6–7.

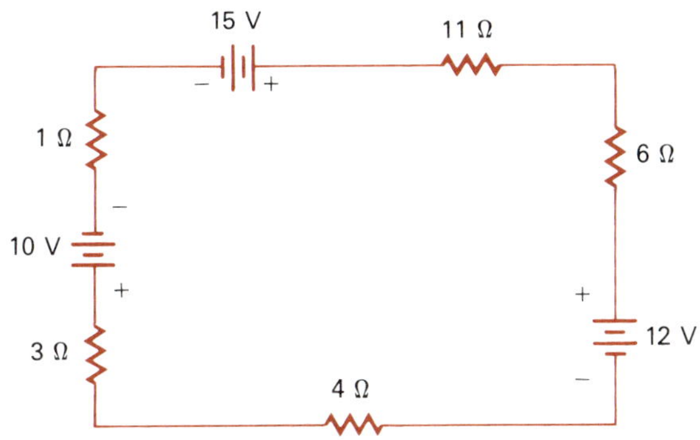

FIGURE 3–37

6. Determine the total current in Figure 3–37.
7. If the 6-Ω resistor in Figure 3–37 was to short, to what level would the circuit current change?

Use Figure 3–38 for problems 8–9.

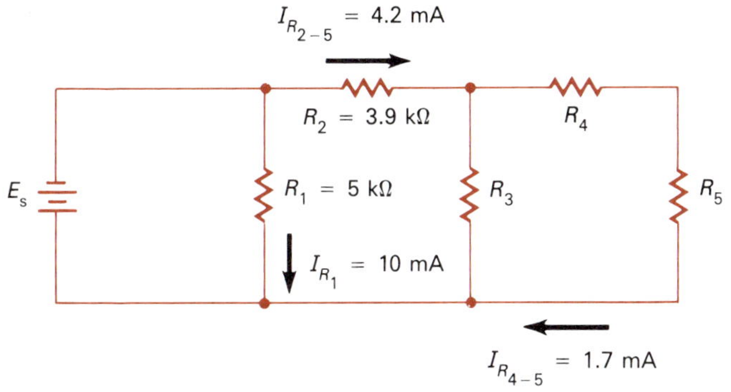

FIGURE 3–38

8. Use Kirchhoff's Current Law to determine the current through R_3 in Figure 3–38.
9. Find total current in Figure 3–38.
10. Two resistors are in parallel. One is 4.7 kΩ and is conducting 12.5 mA. What is the resistive value of the other if it conducts 7.2 mA?

Use Figure 3–39 for problems 11–14.

11. Use Kirchhoff's Current Law to find I_{R_1} in Figure 3–39.
12. What is the value of R_3 in Figure 3–39?
13. Find E_s in Figure 3–39.

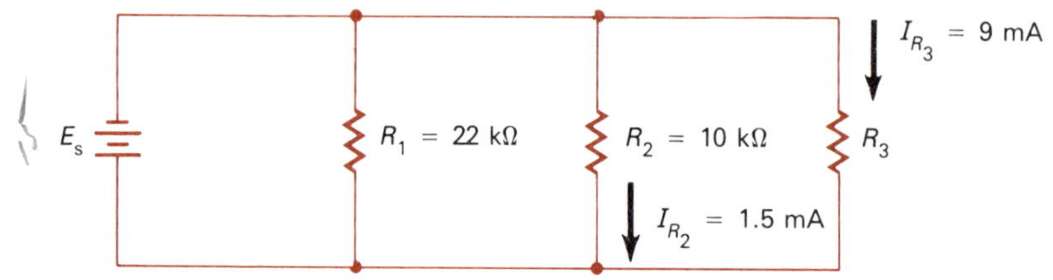

FIGURE 3–39

14. Calculate the equivalent resistance of the three resistors in Figure 3–39.

UNIT 3·4 CIRCUIT ANALYSIS THEOREMS

Objectives:

After studying this unit, you should be able to

- calculate circuit values utilizing the maximum power transfer theorem.
- calculate the Thévenin voltage and resistance for a resistive circuit.
- utilize the superposition theorem in simplifying circuits.
- calculate the Norton current and Thévenin resistance for a resistive circuit.

Maximum Power Transfer Theorem

When a load is connected across a voltage source, there is maximum transfer of power from the load when the resistance of the load is equal to the internal resistance of the source. As an example, consult Figure 3–40 in which

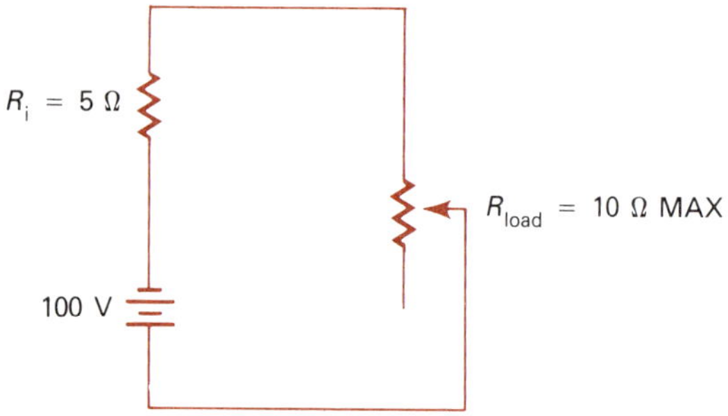

FIGURE 3–40 There will be a maximum transfer of power when the resistance of the load, the variable resistor, is adjusted to 5 ohms, to equal the internal resistance of the source voltage, represented here by R_i.

a variable resistor is connected across a 100-V supply. The internal resistance of the supply is 5 Ω. This resistance is marked R_i and is shown as a separate resistor outside the voltage source. The variable resistor is capable of being set at 2 Ω, 5 Ω, and 10 Ω. Table 3–1 shows what happens when the load resistor is set at these three different values.

This table shows that when the load resistance is 5 Ω, the maximum power is delivered to the load. The maximum in this case is 500 W. Various

TABLE 3–1

Effects of Changing the Load Resistance

E	R_{load}	R_i	R_t	$I = \frac{E}{R_t}$	$P = I^2R_{load}$
100 V	2 Ω	5 Ω	7 Ω	14.29 A	408.408 W
100 V	5 Ω	5 Ω	10 Ω	10.00 A	500.00 W
100 V	10 Ω	5 Ω	15 Ω	6.667 A	444.49 W

circuits (attenuators) are used to make sure that the internal resistance of the voltage source and the load are equal for varying circuit conditions.

The initial reaction to the maximum power transfer theorem is that 50% of the total power dissipated within the power supply appears wasteful. This concept, however, is very important in many electronic applications.

Example: The no-load output voltage (open circuit voltage) of a power supply is 45 V and the internal resistance is 3 Ω. If the supply is to transfer maximum power to a load calculate R_t, I_L, P_L, P_i, and the loaded output voltage (E_o) of the power supply. See Figure 3–41.

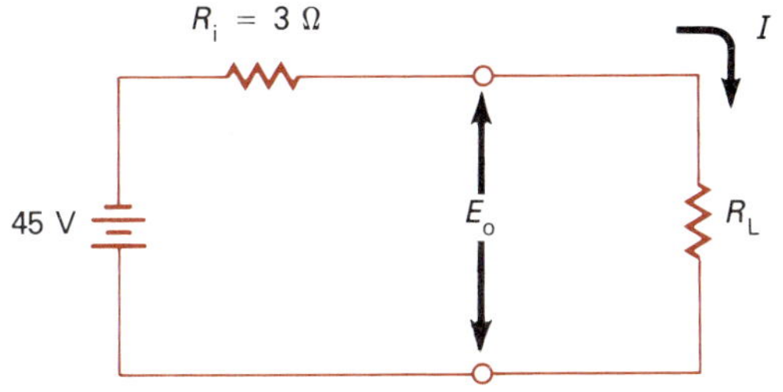

FIGURE 3–41

Solution:

$$R_t = R_i + R_L = 3\ \Omega + 3\ \Omega = 6\ \Omega$$

$$I_L = \frac{E_s}{R_t} = \frac{45\ \text{V}}{6\ \Omega} = 7.5\ \text{A}$$

$$P_L = I^2R_L = (7.5\ \text{A})^2(3\ \Omega) = 168.75\ \text{W}$$

$$\begin{aligned} P_i &= I^2R_i \\ &= (7.5\ \text{A})^2(3\ \Omega) \\ &= 168.75\ \text{W} \end{aligned}$$

$$\begin{aligned} E_o &= IR_L \\ &= 7.5\ \text{A} \times 3\ \Omega \\ &= 22.5\ \text{V} \end{aligned}$$

Thévenin's Theorem

Thévenin's Theorem states that **any two-terminal network containing resistances and voltage sources may be replaced by a single voltage source in series with a single resistance.** The voltage of the single source (E_{TH}) is the open circuit voltage at the network terminals. The resistance (R_{TH}) is the resistance at the network terminals when all of the voltage sources are replaced by their internal resistances.

Example: Find the Thévenin voltage and resistance for the circuit in Figure 3–42A.

Solution: Remove R_L and determine E_{TH} across the network terminals as in Figure 3–42B.

$$\begin{aligned} E_{TH} &= \frac{R_2}{R_1 + R_2} \times E_s \\ &= \frac{18\ \text{k}\Omega}{9\ \text{k}\Omega + 18\ \text{k}\Omega}(6\ \text{V}) \\ &= 4\ \text{V} \end{aligned}$$

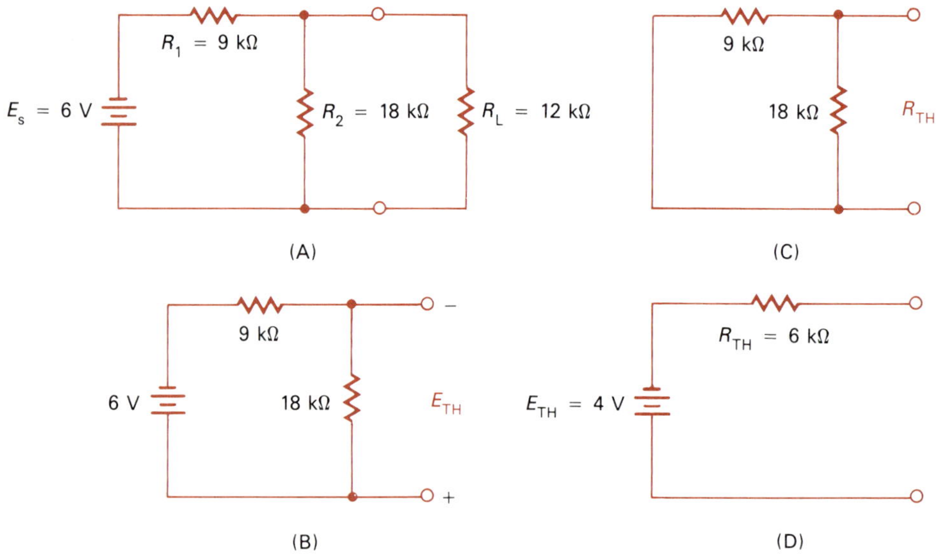

FIGURE 3–42

Replace the source with the internal resistance (typically assumed to be zero) as in Figure 3–42C.

$$R_{TH} = \frac{9\text{ k}\Omega \times 18\text{ k}\Omega}{9\text{ k}\Omega + 18\text{ k}\Omega}$$
$$= 6\text{ k}\Omega$$

Redraw the circuit in the "Thévenized" form as in Figure 3–42D.

In comparing Figures 3–42A and 3–42D, it is important to note that in terms of the load both circuits behave identically. You should calculate the voltage and current values associated with a 12-kΩ load attached to both circuits. These values will be the same.

Example: Determine the Thévenin voltage and resistance for the circuit in Figure 3–43.

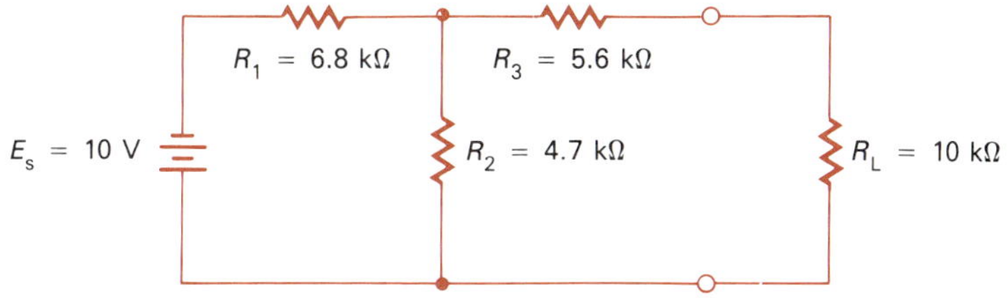

FIGURE 3–43

Solution: As no current flows through R_3 with R_L removed, E_{TH} is the voltage across R_2.

$$E_{TH} = \frac{R_2}{R_1 + R_2} \times E_s$$
$$= \frac{4.7\text{ k}\Omega}{6.8\text{ k}\Omega + 4.7\text{ k}\Omega} \times 10\text{ V}$$
$$= 4.09\text{ V}$$

With E_s replaced with a short R_1 and R_2 are in parallel

$$R_{TH} = R_3 + \frac{R_1 \times R_2}{R_1 + R_2}$$
$$= 5.6\text{ k}\Omega + \frac{6.8\text{ k}\Omega \times 4.7\text{ k}\Omega}{6.8\text{ k}\Omega + 4.7\text{ k}\Omega}$$
$$= 8.38\text{ k}\Omega$$

Superposition Theorem

The superposition theorem is useful when a circuit contains more than one source. The theorem states that **in any network containing more than one voltage source, the current through any branch is the algebraic sum of the currents produced by each source acting independently.**

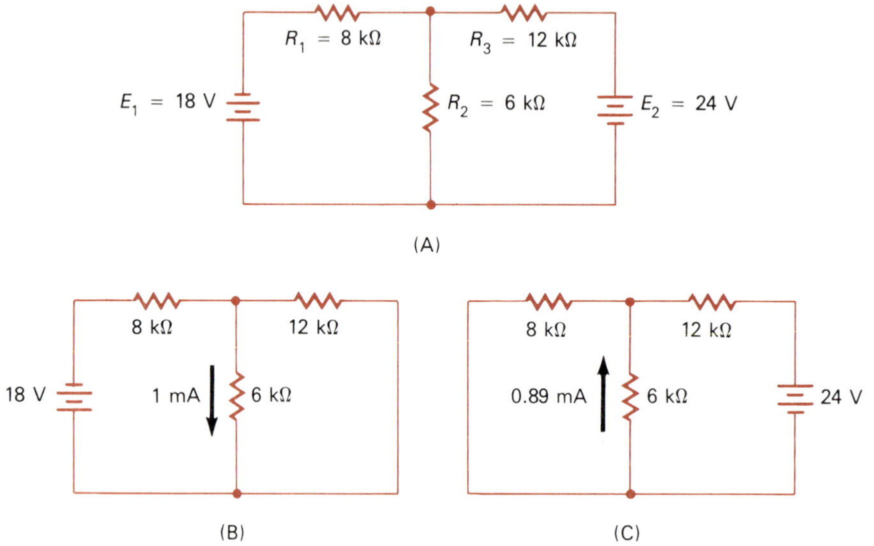

FIGURE 3–44

Example: Use the superposition theorem to solve for the current through R_2 in Figure 3–44A.

Solution: Select one voltage source and replace each of the others with a short. Calculate the value of the specific current. See Figure 3–44B.

$$R_{t_1} = R_1 + \frac{R_2 \times R_3}{R_2 + R_3}$$

$$= 8\text{ k}\Omega + \frac{6\text{ k}\Omega \times 12\text{ k}\Omega}{6\text{ k}\Omega + 12\text{ k}\Omega}$$

$$= 12\text{ k}\Omega$$

$$I_{t_1} = \frac{E_1}{R_{t_1}}$$

$$= \frac{18\text{ V}}{12\text{ k}\Omega}$$

$$= 1.5\text{ mA}$$

$$I_{R_2} = \frac{R_3}{R_2 + R_3} \times I_{t_1}$$

$$= \frac{12\text{ k}\Omega}{6\text{ k}\Omega + 12\text{ k}\Omega} \times 1.5\text{ mA}$$

$$= 1\text{ mA}$$

Place E_2 in the circuit and replace E_1 with a short. Calculate the current through R_2 due to E_2 as in Figure 3–44C.

$$R_{t_2} = R_3 + \frac{R_1 \times R_2}{R_1 + R_2}$$

$$= 12\ \text{k}\Omega + \frac{8\ \text{k}\Omega \times 6\ \text{k}\Omega}{8\ \text{k}\Omega + 6\ \text{k}\Omega}$$
$$= 15.43\ \text{k}\Omega$$

$$I_{t_2} = \frac{E_2}{R_{t_2}}$$
$$= \frac{24\ \text{V}}{15.43\ \text{k}\Omega}$$
$$= 1.56\ \text{mA}$$

$$I_{R_2} = \frac{R_1}{R_1 + R_2} \times I_{t_2}$$
$$= \frac{8\ \text{k}\Omega}{8\ \text{k}\Omega + 6\ \text{k}\Omega} \times 1.56\ \text{mA}$$
$$= 0.89\ \text{mA}$$

As the two currents oppose each other, the algebraic sum is
$I = 1\ \text{mA} - 0.89\ \text{mA}$
$= 0.11\ \text{mA}$, in the direction due to E_1

Example: Find the current through R_2 in Figure 3–45.

Solution: $$R_{t_1} = R_1 + \frac{R_2 \times R_3}{R_2 + R_3}$$
$$= 9\ \text{k}\Omega + \frac{6\ \text{k}\Omega}{2}$$
$$= 12\ \text{k}\Omega$$

$$I_{t_1} = \frac{E_1}{R_{t_1}}$$
$$= \frac{12\ \text{V}}{12\ \text{k}\Omega}$$
$$= 1\ \text{mA}$$

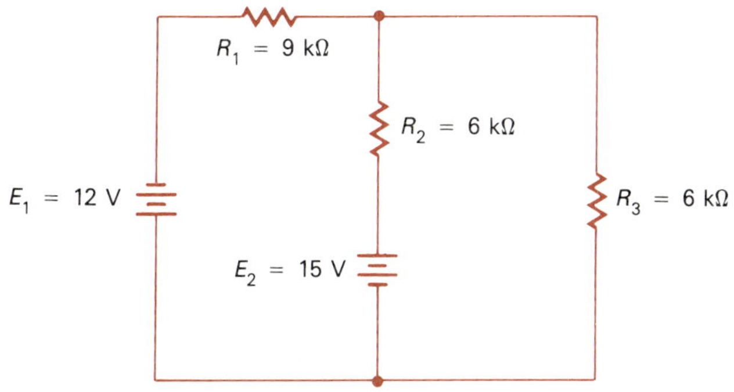

FIGURE 3–45

$$I_{R_2} = \frac{1\text{ mA}}{2} \quad \text{(Equal resistors, current splits)}$$
$$= 0.5\text{ mA} \quad \text{(Current through } R_2 \text{ due to } E_1\text{)}$$

$$R_{t_2} = R_2 + \frac{R_1 \times R_3}{R_1 + R_3}$$
$$= 6\text{ k}\Omega + \frac{9\text{ k}\Omega \times 6\text{ k}\Omega}{9\text{ k}\Omega + 6\text{ k}\Omega}$$
$$= 9.6\text{ k}\Omega$$

$$I_{t_2} = \frac{E_2}{R_{t_2}}$$
$$= \frac{15\text{ V}}{9.6\text{ k}\Omega}$$
$$= 1.56\text{ mA} \quad \text{(Current through } R_2 \text{ due to } E_2\text{)}$$

As both currents are flowing in the same direction the total current is

$$I = 0.5\text{ mA} + 1.56\text{ mA}$$
$$= 2.06\text{ mA}$$

Norton's Theorem

There are some electronic circuits that produce a constant current output as opposed to a constant voltage output. An *ideal current source* is one that produces a constant current regardless of the load resistance. The circuit in Figure 3–46 shows the schematic symbol of a current source delivering a constant

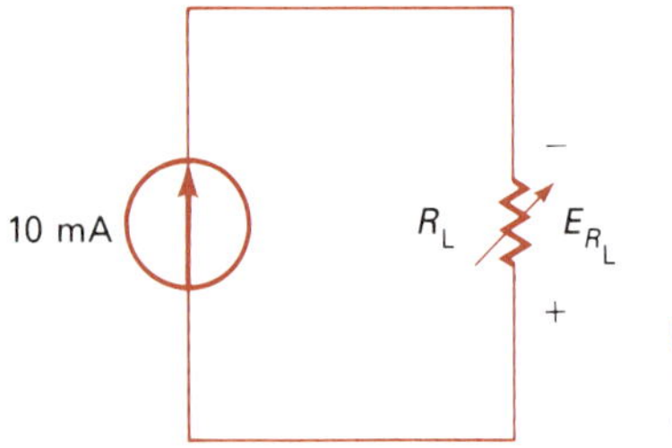

FIGURE 3–46 An ideal current source delivers constant current to a variable load.

10 mA to the load. As R_L is varied the current ideally remains at 10 mA. The voltage developed across the load would, of course, change as the load resistance changed.

Norton's Theorem states that **any two-terminal network containing resistances and current sources may be replaced by a single current source in parallel with a single resistance.** The current of the single source (I_{SL}) is the *shorted-load* current at the network terminals. The resistance (R_{TH}) is the same as the Thévenin resistance.

Example: Find the Norton current source and Thévenin resistance for the circuit in Figure 3–47A.

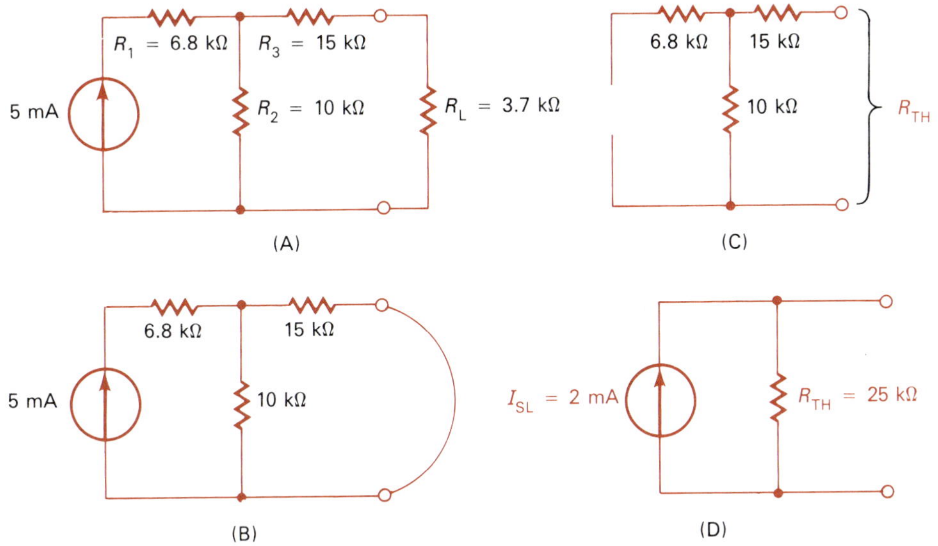

FIGURE 3–47

Solution: Remove and replace R_L with a short as shown in Figure 3–47B. Calculate I_{SL} flowing through the output terminals.

$$I_{SL} = \frac{R_2}{R_2 + R_3} \times I$$
$$= \frac{10\ k\Omega}{10\ k\Omega + 15\ k\Omega} \times 5\ mA$$
$$= 2\ mA$$

Remove the short and replace the current source with an *open* circuit as shown in Figure 3–47C. The resistance across the output terminals is the Thévenin resistance.

$$R_{TH} = R_3 + R_2$$
$$= 15\ k\Omega + 10\ k\Omega$$
$$= 25\ k\Omega$$

Draw the Norton equivalent circuit as shown in Figure 3–47D. In comparing Figure 3–47A and Figure 3–47D it is important to note that in terms of the load both circuits behave identically. You should calculate the voltage and current values associated with a 3.7-kΩ load attached to both circuits. These values will be the same.

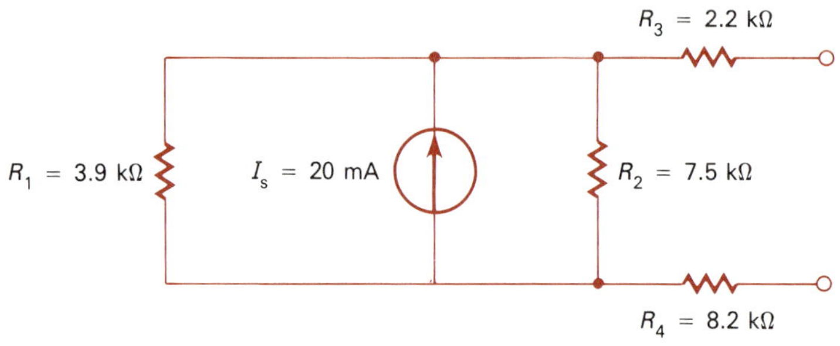

FIGURE 3–48

Example: Find the Norton equivalent circuit values (I_{SL} and R_{TH}) for the circuit in Figure 3–48.

Solution: The resistance to the *right* of the source with a short across the output terminals is

$$R = \frac{R_2 \times (R_3 + R_4)}{R_2 + (R_3 + R_4)}$$
$$= \frac{7.5 \text{ k}\Omega \times (2.2 \text{ k}\Omega + 8.2 \text{ k}\Omega)}{7.5 \text{ k}\Omega + (2.2 \text{ k}\Omega + 8.2 \text{ k}\Omega)}$$
$$= 4.36 \text{ k}\Omega$$

Current to the right is

$$I = \frac{R_1}{R + R_1} \times I_s$$
$$= \frac{3.9 \text{ k}\Omega}{4.36 \text{ k}\Omega + 3.9 \text{ k}\Omega} \times 20 \text{ mA}$$
$$= 9.44 \text{ mA}$$

$$I_{SL} = \frac{R_2}{R_2 + (R_3 + R_4)} \times I$$
$$= \frac{7.5 \text{ k}\Omega}{7.5 \text{ k}\Omega + (2.2 \text{ k}\Omega + 8.2 \text{ k}\Omega)} \times 9.44 \text{ mA}$$
$$= 3.96 \text{ mA}$$

$$R_{TH} = R_3 + \frac{R_2 \times R_1}{R_2 + R_1} + R_4$$
$$= 2.2 \text{ k}\Omega + \frac{7.5 \text{ k}\Omega \times 3.9 \text{ k}\Omega}{7.5 \text{ k}\Omega + 3.9 \text{ k}\Omega} + 8.2 \text{ k}\Omega$$
$$= 12.97 \text{ k}\Omega$$

Multiple Source Circuits

In those instances when a circuit contains resistances, voltage sources, and current sources it is possible to use the superposition theorem to determine the Thévenin equivalent circuit *and* the Norton equivalent circuit. Examine carefully each step in the following example.

Example: Determine the Thévenin equivalent for the circuit in Figure 3–49. Convert the Thévenin equivalent to a Norton equivalent.

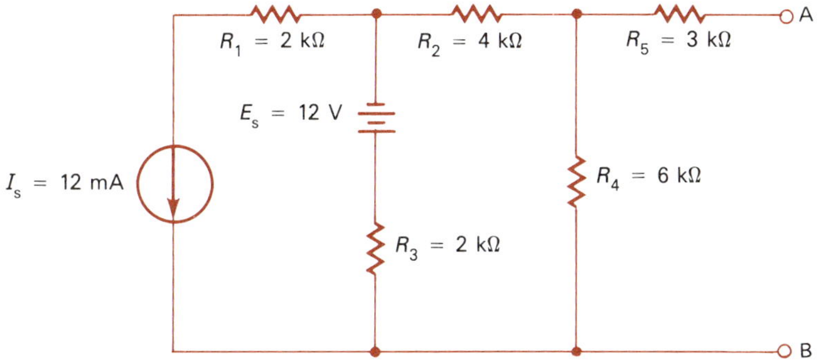

FIGURE 3–49

Solution: Replace I_s with an *open* circuit and determine the polarity and magnitude of the voltage at the network terminals (E_{o_1}) due to E_s. As no current flows through R_5 the output voltage is that across R_4.

$$E_{o_1} = \frac{R_4}{R_2 + R_3 + R_4} \times E_s$$
$$= \frac{6\ \text{k}\Omega}{4\ \text{k}\Omega + 2\ \text{k}\Omega + 6\ \text{k}\Omega} \times 12\ \text{V}$$
$$= 6\ \text{V (A negative with respect to B)}$$

Return I_s to the circuit and replace E_s with a *short* circuit. Again, the voltage at the network terminals (E_{o_2}) is due to the current through R_4 as no current flows through R_5.

$$E_{o_2} = I_{R_4} \times R_4$$
$$= \left[\frac{R_3}{R_3 + (R_4 + R_2)} \times I_s\right] \times R_4$$
$$= \left[\frac{2\ \text{k}\Omega}{2\ \text{k}\Omega + 6\ \text{k}\Omega + 4\ \text{k}\Omega} \times 12\ \text{mA}\right] \times 6\ \text{k}\Omega$$
$$= 2\ \text{mA} \times 6\ \text{k}\Omega$$
$$= 12\ \text{V (B negative with respect to A)}$$

The Thévenin voltage is the algebraic sum of the two voltages. Since the polarities are opposite one of the two must be considered negative.

$$E_{TH} = E_{o_1} + E_{o_2}$$
$$= 6\ \text{V} - 12\ \text{V}$$
$$= -6\ \text{V (The B terminal is negative with respect to A)}$$

To determine the Thévenin resistance replace I_s with an open, E_s with a short, and measure R_{TH} at the terminals.

$$\begin{aligned} R_{TH} &= R_5 + \frac{(R_2 + R_3)(R_4)}{(R_2 + R_3) + R_4} \\ &= 3\ \text{k}\Omega + \frac{(4\ \text{k}\Omega + 2\ \text{k}\Omega)(6\ \text{k}\Omega)}{(4\ \text{k}\Omega + 2\ \text{k}\Omega) + 6\ \text{k}\Omega} \\ &= 6\ \text{k}\Omega \end{aligned}$$

The equivalent Thévenin circuit is given in Figure 3–50A. Note that the polarity of E_{TH} has been drawn so as to be consistent with previous circuits. To convert the Thévenin circuit to a Norton circuit follow the sequence previously described.

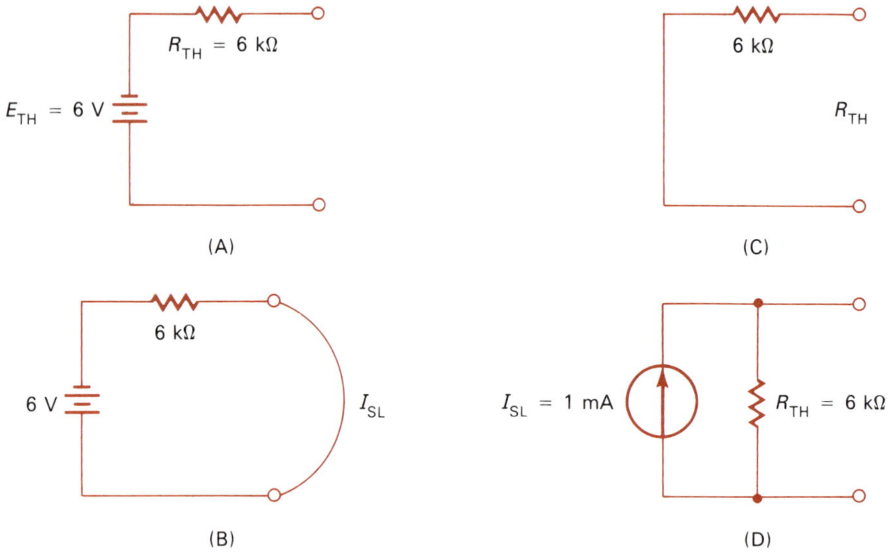

FIGURE 3–50

Short the output terminals and determine I_{SL} as in Figure 3–50B.

$$\begin{aligned} I_{SL} &= \frac{E_{TH}}{R_{TH}} \\ &= 1\ \text{mA} \end{aligned}$$

Remove the short, replace E_{TH} with a short circuit, and determine R_{TH} as in Figure 3–50C.

$R_{TH} = 6\ \text{k}\Omega$

Draw the Norton equivalent circuit as shown in Figure 3–50D.

It is important to note that the circuit in the previous example could have been converted to a Norton circuit, and then to a Thévenin equivalent circuit. You should follow this process to ensure that both approaches yield identical circuits.

EXERCISE 3·4

1. The no-load output of a power supply is 9 V. The supply has an internal resistance of 3 Ω. Determine (a) the maximum power the supply can deliver to a load, (b) the load voltage, and (c) the load current.
2. A 4.5-Ω load dissipates 18 W of power. If this is the maximum power the supply can transfer to the load, calculate the no-load output of the supply.

Use Figure 3–51 for problems 3–4.

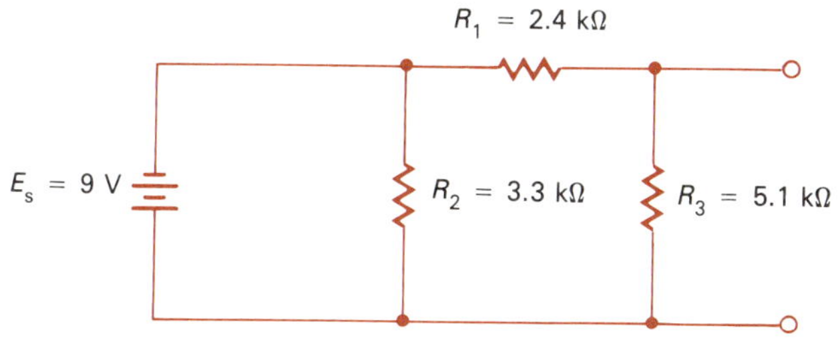

FIGURE 3–51

3. Determine (a) the Thévenin voltage and (b) the Thévenin resistance in Figure 3–51.
4. A 10-kΩ load is connected to the circuit in Figure 3–51. What is the current through this load?

Use Figure 3–52 for problems 5–6.

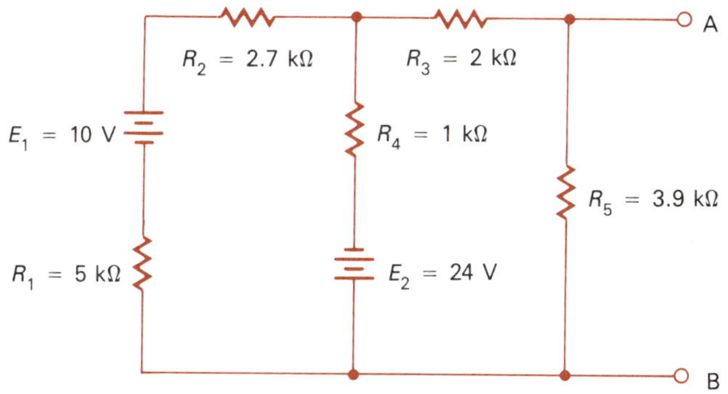

FIGURE 3–52

5. Find (a) E_{TH} and (b) R_{TH} in Figure 3–52.
6. Convert the circuit in Figure 3–52 to a Norton equivalent circuit.

Use Figure 3–53 for problems 7–9.

7. Find (a) I_{SL} and (b) R_{TH} in Figure 3–53.
8. A-1.5 kΩ load is attached to the circuit in Figure 3–53. Find the power dissipated by this load.

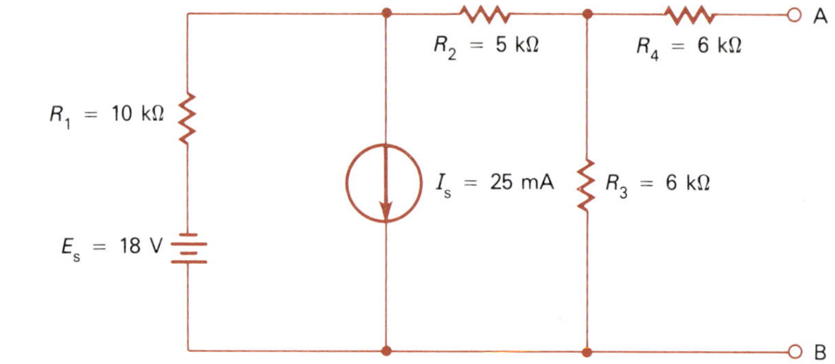

FIGURE 3–53

9. What is the value of E_{TH} in Figure 3–53?
10. Find (a) E_{TH}, (b) R_{TH}, and (c) I_{SL} for the circuit in Figure 3–54.

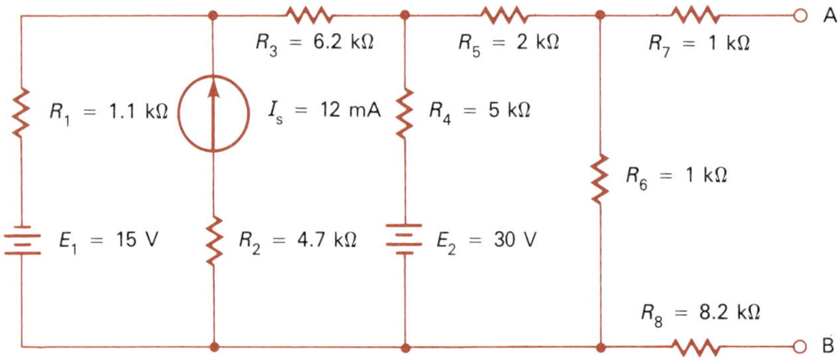

FIGURE 3–54

UNIT 3·5 CAPACITORS AND INDUCTORS IN DIRECT CURRENT CIRCUITS

Objectives:

After studying this unit, you should be able to

- discuss the function of the capacitor and inductor in *dc* circuits.
- determine equivalent capacitance in series and parallel capacitive circuits.
- determine time constants and circuit values in *RC* circuits.
- determine equivalent inductance in series and parallel inductive circuits.
- determine time constants and circuit values in *RL* circuits.

Capacitors in Direct Current Circuits

In the direct current circuits presented thus far in this chapter an assumption of instantaneous current flow has been made. When the switch in Figure 3–55 is closed the current is assumed to flow at the calculated level instantly. This is essentially true for purely resistive circuits. When components such as capacitors and inductors are included in a dc circuit the relationships between voltage, current, and time change.

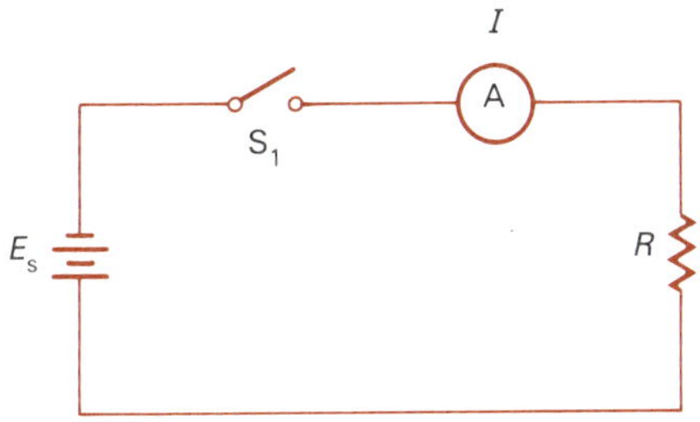

FIGURE 3–55

A capacitor is an electrical device consisting of two metallic conductors (plates) separated from each other by a nonconducting, or insulating, material. A capacitor stores energy in an electric field. The capacitor stores energy when one capacitor plate has an excess of electrons, and the other plate has a deficiency of electrons. The symbols for fixed and variable capacitors are given in Figure 3–56.

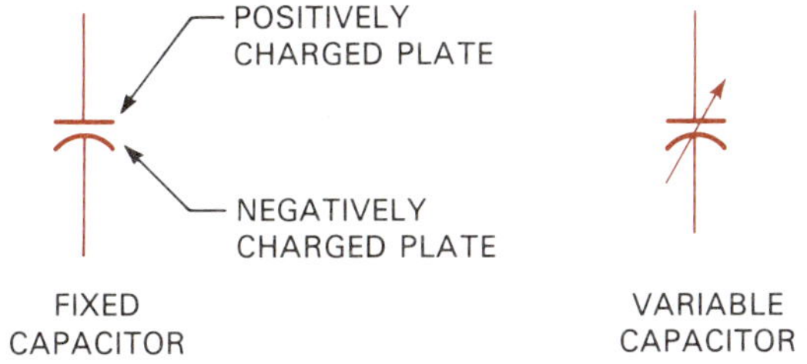

FIGURE 3–56 Schematic symbols for fixed and variable capacitors

Capacitance is determined by the following equation.

$$C = \frac{Q}{V} \qquad (3\text{–}31)$$

where C = capacitance, in farads
V = voltage across the capacitor, in volts
Q = charge on either plate in coulombs (1 coulomb = 1 C, the charge carried by 6.24×10^{18} electrons)

Example: Find the capacitance of a capacitor having a charge of 120 μC, when a voltage of 24 V is applied to it.

Solution:
$$\begin{aligned} C &= \frac{Q}{V} \\ &= \frac{120 \times 10^{-6}\ \text{C}}{24\ \text{V}} \\ &= 5 \times 10^{-6}\ \text{F} \\ &= 5\ \mu\text{F} \end{aligned}$$

Example: How much charge could a 0.2-μF capacitor store when a voltage of 5 V is applied?

Solution:
$$\begin{aligned} C &= \frac{Q}{V} \\ Q &= C \times V \\ &= (0.2 \times 10^{-6}\ \text{F})(5\ \text{V}) \\ &= 1 \times 10^{-6}\ \text{C} \\ &= 1\ \mu\text{C} \end{aligned}$$

The value of a capacitor depends on the following physical parameters:

- The type of dielectric used (K = dielectric constant, in farads/metre)
- The area of the plates (A in square metres)
- The distance between plates (d in metres)

The following equation for a parallel plate capacitor relates these parameters:

$$C = 8.85 \times 10^{-12} \times K \times \frac{A}{d} \tag{3–32}$$

Example: A capacitor has the following parameters. Find the capacitance. $A = 0.1\ \text{m}^2$, $d = 3 \times 10^{-3}\ \text{m}$, $K = 5\ \text{F/m}$ (mica)

Solution:
$$\begin{aligned} C &= (8.85 \times 10^{-12})(5\ \text{F/m})\left(\frac{0.1\ \text{m}^2}{3 \times 10^{-3}\ \text{m}}\right) \\ &= 1.475 \times 10^{-9}\ \text{F} \\ &= 1\ 475\ \text{pF} \end{aligned}$$

Capacitors in Series

Due to their electrical properties capacitors in series react similarly to resistors in parallel. The following equations apply to capacitors in series circuits.

Total capacitance is

$$\frac{1}{C_t} = \frac{1}{C_1} + \frac{1}{C_2} + \cdots + \frac{1}{C_n} \tag{3–33}$$

or,

$$C_t = \frac{1}{\frac{1}{C_1} + \frac{1}{C_2} + \cdots + \frac{1}{C_n}} \tag{3–34}$$

When there are two capacitors the equation is

$$C_t = \frac{C_1 \times C_2}{C_1 + C_2} \qquad (3\text{–}35)$$

When the series capacitors are of equal value the equation is

$$C_t = \frac{C}{n} \qquad (3\text{–}36)$$

where C = value of 1 capacitor
n = the number of capacitors in series

Example: Find the total capacitance in Figure 3–57.

Solution:

$$C_t = \frac{1}{\frac{1}{0.01 \times 10^{-6}} + \frac{1}{0.02 \times 10^{-6}} + \frac{1}{0.05 \times 10^{-6}}}$$
$$= \frac{1}{1 \times 10^8 + 5 \times 10^7 + 2 \times 10^7}$$
$$= \frac{1}{1.7 \times 10^8}$$
$$= 5.882\,4 \times 10^{-9}\ \text{F}$$
$$= 0.006\ \mu\text{F}$$

Example: Determine the total charge stored in Figure 3–57.

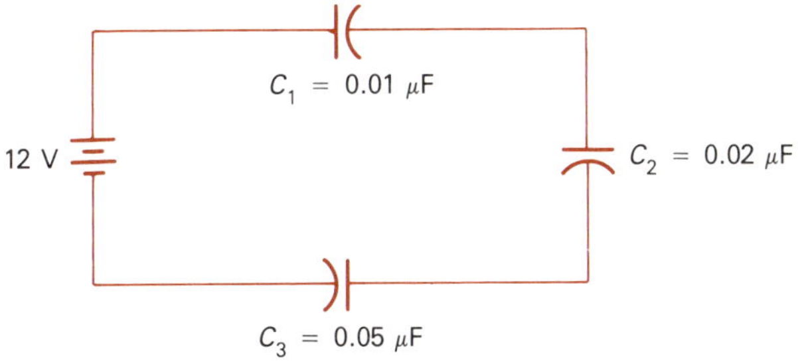

FIGURE 3–57

Solution:

$$C = \frac{Q}{V}$$
$$Q = C \times V$$
$$= 0.006\ \mu\text{F} \times 12\ \text{V}$$
$$= 0.072\ \mu\text{C}$$

Capacitors in Parallel

Due to their electrical properties capacitors in parallel react similarly to resistors in series. The following equation applies to capacitors in parallel circuits.

Total capacitance is

$$C_t = C_1 + C_2 + \cdots + C_n \qquad (3\text{–}37)$$

Example: Determine the total capacitance in Figure 3–58.

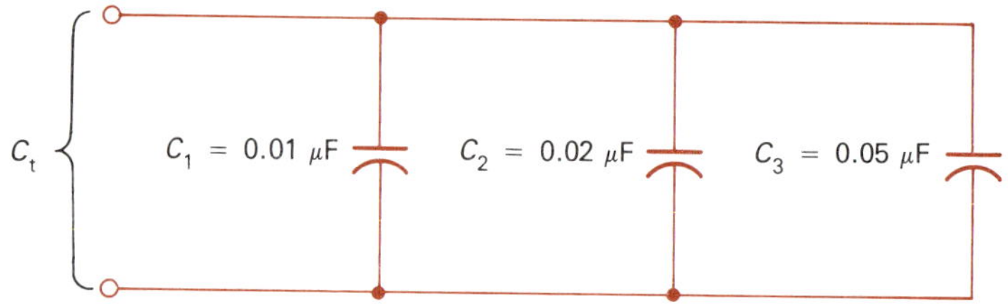

FIGURE 3–58

Solution:

$$\begin{aligned} C_t &= C_1 + C_2 + C_3 \\ &= 0.01\ \mu\text{F} + 0.02\ \mu\text{F} + 0.05\ \mu\text{F} \\ &= 0.08\ \mu\text{F} \end{aligned}$$

RC Circuits

In the circuit of Figure 3–59, the time it takes to charge the capacitor to 63.2% of the final voltage is known as the *time constant.* If you consider that a switch has just been closed, the time constant is found with this formula,

$$T = R \times C \qquad (3\text{–}38)$$

where T is the time constant expressed in seconds, R is the resistance in ohms, and C is the capacitance in farads. These are the basic units and conversions must be made if information is supplied in multiples or submultiples. A capacitor will be fully charged in 5 time constants.

In Figure 3–59 the applied voltage is 200 V, R is 1 kΩ and C is 2 μF. The time constant, then is $T = R \times C$. Converting, $R = 1 \times 10^3\ \Omega$ and $C = 2 \times 10^{-6}$ F. Solving, $T = (1 \times 10^3\ \Omega)(2 \times 10^{-6}\ \text{F}) = 2 \times 10^{-3}\ \text{s} = 0.002$ s or 2 ms. Thus, this capacitor will charge to 63% (63.2% is rounded to 63%) of the applied voltage in 2 ms. At the end of one time constant, that is, at the end of 2 ms the charge across the capacitor will be 63% of 200 V, $0.63 \times 200\ \text{V} = 126$ V.

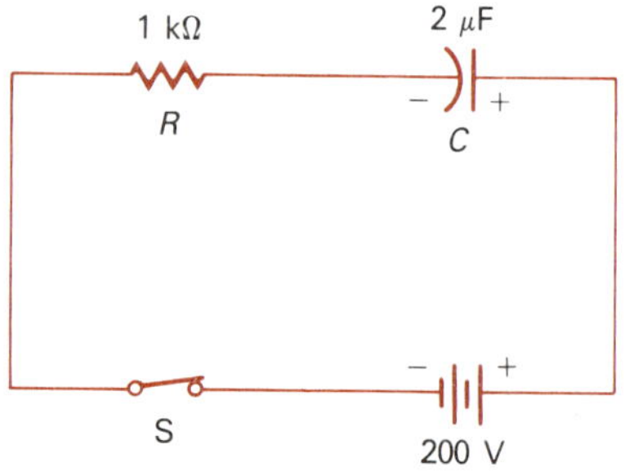

FIGURE 3–59 The time required to charge the capacitor, C, depends on the values of R and C.

The voltage charging across the capacitor is in opposition to the applied voltage. Because of this opposition, the effective applied voltage at the end of one time constant, 2 ms, behaves as though its value was 200 V − 126 V = 74 V.

At the end of another time constant, 4 ms since the switch was closed, the voltage across the capacitor increases by 63% of 74 V, 0.63 × 74 V = 46.62 V. This must be added to the existing charge, 126 V + 46.62 V = 172.62 V. The difference between the 172.62 V charge on the capacitor and the source voltage is now 200 V − 172.62 V = 27.38 V.

At the end of the third time constant, the charge across the capacitor is increased by 63% of 27.38 V, 0.63 × 27.38 V = 17.25 V. Again, this is added to the charge already across the capacitor, 172.62 V + 17.25 V = 189.87 V. A total of 6 ms has now elapsed.

Because 63% is taken of the remainder of applied voltage at each *RC* time (in this case, every 2 ms), it is easy to see that theoretically the voltage can never equal the source voltage, for it is always *approaching* maximum in what is termed an *asymptotic curve*. That is why, in practical electronics work, the value of 5 *RC* time constants is used as the time required for maximum charge. As an exercise, continue with the calculation for 4 *RC* and 5 *RC* times.

Additional time constants can be calculated and ultimately the capacitor will be fully charged, so far as practical work is concerned. At that time, its voltage will be equal and opposite that of the source. No current will flow. The voltage drop across the resistor will then be zero.

The sum of the voltage drops across the resistor and across the capacitor is equal to the source voltage. As the voltage across the capacitor increases, it opposes the flow of current. The reduced movement of current means a smaller voltage drop across the resistor.

Example: The *RC* circuit in Figure 3–60A is a dc circuit and has an applied voltage of 100 V. The value of *R* is 1 MΩ and *C* is 500 pF. What is the amount of charge across the capacitor at the end of two time constants?

Solution:

$$R = 1\ \text{M}\Omega = 1 \times 10^{6}\ \Omega$$
$$C = 500\ \text{pF} = 500 \times 10^{-12}\ \text{F} = 5 \times 10^{-10}\ \text{F}$$
$$T = R \times C = (1 \times 10^{6}\ \Omega)(5 \times 10^{-10}\ \text{F})$$
$$= 5 \times 10^{-4}\ \text{s} = 0.000\,5\ \text{s or } 0.5\ \text{ms}$$

At the end of one time constant, 0.5 ms, the charge across the capacitor is approximately 63% of 100 V, 0.63 × 100 V = 63 V. This voltage, appearing across the capacitor, opposes the source voltage and therefore the effective charging voltage is now 100 V − 63 V = 37 V. To state it in different words, this means that 63 V across the capacitor, opposes 63 V of the source, leaving the source with a net effective charging voltage of only 37 V.

At the end of the second time constant, 1 ms has elapsed and the effective or net charging voltage has decreased by another 63%. That is 0.63 × 37 V = 23.31 V. The amount of charge across the capacitor at the end of two time constants is 63 V + 23.31 V = 86.31 V. The remaining effective charging voltage is 100 V − 86.31 V = 13.69 V. Table 3–2 will aid in impressing this

TABLE 3–2

Effects of Five Time Constants

T(ms)	E_s(V)	E_R(V)	$I = \frac{E_R}{R}$ (μA)	E_C(V)
0.0	100	100.00	100.00	0.00
(1) 0.5	100	37.00	37.00	63.00
(2) 1.0	100	13.69	13.69	86.31
(3) 1.5	100	5.07	5.07	94.93
(4) 2.0	100	1.88	1.88	98.12
(5) 2.5	100	0.70	0.70	99.30

on your memory, and shows why the time in 5 time constants is considered practical.

After approximately five time constants the capacitor in Figure 3–60A is fully charged and the circuit current is almost zero. When the switch is moved to position two, the capacitor acts as a power source and will *discharge* through R_1 and R_2 until the potential difference between the plates is zero.

Example: (a) Calculate the amount of time required to discharge C_1 in Figure 3–60A. (b) Determine the initial discharge current.

Solution: a. $T = R \times C$

$$= (1\ \text{M}\Omega + 1\ \text{M}\Omega)(500\ \text{pF})$$
$$= 2 \times 10^6\ \Omega \times 5 \times 10^{-10}\ \text{F}$$
$$= 1\ \text{ms (With 5 time constants being 5 ms)}$$

b. $I = \dfrac{E_R}{R}$

$$= \frac{100\ \text{V}}{2\ \text{M}\Omega}$$
$$= 50\ \mu\text{A (Decreases to zero in about 5 time constants)}$$

The charging and discharging curves for the circuit in Figure 3–60A are shown in Figure 3–60B.

Inductors in Direct Current Circuits

Whenever current passes through a conductor, a magnetic field exists around the conductor. This field is normally very weak in single, straight line conductors. This magnetic field can be concentrated by coiling the wire so that the magnetic lines of flux are additive. These lines can be concentrated further through the addition of a core as shown in Figure 3–61.

The magnetic field around a current-carrying conductor has the following characteristics:

- As the current in the conductor increases, the lines of force increase.

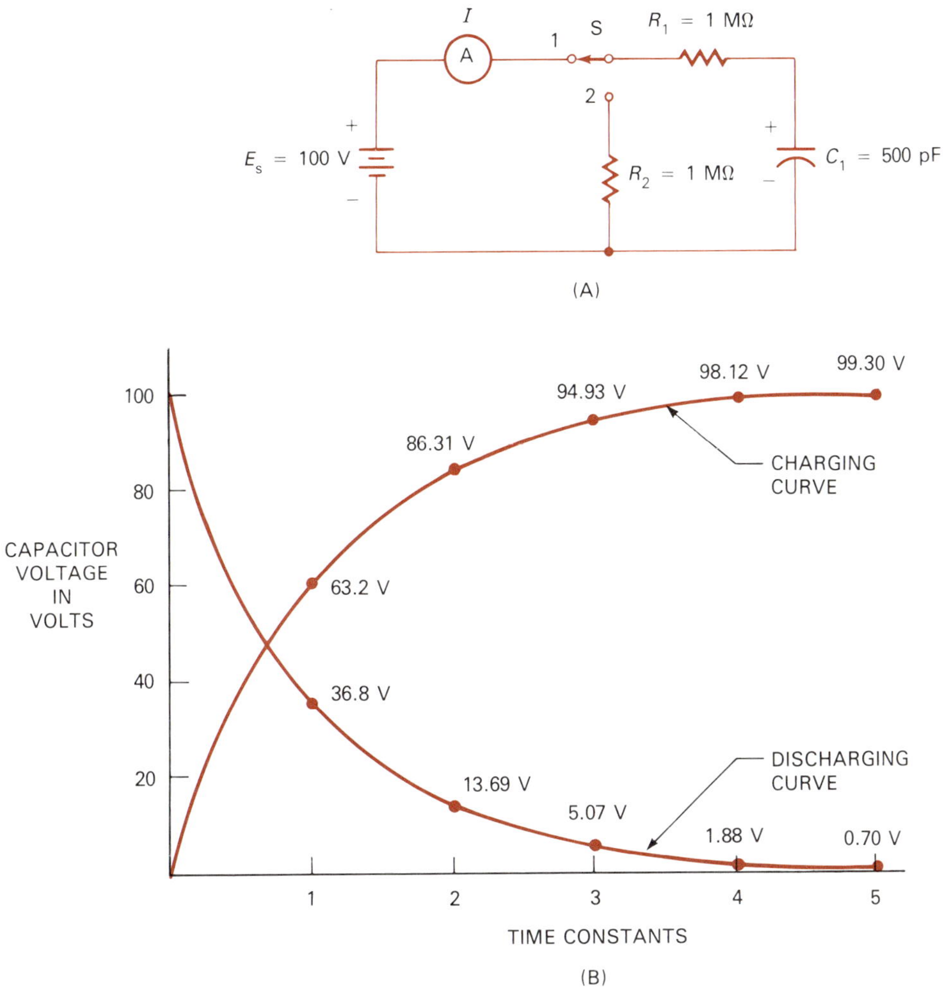

FIGURE 3–60

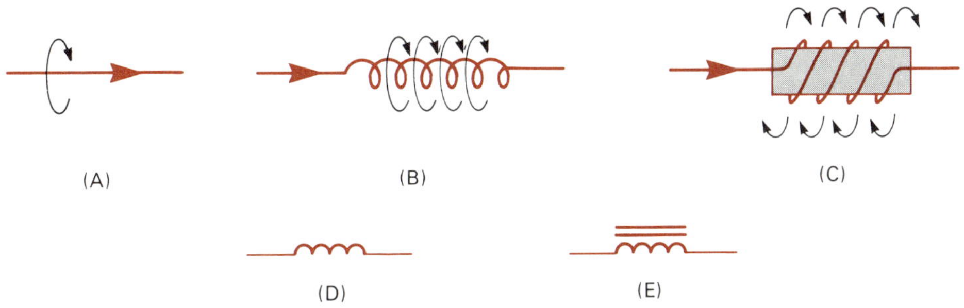

FIGURE 3–61 (A) Single conductor and weak magnetic field (B) Increased field strength in a coil (C) Iron core concentrates field (D) Symbol for air-core inductor (E) Symbol for iron-core inductor

- When the current is steady, the magnetic lines of force remain constant.
- When the current in the conductor decreases, the lines of force decrease.
- When the current becomes zero, the lines of force collapse back toward the center of the conductor and decrease to zero.

As the lines of force move outward from an inductor it is possible that these lines will cut across the windings of another coil. This will induce a voltage in the second coil. The term *mutual inductance* is applied to the process of producing an electromotive force in a secondary winding by changing a current in a primary winding. This concept is the basis of transformers which will be discussed in a later chapter.

Inductors in Series

When two or more inductors are in series, and there is no mutual inductance, the total inductance is given by

$$L_t = L_1 + L_2 + \cdots + L_n \tag{3–39}$$

where L_t is total inductance in henries (H) and L_1 through L_n are the individual inductance values.

In those instances where inductors are physically close, and intercoupling of their magnetic fields exists, the formula to determine total inductance is

$$L_t = L_1 + L_2 \pm 2\,M \tag{3–40}$$

where M is the mutual inductance in henries. In the event the magnetic fields are additive the inductors are connected in series-aiding and the formula is

$$L_t = L_1 + L_2 + 2M \tag{3–41}$$

If the magnetic fields oppose each other the inductors are connected in series-opposing and the formula is

$$L_t = L_1 + L_2 - 2M \tag{3–42}$$

The formula to determine mutual inductance is

$$M = k\sqrt{L_1L_2} \tag{3–43}$$

where k is the coefficient of coupling. This coefficient is the percent of the magnetic lines of one inductor that cut across the other inductor.

Example: Two inductors are connected in series-aiding. Their values are 15 mH and 20 mH. The coefficient of coupling is 0.85. Calculate the total inductance.

Solution: $M = 0.85\sqrt{(15 \times 10^{-3})(20 \times 10^{-3})}$
$= 14.72 \text{ mH}$

$L_t = 15 \text{ mH} + 20 \text{ mH} + 14.72 \text{ mH}$
$= 49.72 \text{ mH}$

Example: The total inductance of two inductors connected in series-opposing is 47.5 mH. Determine the coefficient of coupling if $L_1 = 10$ mH and $L_2 = 62.5$ mH.

Solution: The first step is to determine M.

$$L_t = L_1 + L_2 - 2M$$

$$M = \frac{L_t - L_1 - L_2}{-2}$$

$$M = \frac{47.5 \text{ mH} - 10 \text{ mH} - 62.5 \text{ mH}}{-2}$$

$$M = 12.5 \text{ mH}$$

Solve for k.

$$M = k\sqrt{L_1 L_2}$$

$$k = \frac{M}{\sqrt{L_1 L_2}}$$

$$k = \frac{12.5 \text{ mH}}{\sqrt{10 \text{ mH} \times 62.5 \text{ mH}}}$$

$$k = 0.5 \text{ or } 50\%$$

Inductors in Parallel

As in resistors and capacitors, inductors can also be connected in parallel. Parallel inductors, however, are not used extensively. The formulas and problems presented here will assume parallel connections with no mutual inductance.

Inductors in parallel are similar to resistors in parallel. The general formula for parallel inductors is

$$L_t = \frac{1}{\frac{1}{L_1} + \frac{1}{L_2} + \cdots + \frac{1}{L_n}} \qquad (3\text{–}44)$$

When only two inductors are connected in parallel the equation is

$$L_t = \frac{L_1 \times L_2}{L_1 + L_2} \qquad (3\text{–}45)$$

When any number of inductors with equal values are connected in parallel, the total inductance is

$$L_t = \frac{L}{n} \tag{3–46}$$

where L = inductance of 1 inductor
n = number of inductors in parallel

Example: Two inductors, 0.5 H and 250 mH, are connected in parallel. Calculate the equivalent inductance.

Solution:
$$L_t = \frac{0.5 \text{ H} \times 0.25 \text{ H}}{0.5 \text{ H} + 0.25 \text{ H}}$$
$$= 0.167 \text{ H}$$

RL Circuits

When a circuit, as shown in Figure 3–62, consists of a series coil and resistor, the current flowing in the circuit, when the switch is closed, will not reach an instantaneous peak but will reach its maximum value at a time that is directly proportional to the inductance of the coil and inversely proportional to the resistance. Stated as a formula:

$$T = \frac{L}{R} \tag{3–47}$$

The time constant, T, is expressed in seconds, L is the inductance in henries, and R the resistance in ohms. Multiples and submultiples must be converted prior to being used in the formula. The time constant, T, is the time required for the current to reach 63.2% of its peak value; 63.2% is frequently rounded to 63%.

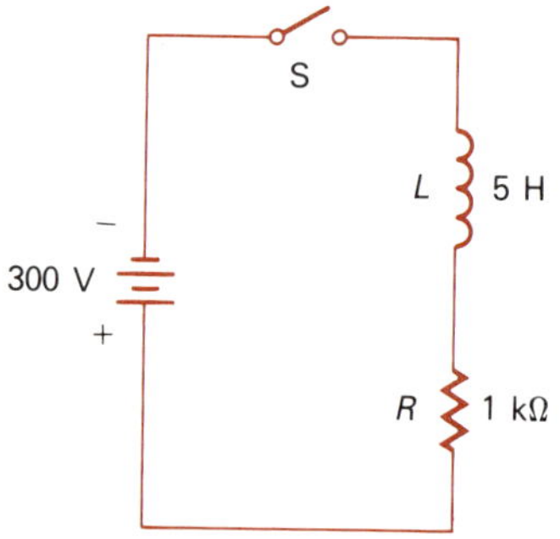

FIGURE 3–62 The time required for the current to reach its maximum value depends on the values of L and R.

In Figure 3–62, a coil having an inductance of 5 H is in series with a 1-kΩ resistor and the applied voltage is 300 V. The resistance of the coil winding is negligible. The time constant, then, is

$$T = \frac{L}{R} = \frac{5 \text{ H}}{1000\ \Omega} = 0.005 \text{ s} = 5 \text{ ms}$$

The maximum possible current can be determined by shorting the coil and considering the resistor as shunted across the voltage source.

$$I = \frac{E}{R} = \frac{300 \text{ V}}{1\ 000\ \Omega} = 0.3 \text{ A} = 300 \text{ mA}$$

At the end of one time constant, or 5 ms, the amount of current flowing is 63% of 300 mA, 0.63 × 300 mA = 189 mA. The difference between the total current capability and the current flowing is now 300 mA − 189 mA = 111 mA.

During the second time constant, 0.63 × 111 mA = 69.93 mA will be added to the original current flow of the first time constant: 189 mA + 69.93 mA = 258.93 mA. Thus, at the end of two time constants, 10 ms, the total current flowing is 258.93 mA out of a possible total of 300 mA.

At the end of 5 time constants, or 25 ms, the current will have increased to 300 mA. There will be 300 V across the resistor and the magnetic field around the inductor will have increased to a specific point and stopped. If, after 5 time constants, the battery could be instantaneously replaced with a short, the field around the coil would collapse. Instead of circuit current immediately dropping to zero, the collapsing magnetic field will continue to generate current. This current will drop to zero after approximately five time constants. Graphs similar to those in Figure 3–60 can be developed for *RL* circuits.

EXERCISE 3·5

1. A capacitor stores 0.5 μC of charge when 50 V are applied. What is the capacitor value?
2. A 200-pF capacitor stores 0.002 4 μC of charge. Determine the difference of potential between the plates.
3. A parallel plate capacitor has a plate area of 0.8 m^2, the distance between the plates is 10×10^{-3} m, and a paraffin coated paper dielectric with a dielectric constant of 2.5 F/m. What is the capacitance?
4. If 20 V are applied to the capacitor in problem 3, determine the amount of charge that would be stored.
5. Three capacitors are connected in series. Find the equivalent capacitance if the values are 35 μF, 50 μF, and 50 μF.
6. A technician needs 5 μF for a test circuit. The only capacitors available are 20 μF. How many must be placed in series to obtain an equivalent capacitance of 5 μF?

Use Figure 3–63 for problems 7–8.

7. Determine the total capacitance in Figure 3–63.
8. If C_3 in Figure 3–63 was to short, find the total capacitance.
9. An 1 800-pF capacitor is connected in parallel with 0.000 9 μF. What is the equivalent capacitance?

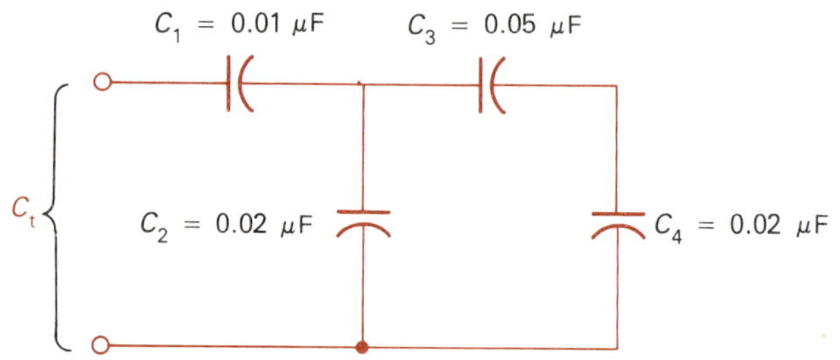

FIGURE 3–63

10. Determine the total capacitance in Figure 3–64.

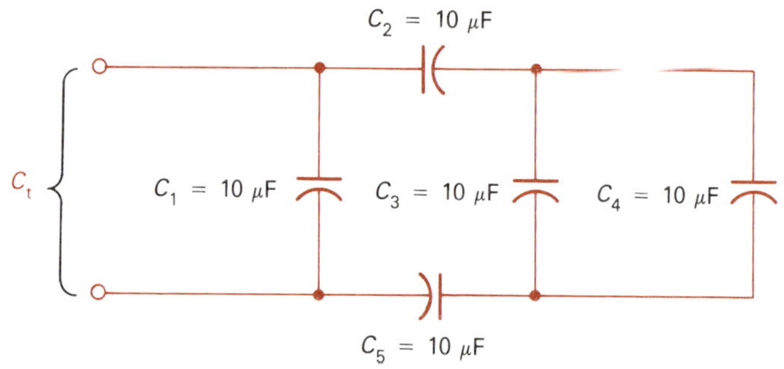

FIGURE 3–64

11. Calculate the time constant for a 0.001-μF capacitor in series with a 50-kΩ resistor.
12. A 10-MΩ resistor and a 15-μF capacitor are connected in series with a 12-V power supply. Determine the voltage across the capacitor after three time constants.
13. A 500-μF capacitor charges through a 10-KΩ resistor and is then discharged through a 100-Ω resistor. Calculate the charging and discharging time constants.
14. Determine the voltage across C_1 in Figure 3–65 after 5 time constants.
15. A180-mH inductor is placed in series with a 0.2-H inductor. Find the equivalent inductance if $M = 0$.
16. Two inductors in series have an equivalent inductance of 100 mH. If the values are 90 mH and 120 mH, determine the mutual inductance.
17. What is the coefficient of coupling in problem 16?
18. Two series-aiding inductors have values of 25 μH and 120 μH. The coefficient of coupling is 0.93. Calculate the total inductance.
19. Two inductors of *equal* value are connected in series. The coefficient of coupling is 0.6 and the mutual inductance is 24 mH. Find the value of each inductor.

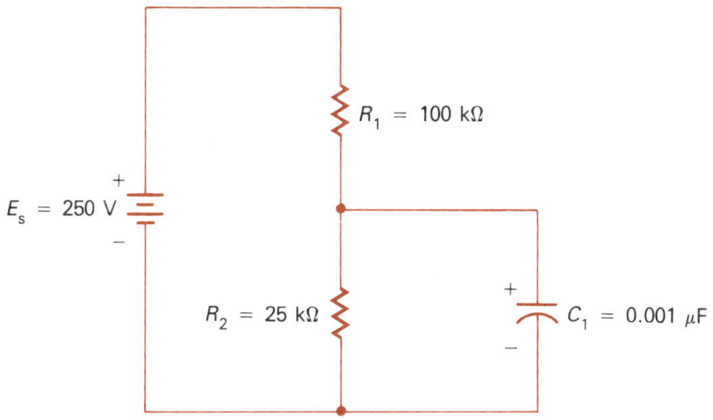

FIGURE 3–65

20. Inductors having values of 1.2 H and 560 mH are connected in parallel. Find the equivalent inductance.
21. Inductors having values of 250 mH, 100 mH, and 0.3 H are connected in parallel. What is the equivalent inductance?
22. What inductance value must be placed in parallel with 12 mH to produce an equivalent inductance of 8 mH?
23. Calculate one time constant for a 10 mH coil in series with a 2.7-kΩ resistor.
24. A 0.1-H inductor is in series with a 100-Ω resistor and a 12-V power supply. What current is flowing after 2 time constants?

UNIT 3·6 INTRODUCTION TO GRAPHS

Objectives:

After studying this unit, you should be able to

- locate points on a Cartesian coordinate system.
- plot a straight line on a graph.
- plot a curved line on a graph.

Types of Graphs

A *graph* is a line, either straight or curved, or a picture, or a geometric construction, showing the relationship between two or more quantities. A graph showing one relationship between frequency and time is a curved line which is called a sine wave. Figure 3–66 is the graph of a pair of sine waves in which wave 2 lags wave 1 by 90°. The two waves are plotted against a time base represented by the horizontal axis.

There are many types of graphs. The one in Figure 3–67 is called a bar graph, because of the way it is constructed. The advantage of a graph of this kind is that it supplies information more quickly and more vividly than a page

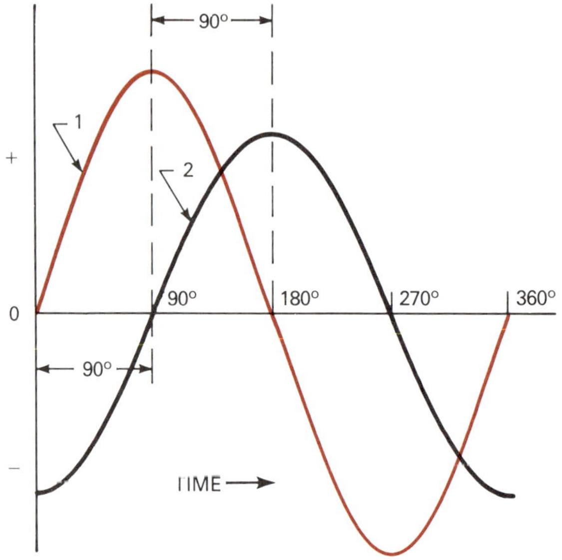

FIGURE 3–66 Graph of a pair of sine waves, wave 1 and wave 2, showing that wave 2 lags wave 1 by 90°

of statistics. Another type of graph is the circle graph, shown in Figure 3–68, in which various sectors are cut to show their relationship to some total.

A coordinate graph, also known as the *Cartesian* or *rectangular coordinate system,* allows relationships to be graphically represented. Coordinate graph

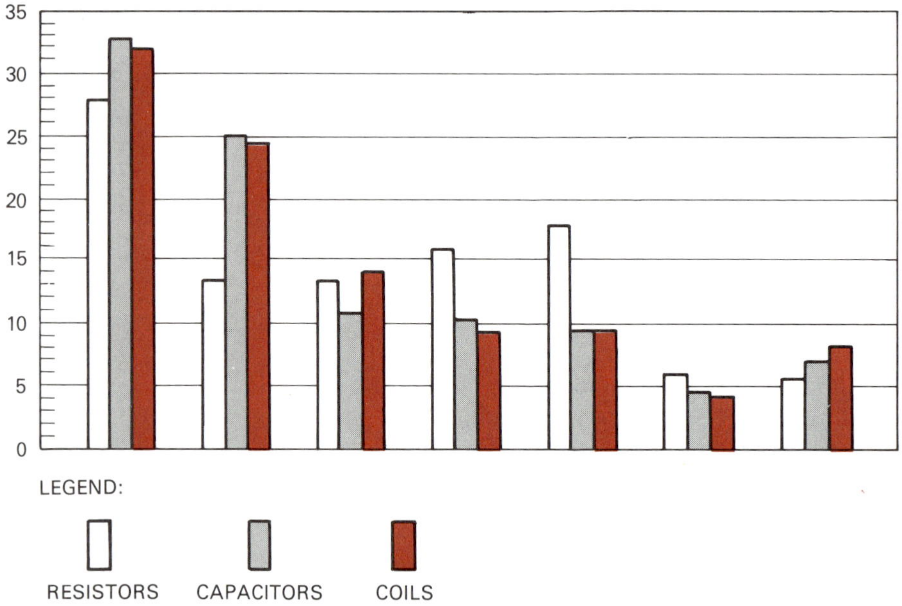

FIGURE 3–67 Example of a bar graph

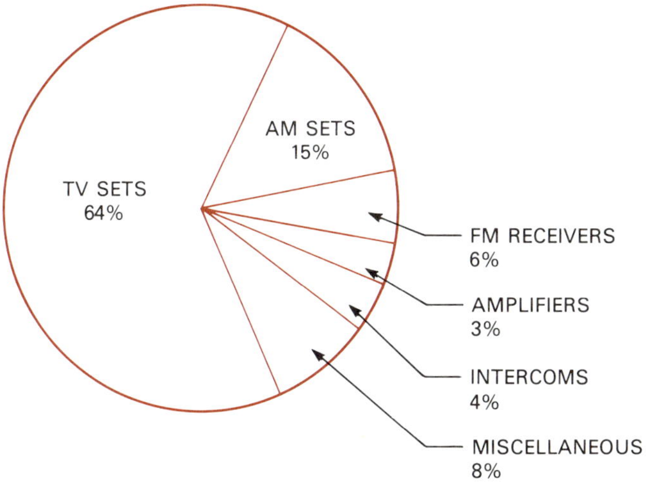

FIGURE 3–68 Circle graph showing types of repairs performed by a service organization

paper, such as the preprinted form of Figure 3–69 consists of a number of equally spaced vertical and horizontal lines. The paper can be divided into 4, 5, 6, 8, 10, or 20 squares to the inch, or 5 squares to the centimeter.

The graph paper can be divided into equal sections or *quadrants,* Figure 3–70, by drawing a pair of straight lines at right angles to each other. The

FIGURE 3–69 Graph paper used to plot points in the Cartesian coordinate system consists of equally spaced vertical and horizontal lines.

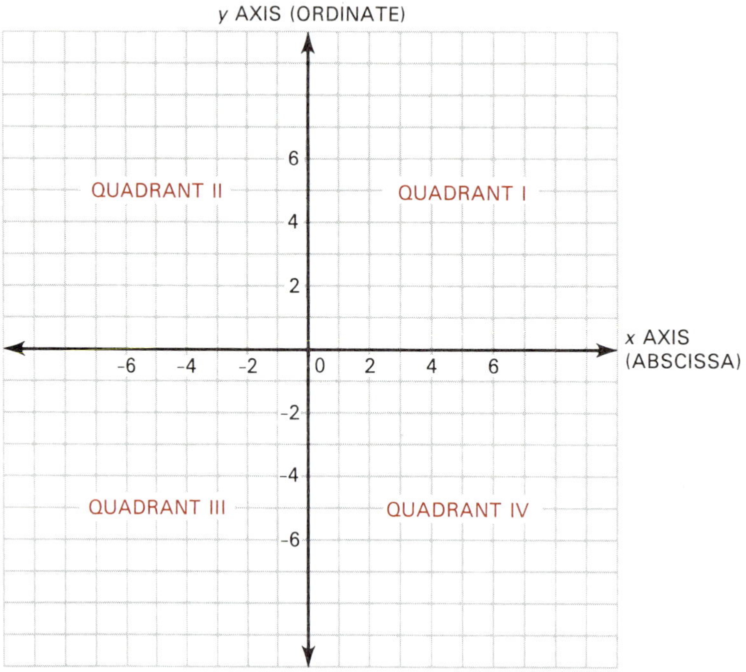

FIGURE 3–70 A pair of axes, x and y, divides the graph into four quadrants.

horizontal line is called the x axis or *abscissa;* the vertical line the y axis or *ordinate.* The two lines divide the sheet into four equal areas or quadrants numbered I, II, III, and IV, in counterclockwise fashion beginning with quadrant I at the upper right. The point of intersection of the two axes is the *origin,* frequently represented by 0.

The horizontal and vertical lines of the graph paper divide the x and y axes into equal segments, each of which can be numbered. Thus, the first point along the x axis, moving to the right from the origin, 0, could be 1, the next point 2, and so on. Moving to the left along the x axis from the origin, each intersection could be marked -1, -2, -3, etc. to distinguish these numbers from those which are at the right of the origin. The y axis can be similarly numbered. All divisions above the origin are positive: 1, 2, 3, etc., while all those below the origin are negative: -1, -2, -3, etc.

Locating a Point

Because the axes are numbered, we now have a method for conveniently locating any point in the space of any quadrant. Thus, $(+3, +4)$ or $(3, 4)$ is a point in the first quadrant. Its precise position can be found by moving $+3$ units along the x axis and then moving up $+4$ units along an imaginary line parallel to the y axis. The two numbers are known as the *coordinates* of the point, with the abscissa always being the first value.

Figure 3–71 shows how points are located in the various quadrants of a

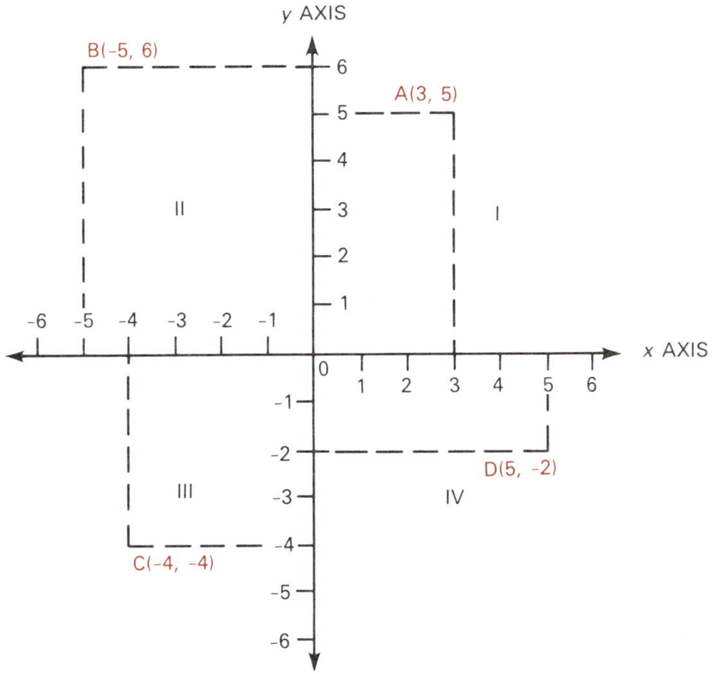

FIGURE 3–71 Coordinate technique for locating points in the various quadrants

graph. In the first quadrant, point A is represented by (3, 5). In the second quadrant, point B is identified by coordinates $(-5, 6)$. In the third quadrant we have point C $(-4, -4)$ while in the fourth quadrant, point D is $(5, -2)$.

Example: What is the position of $(-4, 5)$ and in what quadrant is it located?

Solution: Since the first coordinate is the abscissa and it is negative, move along the x axis to the left of the origin to reach -4. Move up 5 units along a line parallel to the y axis. This will be the position of the point. It will be in the second quadrant.

Reading A Graph

A *linear* graph is a straight line that shows the relationship between two variables. As one of the variables changes, the corresponding point of the second variable will change. Given a value of one variable it is possible to predict the value of the second variable.

The graph in Figure 3–72 shows the relationship between voltage and current in a dc circuit. Note that as voltage increases, that current also increases.

Example: Using the graph in Figure 3–72, find the current when voltage equals 3 V. Also locate the voltage corresponding to a current of 5 mA.

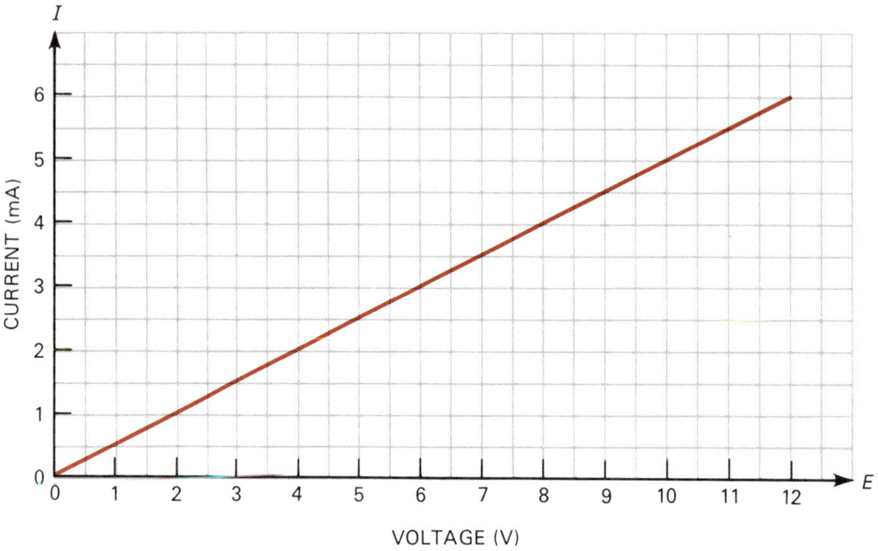

FIGURE 3–72

Solution: The following steps should be followed to obtain the current values:

1. Locate the given point on the appropriate axis.
2. Project vertically from the abscissa (or horizontally from the ordinate) until the graph is intersected.
3. From the intersection point project horizontally to the ordinate (or vertically to the abscissa) and read the value.

In this example the current at 3 V is 1.5 mA and the voltage at 5 mA is 10 V.

Graphs in electronics will not always be linear. Due to the characteristics of some components the graphs are curved or *nonlinear.* The graph in Figure 3–73 shows the increase of voltage across a capacitor as it charges. Note the x variable is marked off in time constants and the y variable in volts.

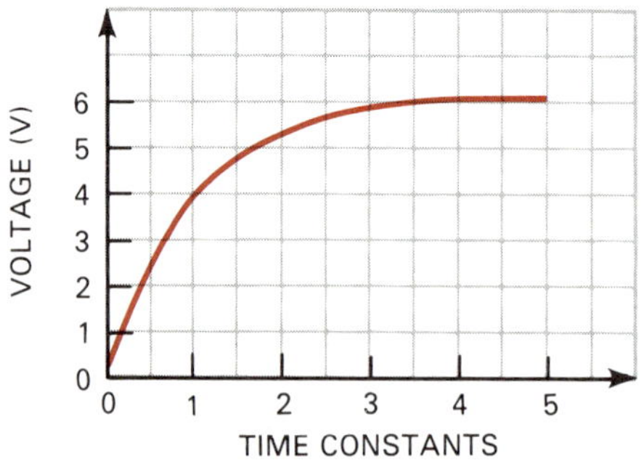

FIGURE 3–73

Example: What is the voltage across the capacitor plates in Figure 3–73 after 1 time constant?

Solution: Approximately 3.8 V.

Plotting a Graph

As graphs are used extensively in electronics, it is important that technicians and engineers be able to construct and draw a graph. The following steps will be helpful in developing useful and accurate graphs.

1. The first step is to collect the data. Assume we wish to graph the relationship between the voltage across, and the current through, a silicon diode. The circuit in Figure 3–74A could be used to collect the necessary information.
2. As the data is collected, a table similar to that in Figure 3–74B should be developed.

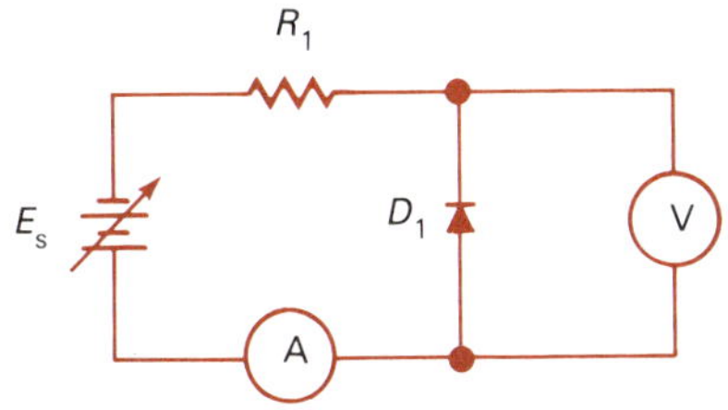

(A)

E	0	0.1 V	0.2 V	0.3 V	0.4 V	0.5 V	0.6 V	0.7 V	0.8 V	0.9 V
I	0	0	0	0	0	2 mA	5 mA	10 mA	18 mA	50 mA

(B)

FIGURE 3–74

3. The next step is to draw the ordinate and abscissa. A majority of the graphs in electronics are plotted in the first quadrant. When this is the case, then draw only the first quadrant. If the graph includes negative values then all four quadrants may need to be drawn.
4. Once the base lines are drawn it will be necessary to select appropriate scales for each. Problems stating that variable *A* is to be plotted *against* variable *B* should have the scale for variable *A* on the abscissa and the scale for *B* on the ordinate. The voltage values in Figure 3–74B range from 0 to 0.9 V. With this in mind, we might mark the abscissa in tenths of a volt. Note that voltage is usually placed on the *x* axis. The current values range from 0 to 50 mA. Marking the ordinate for each milliampere may create a problem in that the scale would be too lengthy and difficult to read. Using 5 mA marks would create ten divisions, the same as the voltage scale. Once the divisions along each

axis are marked, each scale must be identified in terms of the units of measurement. In Figure 3–74, these would be voltage in volts and current in milliamperes.

5. The next step is to plot the data points. The first point to be plotted in Figure 3–74 is (0, 0), or the origin. The next four points are plotted along the abscissa. To plot the sixth point (0.5, 2) locate 0.5 V on the voltage axis, 2 mA on the current axis, and place a small point at the intersection of these two values.
6. After all data are plotted, a smooth line is drawn to connect the points. When points do not fall on the line, the line is drawn so that as many of the points are as close to the line as possible.

The data in Figure 3–74 has been plotted in Figure 3–75.

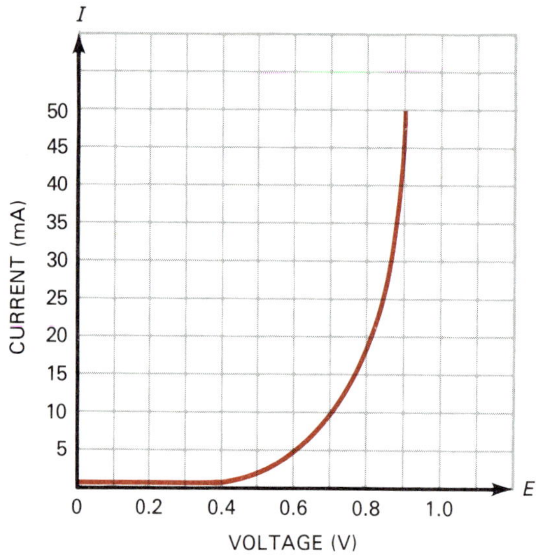

FIGURE 3–75

EXERCISE 3·6

In problems 1–10, determine in which quadrant the given point is located.

1. (2, 0)
2. (−3, 6)
3. (4, −2.7)
4. $(-\frac{1}{2}, -\frac{3}{4})$
5. (3, 3)
6. (−3, −3)
7. (0, −6)
8. (5, 1)
9. (48, −9)
10. (0, 0)

Problems 11–15 are based upon the zener diode characteristic curve in Figure 3–76.

11. What is the approximate voltage when the forward current is 10 mA?

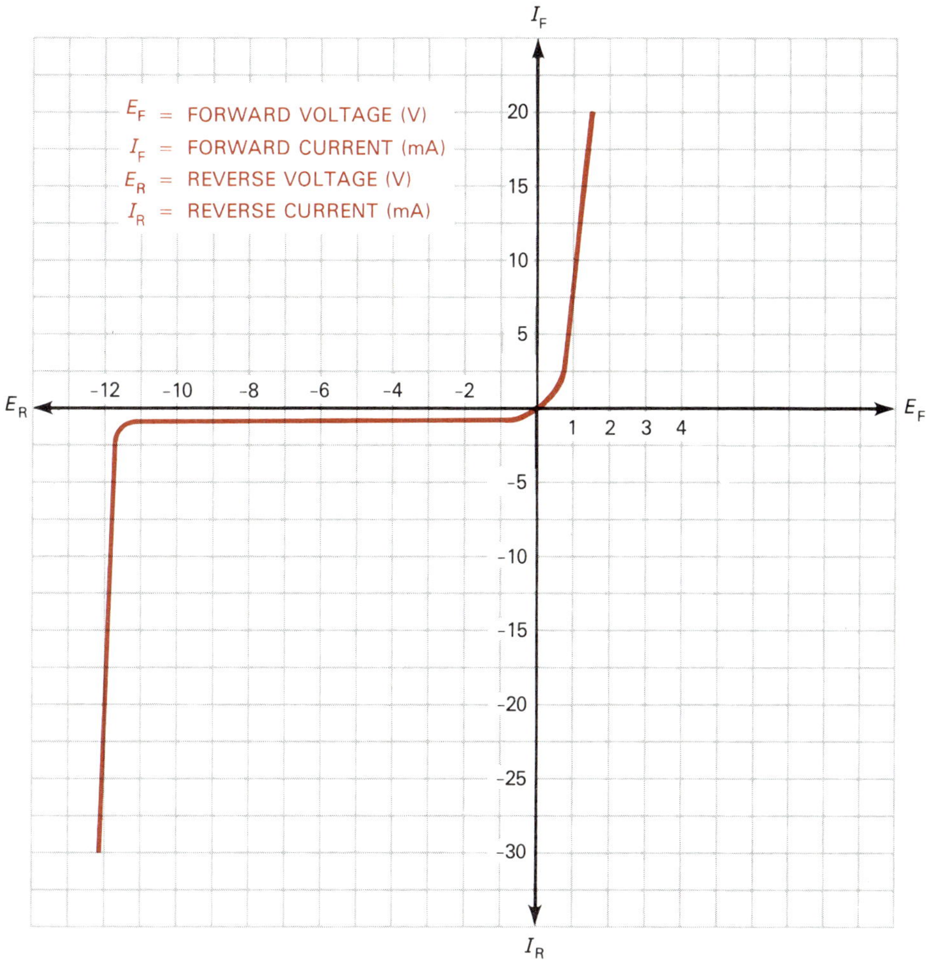

FIGURE 3–76 Zener diode characteristic curve

12. Find the approximate current when the reverse voltage is −6 V.
13. At what voltage does the zener breakdown and conduct heavily in the reverse direction?
14. What is the voltage when the zener current is −30 mA?
15. Find the zener current when the voltage is 0.5 V.

Problems 16–21 are based upon the family of characteristic curves for a common emitter transistor circuit shown in Figure 3–77.

16. What is the approximate value of I_c for a base current (I_b) of 100 μA?
17. A change of 50 μA of base current causes how much of a change in I_c?
18. Calculate the approximate resistance from collector to emitter for E_{ce} = 10 V and I_b = 250 μA. Note: $R_{ce} = E_{ce} \div I_c$

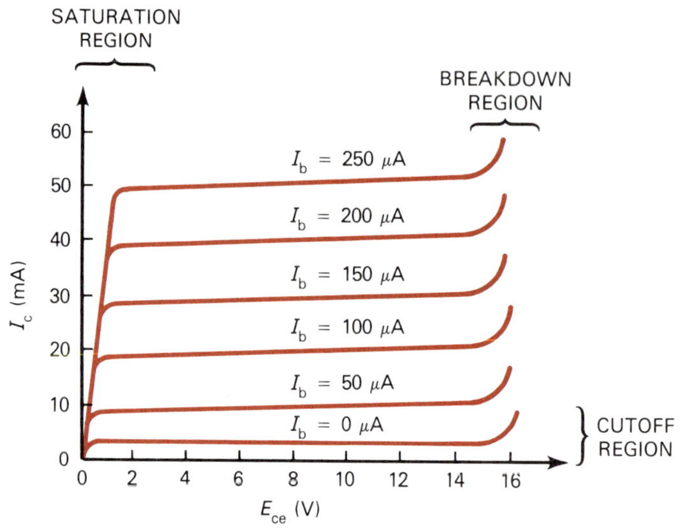

FIGURE 3–77

19. Calculate the approximate resistance from collector to emitter for $E_{ce} = 10$ V and $I_b = 50$ μA. Note: $R_{ce} = E_{ce} \div I_c$
20. Compute the current gain (β) of the circuit for $I_b = 100$ μA. Note: $\beta \approx I_c \div I_b$
21. Calculate the current gain (β) of the circuit for $I_b = 250$ μA. Note: $\beta \approx I_c \div I_b$

In problems 22–25 use the data provided to plot a graph.

22. The area of a square as the length of the side is changed. The side length should be plotted against the area of the square.

Side Length (inches)	0	1	2	3	4	5
Area (square inches)	0	1	4	9	16	25

23. The volume of a cylinder is given by the formula $V = \pi r^2 h$, where V = volume in cubic inches, r = radius in inches, and h = height in inches. Assuming the height remains constant at 6 inches, plot the radius against volume for radius values of 1″ through 5″.
24. Plot the output of a filter circuit. Frequency should be plotted against output voltage.

Frequency (kHz)	0	10	20	30	40	50	60	70	80	90	100
Voltage (volts)	0	3	9	12	12	12	12	12	9	3	0

25. Plot the current-voltage characteristics of a tunnel diode. Voltage is to be plotted against current.

Voltage (mV)	Current (mA)
0	0
50	15
100	35
150	50
200	40
250	25
300	15
350	10

Voltage (mV)	Current (mA)
400	8
450	6
500	8
550	10
600	15
650	20
700	40
750	75

UNIT 3·7 PROBLEM REVIEW

1. A 180-Ω resistor is connected to 15 V. What is the current flowing through the resistor?
2. A 10-Ω load conducts 138 mA of current. What is the power dissipated by the load?
3. A 100-W bulb is connected to a 120-V source. Determine the bulb's resistance to current flow.
4. The power rating of a direct current motor is 150 W. If the motor's resistance is 3.84 Ω calculate the applied voltage.

Use Figure 3–78 for problems 5–8.

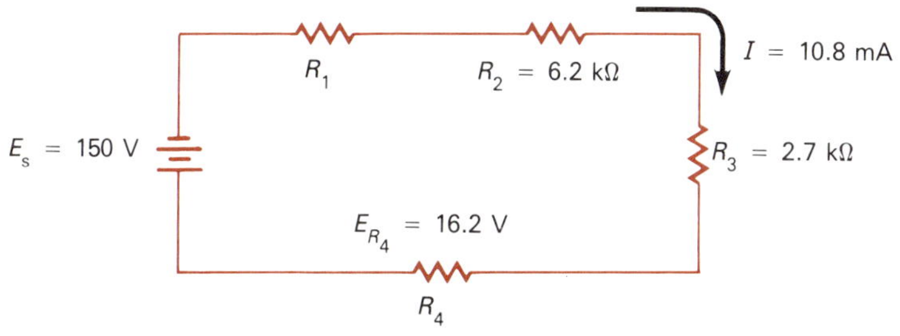

FIGURE 3–78

5. What is the value of R_4 in Figure 3–78?
6. Find R_t in Figure 3–78.
7. What is E_{R_1} in Figure 3–78?
8. What is the total power dissipated by the circuit in Figure 3–78?
9. Find the equivalent resistance of 68 kΩ and 47 kΩ connected in parallel.
10. Three parallel resistors are connected to 24 V. Find the total power dissipation if the values are 1.2 kΩ, 3 kΩ, and 2.7 kΩ.

11. Two resistors of equal value are connected in parallel to a 10-V source and conduct a total current of 0.5 mA. Find the value of each resistor.

Use Figure 3–79 for problems 12–14.

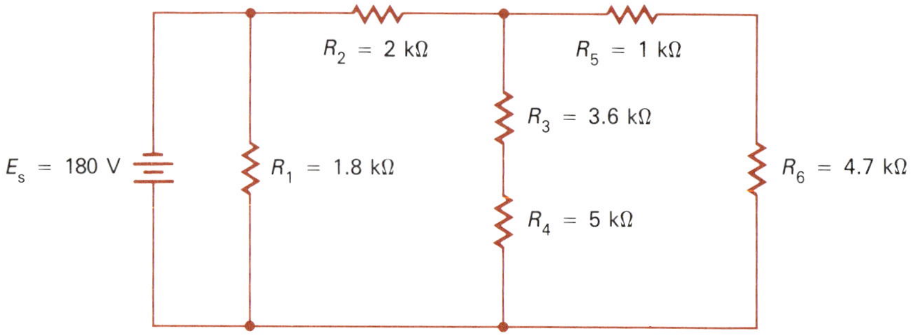

FIGURE 3–79

12. What is the total resistance of the circuit in Figure 3–79?
13. Calculate E_{R4} in Figure 3–79.
14. What is I_{R6} in Figure 3–79?

Use Figure 3–80 for problems 15–17.

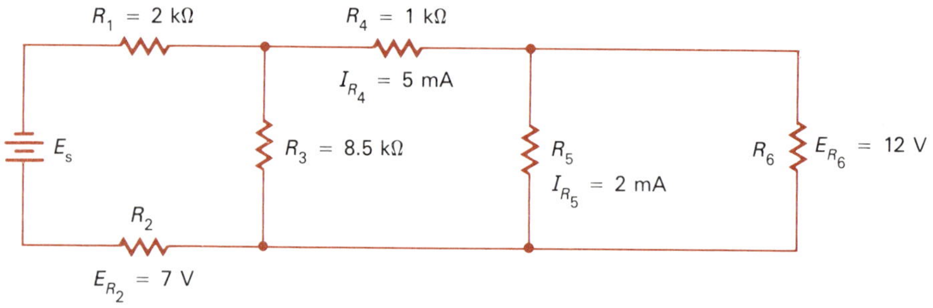

FIGURE 3–80

15. What are the values of (a) R_2, (b) R_5, and (c) R_6 in Figure 3–80?
16. Calculate (a) I_{R2}, (b) I_{R3}, and (c) I_t in Figure 3–80.
17. What are the values of (a) R_t and (b) E_s in Figure 3–80?
18. A capacitor stores 0.008 mC of charge when a voltage of 10 V is applied. What is the capacitor's value?
19. A 0.05-μF capacitor is placed in a circuit operating at 15 V. How much charge does the capacitor store?
20. A 25-μF capacitor stores 0.125 mC of charge. Determine the difference of potential between the plates.
21. The following parameters apply to a specific parallel plate capacitor: $A = 0.08\ \text{m}^2$, $d = 0.1 \times 10^{-3}$ m, and $K = 5$ F/m. What is the capacitor value?

22. Three capacitors are connected in series. Find the equivalent capacitance if the values are 0.05 μF, 0.02 μF, and 1 μF.

Use Figure 3–81 for problems 23–24.

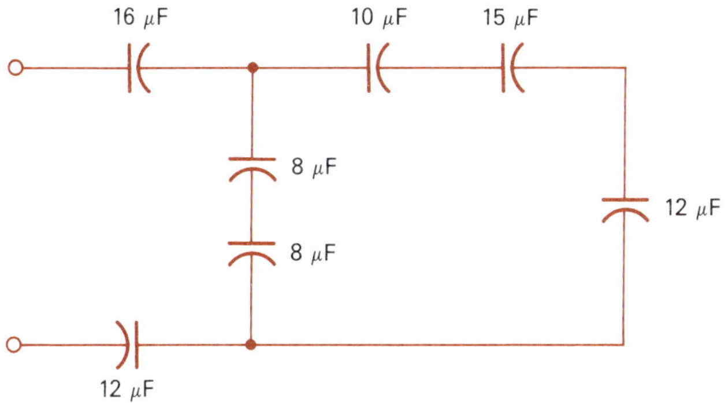

FIGURE 3–81

23. Find the total capacitance of the circuit in Figure 3–81.
24. If one of the 8-μF capacitors in Figure 3–81 were to open, what would the total capacitance be?
25. Determine one time constant for a 120-pF capacitor in series with 5 MΩ.
26. A 0.01-μF capacitor is in series with a resistor. What is the resistor value if 5 time constants total 5 ms?
27. Two 10-μF capacitors are wired in parallel and then connected in series with a 50-kΩ resistor. Calculate (a) one time constant, and (b) the voltage across the resistor after three constants. The applied voltage to the *RC* circuit is 24 V.
28. Two inductors are connected in series. Find the equivalent inductance if $L_1 = 100$ mH, $L_2 = 150$ mH, and $M = 0$.
29. Two inductors are connected in series-opposing. Their values are 75 mH and 200 mH. Find the equivalent inductance if the coefficient of coupling is 0.8.
30. A circuit contains two inductors connected in series-aiding. If $L_t =$ 200 mH, $M = 25$ mH, and one of the inductors has a value of 120 mH, find the value of the second inductor.
31. Find the coefficient of coupling for the circuit described in problem 30.
32. Find the equivalent inductance of 0.8-H and 1 500-mH inductors connected in parallel.
33. One time constant for a 0.5-mH inductor in series with a resistor is 0.02 ms. Find the value of the resistor.
34. Calculate the current flow in Figure 3–82 after 4 time constants.

Use Figure 3–83 for problems 35–36.

35. Determine the total current in Figure 3–83.

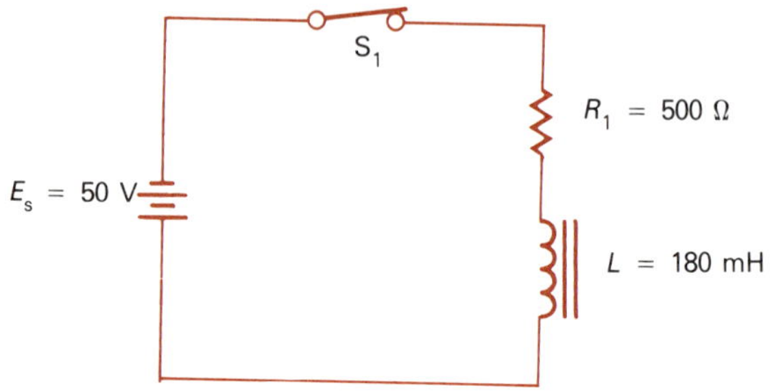

FIGURE 3–82

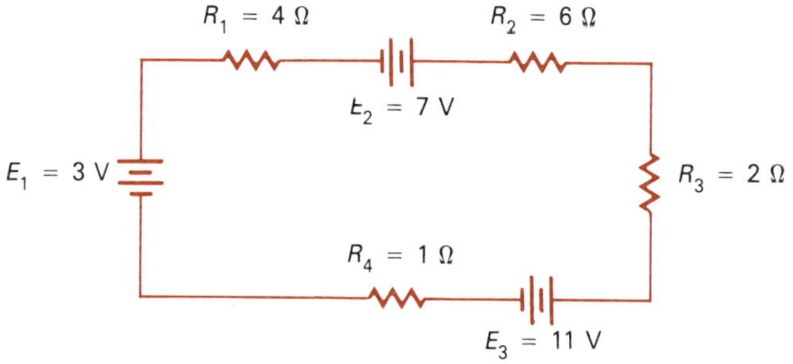

FIGURE 3–83

36. Calculate I_t in Figure 3–83 if the polarity of E_3 is reversed.
37. Use the voltage divider rule to determine the resistance of R_1 in Figure 3–84.

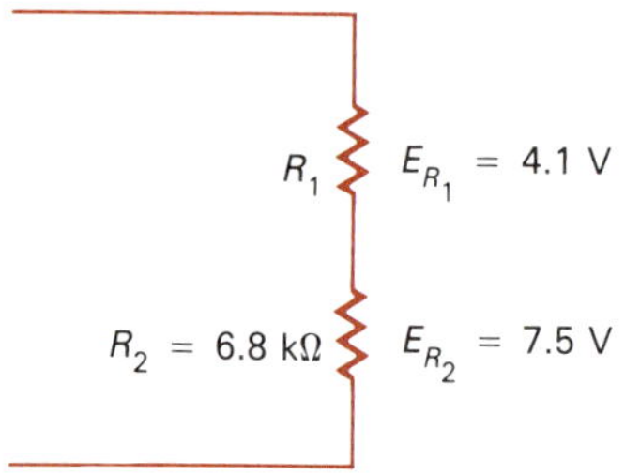

FIGURE 3–84

Use Figure 3–85 for problems 38–39.

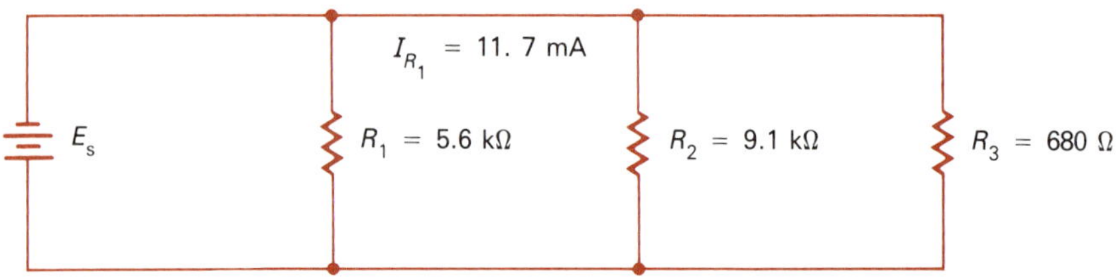

FIGURE 3–85

38. Use the current divider rule to determine (a) I_{R_2} and (b) I_{R_3} in Figure 3–85.
39. What are (a) the source voltage and (b) the total resistance in Figure 3–85?

Use Figure 3–86 for problems 40–44.

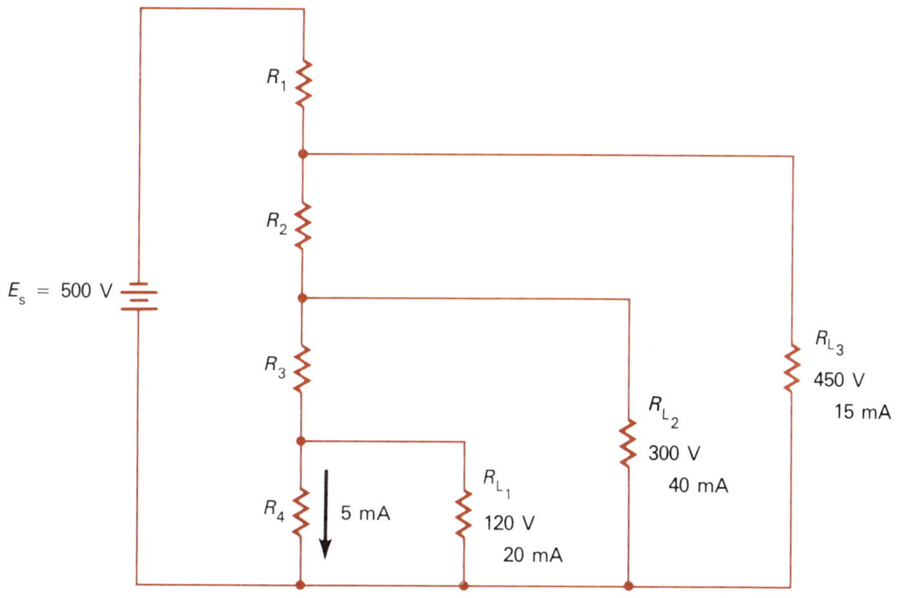

FIGURE 3–86

40. What is the value of R_4 in Figure 3–86?
41. Calculate the current passing through R_2 in Figure 3–86.
42. What are the values of (a) R_1, (b) R_2, and (c) R_3 in Figure 3–86?
43. Determine the total resistance seen by the source voltage in Figure 3–86.
44. If R_{L_1} in Figure 3–86 was to short (zero resistance), determine (a) E_{R_3}, (b) $I_{R_{L_2}}$, and (c) $E_{R_{L_3}}$.

Use Figure 3–87 for problems 45–47.

45. Find the value of R_x in Figure 3–87.
46. If R_3 in Figure 3–87 had been 11.7 kΩ, what would be the value of R_x?
47. Given the values in Figure 3–87, determine the total current delivered by the supply when the bridge is balanced.

Use Figure 3–88 for problems 48–50.

48. The full scale current (I_m), of the meter movement in Figure 3–88, is 0.25 mA. The resistance of the meter (R_m) is 10 Ω. Determine the shunt resistance (R_s) if the meter is to measure 10 mA.
49. Using Figure 3–88, calculate R_m if R_s = 0.8 Ω, I_m = 1 mA, and I_t = 5 mA.

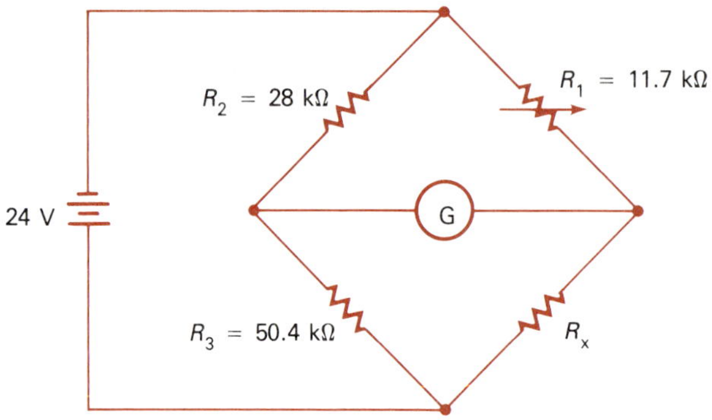

FIGURE 3–87

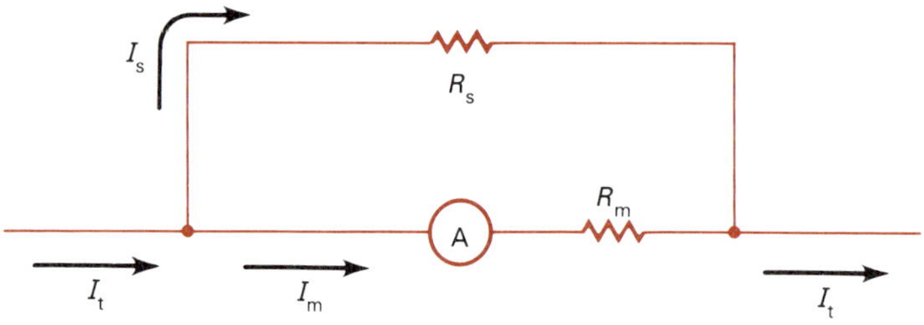

FIGURE 3–88

50. Calculate I_t in Figure 3–88 if I_m = 0.5 mA, R_m = 50 Ω, and R_s = 1.724 Ω.
51. The internal resistance of a power supply is 12 Ω. A load is attached such that maximum power is transferred to the load. The load current is 1.5 A. Determine the no-load output voltage of the supply.
52. The no-load output voltage of a power supply is 48 V. The internal resistance is 5.85 Ω. Calculate the load current when maximum power is being transferred to a load.

Use Figure 3–89 for problems 53–54.

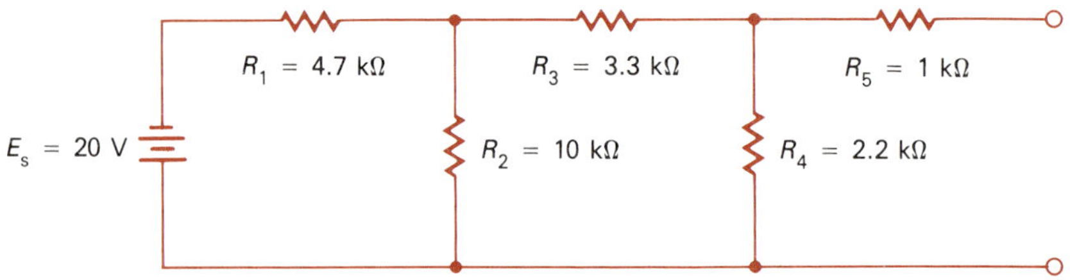

FIGURE 3–89

53. Calculate (a) E_{TH} and (b) R_{TH} for the circuit in Figure 3–89.
54. Determine the Norton current (I_{SL}) for Figure 3–89.

Use Figure 3–90 for problems 55–56.

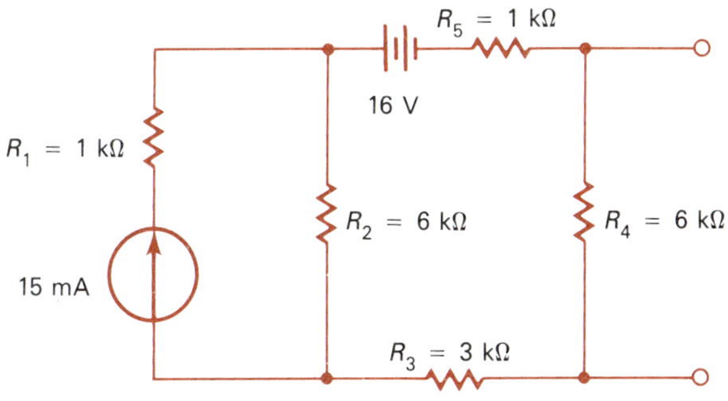

FIGURE 3–90

55. Calculate (a) E_{TH}, (b) R_{TH}, and (c) I_{SL} for the circuit in Figure 3–90.
56. What load, when connected to the circuit in Figure 3–90, causes the supply to transfer the maximum power to the load?

Use Figure 3–91 for problems 57–58.

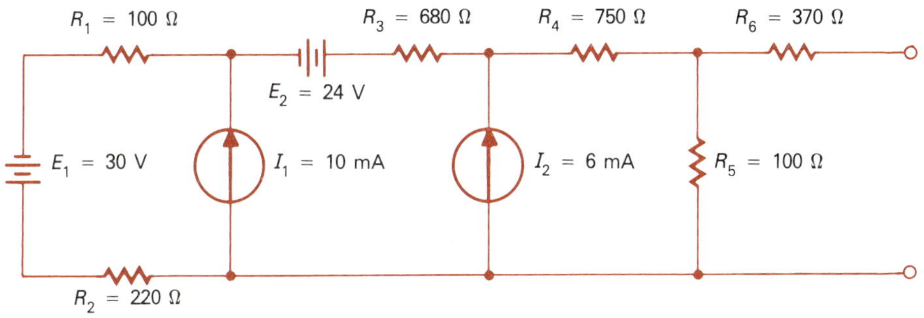

FIGURE 3–91

57. Determine (a) I_{SL} and (b) R_{TH} for Figure 3–91.
58. Convert the circuit in Figure 3–91 to a Thévenin circuit.
59. Determine in which quadrant the following points are located.
    a. $(5, -3)$
    b. $(7, 2.6)$
    c. $(-0.4, -1.8)$
    d. $(-9, 11.3)$
60. Determine in which quadrant the following points are located.
    a. $(0, 14)$
    b. $(-10, -10)$
    c. $(3.6, -2)$
    d. $(-2, 3.6)$

Figure 3–92 indicates that a sine wave of the voltage in an ac circuit increases and decreases during 360° rotation. Use Figure 3–92 for problems 61–64.

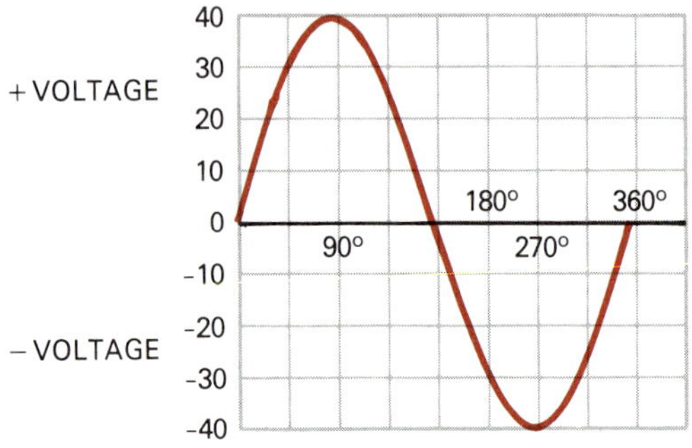

FIGURE 3–92

61. Find the voltage at 270°.
62. At approximately which degree points is the output +20 V?
63. Find the output voltage at 180°.
64. What is the approximate output voltage at 225°?

CHAPTER 4

ALGEBRA FOR ELECTRONICS

Algebra is a problem-solving tool which is no more difficult than ordinary arithmetic. Algebra is a shorthand technique which uses letters of the alphabet as a means of abbreviating what would otherwise be a lengthy and hard-to-handle statement. A description such as "voltage is equal to the product of current and resistance" can be stated as $E = I \times R$, an algebraic *formula* which uses letters to replace a statement. These letters can be manipulated—transposed—to give new formulas, something that is either impossible or difficult to do with complete sentences.

UNIT 4·1 THE LANGUAGE OF ALGEBRA

Objectives:

After studying this unit, you should be able to

- identify numerical and literal expressions.
- identify similar and dissimilar terms.
- identify monomials, binomials, and polynomials.
- apply the basic laws of algebra.
- evaluate basic algebraic expressions.

Formulas

A *formula* uses letters to express the relationship of two or more quantities, or algebraic expressions. In a formula such as Ohm's Law, each of the letters in the formula is a quantity: $E = I \times R$ consists of three quantities, E, I, and R. In this formula, and in all others, the quantities are represented by letters. Information is generally supplied for the quantities on the right side of the equals sign. In an electronics formula such as $P = I^2 \times R$, the values of current and resistance are either known or can be obtained in some way. The value of the unknown P will be determined just as soon as numerical values of I and R are substituted into the formula. By convention the unknown quantity is usually placed on the left side of the equals sign.

A formula is an equation. An *equation* indicates that algebraic expressions are equal to each other, as shown by their separation with an equals sign. A formula such as $P = E \times I$ is an equation. The algebraic expression, quantity, or term on the left, represented by P is equal to the product of the quantities (E and I) on the right. The two quantities, E and I, on the right are an algebraic statement or expression.

Algebraic Expressions

Just as you use words and phrases in English, algebra uses expressions and terms. An algebraic expression is the representation of any quantity in algebraic signs and symbols. For example, $3x - 9$ is an algebraic expression. An algebraic expression can be considered as an instruction and in the example just supplied we are told to subtract 9 from $3x$, whatever $3x$ might be. Terms are components of algebraic expressions which are separated by plus or minus signs. In this example, $3x$ is a term and so is 9. An algebraic expression can consist of one or more terms.

An algebraic expression can consist entirely of numbers. Thus, $9 - (7 + 4)$ is an algebraic expression. Since it contains numbers only, it is known as a *numerical* algebraic expression. The opposite of a numerical expression would be one containing letters only, such as $x + (y - z)$, and this is known as a *literal* algebraic expression.

An algebraic expression can contain two or more terms which are separated by plus or minus signs. The expression $4a + 3b - 2c$ contains three terms, $4a$, $3b$, and $2c$. If the terms have the same letters and exponents, such as $4a^2b$, $10a^2b$, and $30a^2b$, they are called *similar terms.* Terms that do not contain the same letters and exponents are *dissimilar terms:* $3ay^3$, $4ax^2$, and $12\ mn^5$ are dissimilar terms.

An algebraic expression with a single term is known as a *monomial,* e.g., $4abc$ is a monomial. So is $3a^2b^2c$. If the expression has two terms, it is referred to as a *binomial:* $x^3 - 3ay^2$ is a binomial. If it has three terms it is called a *trinominal:* $x^3 - y^2 + z$ is a trinomial. Generally, though, any algebraic expression having two or more terms is identified as a *polynomial.*

Coefficients

The numerical part of an algebraic term is frequently referred to as a *coefficient.* In a term such as $9a^2b$, the number 9 is the numerical portion of the expression. Technically speaking, however, each component of an algebraic expression can be regarded as a coefficient. Thus, in $9a^2b$, 9 is a numerical coefficient of a^2b; a^2 is a coefficient of $9b$; and b is a coefficient of $9a^2$. However, throughout this text, the word *coefficient* is used only in connection with the numerical part of algebraic expressions.

Example: In each of the following expressions determine whether the terms are similar or dissimilar, and whether the expression is a monomial, binomial, or polynomial.

a. $3x^2a + 27$
b. $17x^2y^{-4}$
c. $16x^3 - 11x^2 + 7x^3 - 21$

Solution: a. binomial; dissimilar
b. monomial
c. polynomial, two similar terms ($16x^3$ and $7x^3$)

Grouping Signs

The use of addition, subtraction, multiplication, and division signs in algebra may not always mean that the instructions represented by these signs will be understood. A statement such as $6n$ is clear and distinct; it means that 6 is to be multiplied by n, with the implication that n will ultimately be replaced by a number or possibly an expression that conveys more understanding than just the letter n. Parentheses ( ) are used to clarify algebraic expressions. The purpose of parentheses is to indicate that the terms they enclose are to be considered as a single quantity.

In addition to the parentheses, there are other grouping signs, including the brackets [ ], the braces { } and the vinculum ——. Technically, the radical sign is represented by $\sqrt{}$. The vinculum is added to make sure that all the terms beneath the vinculum are regarded as a group. All of these grouping signs have the same meaning.

The purpose of grouping signs is to make sure that the instructions conveyed by an algebraic expression are clear. However, when the work done by the algebraic symbols, such as +, or −, is completed, the grouping symbols can be removed.

Basic Laws of Algebra

Ohm's Law governs the relationships among voltage, current, and resistance in electronic circuits. Similarly, operations with numbers are governed by a series of basic laws.

The first of these is the *commutative law of addition.* This law states that the sum of two or more numbers is the same, regardless of the order in which the numbers are added. The commutative law of addition is expressed as

$$a + b = b + a \qquad (4\text{–}1)$$

The *commutative law of multiplication* states that the product of two or more numbers is the same, regardless of the order in which the numbers are multiplied. This law is written as

$$ab = ba \qquad (4\text{–}2)$$

The third law is the *associative law of addition.* This law states that the sum of three or more numbers is the same, regardless of how the numbers are grouped. This law is expressed as

$$a + (b + c) = (a + b) + c \tag{4–3}$$

The *associative law of multiplication* states that the product of three or more numbers is the same, regardless of how the numbers are grouped. This law is written as

$$a(bc) = (ab)c \tag{4–4}$$

The final law presented here is the *distributive law.* This very important law states that the product of a number (a) and the sum of two or more other numbers ($b + c$) is equal to the sum of the products of each number and the first number ($ab + ac$). This law is written as

$$a(b + c) = ab + ac \tag{4–5}$$

Example: Use the distributive law to simplify $3(7 + 4 - 6)$.

Solution:

$$\begin{aligned} 3(7 + 4 - 6) &= (3)(7) + (3)(4) - (3)(6) \\ &= 21 + 12 - 18 \\ &= 33 - 18 \\ &= 15 \end{aligned}$$

Translating English into Algebraic Symbols

In electronics problems word statements must be translated or converted into their algebraic symbols to find solutions or unknown values.

A phrase such as **the sum of** means that two or more quantities are to be added. "The sum of the voltage drops" is translated as $E_1 + E_2 + E_3 \ldots .$

A sentence containing a phrase such as **the product of** means multiplication is required. "The product of current and resistance" is translated as $I \times R$.

Difference means subtraction. A voltage difference means one voltage subtracted from another. "The voltage difference between E_1 and E_2" is translated as $E_1 - E_2$. Sometimes the phrase "potential difference" is used. The potential difference across the output of a power supply is the reference voltage subtracted from the maximum voltage. If the output of a power supply is 300 V and the reference point is called either zero or ground, the potential difference is 300 V − 0 V = 300 V. Quite often, potential difference is indicated as a single value. In the example just supplied, the potential difference might be referred to as 300 V.

The expression, **the quantity,** indicates two or more terms are enclosed in parentheses. A statement such as "the quantity x minus two times the quantity x plus seven" can be written as $(x - 2)(x + 7)$.

The phrase, **the ratio of,** indicates division. A statement such as "the ratio of E_1 to E_2" can be written as

$$\frac{E_1}{E_2}$$

The phrases, **is the same as,** or **is identical to,** are indicative of an equals sign. A statement such as "E_1 is identical to E_2" means $E_1 = E_2$.

Signed Numbers

When using literal expressions, it is necessary to be able to perform basic operations on these expressions. Use these rules when performing the basic operations involving signed numbers.

To add signed numbers with like signs, add the numbers as in arithmetic and place the common sign before the sum. For example, $(-3) + (-3) = -6$.

To add signed numbers with unlike signs, subtract the numbers as in arithmetic and place the sign of the larger number before the answer. For example, $(+2) + (-5) = -3$.

To subtract signed numbers, change the sign of the number being subtracted and proceed as in addition. For example, $(+2) - (+4) = (+2) + (-4) = -2$.

To multiply or divide signed numbers with like signs, multiply or divide as in arithmetic and place a plus sign before the answer. For example, $(-4)(-3) = +12$ or 12.

To multiply or divide signed numbers with unlike signs, multiply or divide as in arithmetic and place a minus sign before the answer. For example, $(-12) \div (+4) = -3$.

Evaluating Algebraic Expressions

To check algebraic solutions, the numerical values are substituted in the algebraic expression. An expression, such as $5a^2b - 14$ may be reduced to a single numerical value once the values of a and b are known.

Example: Evaluate $5a^2b - 14$ if $a = 2$ and $b = 5$.

Solution: $5(2^2)(5) - 14 = 86$

Example: Simplify the expression "the quantity $4a$ plus $7b$ is divided by the quantity $5a^2$ minus b" given that $a = 4$ and $b = 6.5$.

Solution: $\dfrac{4a + 7b}{5a^2 - b} = \dfrac{4(4) + 7(6.5)}{5(4^2) - 6.5} = 0.837$

Example: Simplify $3a^2b - (4a + 7b) - 5a^3b^4 + 6ab^5$ given that $a = -5$ and $b = 3$.

Solution: $3(-5^2)(3) - [(4)(-5) + (7)(3)] - 5(-5^3)(3^4) + 6(-5)(3^5) = 225 - 1 + 50\,625 - 7\,290 = 43\,559$

EXERCISE 4·1

Simplify each of the expressions in problems 1–5.

1. $12 - 6(4 + 7) - 2$
2. $-8 - (-9 + 3 \times 2) + 10$

3. $-4(6 - 4) + (5 - 1)(-7)$
4. $[-3(6 - 2) + 3][(-4 + 7) 3]$
5. $2[(7 + 2 \times (-3) + 5)] - 18$
6. What are the numerical coefficients in $16x^4y^2 - 3x^3y^3 - x^2y$?
7. Is the expression in problem 6 a monomial, binomial, or polynomial?
8. Write an algebraic expression for the following statement: "The sum of a number squared and five is divided by the difference of the same number and another number which is cubed."
9. Write an expression for this statement: "The product of three times one number and another number squared is divided by the difference of the two numbers."
10. Simplify: $4a^2b^3y - 12a$ if $a = 2$, $b = -4$, $y = 0.5$
11. Simplify: $5x^2 - (3x^2 + 7)$ if $x = 8$
12. Simplify: $6ax^4 + (7x^2)(2)$ if $a = -10$ and $x = 4\frac{1}{2}$
13. Simplify: $25 - 25x^2 + 17x$ if $x = -3$
14. Simplify: $\frac{165}{x^2} - (x^2 + 7y) + 14y^3$ if $x = 5$ and $y = 3$
15. Simplify: $\left[\frac{5a^2b^3 - 12a^3}{b^4} - (3a^2 - b)\right] + ab$

    if $a = -6$ and $b = 7$

UNIT 4·2 EXPONENTS AND RADICALS

Objectives:

After studying this unit, you should be able to

- apply the basic rules of exponents.
- apply the basic rules of radicals.
- simplify expressions containing exponents and radicals.

Multiplying Exponential Numbers

Exponents and radicals were briefly discussed in Chapter 2. As they play an important part in the study of algebra, the basic rules of exponents and radicals are presented here.

The product of exponential factors with the same base is that base raised to a power that is the *sum* of the individual exponents.

$$x^m \cdot x^n = x^{m+n} \tag{4–6}$$

Example: Simplify each of the following.

a. $5^2 \cdot 5^3$
b. $12^3 \cdot 12^4 \cdot 12^1$
c. $x^2 \cdot x^3 \cdot x^7$
d. $10^1 \cdot 10^2 \cdot 10^3$

Solution: a. $5^2 \cdot 5^3 = 5^{2+3} = 5^5$
b. $12^3 \cdot 12^4 \cdot 12^1 = 12^{3+4+1} = 12^8$
c. $x^2 \cdot x^3 \cdot x^7 = x^{2+3+7} = x^{12}$
d. $10^1 \cdot 10^2 \cdot 10^3 = 10^{1+2+3} = 10^6$

Raising Exponential Numbers to Powers

When a term with an exponent is raised to a power, the resulting exponent is the *product* of the two exponents.

$$(x^m)^n = x^{m \cdot n} = x^{mn} \qquad (4\text{–}7)$$

Example: Simplify each of the following.
a. $(3^2)^3$
b. $(5^5)^2$
c. $(a^3)^7$
d. $(8^4)^4$

Solution: a. $(3^2)^3 = (3^2)(3^2)(3^2) = 3^{2+2+2} = 3^6$
b. $(5^5)^2 = 5^{5 \cdot 2} = 5^{10}$
c. $(a^3)^7 = a^{3 \cdot 7} = a^{21}$
d. $(8^4)^4 = 8^{4 \cdot 4} = 8^{16}$

When factors within parentheses are raised to a power, each factor is raised to that power.

$$(xyz)^m = x^m y^m z^m \qquad (4\text{–}8)$$

Example: Simplify each of the following.
a. $(4a)^2$
b. $(6ab^2c^3)^4$
c. $(x^3yz^4)^9$
d. $(16ab^3)^5$

Solution: a. $(4a)^2 = 4^2a^2 = 16a^2$
b. $(6ab^2c^3)^4 = 6^4a^4b^8c^{12} = 1\ 296a^4b^8c^{12}$
c. $(x^3yz^4)^9 = x^{27}y^9z^{36}$
d. $(16ab^3)^5 = 16^5a^5b^{15} = 1\ 048\ 576a^5b^{15}$

When division occurs within parentheses, and the quantity is raised to a power, each factor in the numerator and denominator is raised to that power.

$$\left(\frac{ax}{by}\right)^m = \frac{a^m x^m}{b^m y^m} \qquad (4\text{–}9)$$

Example: Simplify each of the following terms.

a. $\left(\dfrac{3a^2}{b}\right)^3$ c. $\left(\dfrac{x^3a^5b^4}{y^2z^6}\right)^2$

b. $\left(\dfrac{5x^2y}{4a}\right)^4$ d. $\left(\dfrac{6a^9b^2}{x^3y}\right)^5$

Solution: a. $\left(\dfrac{3a^2}{b}\right)^3 = \left(\dfrac{3a^2}{b}\right)\left(\dfrac{3a^2}{b}\right)\left(\dfrac{3a^2}{b}\right) = \dfrac{27a^6}{b^3}$

b. $\left(\dfrac{5x^2y}{4a}\right)^4 = \dfrac{625x^8y^4}{256a^4}$

c. $\left(\dfrac{x^3a^5b^4}{y^2z^6}\right)^2 = \dfrac{x^6a^{10}b^8}{y^4z^{12}}$

d. $\left(\dfrac{6a^9b^2}{x^3y}\right)^5 = \dfrac{7\,776a^{45}b^{10}}{x^{15}y^5}$

Dividing Exponential Numbers

The quotient of terms with the same base is that base raised to a power that is the *difference* of the individual exponents.

$$\frac{x^m}{x^n} = x^{m-n} \qquad (4\text{–}10)$$

If m is greater than n ($m > n$) then the resultant is in the numerator. If $n > m$ then the resultant is in the denominator, or is expressed as a negative exponent. If $m = n$ then the exponent is zero and the resultant is 1, $x^0 = 1$, where $x \neq 0$.

Example: Simplify each of the following.

a. $\dfrac{a^4}{a^2}$

b. $\dfrac{x^5}{3x}$

Solution: a. $\dfrac{a^4}{a^2} = \dfrac{a \cdot a \cdot a \cdot a}{a \cdot a} = \dfrac{a^2}{1} = a^2$

b. $\dfrac{x^5}{3x} = \dfrac{x^{5-1}}{3} = \dfrac{x^4}{3}$

Example: Simplify each of the following.

a. $\dfrac{b^3}{b^5}$

b. $\dfrac{5a^7}{a^{12}}$

c. $\dfrac{x^4}{x^7}$

Solution: a. $\dfrac{b^3}{b^5} = \dfrac{b \cdot b \cdot b}{b \cdot b \cdot b \cdot b \cdot b} = \dfrac{1}{b^2} = b^{-2}$

b. $\dfrac{5a^7}{a^{12}} = \dfrac{5}{a^{12-7}} = \dfrac{5}{a^5} = 5a^{-5}$

c. $\dfrac{x^4}{x^7} = x^{4-7} = x^{-3} = \dfrac{1}{x^3}$

Fractional Exponents or Radicals

Fractional exponents were presented earlier in the text. By definition:

$$\sqrt[m]{x} = x^{1/m} \qquad \textbf{(4–11)}$$

Calculator Tip

The use of the y^x function was discussed in Chapter 2. This function will normally accept decimal exponents, which represent specific roots. Examine two methods for determining the square root of 144.

1. $\sqrt{144}$: 144 [$\sqrt{x}$] and 12 is displayed
2. $144^{1/2} = 144^{0.5} =$ 144 [y^x] .5 [=] 12

Example: Simplify each of the following.

a. $\sqrt[3]{64}$
b. $625^{1/4}$
c. $\sqrt[8]{6\,561}$
d. $387^{0.5}$
e. $\sqrt[4]{1\,000}$

Solution: a. $\sqrt[3]{64} = 64^{1/3} = 4$
b. $625^{1/4} = 5$
c. $6\,561^{1/8} = 3$
d. $387^{1/2} = 19.67$
e. $1\,000^{1/4} = 5.62$

Since a radical is a fractional exponent, the same rules apply for radicals as for exponents. **When factors within a grouping symbol are raised to a fractional power, each factor is raised to that fractional power.**

$$\sqrt[m]{xy} = \sqrt[m]{x} \cdot \sqrt[m]{y} \qquad \textbf{(4–12)}$$

Calculator Tip

Simplify the following expression if $a = 4$ and $x = 8$.

$$\sqrt[5]{32a^3x^6}$$

1. Removing the radical sign gives
$(32^{1/5})(a^3)^{1/5}(x^6)^{1/5}$ or $(32^{1/5})(a^{3/5})(x^{6/5})$
or $(32^{0.2})(a^{0.6})(x^{1.2})$
and the sequence is
32 [yˣ] 0.2 [×] 4 [yˣ] 0.6 [×] 8 [yˣ] 1.2 [=] 55.72
2. A more direct sequence might be
32 [×] 4 [yˣ] 3 [×] 8 [yˣ] 6 [=] [INV] [yˣ] 5 [=] 55.72

Example: Simplify each of the following.

a. $\sqrt[3]{8a^6}$

b. $\sqrt[5]{243x^4}$

c. $\sqrt{49a^8b^3}$

Solution: a. $\sqrt[3]{8a^6} = \sqrt[3]{8}\,\sqrt[3]{a^6} = 8^{1/3}\,(a^6)^{1/3} = 2a^2$

b. $\sqrt[5]{243x^4} = (243^{1/5})(x^4)^{1/5} = 3x^{4/5} = 3\sqrt[5]{x^4}$

c. $\sqrt{49a^8b^3} = (49^{1/2})(a^8)^{1/2}\,(b^3)^{1/2} = 7a^4b^{3/2} = 7a^4\sqrt{b^3}$

When a quotient is raised to a fractional power, each factor in the numerator and denominator is raised to that fractional power.

$$\sqrt[m]{\frac{x}{y}} = \frac{\sqrt[m]{x}}{\sqrt[m]{y}} \qquad \textbf{(4–13)}$$

Example: Simplify each of the following.

a. $\sqrt{\dfrac{64x^4}{9y^2}}$

b. $\sqrt[4]{\dfrac{256a^6b^8}{16x^{10}}}$

Solution: a. $\sqrt{\dfrac{64x^4}{9y^2}} = \dfrac{\sqrt{64x^4}}{\sqrt{9y^2}} = \dfrac{64^{1/2}(x^4)^{1/2}}{9^{1/2}(y^2)^{1/2}} = \dfrac{8x^2}{3y}$

b. $\sqrt[4]{\dfrac{256a^6b^8}{16x^{10}}} = \dfrac{256^{1/4}(a^6)^{1/4}(b^8)^{1/4}}{16^{1/4}\,(x^{10})^{1/4}} = \dfrac{4b^2\sqrt{a^3}}{2\sqrt{x^5}}$

When a term with a fractional exponent is raised to a fractional power, the resulting exponent is the product of the two fractional exponents.

$$\sqrt[n]{\sqrt[m]{x}} = \sqrt[mn]{x} \qquad (4\text{–}14)$$

Example: Simplify each of the following.

a. $\sqrt[3]{\sqrt{729}}$

b. $\sqrt{\sqrt{a^{12}}}$

c. $\sqrt[5]{\sqrt[3]{14\,756}}$

Solution: a. $\sqrt[3]{\sqrt{729}} = \sqrt[6]{729} = 729^{1/6} = 3$

b. $\sqrt{\sqrt{a^{12}}} = \sqrt[4]{a^{12}} = (a^{12})^{1/4} = a^3$

c. $\sqrt[5]{\sqrt[3]{14\,756}} = \sqrt[15]{14\,756} = (14\,756)^{1/15} = 1.896$

Example: Using the rules just presented simplify each of the following expressions by removing all grouping signs, radical signs, and using positive exponents only.

a. $\sqrt{\dfrac{2x^2(8x^2)}{4a^4b^3}}$

b. $\dfrac{81^{-1/2}x^{-2}a^3}{9x^3a^{-5}}$

c. $\left[\dfrac{15^{-1}a^{-3}b^3c^{-5}}{60a^{-3}b^4c}\right]^{-1/2}$

Solution: a. $\sqrt{\dfrac{2x^2(8x^2)}{4a^4b^3}} = \sqrt{\dfrac{16x^4}{4a^4b^3}}$

$= \sqrt{\dfrac{4x^4}{a^4b^3}}$

$= \dfrac{4^{1/2}(x^4)^{1/2}}{(a^4)^{1/2}(b^3)^{1/2}}$

$= \dfrac{2x^2}{a^2b^{3/2}}$

b. $\dfrac{81^{-1/2}x^{-2}a^3}{9x^3a^{-5}} = \dfrac{a^5a^3}{(9)(81^{1/2})(x^2)(x^3)}$

$= \dfrac{a^8}{81x^5}$

c. $\left[\frac{15^{-1}a^{-3}b^{3}c^{-5}}{60a^{-3}b^{4}c}\right]^{-1/2} = \left[\frac{a^{3}b^{3}}{(60)(15)(a^{3})(b^{4})(c)(c^{5})}\right]^{-1/2}$

$= \left[\frac{a^{3}b^{3}}{900a^{3}b^{4}c^{6}}\right]^{-1/2}$

$= \left[\frac{1}{900bc^{6}}\right]^{-1/2}$

$= [900bc^{6}]^{1/2}$

$= 30b^{1/2}c^{3}$

EXERCISE 4·2

Simplify each of the following expressions by removing all grouping signs, radical signs, and using positive exponents only.

1. $(4ab^{2})^{2}$
2. $5(2x^{4}y)^{3}$
3. $\frac{(a^{3}b^{2}c^{-2})^{3}}{c^{4}a^{2}}$
4. $(3a^{-2}b^{2}x^{-4})^{-5}$
5. $(15^{0})^{2}$
6. $(45.697\ 6)^{0.25}$
7. $\sqrt[5]{144^{3}}$
8. $\sqrt[4]{81x^{8}b^{4}c^{2}}$
9. $\sqrt{\frac{(x^{2}y^{3}z^{4})^{-2}}{16x^{6}}}$
10. $\left[\frac{(9a^{-5}b^{2}c^{-3})^{2}}{(3b^{-4}c^{3})^{-2}}\right]^{3}$
11. $[(5a^{2}x)^{-3}]^{-2}$
12. $\frac{3xy^{-3}}{(4x^{2}y^{-3})^{2}}$
13. $25(5a^{2}b^{4})^{-2}$
14. $(\sqrt[9]{4\ 278})^{-3}$
15. $\frac{1}{(12x^{-2}y^{3})^{-3}}$
16. $\sqrt{\frac{(3a^{5}y^{3})^{4}}{a^{6}y^{-2}}}$
17. $-(5x^{4}a^{-1}b)^{2}$
18. $[(107\ 283)^{0.1}]^{2}$
19. $\frac{10}{\left[\frac{25a^{4}y^{2}}{10a^{3}y}\right]^{-2}}$
20. $\sqrt[5]{243a^{35}b^{10}y^{2}}$
21. $\left(\frac{b^{-m}a^{n}c}{b^{n}a^{n}c^{m}}\right)$
22. $\sqrt[m]{a^{-2}b^{3}x^{4}}$
23. $\left(\frac{14^{2}b^{3}x^{-3}}{7b^{-4}}\right)^{2}$
24. $\left[\frac{(1.47 \times 10^{5})^{4}\ (2.47 \times 10^{-6})^{10}}{0.8 \times 10^{-3}}\right]^{-2}$
25. $\left(\sqrt[16]{3.14 \times 10^{23}}\right)^{-3}$
26. $\left[\frac{1}{(7x^{-4}y^{2}b)^{-3}}\right]^{2}$
27. $(1.98 \times 10^{-6})^{4}$
28. $\left(\frac{\sqrt[7]{(1\ 702)^{2}}}{\sqrt[4]{625}}\right)^{5}$
29. $\left[\frac{32a^{4}b^{3}a^{-3}x^{-4}}{(4a^{3}bx)^{2}\ (12a^{4}b)}\right]^{-3}$
30. $\frac{(2y^{-3}x^{2}b)^{4}\ (12x^{-3}b^{5})^{-2}}{(-4y^{3}x^{-2}b^{5})^{5}\ (x^{3}y^{3}b^{3})^{-3}}$

UNIT 4·3 ADDITION AND SUBTRACTION OF ALGEBRAIC EXPRESSIONS

Objectives:

After studying this unit, you should be able to

- add literal algebraic expressions.
- subtract literal algebraic expressions.

Addition of Monomials

In arithmetic, we have ten different symbols all of which can be added to each other. In algebra, only similar terms can be added. In a problem such as

$$\begin{array}{r} a \\ 2a \\ +4a \\ \hline 7a \end{array}$$

each algebraic symbol, a in this example, has a numerical coefficient expressed or implied. The first letter a has a coefficient of 1 and can be written as $1a$. The number 1 is ordinarily omitted. The other coefficients, 2 and 4, are expressed or written.

The previous problem shows vertical addition. Vertical addition is customarily used in arithmetic while horizontal addition seems to be the preferred form in algebra, although there is no reason why both cannot be used. In the horizontal form the problem would be:

$$a + 2a + 4a = 7a$$

In algebraic addition, it is just the coefficients that are added, while the algebraic symbol is simply brought along into the final answer. In a problem such as

$$\begin{array}{r} 245c \\ 112c \\ +302c \\ \hline 659c \end{array}$$

the answer or sum is 659 times c, or $659 \times c$. The answer can also be resolved further into powers of 10 by writing it as

$$(6 \times 10^2) + (5 \times 10^1) + (9 \times 10^0) \times c^1$$

Since, in algebraic addition, only like symbols can be added, a problem such as finding the sum of a, b, and c can only be written as

$$a + b + c$$

Example: Add these monomials: $6a^2x$, a^2x, and $14a^2x$

Solution: Since all 3 terms are similar the numerical coefficients may be added.

$$6a^2x + 1a^2x + 14a^2x = 21a^2x$$

Example: Add these monomials: $3b^2c + 4bc^2 + 18b^2c$

Solution: Only two of these terms are similar, therefore

$$3b^2c + 4bc^2 + 18b^2c = 21b^2c + 4bc^2$$

It is important to note in the previous example that further simplification of the expression is not now possible as the two expressions are dissimilar. Once the values of b and c are known the expressions may be evaluated.

Addition of Polynomials

Addition of polynomials is similar to that of monomials. Again, only like terms may be added. With this in mind the polynomials are arranged in columns with like terms in the same columns. The columns are then added. Carefully follow this example.

Example: Add $(6y^2x + 3a^2b^2 + 5x^2y + 12)$ and $(x^2y + 7 + 4a^2b^2 + 3y^2x)$.

Solution: Arrange similar terms in columns and then add within each column.

$$\begin{array}{r} 6y^2x + 3a^2b^2 + 5x^2y + 12 \\ (+)\,3y^2x + 4a^2b^2 + x^2y + 7 \\ \hline 9y^2x + 7a^2b^2 + 6x^2y + 19 \end{array}$$

Example: Add $(9a^3b + 4ab^2 + x^2y + 11)$ and $(4ab^2 + 3x^2y + 13 + 5a^2b)$.

Solution: Note in this example that one term in each expression is not similar to any other term.

$$\begin{array}{r} 9a^3b + 4ab^2 + x^2y + 11 \qquad\quad \\ (+)\qquad\quad 4ab^2 + 3x^2y + 13 + 5a^2b \\ \hline 9a^3b + 8ab^2 + 4x^2y + 24 + 5a^2b \end{array}$$

Subtraction of Monomials and Polynomials

The subtraction of monomials and polynomials is similar to addition, although care must be exercised to ensure that each term carries the appropriate sign.

Example: Subtract $4a^2b^2$ from $10a^2b^2$.

Solution:

$$\begin{array}{r} 10a^2b^2 \\ -\ 4a^2b^2 \\ \hline 6a^2b^2 \end{array}$$

Example: Subtract $(5xy^2 + 7xy - 6x + 4)$ from $(7xy^2 + 12x - 11 - 18xy)$.

Solution: Write the expressions in a column format:

$$\begin{array}{rrrrr} (7xy^2 & + 12x & - 11 & - 18xy) \\ -(5xy^2 & - 6x & + 4 & + 7xy) \\ \hline \end{array}$$

Remove the parentheses by changing *each* sign in the subtrahend and then combine the terms.

$$\begin{array}{lrrrr} & 7xy^2 & + 12x & - 11 & - 18xy \\ (+) & -5xy^2 & + 6x & - 4 & - 7xy \\ \hline & 2xy^2 & + 18x & - 15 & - 25xy \end{array}$$

Example: Subtract $(-12x^2y + xy - 14c^2 + x^3c^2)$ from $(3x^3c^2 - 5xy - x^2y - 14c^2)$.

Solution:

$$\begin{array}{lrrrr} & 3x^3c^2 & - 5xy & - x^2y & - 14c^2 \\ (+) & -x^3c^2 & - xy & + 12x^2y & + 14c^2 \\ \hline & 2x^3c^2 & - 6xy & + 11x^2y & + 0 \end{array}$$

Example: Simplify the following expressions.
$(25x^2y - 7xy) - [(12x^2y + 10xy^2 - 8x^2y) - (14xy^2 + 6xy)]$

Solution:
$(25x^2y - 7xy) - [(12x^2y + 10xy^2 - 8x^2y) - (14xy^2 + 6xy)]$
$(25x^2y - 7xy) - [4x^2y + 10xy^2 - 14xy^2 - 6xy]$
$(25x^2y - 7xy) - [4x^2y - 4xy^2 - 6xy]$
$25x^2y - 7xy - 4x^2y + 4xy^2 + 6xy$
$21x^2y - xy + 4xy^2$

EXERCISE 4·3

Simplify each of the following expressions.

1. $15x - (4x + x) + 3x$
2. $6a^2 + (8a - 3a^2 + 7)$
3. $(2x^2y - 4x) - (5x + 3x^2y)$
4. $(4b^2 + c) - 2c - (3b^2 + c - 12) - b^2$
5. $5y - [3y - (5y + 6x) + 7x - (x - y)]$
6. $7w - \{-9x - [4wx - 5w + x - (-3x + w)]\}$
7. $(x^2y^2 + 3x^2y + 6xy^2 - 12) - (4x^2y^2 - 7x^2y + xy^2 - 12)$
8. $(8ab^2 - 2ab) - (-7ba + 14ab^2) + (-b^2a + 7ab)$
9. $-(25xy^2z + 112x^2yz^2 - 13xyz^2) - [48x^2yz^2 - (6xyz^2 - 19x^2yz^2)] +$
   $(-x^2yz^2 + 7xy^2z)$
10. $(12m - 16n + 7) - (4n + 2) + (-5m - 3n - 12) - (m - n)$
11. Simplify this expression if $x = 5$, $y = -3$, and $z = 8$.
    $(5z^2x - 2y + 7) - (4x^2y^2 - z - 47) + (-y + z^2x - z)$
12. Simplify this expression if $a = 4$, $b = 2$, and $c = -7$.
    $-(\sqrt{ab^2} + 3ab - c) + (-c + 5ab) + c + ab$
13. Simplify this expression if $x = -9$, $y = -3$, $z = 10$.
    $x^2y^2z^2 - [(9x^2y + 13yz) - (yz + 7x^2y) + (-8yz + 6x^2y)]$

14. $12m - [-12m + (12m - m) - (-12m)]$
15. $-(11ab^2 + 16a^2b) - (-5a^2b + ab^2) + (4ab^2 + 3ab^2)$
16. Simplify this expression if $m = -6$ and $n = 6$.
$-(18m^2n^2 + m^2n - 10mn^2) + (7m^2n^2 - 24m^2n - mn^2)$
17. Subtract $(-3gh^2 + 12gh - 9)$ from $(7gh^2 - 12gh + 42)$.
18. $5m - (m + n - 12)^0 + (6m - 3n + 7) - 14n + 12$
19. $-14y^2z^3 + (7y^2z^3 - 12y^2z^2 + 18yz) - (-x^2z^2 + yz - 10y^2z^3)$
20. Simplify this expression if $a = 7$, $b = -2$, $c = 5$ and $d = -8$.
$-a^2b^2c + (ab^2cd - c^2d - a^2b^2c) - (dc^2 + ab^2cd) + (-a^2b^2c + a^2b^2c)$

UNIT 4·4 MULTIPLICATION AND DIVISION OF ALGEBRAIC EXPRESSIONS

Objectives:

After studying this unit, you should be able to

- multiply literal algebraic expressions.
- divide literal algebraic expressions.

Multiplication of Monomials

The following examples will demonstrate the process by which monomials are multiplied.

Example: Multiply $6ab$ and $2a^2b$.

Solution: In multiplication, those factors with the same base are combined through the addition of their respective exponents.

$(6ab)(2a^2b)$	(The initial problem)
$12aba^2b$	(The numerical coefficients are multiplied.)
$12a^{1+2}b^{1+1}$	(The exponents are summed.)
$12a^3b^2$	(Final expression)

Example: Multiply $-8x^4y^3z$ and $5xy^3z^2$.

Solution: $(-8x^4y^3z)(5xy^3z^2) = -40x^5y^6z^3$

Example: Simplify $(15m^3n^{-2})(7m^2n)(m^{-6}n^5)$.

Solution: $(15m^3n^{-2})(7m^2n)(m^{-6}n^5) = 105m^{-1}n^4 = \dfrac{105n^4}{m}$

Example: Simplify $(6x^2y^3z)(5xy^4z^2)^3$.

Solution: $(6x^2y^3z)(5xy^4z^2)^3 = (6x^2y^3z)(125x^3y^{12}z^6) = 750x^5y^{15}z^7$

Multiplication of a Monomial and a Polynomial

The following examples will demonstrate the process by which a monomial is multiplied by a polynomial.

Example: Multiply $3x$ and the quantity $(x + y)$.

Solution:

$3x(x + y)$ (The initial problem)

$3x(x + y)$ (The monomial is multiplied by *each* term in the binomial.)

$3x^2 + 3xy$ (The final expression)

Example: Multiply $-4a^2b$ and the quantity $(7a^2b^2 + 3a^2b - 8ab^2 - 7)$.

Solution:

$$(-4a^2b)(7a^2b^2 + 3a^2b - 8ab^2 - 7)$$
$$= (-4a^2b)(7a^2b^2) + (-4a^2b)(3a^2b) - (-4a^2b)(8ab^2) - (-4a^2b)(7)$$
$$= -28a^4b^3 - 12a^4b^2 + 32a^3b^3 + 28a^2b$$

Example: Simplify $(2m^2n)^3\,(4m^2n^2 - mn^2 + 3m^2n^4)$.

Solution:

$$(2m^2n)^3\,(4m^2n^2 - mn^2 + 3m^2n^4)$$
$$= (8m^6n^3)(4m^2n^2 - mn^2 + 3m^2n^4)$$
$$= (8m^6n^3)(4m^2n^2) - (8m^6n^3)(mn^2) + (8m^6n^3)(3m^2n^4)$$
$$= 32m^8n^5 - 8m^7n^5 + 24m^8n^7$$

Multiplication of Polynomials

The following examples will demonstrate the process by which a polynomial is multiplied by a polynomial.

Example: Multiply $(m + n)$ and $(3m^2n - m)$.

Solution:

$(m + n)(3m^2n - m)$ (The initial problem)

$(m + n)(3m^2n - m)$ (*Each* term in the first expression is multiplied by *each* term in the second expression.)

$(m)(3m^2n) - (m)(m) + (n)(3m^2n) - (n)(m)$

$3m^3n - m^2 + 3m^2n^2 - mn$ (The final expression)

Example: Simplify $(4a^2b + b^2)(6a^2b^2 - 3ab + 4)$.

Solution:

$$(4a^2b + b^2)(6a^2b^2 - 3ab + 4)$$
$$= (4a^2b)(6a^2b^2) - (4a^2b)(3ab) + (4a^2b)(4) + (b^2)(6a^2b^2) - (b^2)(3ab) + (b^2)(4)$$
$$= 24a^4b^3 - 12a^3b^2 + 16a^2b + 6a^2b^4 - 3ab^3 + 4b^2$$

Example: Simplify $(x + 4y)^2(3x - y)$.

Solution:

$$(x + 4y)^2(3x - y)$$
$$= [(x + 4y)(x + 4y)](3x - y)$$
$$= [(x)(x) + (x)(4y) + (4y)(x) + (4y)(4y)](3x - y)$$
$$= (x^2 + 4xy + 4xy + 16y^2)(3x - y)$$
$$= (x^2 + 8xy + 16y^2)(3x - y)$$
$$= (3x)(x^2) + (3x)(8xy) + (3x)(16y^2) - (y)(x^2) - (y)(8xy) - (y)(16y^2)$$
$$= 3x^3 + 24x^2y + 48xy^2 - x^2y - 8xy^2 - 16y^3$$
$$= 3x^3 + 23x^2y + 40xy^2 - 16y^3$$

Division of Monomials

The following examples will demonstrate the use of the laws of exponents to divide monomials.

Example: Divide $18x^4y^2z$ by $3x^6yz^3$.

Solution: $\dfrac{18x^4y^2z}{3x^6yz^3} = \dfrac{6y}{x^2z^2}$

Example: Divide $(8a^{-4}bc^2)^2$ by $(2a^{-2}c)^5$.

Solution:

$$\frac{(8a^{-4}bc^2)^2}{(2a^{-2}c)^5}$$

$$= \frac{64a^{-8}b^2c^4}{32a^{-10}c^5}$$

$$= \frac{64a^{10}b^2c^4}{32a^8c^5}$$

$$= \frac{2a^2b^2}{c}$$

Example: Simplify $\dfrac{(24x^2y^3z)(6y^4z^2)^2}{(3x^2yz^4)^3}$.

Solution:

$$\frac{(24x^2y^3z)(6y^4z^2)^2}{(3x^2yz^4)^3}$$

$$= \frac{(24x^2y^3z)(36y^8z^4)}{27x^6y^3z^{12}}$$

$$= \frac{864x^2y^{11}z^5}{27x^6y^3z^{12}}$$

$$= \frac{32y^8}{x^4z^7}$$

Division of a Polynomial by a Monomial

In order to divide a polynomial by a monomial, *each* term of the polynomial is divided by the monomial.

Example: Divide $(24x^4a^3 + 8x^2a^2)$ by $2xa$.

Solution:

$$\frac{24x^4a^3 + 8x^2a^2}{2xa}$$

$$= \frac{24x^4a^3}{2xa} + \frac{8x^2a^2}{2xa}$$

$$= 12x^3a^2 + 4xa$$

Example: Divide $(48m^3n^6 - 20mn^8)$ by $4m^2n^7$.

Solution:

$$\frac{48m^3n^6 - 20\,mn^8}{4m^2n^7}$$
$$= \frac{48m^3n^6}{4m^2n^7} - \frac{20mn^8}{4m^2n^7}$$
$$= \frac{12m}{n} - \frac{5n}{m}$$

Example: Divide $(18a^2b^{-2}c + 24b^4c^5 - 4a^4bc^{-5})$ by $12a^3b^2c$.

Solution:

$$\frac{18a^2b^{-2}c + 24b^4c^5 - 4a^4bc^{-5}}{12a^3b^2c}$$
$$= \frac{18a^2b^{-2}c}{12a^3b^2c} + \frac{24b^4c^5}{12a^3b^2c} - \frac{4a^4bc^{-5}}{12a^3b^2c}$$
$$= \frac{3}{2ab^4} + \frac{2b^2c^4}{a^3} - \frac{a}{3bc^6}$$

Example: Simplify this algebraic expression.

$$\frac{(5x^2y)^3(x + y)}{(5x)(5xy^3)}$$

Solution:

$$\frac{(5x^2y)^3(x + y)}{(5x)(5xy^3)}$$
$$= \frac{(125x^6y^3)(x + y)}{25x^2y^3} \quad \text{or} \quad = \frac{125x^6y^3\,(x + y)}{25x^2y^3}$$
$$= 5x^4(x + y) \qquad\qquad = \frac{125x^7y^3 + 125x^6y^4}{25x^2y^3}$$
$$= 5x^5 + 5x^4y \qquad\qquad = 5x^5 + 5x^4y$$

Note in the previous example that the numerator was considered a monomial, therefore allowing direct division by the denominator. This is possible because the expression $(x + y)$ is a separate factor being multiplied by the other factors. It is also correct to simplify the numerator, which becomes a binomial, and then perform the division.

Division of a Polynomial by a Polynomial

The process of dividing a polynomial by a polynomial is similar to that of long division. The following rules and examples will demonstrate this process.

Example: Divide $(10x + x^2 + 24)$ by $x + 4$.

Solution:

1. Arrange both expressions in descending order according to the unknown and the associated exponents:
   $x^2 + 10x + 24 \div x + 4$
2. Place both expressions in the standard division format:

$$x + 4\,\overline{)\,x^2 + 10x + 24}$$

3. Divide the first term of the dividend by the first term of the divisor.

$$x + 4 \overline{\smash{)}\, x^2 + 10x + 24} \quad \text{quotient: } x$$

4. Multiply this first term in the quotient by each term in the divisor. Place these products under similar terms in the dividend and subtract.

$$\begin{array}{r l} & x \\ x + 4 \,\big) & x^2 + 10x + 24 \\ & -(x^2 + 4x) \\ \hline & 6x \end{array}$$

(Subtract, change signs and add)

5. Bring down the next term in the dividend and divide the first term of this new expression by the first term of the divisor.

$$\begin{array}{r l} & x + 6 \\ x + 4 \,\big) & x^2 + 10x + 24 \\ & -(x^2 + 4x) \\ \hline & 6x + 24 \end{array}$$

6. Multiply this second term in the quotient by the divisor. Place these products under similar terms and subtract.

$$\begin{array}{r l} & x + 6 \\ x + 4 \,\big) & x^2 + 10x + 24 \\ & -(x^2 + 4x) \\ \hline & 6x + 24 \\ & -(6x + 24) \\ \hline \end{array}$$

(No remainder)

7. To check this answer, the quotient can be multiplied by the divisor. This should result in a product identical to the dividend.

$$(x + 4)(x + 6)$$
$$= x^2 + 6x + 4x + 24$$
$$= x^2 + 10x + 24$$

Example: Divide $3a^3 - 16a^2 - 29a - 50$ by $a - 7$.

Solution:

$$\begin{array}{r l} & 3a^2 + 5a + 6 \\ a - 7 \,\big) & 3a^3 - 16a^2 - 29a - 50 \\ & -(3a^3 - 21a^2) \\ \hline & 5a^2 - 29a \\ & -(5a^2 - 35a) \\ \hline & 6a - 50 \\ & -(6a - 42) \\ \hline & -8 \end{array}$$

(Remainder)

Answer: $3a^2 + 5a + 6$ + a remainder of -8.

Check: $(a - 7)(3a^2 + 5a + 6) - 8$
$= (3a^3 + 5a^2 + 6a - 21a^2 - 35a - 42) - 8$
$= 3a^3 - 16a^2 - 29a - 50$

EXERCISE 4·4

Simplify each of the following problems. Answers should contain only positive exponents and should not contain parentheses.

1. $(12x^2y)(3x^4y^2)$
2. $(-9a^{-4}b^3)(6ab^{-2})(5a^4b^4)$
3. $(-3x^4y^{-2})^{-4}$
4. $[(m^{-3}n^4)(m^{-2}n^2)]^{-4}$
5. $4b^2(3b - 9b^4)$
6. $5xy^{-3}(-8x^2y^2 + 4xy^2 + x^2y - 12)$
7. $(a - b)(a + b)^2$
8. $x^2y^3(18x^{-2} - 2x^2y^2 + 7y^5)$
9. $(8a^2b)^{-1}(16a^4b^3 + 12ab^{-5})$
10. $(3a - b)(6a^2b^2 - 11a^2b + 7ab^2 - 19)$
11. $(5a - 2b)(5a + 2b)$
12. $(7x^2y)^2(4xy^2 - 2xy)^2$
13. $\dfrac{(27x^4b^{-3}a^2)(16x^{-5}ba)^2}{(3xba)^3}$
14. $\left[\dfrac{6a^{-4}b^3c^{-2}}{(4ab^{-3}c^4)^{-2}}\right]^3$
15. $(3x^3y^5)^2\dfrac{48x^4y^2z}{12x^{-3}z^{-5}}$
16. $\dfrac{(4a - b)^2}{2ab}$
17. $\dfrac{132x^6y^3 - (4x^2 - y^3)^2}{12x^2y^2}$
18. $\dfrac{(7x - 3y)(x^2 + y^2)}{x^4y}$
19. $\dfrac{(2ab)(4a - b)(3a + 2b)}{(6b^2)(4ab^3)}$
20. $\dfrac{(5xy)^2(2x + y^2)(2y + x^2)}{(5x^2y^2)^2}$
21. $\dfrac{3x^2 - 19x - 14}{x - 7}$
22. $\dfrac{5a^3 + 8a^2 - 58a - 40}{a + 4}$
23. $\dfrac{24a^5 - 74a^4 + 34a^3 - 34a^2 + 16a - 2}{6a - 2}$
24. $\left(\dfrac{8x^3 - 28x^2y + 8xy^2 + 12y^3}{2x + y}\right) \div (4x - 4y)$

UNIT 4·5 SPECIAL PRODUCTS AND FACTORING

Objectives:

After studying this unit, you should be able to

- recognize special algebraic products.
- utilize special products to factor algebraic expressions.

Special Products

In the study of algebra there are special products which occur frequently. Understanding and remembering these relationships will prove extremely useful, not only in simplifying algebraic expressions, but also in factoring them. Each of these will be presented, and examples of their use given. You should make every effort to commit these to memory.

Product of a Monomial and Binomial

$$m(x + y) = mx + my \qquad (4\text{–}15)$$

Example: Simplify (a) $-5(x + y)$, (b) $4a(a - b)$, and (c) $3x(x^2 - x)$.

Solution:
a. $-5(x + y) = -5x - 5y$
b. $4a(a - b) = 4a^2 - 4ab$
c. $3x(x^2 - x) = 3x^3 - 3x^2$

A Binomial Squared

$$(x + y)^2 = (x + y)(x + y) = x^2 + xy + yx + y^2 = x^2 + 2xy + y^2 \quad (4\text{–}16)$$

$$(x - y)^2 = (x - y)(x - y) = x^2 - xy - yx + y^2 = x^2 - 2xy + y^2 \quad (4\text{–}17)$$

To square a binomial, square the first term, add two times the product of the terms, and then add the square of the second term.

Example: Simplify each of the following.
a. $(x + 4)^2$
b. $(3x - 2y)^2$
c. $(5a - b)^2$
d. $(-3 + y)^2$

Solution:
a. $(x + 4)^2 = (x + 4)(x + 4) = x^2 + 4x + 4x + 16 = x^2 + 8x + 16$
b. $(3x - 2y)^2 = (3x - 2y)(3x - 2y) = 9x^2 - 6xy - 6xy + 4y^2 = 9x^2 - 12xy + 4y^2$
c. $(5a - b)^2 = (5a - b)(5a - b) = 25a^2 - 5ab - 5ab + b^2 = 25a^2 - 10ab + b^2$
d. $(-3 + y)^2 = (-3 + y)(-3 + y) = 9 - 3y - 3y + y^2 = 9 - 6y + y^2$

The Product of the Sum and Difference of Two Numbers

$$(x + y)(x - y) = x^2 - xy + xy - y^2 = x^2 - y^2 \quad (4\text{–}18)$$

Note that this product results in the difference of the two terms squared.

Example: Simplify each of the following.
a. $(x + 4)(x - 4)$
b. $(3a + 3b)(3a - 3b)$
c. $(2 + 5x)(2 - 5x)$
d. $(\sqrt{x} + \sqrt{y})(\sqrt{x} - \sqrt{y})$

Solution:
a. $(x + 4)(x - 4) = x^2 - 4x + 4x - 4^2 = x^2 - 16$
b. $(3a + 3b)(3a - 3b) = 9a^2 - 9b^2$
c. $(2 + 5x)(2 - 5x) = 4 - 25x^2$
d. $(\sqrt{x} + \sqrt{y})(\sqrt{x} - \sqrt{y}) = x - y$

The Product of Two Binomials with No Similar Terms

$$(x + y)(m + n) = xm + xn + ym + yn \quad (4\text{–}19)$$

Example: Simplify each of the following.

a. $(x + 5)(y - 7)$
b. $(3 - x)(a + 4)$
c. $(4a - 6)(2 - b)$

Solution: a. $(x + 5)(y - 7) = xy - 7x + 5y - 35$
b. $(3 - x)(a + 4) = 3a + 12 - ax - 4x$
c. $(4a - 6)(2 - b) = 8a - 4ab - 12 + 6b$

The Product of Two Binomials with One Similar Term.

$$(x + m)(x + n) = x^2 + xn + xm + mn = x^2 + x(m + n) + mn \quad (4\text{–}20)$$

Example: Simplify each of the following.

a. $(x + 5)(x - 3)$
b. $(2m - a)(2m + b)$
c. $(x - 1)(x - 2)$
d. $(6 - x)(6 - y)$

Solution: a. $(x + 5)(x - 3) = x^2 + x(5 - 3) - 15 = x^2 + 2x - 15$
b. $(2m - a)(2m + b) = 4m^2 + 2m(-a + b) - ab =$
$4m^2 - 2ma + 2mb - ab$
c. $(x - 1)(x - 2) = x^2 + x(-1 - 2) + 2 = x^2 - 3x + 2$
d. $(6 - x)(6 - y) = 36 + 6(-x - y) + xy = 36 - 6x - 6y + xy$

The Binomial Cubed

$$\begin{aligned}(x + y)^3 &= (x + y)(x + y)(x + y)\\ &= (x + y)(x^2 + 2xy + y^2)\\ &= x^3 + 2x^2y + xy^2 + x^2y + 2xy^2 + y^3\\ &= x^3 + 3x^2y + 3xy^2 + y^3\\ (x - y)^3 &= (x - y)(x - y)(x - y)\\ &= (x - y)(x^2 - 2xy + y^2)\\ &= x^3 - 2x^2y + xy^2 - x^2y + 2xy^2 - y^3\\ &= x^3 - 3x^2y + 3xy^2 - y^3\end{aligned}$$

To cube a binomial, cube the first term, add three times the first term squared times the second term, add three times the second term squared times the first term, and add the second term cubed.

$$(x + y)^3 = x^3 + 3x^2y + 3xy^2 + y^3 \quad (4\text{–}21)$$

$$(x - y)^3 = x^3 - 3x^2y + 3xy^2 - y^3 \quad (4\text{–}22)$$

Example: Simplify each of the following.

a. $(x + 4)^3$
b. $(3a + 2)^3$
c. $(1 - x)^3$

Solution: a. $(x + 4)^3 = x^3 + (3)(x^2)(4) + (3)(4^2)(x) + 4^3$
$= x^3 + 12x^2 + 48x + 64$

b. $(3a + 2)^3 = (3a)^3 + (3)(3a)^2(2) + (3)(2^2)(3a) + 2^3$
$= 27a^3 + 54a^2 + 36a + 8$

c. $(1 - x)^3 = 1^3 + (3)(1^2)(-x) + (3)(-x^2)(1) + (-x^3)$
$= 1 - 3x + 3x^2 - x^3$

Factoring Common Terms

When simplifying algebraic expressions it is often useful to break the expression into prime *factors*. The factor evenly divides the original expression. By factoring, it is possible to reverse the process described in the special products just presented. Being able to identify special products, and factoring them out of an expression, will be extremely useful in solving equations.

The first step in factoring an expression is to inspect each term and factor out any factor(s) common to each. These common factors could be numerical coefficients or literal factors. Carefully follow these examples.

Example: Factor $8x^4yz - 12x^3y^3 + 20x^2y^2z^3$.

Solution: As the terms are inspected each has a numerical coefficient, an x, and a y. Each numerical coefficient is factored into its prime factors,

$8 = 2 \times 2 \times 2$
$12 = 2 \times 2 \times 3$
$20 = 2 \times 2 \times 5$

The factors common to each coefficient are 2×2, or 4. Each term in the original expression is divided by 4, resulting in

$$\frac{8x^4yz}{4} - \frac{12x^3y^3}{4} + \frac{20x^2y^2z^3}{4} = 4(2x^4yz - 3x^3y^3 + 5x^2y^2z^3)$$

In terms of x and y, each is factored out only the number of times it is present in *each* term. In the case of x, since x^2 appears in the third term, only x^2 can be factored out of all three terms. As y is raised to the first power, only y can be factored out of all three terms.

$$4\left(\frac{2x^4yz}{x^2y} - \frac{3x^3y^3}{x^2y} + \frac{5x^2y^2z^3}{x^2y}\right)$$

$= 4x^2y\ (2x^2z - 3xy^2 + 5yz^3)$

As no other factors are common to the three terms, the expression has been factored. You should verify this answer by multiplying the monomial by the polynomial to give the original expression.

Example: Factor $6a^4b^2c + 3a^5b^3c^2 - 15a^4bc$.

Solution: The common terms are 3, a^4, b, and c, giving

$$\frac{6a^4b^2c}{3a^4bc} + \frac{3a^5b^3c^2}{3a^4bc} - \frac{15a^4bc}{3a^4bc}$$

$= 3a^4bc\ (2b + ab^2c - 5)$

Factoring Special Products

Once common factors have been factored out further simplification may still be possible through recognition of special products. Establishing a strict procedure for factoring is difficult due to the numerous combinations of terms. It is often helpful to ask a series of questions and, depending upon the answers, factor accordingly.

Is the expression a binomial, trinomial, or polynomial? If the response is binomial, then is it the product of the sum and difference of two numbers? Inspect these examples.

Example: Factor $5a^2 - 45$.

Solution: Factor all common factors:
$5(a^2 - 9)$
Does each term have a square root?
$\sqrt{a^2} = a, \ \sqrt{9} = 3$
Writing the sum and difference gives:
$5(a + 3)(a - 3)$

Example: Factor $49 - x^2$.

Solution: $49 - x^2 = (7 - x)(7 + x)$

Assuming the answer to the original question was trinomial, then the alternatives would include a binomial squared and the product of two binomials with one similar term.

The key in determining if the expression is a binomial squared is to examine the first and third terms. Both of these will be positive and perfect squares. If this is the case, then take the square roots to determine the two terms of the binomial. Determination of the signs can be made by inspecting the sign of the second term of the trinomial. If this sign is positive then the two terms are either both positive or both negative. If this sign is negative then one term is positive and one is negative.

Example: Factor $a^2 + 12a + 36$.

Solution: Find the square roots of the first and third terms.
$\sqrt{a^2} = a$ and $-a$
$\sqrt{36} = 6$ and -6
Place the factors in the parentheses.
$(a \quad 6)\ (a \quad 6)$
Since the second term is positive then the expressions are
$(a + 6)(a + 6)$ or $(-a - 6)(-a - 6)$
Note: Both of these expressions are equal. Examine the following comparison.
$(-a - 6) = -1(a + 6)$, and

$$
\begin{aligned}
(-a - 6)(-a - 6) &= -1(a + 6) \times -1(a + 6) \\
&= -1 \times -1 \times (a + 6)(a + 6) \\
&= 1 \times (a + 6)(a + 6) \\
&= (a + 6)(a + 6)
\end{aligned}
$$

Example: Factor $32x^3 - 16x^2y + 2xy^2$.

Solution: Factor common factors:

$2x(16x^2 - 8xy + y^2)$

Taking square roots yields

$2x(4x \quad y)(4x \quad y)$

Since the second term is negative

$2x(4x - y)(4x - y) = 2x(4x - y)^2$

Example: Factor $9x^4 - 6x^2 + 1$.

Solution: $(3x^2 - 1)(3x^2 - 1)$

If, in examining the first and third terms of the trinomial, one is not a perfect square then the expression may be the product of two binomials with one similar term. Re-examine the standard format for this product:

$$(x + m)(x + n) = x^2 + x(m + n) + mn$$

Note that the first term is easily identified, as it will be a perfect square. To determine the values of m and n inspect the third term. This value must be the product of m and n. The numerical coefficient of the second term is the sum of m and n. Follow this example:

Example: Factor $y^2 + 9y + 18$.

Solution: Set up the standard format and insert the $\sqrt{y^2}$

$(y \quad)(y \quad)$

The product of m and n must be 18, and the sum is 9. Factoring 18 gives:

$18 = 1 \times 18$

$\quad = 2 \times 9$

$\quad = 3 \times 6$

and $1 + 18 = 19$

$\quad 2 + 9 = 11$

$\quad 3 + 6 = 9$

so that $m = 3$ and $n = 6$,

$(y \quad 3)(y \quad 6)$

To determine the signs follow this reasoning:

- If mn is positive, then m and n are both positive or both negative.
- If mn is negative, then m or n is negative, but not both. Therefore:
- If mn is positive, and the second term is positive, then m and n are positive. The reverse is true if the second term is negative.
- If mn is negative then the magnitude of the second term will indicate which term is positive and which is negative.

So the resultant in this example is:

$(y + 3)(y + 6)$

Example: Factor $2a^2 + 2a - 24$.

Solution: Factor common factors
$2(a^2 + a - 12)$
Standard format
$2\,(a \quad)(a \quad)$
Factor -12:
$-12 = 1 \times -12$ or -1×12
$-12 = 2 \times -6$ or -2×6
$-12 = 3 \times -4$ or -3×4
Of the six combinations above only one sums to $+1$, therefore the factored expression is
$2(a - 3)(a + 4)$

Example: Factor $-x^2 - 3x + 28$.

Solution: Since there are no common factors place the unknown in the standard format. Obviously one x must be negative.
$(x \quad)(-x \quad)$
Factor 28
$28 = 1 \times 28$ or -1×-28
$28 = 2 \times 14$ or -2×-14
$28 = 4 \times 7$ or -4×-7
The sum (or difference) of the two numbers is -3. Keeping in mind that multiplication by $-x$ will change the sign of the number in the other parentheses, the logical choice is 4 and 7. Will it make a difference where these numbers are placed? Try both possibilities:
$(x + 4)(-x + 7) = -x^2 + 3x + 28$ (incorrect)
$(x + 7)(-x + 4) = -x^2 - 3x - 28$ (correct)

The previous examples have demonstrated methods of factoring trinomial expressions. In the event the expression has four terms, then among the factorable alternatives are the product of two binomials with no common terms, and a binomial cubed.

The factoring of the product of two binomials with no common terms can usually be completed by inspection of the terms.

Example: Factor $mn - 3m + 7n - 21$.

Solution: Set up the standard parentheses. It is obvious that m and n are in different binomials and are either both positive or both negative.
$(m \quad)(n \quad)$
The numbers 7 and 3 are the two other terms in the binomials. The term $-3m$ indicates that the three is negative and in the binomial with n.
$(m + 7)(n - 3)$
To check, $(m + 7)(n - 3) = mn - 3m + 7n - 21$

Example: Factor $-8x^2y + 8x^2 - 16xy + 16x$.

Solution: Factoring common terms gives

$8x(-xy + x - 2y + 2)$

Inspection indicates that the numerical terms are 1 and 2.

$8x(x \quad 2)(y \quad 1)$

Since xy is negative, the 1 and 2 are positive, and if $1x$ is positive then y must be negative

$$8x(x + 2)(-y + 1) = 8x(-xy + x - 2y + 2)$$
$$= -8x^2y + 8x^2 - 16xy + 16x$$

The factoring of a binomial cubed is similar to that of a binomial squared. Place the terms in descending order according to exponent and take the cube root of the first and last term. Use the respective signs.

Example: Factor $a^3 + 15a^2 + 75a + 125$.

Solution: Find the cube roots of the first and fourth terms:

$\sqrt[3]{a^3} = a$

$\sqrt[3]{125} = 5$

The answer is $(a + 5)^3$

Example: Factor $27 - 27x + 9x^2 - x^3$.

Solution: Find the cube roots of the first and fourth terms:

$\sqrt[3]{27} = 3$

$\sqrt[3]{-x^3} = -x$

The answer is $(3 - x)^3$

Recognizing special products and factoring are often very useful in the simplification of algebraic expressions. This will become more apparent in the study of algebraic fractions and equations. It should also be pointed out that not all algebraic expressions can be factored. After applying the techniques presented in this unit one may determine that a particular expression cannot be factored, or simplified, without additional information.

EXERCISE 4·5

Simplify each of the following.

1. $3a(x - 4)$
2. $-6b(b - c)$
3. $y(-3y^4 + 2y^2 - y + 1)$
4. $12z(x - 2z)$
5. $4x(1 - x)^2$
6. $5y^4(y^3 + y^2)^2$
7. $a(-a - b)^2$
8. $-8m(n^3 - m^3)^2$
9. $(y + 7)(y - 7)$
10. $12x(x - 1)(x + 1)$
11. $(5x^2 + 9)(5x^2 - 9)$
12. $[(x + 3) + (a - 5)][(x + 3) - (a - 5)]$
13. $8(7 - x)(y + 4)$
14. $(x + 4)(y - 4)$
15. $(6x^2 + y)(3x - 10)$
16. $12b^2(b^4 + b^3)(b^2 - b)$
17. $(a - 2)(a + 7)$
18. $(6x - 3y)(3y + x)$

19. $(m + 3)(3 - m)$
20. $4(a + 10)(a + 18)$
21. $(x - 4)^2(x - 4)$
22. $(3 - m)^3$
23. $12(3a + 4)^3$
24. $2[6x(x - 1)]^3$
25. $(x^3 - x^2)^3$

Factor each of the following.

26. $4m^3n^2 - 12m^4n - 6m^2n^2$
27. $(3a^4b^2c)^3 + 9a^3b^2c^2 - 15a^5b^3c^2$
28. $48a^4 - 3a^2$
29. $3x^6 - 3x^4$
30. $81m^4n^2 - 1$
31. $x^2 - 1$
32. $x^2 + 4x + 4$
33. $49 - 14x + x^2$
34. $18a^2 - 12a + 2$
35. $64b^3 + 48b^2 + 9b$
36. $3y^4 - 6y^3 + 3y^2$
37. $x^2 + 2x - 3$
38. $-w^2 + 2w + 8$
39. $(5a^2)^2 - 15a^2 - 18$
40. $-2x^2y - 10xy - 12y$
41. $ab - a - b + 1$
42. $24 - 6x + 4y - xy$
43. $b^3 + 3b^2 + 3b + 1$
44. $x^3 - 12x^2 + 48x - 64$
45. $27x^6 + 27x^5 + 9x^4 + x^3$

UNIT 4·6 ADDITION AND SUBTRACTION OF ALGEBRAIC FRACTIONS

Objectives:

After studying this unit, you should be able to

- simplify algebraic fractions.
- add algebraic fractions.
- subtract algebraic fractions.

Adding and Subtracting Fractions

The addition and subtraction of algebraic fractions is similar to the addition and subtraction of fractions in arithmetic. In order to demonstrate this similarity review the following example.

Example: Combine $\frac{3}{4} + \frac{7}{12} - \frac{1}{16}$.

Solution: The first step is to determine the lowest common denominator (LCD) that the individual fractions have in common. Each denominator is separated into prime factors as follows:

$4 = 2 \times 2 \qquad\qquad = 2^2$

$12 = 2 \times 2 \times 3 \qquad = 2^2 \times 3^1$

$16 = 2 \times 2 \times 2 \times 2 = 2^4$

The LCD is the product of each factor raised to the highest power which occurs in any one expression. In this example there are two factors, with the highest power of 2 being the fourth power and the highest power of 3 being the first power. The LCD is:

$(2^4)(3^1) = 16 \times 3 = 48$

Converting each fraction to the LCD gives:

$$\frac{3}{4} \times \frac{48}{48} = \frac{36}{48}$$

$$\frac{7}{12} \times \frac{48}{48} = \frac{28}{48}$$

$$\frac{1}{16} \times \frac{48}{48} = \frac{3}{48}$$

Combining fractions gives:

$$\frac{36}{48} + \frac{28}{48} - \frac{3}{48} = \frac{61}{48} = 1\frac{13}{48}$$

Adding and Subtracting Algebraic Fractions

Addition and subtraction of literal algebraic fractions follows the same process. Carefully examine each step in the following examples.

Example: Add $\frac{a}{b}$ and $\frac{x}{y}$.

Solution: With only one term in each denominator the LCD is by. As the final result will have by as a denominator, the numerator terms are found by multiplying each fraction by the LCD.

$$by\left(\frac{a}{b} + \frac{x}{y}\right) = \frac{aby}{b} + \frac{xby}{y} = ay + xb$$

The sum of the two fractions is:

$$\frac{a}{b} + \frac{x}{y} = \frac{ay + xb}{by}$$

Example: Simplify $\frac{5}{a + b} - \frac{3}{a - b} + \frac{a}{a^2 - b^2}$.

Solution:

$$\frac{5}{a + b} - \frac{3}{a - b} + \frac{a}{a^2 - b^2}$$

$$= \frac{5}{a + b} - \frac{3}{a - b} + \frac{a}{(a + b)(a - b)}$$

$$= (a + b)(a - b)\left[\frac{5}{a + b} - \frac{3}{a - b} + \frac{a}{(a + b)(a - b)}\right]$$

$$= \frac{5(a - b) - 3(a + b) + a}{(a + b)(a - b)}$$

$$= \frac{5a - 5b - 3a - 3b + a}{(a + b)(a - b)}$$

$$= \frac{3a - 8b}{(a + b)(a - b)}$$

Example: Simplify $\frac{8x^2}{x^2 + x - 6} + \frac{x}{x^2 + 6x + 9} - \frac{(x - 2)^2}{x - 1}$.

Solution: $\dfrac{8x^2}{x^2 + x - 6} + \dfrac{x}{x^2 + 6x + 9} - \dfrac{(x - 2)^2}{x - 1}$

$$= \frac{8x^2}{(x + 3)(x - 2)} + \frac{x}{(x + 3)^2} - \frac{(x - 2)^2}{(x - 1)}$$

$$= (x + 3)^2(x - 2)(x - 1)\left[\frac{8x^2}{(x + 3)(x - 2)} + \frac{x}{(x + 3)^2} - \frac{(x - 2)^2}{(x - 1)}\right]$$

$$= \frac{8x^2(x + 3)(x - 1) + x(x - 2)(x - 1) - (x - 2)^2(x + 3)^2(x - 2)}{(x + 3)^2(x - 2)(x - 1)}$$

$$= \frac{-x^5 + 8x^4 + 32x^3 - 37x^2 - 58x + 72}{(x + 3)^2(x - 2)(x - 1)}$$

EXERCISE 4·6

Perform the indicated operations in each of the following problems.

1. $\dfrac{3}{8} + \dfrac{2}{3} - \dfrac{5}{6}$
2. $4\dfrac{1}{3} - \dfrac{4}{15} - \dfrac{23}{30}$
3. $\dfrac{3x}{a} - \dfrac{5x}{b}$
4. $\dfrac{12b}{x - 1} - \dfrac{16b}{1 - x}$
5. $\dfrac{4y}{y^2 - 1} + \dfrac{y + 1}{y - 1}$
6. $\dfrac{x^2 + x}{x^2 - 4x - 5} - \dfrac{x^2 + 6x + 8}{x^2 + x - 2} - (x + 4)$
7. $\dfrac{3a - b}{a^2 + 2ab + b^2} + \left[\dfrac{2a - 4b}{a - b}\right]^2$
8. $1 - \dfrac{6m}{m^2 + 9} - \dfrac{m - 3}{m + 3}$
9. $\dfrac{6x^2}{5(x - 2)} + \dfrac{x(x - 1)}{2x^2 - 8} - \dfrac{1}{x^2 - 4x + 4}$
10. $\dfrac{(y + 2)(y - 3)}{y^2 - 5y + 6} - (y + 2)$
11. $\dfrac{a(a + 4)^2}{3a^3 + 9a^2 - 12a} - \dfrac{a - 1}{a + 4} + 1$
12. $\dfrac{1}{16 - x^2} + \dfrac{4x^2 + 16x}{2x^2 - 2}$
13. $1 - \dfrac{3(m + 2)^2}{3m^2 - 12m - 36} + \dfrac{m - 6}{m + 2}$
14. $\dfrac{5}{a - 7} + \dfrac{7}{a - 5} - \dfrac{a - 7}{a - 5}$

15. $\dfrac{x^2 - 2x + 1}{x^2 + 2x + 1} + \dfrac{x + 1}{x - 1}$

16. $\dfrac{4b + 2}{2b^2 - 7b - 4} + \dfrac{b}{b^2 - 16} - \dfrac{3b + 6}{3b^2 + 6b}$

17. $\dfrac{x - y}{x + y} + \dfrac{y - x}{x - y} + \dfrac{x + y}{y - x}$

18. $\dfrac{3n^2}{m^2} - \dfrac{4n}{2mn} - \dfrac{m}{n}$

19. $\dfrac{a^2 + 3a + 2}{4a^2 + 3a - 1} - \dfrac{1}{4a - 4}$

20. $\dfrac{2x - 1}{x + 2} - \dfrac{x + 2}{x - 1} + \dfrac{3x - 5}{-3x + 5} + 1$

UNIT 4·7 MULTIPLICATION AND DIVISION OF ALGEBRAIC FRACTIONS

Objectives:

After studying this unit, you should be able to

- multiply algebraic fractions.
- divide algebraic fractions.

Multiplying Algebraic Fractions

The multiplication and division of algebraic fractions is similar to multiplying and dividing fractions in arithmetic. In order to multiply algebraic fractions simply multiply the numerators and then multiply the denominators.

Example: Multiply $\dfrac{3x^2}{5b}$ times $\dfrac{b}{4x}$.

Solution: $\dfrac{3x^2}{5b} \cdot \dfrac{b}{4x} = \dfrac{3x^2b}{5b4x} = \dfrac{3x^2b}{20bx} = \dfrac{3x}{20}$

Example: Multiply $\dfrac{y^2 - 2y - 8}{y + 1}$ times $\dfrac{y^2 - 1}{y^2 + 3y + 2}$.

Solution:

$$\frac{y^2 - 2y - 8}{y + 1} \cdot \frac{y^2 - 1}{y^2 + 3y + 2}$$

$$= \frac{(y + 2)(y - 4)(y + 1)(y - 1)}{(y + 1)(y + 2)(y + 1)}$$

$$= \frac{(y - 4)(y - 1)}{(y + 1)}$$

Example: Multiply $\dfrac{x^2 + 2x - 15}{x^2 + 9x + 20}$ times $\dfrac{x^2 + 5x + 4}{x^2 - 2x - 3}$.

Solution:
$$\frac{x^2 + 2x - 15}{x^2 + 9x + 20} \cdot \frac{x^2 + 5x + 4}{x^2 - 2x - 3}$$
$$= \frac{(x + 5)(x - 3)(x + 4)(x + 1)}{(x + 5)(x + 4)(x - 3)(x + 1)}$$
$$= 1$$

Dividing Algebraic Fractions

In order to divide algebraic fractions the denominator is inverted and multiplied by the numerator. Typically, any factoring is completed prior to inverting and multiplying.

INVERT THEN MULTIPLY

Example: Divide $\frac{x^2 - 1}{3x}$ by $\frac{x^2 - 2x + 1}{x}$.

Solution:
$$\frac{x^2 - 1}{3x} \div \frac{x^2 - 2x + 1}{x}$$
$$= \frac{(x + 1)(x - 1)}{3x} \div \frac{(x - 1)(x - 1)}{x}$$
$$= \frac{(x + 1)(x - 1)}{3x} \cdot \frac{x}{(x - 1)(x - 1)}$$
$$= \frac{x + 1}{3(x - 1)}$$

Example: Simplify $\left[\frac{x^2 - y^2}{2x^2 - 2xy} \cdot \frac{x + y}{x^2 - 2xy + y^2}\right] \div \frac{4x^3 + 8x^2y + 4xy^2}{x - y}$.

Solution: The first step is to complete any factoring:
$$\left[\frac{(x + y)(x - y)}{2x(x - y)} \cdot \frac{(x + y)}{(x - y)(x - y)}\right] \div \frac{4x(x + y)(x + y)}{(x - y)}$$
$$= \frac{(x + y)^2}{2x(x - y)^2} \cdot \frac{x - y}{4x(x + y)^2}$$
$$= \frac{1}{8x^2(x - y)}$$

EXERCISE 4·7

Perform the indicated operations in the following problems. Answers should contain no parentheses.

1. $\frac{12x^4b^2}{c^4} \cdot \frac{b^4}{3x^3c}$
2. $\frac{2x(x - 3)}{x^2 - 2x - 3} \cdot \frac{x + 1}{x}$
3. $\frac{4a^4x}{b^2c} \cdot \frac{1}{bx^3a^2} \cdot 9b^3x$
4. $\frac{x + y}{x - y} \cdot \frac{x^2 - y^2}{x}$
5. $\left[\frac{a}{b} + \frac{a}{c}\right]2bc$

6. $\left(1 + \dfrac{2x}{x - 1}\right)\left(\dfrac{x + 1}{2x - 1}\right)$

7. $\dfrac{4x^2 - 16}{x + 2} \cdot \dfrac{x - 1}{x - 2}$

8. $\dfrac{3x + 6}{y + 2} \cdot \dfrac{5}{x + 2}$

9. $\dfrac{a^2 + 5a - 14}{a^2 - 7a + 12} \cdot \dfrac{a^2 - 6a + 8}{a^2 - 5a + 6}$

10. $\left(\dfrac{1}{x} + \dfrac{3x - 2}{3y}\right)\left(\dfrac{4y}{x}\right)$

11. $\dfrac{2y^3 - 2y^2}{y + 4} \div \dfrac{2y^2}{y^2 + 3y - 4}$

12. $\dfrac{2a^2 - a - 6}{a^2 - a - 2} \div \dfrac{a - 2}{a + 1}$

13. $\left(\dfrac{1}{a + 1} + \dfrac{a + 1}{a - 1}\right) \div \dfrac{1}{a^2 - 1}$

14. $\dfrac{1}{x - 1} \div \dfrac{1}{x^2 + 4x - 5}$

15. $\dfrac{3b^3 - 15b^2 - 42b}{b^2 + 2b} \div \dfrac{2b^2 - 3b - 14}{b + 2}$

16. $\left[\dfrac{x}{x + 1} + \dfrac{x}{x + 2}\right] \div \dfrac{x + 1}{x + 2}$

17. $\left[\left(\dfrac{1}{a} + \dfrac{1}{b}\right)\dfrac{b}{a}\right] \div \dfrac{a}{b}$

18. $\dfrac{x^2 - 4x + 3}{x^2 + 2x - 3} \cdot \dfrac{x^2 + 7x + 6}{x^2 + 9x + 20} \cdot \dfrac{x^2 + 8x + 15}{x^2 - 1}$

19. $\left[\dfrac{y^4 - 1}{2y + 2} \div \dfrac{y^2 - 3y + 2}{y^2}\right] \cdot \dfrac{y - 2}{y - 1}$

20. $\dfrac{a^2 - 6a + 9}{a^2 + 2a - 3} \div \left[\dfrac{a^2 + a - 12}{a^2 - 1} \div \dfrac{2a^2 + 7a - 4}{6a^2 + a - 2}\right]$

UNIT 4·8 LOGARITHMS

Objectives:

After studying this unit, you should be able to

- perform conversions between exponential and logarithmic equations.
- determine the common logarithm of a number.
- determine the antilogarithm of a number.
- solve problems involving decibels.
- determine the natural logarithm of a number.
- determine the number equivalent to a natural logarithm.
- use natural logarithms to solve electronics problems.

Logarithmic and Exponential Equations

Numbers can be expressed in various ways. A number such as 6 might be expressed as a product, that is, the result of multiplying 3 by 2. A number such as 100 can be expressed as 10 raised to the second power, or 10^2. If you have a number such as 1 000, you can express this as an exponential number. Using the base 10, 1 000 can be written as 10^3 since $10^3 = 1\ 000$. The exponent, 3, is known as the *logarithm* of the number. Calling a number a logarithm is synonymous with naming it an exponent. "The logarithm of 1 000 is 3" is just another way of saying that 3 is the exponent that will produce the number 1 000 with a base of 10.

The logarithm (abbreviated log) is the exponent of 10 when the number is expressed in exponential form with a base of 10. The log of 10 000 is 4 or the exponent required to produce 10 000 using a base of 10. Similarly, the log of 100 000 is 5 and the log of 1 000 000 is 6 using the base 10.

Although base 10 has been specified in these examples, there are other bases. Thus it is possible to have a number such as $2^4 = 2 \times 2 \times 2 \times 2 = 16$. To avoid confusion this log is written with the base as a subscript in this way: $\log 16_2 = 4$. As another example: $5^3 = 5 \times 5 \times 5 = 125$. This is written $\log 125_5 = 3$, or the log of 125 when using 5 as a base is 3. The log of a number is an exponent and we generally assume 10 is the base. The $\log 100_{10} = 2$, however base 10 is generally written without the identifying subscript, $\log 100 = 2$.

A statement such as $10^z = 100$ is an *exponential equation*. Similarly, $\log 100 = 2$ is also an equation, but since the log of a number is involved, it is referred to as a *logarithmic equation*.

A basic exponential equation can be written:

$$b^y = x \tag{4–23}$$

where b = the base number
y = the exponent by which the base is raised
x = the product of the base b being used as a factor y times

The logarithmic equation for this basic exponential equation is

$$y = \log_b x \tag{4–24}$$

where y is the logarithm of x in the base b.

Example: Write the logarithmic equation for each of the following exponential equations.

a. $6^3 = 216$
b. $2^7 = 128$
c. $8^{2/3} = 4$
d. $5^{-2} = 0.04$
e. $(1/4)^{-5/2} = 32$
f. $10^5 = 100\ 000$

Solution:
a. $\log_6 216 = 3$, which is equivalent to saying that the logarithm of 216 in the base 6 is 3.
b. $\log_2 128 = 7$
c. $\log_8 4 = 2/3$
d. $\log_5 0.04 = -2$
e. $\log_{1/4} 32 = -5/2$
f. $\log 100\ 000 = 5$, note in the base 10 that the base subscript is not written.

Example: Write the exponential equation for each of the following logarithmic equations.

a. $\log_7 343 = 3$
b. $\log_2 128 = 7$
c. $\log_{625} 0.2 = -1/4$
d. $\log_{1/5} 25 = -2$
e. $\log_4 512 = 4.5$
f. $\log 63.095\ 734 = 1.8$

Solution:
a. $7^3 = 343$
b. $2^7 = 128$
c. $625^{-1/4} = 0.2$
d. $(1/5)^{-2} = 25$
e. $4^{4.5} = 512$
f. $10^{1.8} = 63.095\ 734$

Common Logarithms

A table of logarithms can be derived from a table of powers of 10.

$$
\begin{aligned}
10^0 &= 1 & \log 1 &= 0 \\
10^1 &= 10 & \log 10 &= 1 \\
10^2 &= 100 & \log 100 &= 2 \\
10^3 &= 1\ 000 & \log 1\ 000 &= 3 \\
10^4 &= 10\ 000 & \log 10\ 000 &= 4 \\
10^5 &= 100\ 000 & \log 100\ 000 &= 5 \\
10^6 &= 1\ 000\ 000 & \log 1\ 000\ 000 &= 6
\end{aligned}
$$

A system of logs in which 10 is the base is sometimes called a *decimal log* system, or a system of *common logs.* Table 4–1 lists logs for numbers ranging from 1 to 100. This table shows that the log of 1 is 0 and the log of 100 is 2. To find the log of any number between 1 and 100, locate the number in the No. column and immediately to the right is the log. The log of 16, for example, is 1.204 which means $10^{1.204} = 16$. Similarly, the log of 75 is 1.875, or, $10^{1.875} = 75$.

Although the table seems limited to the logs of the numbers from 1 to 100, it can be used to supply the logs of other numbers. The log of 5, for example, is 0.699. Locate the log of 50 and you will see that it is 1.699. The

TABLE 4–1

Logarithms from 1 to 100

No.	Log	No.	Log	No.	Log	No.	Log	No.	Log
1	0.000	21	1.322	41	1.613	61	1.785	81	1.908
2	0.301	22	1.342	42	1.623	62	1.792	82	1.914
3	0.477	23	1.362	43	1.633	63	1.799	83	1.919
4	0.602	24	1.380	44	1.643	64	1.806	84	1.924
5	0.699	25	1.397	45	1.653	65	1.813	85	1.929
6	0.778	26	1.415	46	1.663	66	1.820	86	1.934
7	0.845	27	1.431	47	1.672	67	1.826	87	1.940
8	0.903	28	1.447	48	1.681	68	1.833	88	1.944
9	0.954	29	1.462	49	1.690	69	1.839	89	1.949
10	1.000	30	1.477	50	1.699	70	1.845	90	1.954
11	1.041	31	1.491	51	1.708	71	1.851	91	1.959
12	1.079	32	1.505	52	1.716	72	1.857	92	1.964
13	1.114	33	1.518	53	1.724	73	1.863	93	1.968
14	1.146	34	1.531	54	1.732	74	1.869	94	1.973
15	1.176	35	1.544	55	1.740	75	1.875	95	1.978
16	1.204	36	1.556	56	1.748	76	1.881	96	1.982
17	1.230	37	1.568	57	1.756	77	1.886	97	1.987
18	1.255	38	1.580	58	1.763	78	1.892	98	1.991
19	1.279	39	1.591	59	1.771	79	1.898	99	1.996
20	1.301	40	1.602	60	1.778	80	1.903	100	2.000

only difference between these two logs is the number preceding the decimal point. Every log consists of two parts: an integral number preceding the decimal point, known as the *characteristic,* and the decimal number following it, called the *mantissa.* The characteristic is easy to determine, for it is always 1 less than the number of digits to the left of the decimal point in the number whose log is being found. The mantissa, the portion of the log following the decimal point, may be supplied by a table.

With this information on hand, we can now find the log of numbers greater than 100. For example, what is the log of 600? Since 600 consists of three digits, the first number of the logarithm will be 2 followed by a decimal point. Locate the number 6 in Table 4–1. The log is 0.778. The log of 600, then, is 2.778. To find the log of 2 700, find 27 in Table 4–1. The log of 27 is 1.431. Since the number 2 700 has four digits, the first number of the log is 3. The log of 2 700 is 3.431.

Every number has a log, that is, every number has an exponent using 10 as a base. A number such as 1 000 has a log of 3, or, $10^3 = 1\ 000$. Similarly, a number such as 100 has a log of 2. There are many numbers between 100 and 1 000. You can start at 101, continue on to 102, 103, 104, etc. Each of these numbers has a log, that is, each number has an exponent with 10 as the base. If $10^2 = 100$ and $10^3 = 1\ 000$, then a number such as 685 must have a log somewhere between 2 and 3. The limitations of Table 4–1 become obvious when looking for the log of a number such as 685. For a more complete table of mantissa values, refer to the log table of mantissa values in the Appendix. This table supplies only the mantissa portion of a log since the characteristic is easily determined.

Using the log table found in the Appendix, find the log of 313. The characteristic is 2. To find the mantissa, locate 31 in the N column in the table. Move to the right to column 3. The mantissa is given as 4955. The complete log is 2.495 5 or $10^{2.4955} = 313$.

The mantissa is always the same for any particular number group, regardless of the positioning of the decimal point. The mantissa, for example, is the same for 1 570, 157, 15.7, and 1.57. The logs of such numbers differ only in their characteristics. The mantissa for all these numbers is 195 9.

Number	**Log**
1 570	3.195 9
157	2.195 9
15.7	1.195 9
1.57	0.195 9

Example: Find the log of 549.

Solution: Since the number has three digits, the characteristic is 2. Using the log table in the Appendix, find the mantissa by locating 54 in the N column and then moving across to the last column, the 9 column, on the right. The mantissa is 739 6. The log of 549 is 2.739 6.

Logarithms can be an extremely useful and time saving tool in some mathematical calculations. Prior to the advent of electronic calculators these computations were made using tables such as the one just presented. Modern scientific calculators have the capability of displaying the logarithm of a number by simply depressing the "log" key.

Calculator Tip

Most scientific calculators use the [log] key to display the common logarithm of positive numbers greater than zero. The following sequences will produce the logarithms of 600, 2 700, 3.26, and 4.5×10^8.

600 [log] and 2.778 151 3 is displayed
2 700 [log] and 3.431 363 8 is displayed
3.26 [log] and 0.513 217 6 is displayed
4.5 [EE] 8 [log] and 8.653 212 5 is displayed

Logs of Numbers Smaller than 1

To determine the characteristic of a number smaller than 1, we can start with numbers larger than 1 and proceed in reverse order.

Number	Characteristic
10 000	4
1 000	3
100	2
10	1
1	0
0.1	−1
0.01	−2
0.001	−3
0.000 1	−4

There are two facts that are apparent about the characteristics of the logs of numbers less than 1. The characteristic is equal to the number of places to the first significant figure and is preceded by a minus sign. The characteristic of the log of a decimal number such as 0.008 467 is −3 since there are three decimal places from the decimal point to the first significant figure. As a further example, the characteristic of the log of 0.000 078 9 is −5 since there are five decimal places to the right of the decimal point to the first significant figure. For a number such as 0.589 78, the characteristic of the log is −1 since the first significant figure follows the decimal point immediately.

When finding the log of a number smaller than 1, that is, a decimal fraction, the mantissa is obtained in the same way as for numbers larger than 1. The mantissa is always positive, even when the characteristic is negative. Thus the log of 0.157 is actually $-1 + 0.195\ 9$. To avoid arithmetic difficulties involved in using a log consisting of a negative and a positive portion, the logs of decimal numbers can be written in the form $9.195\ 9 - 10$, and the log of 0.015 7 can be written as $8.195\ 9 - 10$.

Example: Find the log of 0.456.

Solution: Since the first significant figure follows the decimal point, the characteristic is -1. The mantissa, taken from the log table in the Appendix, is 659 0. The log of 0.456 is $9.659\ 0 - 10$.

The typical method of expressing logarithms for numbers smaller than 1 is to write the mantissa for that number, and then use the characteristic to indicate the location of the decimal point. However, electronic calculators complete the subtraction explicit in the logarithmic statement.

Calculator Tip

Use your calculator to determine the log of 0.456. The sequence and answer are:

0.456 [log] and $-0.341\ 035\ 2$ is displayed

In finding the log of 0.456, the log derived from the table is $9.659\ 0 - 10$, and the calculator indicates the log for the same number is $-0.341\ 035\ 2$. Note that these two expressions are equivalent:

$$9.569\ 0 - 10 = -0.341\ 0$$

The calculator expresses the logarithm as the characteristic subtracted from the mantissa. Closely study the examples in Table 4–2.

You must be aware that the table and calculator expressions are equivalent. In this text the calculator expression for numbers smaller than 1 will be used.

Antilogarithms

A problem in electronics can sometimes supply an answer in log form, making it necessary to find the number corresponding to the log. This number is known as the *antilogarithm* or *antilog*. The antilog of 2 is 100. The antilog of 3 is 1 000. The log table in the Appendix can be used to find the answer when a solution is expressed in log form.

TABLE 4–2

Equivalent Table and Calculator Log Values

Decimal Number	Characteristic	Table Mantissa	Log Expression	Calculator Expression
26 500	4	423 2	4.423 2	4.423 246
2 650	3	423 2	3.423 2	3.423 246
265	2	423 2	2.423 2	2.423 246
26.5	1	423 2	1.423 2	1.423 246
2.65	0	423 2	0.423 2	0.423 246
0.265	−1	423 2	9.423 2 − 10	−0.576 754
0.026 5	−2	423 2	8.423 2 − 10	−1.576 754
0.002 65	−3	423 2	7.423 2 − 10	−2.576 754

Example: The solution to an electronics problem, appearing in log form, is 2.605 3. Find the number corresponding to this log.

Solution: Locate 605 3 in the log table in the Appendix. This mantissa appears across from 40 in the N column and under column 3. Since the characteristic is 2, the antilog has three places. The antilog of 2.605 3 is 403. As a check on the accuracy of the answer, reverse the procedure and find the log of 403.

Calculator Tip

Given a logarithm consisting of a characteristic and mantissa, your calculator will display the decimal number in the base 10 equivalent to that logarithm. Determining the antilog of a value is often difficult when using tables. When the exact value is not listed, *interpolation* must be used. This time consuming, and often cumbersome procedure, is unnecessary when using a calculator.

A typical sequence to find antilogs is presented here. This sequence will be used to find the antilogs of 5.362 9, 3.017 5, 0.78, −2.164 729, and 12.148 7.

5.362 9	INV	log	and 230 621.61	is displayed
3.017 5	INV	log	and 1 041.118 1	is displayed
0.78	INV	log	and 6.025 595 9	is displayed
−2.164 729	INV	log	and 0.006 843 4	is displayed
12.148 7	INV	log	and $1.408\ 315\ 6 \times 10^{12}$	is displayed

The Loudness Indicator

Logarithms, at one time, were used extensively to make the solving of complex problems easier and quicker. Electronic calculators are now capable of performing these computations without the numbers first being converted to logarithms, although for a technician, there are applications of logarithms in the electronics field which make a basic understanding of this process important.

One of the ways of indicating the output of a power amplifier is to supply information in terms of watts. Amplifiers are referred to as 50-W units, 100-W units, 200-W units and so on. However, while the power rating is satisfactory for identifying the maximum amount of electrical power output of an amplifier, the power rating does not indicate loudness. With a 10-W amplifier, an increase of power to 12 W, a 20% increase, might be barely noticeable.

A preferred method for indicating loudness is through the use of a unit known as a *decibel* (dB). The basic unit is the bel (B). The decibel is a submultiple, equivalent to a tenth of a bel.

The decibel is a logarithmic unit and is used to indicate the ratio of two quantities, such as voltages or power. The bel, the fundamental unit, is the common log of the ratio of two powers. Expressed in terms of a formula:

$$B = \log \frac{P_2}{P_1} \qquad (4\text{–}25)$$

where P_1 = the power input
P_2 = the power output

The value of P_2 can be either greater or smaller than P_1. If P_2 is larger than P_1, the number of bels will be positive, representing a power gain or increase. Conversely, if P_2 is smaller than P_1, a minus sign is placed in front of the result to indicate a decrease in power.

Since P_1 and P_2 can be expressed in watts, submultiples such as milliwatt and microwatt can also be used as well as the multiple, the kilowatt. However, P_1 and P_2 must be expressed in the same multiple or submultiple. If P_1, for example, is in microwatts, P_2 must also be in microwatts. When different submultiples are used, convert from either one to the other.

The bel is too large a unit to use conveniently, so a submultiple, the decibel, is used in its place. The formula for the ratio of two powers becomes:

$$dB = 10 \log \frac{P_2}{P_1} \qquad (4\text{–}26)$$

Since the decibel in the formula is only a tenth of a bel, the number 10 is used on the right side to indicate that ten times as many units are now required.

Example: A power amplifier has an input of 1 W and an output of 100 W. What is the power gain of this amplifier in decibels?

Solution: Let P_2 represent the output power; P_1 is the input power. Then:

$$
\begin{aligned}
dB &= 10 \log \frac{P_2}{P_1} \\
&= 10 \log \frac{100\text{ W}}{1\text{ W}} \\
&= 10 \log 100 \\
&= 10(2) \\
&= 20\text{ dB}
\end{aligned}
$$

It requires an increase of 2 dB to 3 dB in sound power output to become noticeable. A rather substantial increase in power output may not be noticed by a listener.

Example: Assume that the amplifier described in the previous example increases its power output to 125 W, but still uses this same power input. Will the change in power output be noticed by a listener?

Solution: Find the output in decibels.

$$
\begin{aligned}
dB &= 10 \log \frac{P_2}{P_1} \\
&= 10 \log \frac{125\text{ W}}{1\text{ W}} \\
&= 10 \log 125 \\
&= 10(2.096\ 91) \\
&= 20.969\ 1\text{ dB}
\end{aligned}
$$

For 100-W output, the conversion to decibels resulted in 20 dB. With an increase in power output to 125 W, the output in decibels increased to only 20.969 dB. This increase in power output would <u>not</u> be noticed by a listener since the power must increase by several decibels before a change will be noticed.

Power Gain

The ratio of the output power of an amplifier to its input power is known as the power gain. If an amplifier has an output of 65 W and input of 1 W the power gain is

$$\frac{65\text{ W}}{1\text{ W}} = 65$$

The answer does not have a unit such as the ohm, volt, or the watt. It is simply an arithmetic ratio between two powers. For this amplifier, we can say that in terms of power the output is 65 times as great as the input.

Example: A certain amplifier delivers 625 mA across an 8-Ω load. The amplifier has an input of 80 mW. What is the power gain of the amplifier?

Solution: The output power must first be calculated by using one of the power laws.

$$P = I^2 \times R$$

$$P_2 = 0.625\ \text{A} \times 0.625\ \text{A} \times 8\ \Omega = 3.125\ \text{W}$$

$$\text{Power gain} = \frac{P_2}{P_1} = \frac{3.125\ \text{W}}{0.080\ \text{W}} = 39.06$$

The power gain can be substituted in the formula for finding the decibel power gain or loss of an amplifier.

$$\begin{aligned} dB &= 10 \log \frac{P_2}{P_1} \\ &= 10 \log \text{power gain} \end{aligned}$$

Example: A solid-state amplifier has a power gain of 60. What is the power gain in decibels?

Solution:

$$\begin{aligned} dB &= 10 \log \text{power gain} \\ &= 10 \log 60 \\ &= 10(1.778\ 2) \\ &= 17.782\ \text{dB} \end{aligned}$$

Sometimes an electronics problem involving power gain or power loss of an amplifier or network will be phrased in terms of decibels rather than watts.

Example: What must be the power gain of an amplifier if the output is to be 30 dB?

Solution:

$$\begin{aligned} dB &= 10 \log \text{power gain} \\ 30 &= 10 \log \text{power gain} \\ \frac{30}{10} &= \log \text{power gain} \\ \text{antilog } 3 &= \text{power gain} \\ \text{power gain} &= 1\ 000 \end{aligned}$$

In the previous example, the amplifier has a power gain of 1 000. If the input is 1 W, the output must be 1 000 W (1 kW). If the input is 1 mW, the output must be 1 W. An amplifier having an output of 1 kW may have the same power gain as an amplifier having an output of 1 W.

To compare the effect of a change in the power gain on the decibel output, consider the amplifier in the previous example and an amplifier having a power gain of 100.

$$\begin{aligned} dB &= 10 \log \text{power gain} \\ &= 10 \log 100 \\ &= 10(2) \\ &= 20\ \text{dB} \end{aligned}$$

Thus, an amplifier with a power gain of 100 has an output of 20 dB; one with a power gain of 1 000 has an output of 30 dB. It takes 10 times as much power gain to produce an increase of 10 dB in the output.

In some instances, a particular amount of power output is required. The question, then, is the power input required. This can be calculated if the power gain of the amplifier is known.

Example: An audio power amplifier has a gain of 40 dB with an output measured as 10 V across a resistor whose current flow is metered at 600 mA. What power input is required?

Solution: The power output can be calculated from the formula:

$$P = E \times I = 10\text{ V} \times 0.6\text{ A} = 6\text{ W}$$

$$dB = 10 \log \frac{P_2}{P_1}$$

$$40\text{ dB} = 10 \log \frac{6\text{ W}}{P_1}$$

$$4 = \log \frac{6}{P_1} \qquad \text{(Dividing both sides by 10)}$$

$$10\ 000 = \frac{6}{P_1} \qquad \text{(Taking the antilog of 4)}$$

$$P_1 = 0.000\ 6\text{ W or } 0.6\text{ mW}$$

Power Loss

Not all circuits supply a power gain. In a passive network such as an attenuator, power out of the attenuator is less than the power input. A long transmission line can also have a power loss. The same formula for calculating power gain can be used for finding power loss.

Example: A resistive network requires an input of 40 W. The power dissipated in the form of heat in the network is 39 W. The amount of useful power is 1 W. What is the power loss in decibels?

Solution:

$$dB = 10 \log \frac{P_2}{P_2}$$

$$= 10 \log \frac{1\text{ W}}{40\text{ W}}$$

$$= 10 \log 0.025$$

$$= (10)(-1.602)$$

$$= -16.02\text{ dB}$$

Power Reference Level

When power gain or loss is to be calculated, both power input and output must be known. Conceivably, an amplifier with a power output of 50 W could have a higher power gain than one that delivers 60 W. To be able to make a comparison between two amplifiers, some reference level is required. There are a number of such reference levels in use, but one of the more common is

6 mW or 0.006 W. This is regarded as 0 dB. An amplifier using this reference and having a power gain of 40 dB would have a power gain of 40 dB above 0 dB or above 6 mW. It is important to specify the reference level since several different ones are used.

Example: What is the power output in milliwatts of an amplifier rated at 15 dB, assuming its zero reference level is 6 mW?

Solution:

$$dB = 10 \log \frac{P_2}{P_1}$$

$$15 = 10 \log \frac{P_2}{0.006}$$

$$1.5 = \log \frac{P_2}{0.006}$$

$$31.6 = \frac{P_2}{0.006}$$

$$P_2 = (31.6)(0.006)$$

$$P_2 = 0.189\,6 \text{ W or } 189.6 \text{ mW}$$

To check, reverse the problem-solving technique.

$$db = 10 \log \frac{189.6 \text{ mW}}{6 \text{ mW}}$$

$$= 10 \log 31.6$$

$$= (10)(1.5)$$

$$= 15 \text{ dB}$$

Reference levels, such as 6 mW can be used for amplifiers, in which case there is a power gain, or for devices in which there is a power loss. A microphone, for example, with an output of −60 dB can be assumed to be 60 dB below reference level. An attenuator of −35 dB is 35 dB below the zero reference level of 6 mW.

If the same reference level is used for all components, whether they have a power gain or loss, then the total output in decibels is simply the sum of their individual gains or losses in decibels.

Example: A microphone with an output of −40 dB is fed into a preamp which has a power gain of 40 dB. The output of the preamp is controlled by an attenuator network having a power loss of 20 dB. The attenuator is connected to a power amplifier whose power gain is rated at 50 dB. What is the power gain, in decibels, of this system if the same reference level is used for all components?

Solution: $-40 \text{ dB} + 40 \text{ dB} - 20 \text{ dB} + 50 \text{ dB} = 30 \text{ dB}$

Voltage and Current Ratios

Since voltage and current are so closely related to power,

$$P = I^2R \text{ and } P = \frac{E^2}{R}$$

the power gain of an amplifier or the power loss of a network, in decibels, can be calculated when voltage or current values are known. For voltage, the formula for calculating decibels is

$$dB = 20 \log \frac{E_2\sqrt{R_1}}{E_1\sqrt{R_2}}$$

In this formula, E_2 is the output voltage, E_1 the input voltage, R_1 is the input resistance (or impedance) while R_2 is the output resistance (or impedance). If R_1 is equal to R_2, then these two terms will cancel and the formula will be simplified to:

$$dB = 20 \log \frac{E_2}{E_1}$$

The formula used for calculating decibels when current ratios are involved is

$$dB = 20 \log \frac{I_2\sqrt{R_2}}{I_1\sqrt{R_1}}$$

Again, if R_2 and R_1 are equal, they cancel, and the formula simplifies to:

$$\text{dB} = 20 \log \frac{I_2}{I_1}$$

Example: A solid-state amplifier is capable of delivering a sine wave output of 300 V when the input is driven by 1 V. The input resistance is 400 Ω while the higher impedance output is 10 kΩ. What is (a) the power gain of this amplifier, in decibels, and (b) the power gain of the amplifier?

Solution: a.

$$\begin{aligned} dB &= 20 \log \frac{E_2\sqrt{R_1}}{E_1\sqrt{R_2}} \\ &= 20 \log \frac{300 \text{ V} \sqrt{400\ \Omega}}{1 \text{ V} \sqrt{10\ 000\ \Omega}} \\ &= 20 \log \frac{300 \times 20}{100} \\ &= 20 \log \frac{6\ 000}{100} \\ &= 20 \log 60 \\ &= (20)(1.778\ 2) \\ &= 35.564 \text{ dB gain} \end{aligned}$$

b. To calculate the power gain:

$$P = \frac{E^2}{R}$$

for the output:

$$P_2 = \frac{(300\ \text{V})^2}{10\ 000\ \Omega} = 9\ \text{W}$$

for the input:

$$P_1 = \frac{(1\ \text{V})^2}{400\ \Omega} = 0.002\ 5\ \text{W}$$

$$\text{power gain} = \frac{P_2}{P_1} = \frac{9\ \text{W}}{0.002\ 5\ \text{W}} = 3\ 600$$

To check:

$$\begin{aligned} dB &= 10 \log \text{power gain} \\ &= 10 \log 3\ 600 \\ &= 10(3.556\ 3) \\ &= 35.563\ \text{dB} \end{aligned}$$

When using current or voltage, it is often easier to calculate the power gain and then the decibel gain since the formulas for calculating the decibel gain involve square root. However, one can be used as a check on the other to verify the accuracy of the work.

Volume Units

In addition to being expressed in terms of decibels, power ratios are sometimes specified in *volume units,* abbreviated as VU. When using volume units, a zero reference of 1 mW (0.001 W) is understood unless otherwise indicated. The formula for volume units is

$$VU = 10 \log \frac{P_2}{0.001}$$

The fraction can be eliminated by multiplying numerator and denominator by 1 000.

$$VU = 10 \log 1\ 000(P_2)$$

Example: What is the gain in volume units of an amplifier whose output is 15 W?

Solution:

$$\begin{aligned} VU &= 10 \log 1\ 000(P_2) \\ &= 10 \log (1\ 000)(15) \\ &= 10 \log 15\ 000 \\ &= (10)(4.176\ 1) \\ &= 41.761\ \text{VU} \end{aligned}$$

The reference level for electronic instruments is 1 mW. When this reference is used the gain is measured in units known as dBm.

Example: The output of an amplifier is 4.5 W. What is the output in decibels above 1 milliwatt (dBm)?

Solution:

$$\begin{aligned} dBm &= 10 \log \frac{P_2}{1 \text{ mW}} \\ &= 10 \log \frac{4.5 \text{ W}}{1 \text{ mW}} \\ &= 10 \log 4\,500 \\ &= 10(3.653\,2) \\ &= 36.532 \text{ dBm} \end{aligned}$$

Natural Logs

A commonly used logarithm base is the number whose approximate value is 2.718 28. Logs using this number as a base are called *natural* or *Naperian logs,* compared to common logs which have 10 as a base. This number base is generally represented by the letter e.

A common log can be written as $\log_{10}$ or simply as log. A natural log can be written as $\log_e$ or simply as ln. There is a simple arithmetic relationship between natural and common logs. The common log is 0.434 3 times the natural log. To convert from natural logs to common logs, multiply the natural log by 0.434 3. To convert from common logs to natural logs, multiply the common log by 2.302 6, which is the reciprocal of 0.434 3. Thus:

$$\log a = 0.434\,3 \ln a \qquad (4\text{–}27)$$

where the letter a represents any number. Similarly:

$$\ln a = 2.302\,5 \log a \qquad (4\text{–}28)$$

Example: What is the numeric value of ln 500?

Solution: This value can be found in a table of natural logarithms. Using the conversion factor, however, makes it possible to use the table of common logs for both.

$$\begin{aligned} \ln 500 &= 2.302\,6 \log 500 \\ &= 2.302\,6(2.699\,0) \\ &= 6.214\,717 \end{aligned}$$

The basic exponential equation for natural logarithms can be written as

$$e^y = x \qquad (4\text{–}29)$$

where e = the base number (2.718 28)
y = the exponent to which the base is raised
x = the product of the base e being used as a factor y times

The natural logarithmic equation for this basic exponential equation is

$$y = \log_e x = \ln x \qquad (4\text{–}30)$$

where y is the natural logarithm of x.

Example: Write the natural logarithmic equation for each of the following exponential equations.

a. $e^2 = 7.389$
b. $e^{-1.5} = 0.223$
c. $e^{12} = 1.628 \times 10^5$
d. $e^{-2/3} = 0.513$

Solution: a. $\log_e 7.389 = 2$, which is equivalent to saying that the natural logarithm of 7.389 is 2. This could also be written as $\ln 7.389 = 2$.
b. $\ln 0.223 = -1.5$
c. $\ln 1.628 \times 10^5 = 12$
d. $\ln 0.513 = -0.667$

Example: Write the exponential equation for each of the following natural logarithmic equations.

a. $\ln 200 = 5.298$
b. $\ln 0.986 = -1.4 \times 10^{-2}$
c. $\ln 2.8 \times 10^7 = 17.148$
d. $\ln 3.7 \times 10^{-3} = -5.6$

Solution: a. $e^{5.298} = 200$
b. $e^{-1.4 \times 10^{-2}} = 0.986$
c. $e^{17.148} = 2.8 \times 10^7$
d. $e^{-5.6} = 3.7 \times 10^{-3}$

Calculator Tip

Just as your calculator will compute common logarithms, it will also compute natural logarithms. Find the natural logarithm ($y = \ln x$) for each of these numbers: 0.26, 1.5, 376.23, 47 000, and 5.8×10^{17}.

0.26	[ln x]	and −1.347 073 6 is displayed
1.5	[ln x]	and 0.405 465 1 is displayed
376.23	[ln x]	and 5.930 200 7 is displayed
47 000	[ln x]	and 10.757 903 is displayed
5.8×10^{17}	[ln x]	and 40.901 804 is displayed

What decimal numbers have natural logarithms ($e^y = x$) equivalent to -7.26, 0.983, 16, and 24.173 8?

$e^{-7.26}$ INV ln x and 7.031×10^{-4} is displayed

$e^{0.983}$ INV ln x and 2.672 461 6 is displayed

e^{16} INV ln x and 8.886×10^{6} is displayed

$e^{24.1738}$ INV ln x and 3.151×10^{10} is displayed

Practical applications of natural logarithms occur frequently in the study of capacitors and inductors. The following examples demonstrate the use of natural logs.

Example: The instantaneous voltage across a charging capacitor is given as

$$v_C = E(1 - e^{-t/RC})$$

where v_C = instantaneous voltage in volts
E = applied voltage in volts
t = charging time in seconds
R = series resistance in ohms
C = capacitor value in farads
RC = time constant in seconds

Find the voltage across C_1 in Figure 4–1 after switch S_1 has been closed 2 ms.

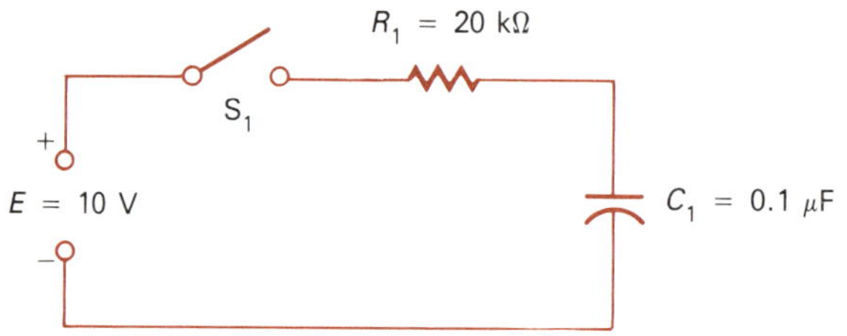

FIGURE 4–1

Solution:

$$\begin{aligned} RC &= (20 \times 10^3\ \Omega)(0.1 \times 10^{-6}\ \text{F}) \\ &= 2 \times 10^{-3}\ \text{seconds} \end{aligned}$$

$$\begin{aligned} v_C &= E(1 - e^{-t/RC}) \\ &= 10\ \text{V}\ (1 - e^{-2 \times 10^{-3}\text{s}/2 \times 10^{-3}\text{s}}) \\ &= 10\ \text{V}\ (1 - e^{-1}) \\ &= 10\ \text{V}\ (1 - 0.368) \\ &= 10\ \text{V}\ (0.632) \\ &= 6.32\ \text{V} \end{aligned}$$

Note that after one time constant the capacitor charges to 63.2% of the applied voltage.

Example: Find the voltage across C_1 in Figure 4–1 after (a) 5 ms and (b) 10 ms.

Solution:

a.
$$\begin{aligned} v_C &= E(1 - e^{-t/RC}) \\ &= 10\ \text{V}\ (1 - e^{-5 \times 10^{-3}\text{s}/2 \times 10^{-3}\text{s}}) \\ &= 10\ \text{V}\ (1 - e^{-2.5}) \\ &= 10\ \text{V}\ (1 - 0.082) \\ &= 10\ \text{V}\ (0.918) \\ &= 9.18\ \text{V} \end{aligned}$$

b.
$$\begin{aligned} v_C &= E(1 - e^{-t/RC}) \\ &= 10\ \text{V}\ (1 - e^{-10 \times 10^{-3}\text{s}/2 \times 10^{-3}\text{s}}) \\ &= 10\ \text{V}\ (1 - e^{-5}) \\ &= 10\ \text{V}\ (1 - 0.007) \\ &= 10\ \text{V}\ (0.993) \\ &= 9.93\ \text{V} \end{aligned}$$

Example: The instantaneous current passing through the inductor in Figure 4–2 after the switch is closed is given as

$$i_L = \frac{E}{R}(1 - e^{-t/(L/R)})$$

where i_L = instantaneous current in amperes
E = applied voltage in volts
R = series resistance in ohms
t = charging time in seconds
L = inductor value in henries
L/R = time constant in seconds

Find the current in Figure 4–2 after 200 μs.

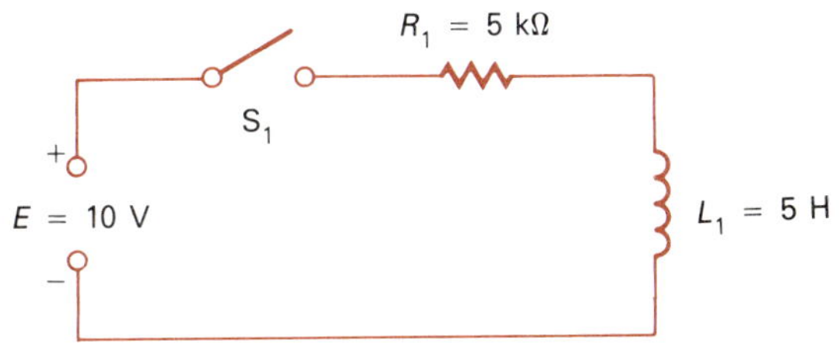

FIGURE 4–2

Solution:

$$\begin{aligned} L/R &= 5\ \text{H} \div 5 \times 10^3\ \Omega \\ &= 1 \times 10^{-3}\ \text{seconds} \end{aligned}$$

$$\begin{aligned} i_L &= \frac{E}{R}(1 - e^{-t/(L/R)}) \\ &= \frac{10\ \text{V}}{5\ \text{k}\Omega}(1 - e^{-200 \times 10^{-6}\text{s}/1 \times 10^{-3}\text{s}}) \\ &= 2\ \text{mA}(1 - e^{-0.2}) \end{aligned}$$

$= 2\text{ mA}(1 - 0.819)$
$= 2\text{ mA}(0.181)$
$= 0.363\text{ mA}$

Example: Find the circuit current in Figure 4–2 after 3 ms.

Solution:
$$i_L = \frac{E}{R}(1 - e^{-t/(L/R)})$$
$$= 2\text{ mA}(1 - e^{-3\times10^{-3}\text{s}/1\times10^{-3}\text{s}})$$
$$= 2\text{ mA}(1 - e^{-3})$$
$$= 2\text{ mA}(1 - 0.05)$$
$$= 2\text{ mA}(0.95)$$
$$= 1.9\text{ mA}$$

EXERCISE 4·8

1. Write the logarithmic equation for each of these exponential equations.
   a. $7^4 = 2\,401$
   b. $(1/8)^{-4} = 4\,096$
   c. $4^{3/4} = 2.828$
   d. $5^{3/5} = 2.627$
2. Write the loagrithmic equation for each of these exponential equations.
   a. $(3/4)^{-4/3} = 1.468$
   b. $256^{3/8} = 8$
   c. $10^7 = 10\,000\,000$
   d. $100^3 = 1\,000\,000$
3. Write the exponential equation for each of the following logarithmic equations.
   a. $\log_9 6\,561 = 4$
   b. $\log_{0.2} 125 = -3$
   c. $\log_6 14.697 = 3/2$
   d. $\log_{15} 25.782 = 1.2$
4. Write the exponential equation for each of the following logarithmic equations.
   a. $\log_{4/9} 1.5 = -1/2$
   b. $\log_4 1\,024 = 5$
   c. $\log 3.981 = 0.6$
   d. $\log_{0.04} 5 = -0.5$
5. Use your calculator to find the common logarithm for each of these numbers. Round each mantissa to four decimal places.
   a. 765 c. 218.987 e. 0.925 g. 0.000 625
   b. 37 d. 100 000 f. 720.084 h. $(1.28)^6$
6. Use your calculator to find the common logarithm for each of these numbers. Round each mantissa to four decimal places.
   a. 157 624.13 c. $(4.2)^4$
   b. 1.3×10^{-5} d. 7.213

e. 0.006 g. 0.1

f. 8.5×10^9 h. 6.28×10^{-5}

7. Use your calculator to find the decimal number corresponding to each of these common logarithms. Round answers to four decimal places.

   a. 1.372 9 e. 0.107 6

   b. 4.065 0 f. 2.399 1

   c. −3.172 8 g. −5

   d. 8 h. −1.729 4

8. Use your calculator to find the decimal number corresponding to each of these common logarithms. Round answers to four decimal places.

   a. −0.983 e. 1.75

   b. 10.16 f. 7.143

   c. 6.072 93 g. 0.001 257

   d. −4 h. −3.062

9. An amplifier has a power gain of 500. What is the bel power gain?
10. What is the decibel power gain for the amplifier in problem 9?
11. The input power to an amplifier is 0.2 W. The output power is measured at 50 W. Calculate the power gain in decibels.
12. What is the power gain in decibels of an amplifier if the input is 600 mW and the output 5 W?
13. An amplifier delivers a sine wave current of 450 mA across a 16-Ω resistive load. The input is driven by 100 mW. What is the decibel power gain of this amplifier?
14. What is the power gain in decibels for an amplifier when the output power is 20 W and the input 1 mW?
15. An amplifier having identical input and output impedances has a 3 mV input signal with 4 V appearing across the output load. What is the gain of this amplifier in decibels?
16. The input power to a circuit is 125 mW. Find the power gain in decibels if the output power is 100 mW.
17. The output power of a circuit is 1/2 of the input power. Express this gain in decibels.
18. An amplifier has a decibel gain of 20 dB. What is the circuit's power gain?
19. An amplifier is to have a 40 dB gain. What is the power output if the input is 120 μW?
20. The output power of a circuit is 720.8 mW. Find the input power if the circuit gain is −2.64 dB.
21. The input and output impedances of a circuit are identical. Using an oscilloscope the input and output voltages are found to be 450 μV and 2.2 V respectively. What is the circuit's power gain in decibels?
22. The input current and impedance of a circuit are 350 μA and 10 kΩ respectively. The circuit's power gain in decibels is −6.902 dB. Find the output power.

23. Given the circuit described in problem 22, find the output power if the power gain had been 3 dB.
24. What is the gain in volume units of an amplifier whose output is 10 W?
25. The output of an amplifier is 850 mW. Express the output in dBm units.
26. Use your calculator to find the natural logarithms for the numbers given in problem 5. Round each mantissa to four places.
27. Use your calculator to find the natural logarithms for the numbers given in problem 6. Round each mantissa to four places.
28. Assume that the values given in problem 7 are natural logarithms. Use your calculator to determine the decimal equivalents to four decimal places.
29. Assume that the values given in problem 8 are natural logarithms. Use your calculator to determine the decimal equivalents to four decimal places.
30. Finding insulation resistance (R) by the leakage method is given by the following formula. Using the values provided calculate the resistance in ohms.

$$R = 10^6 \times \frac{t}{c} \times \frac{1}{\log_e (V_o/V)}$$

where $t = 180$
$c = 0.075$
$V_o = 150$
$V = 116$

Use Figure 4–3 for problems 31–33.

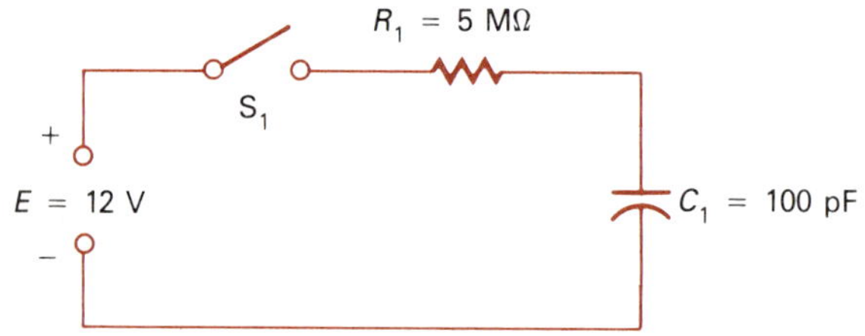

FIGURE 4–3

31. Find the voltage across C_1 in Figure 4–3 after 1 ms.
32. What is the voltage across C_1 in Figure 4–3 after 2.45 ms?
33. Resistor R_1 in Figure 4–3 is increased to 100 MΩ. What is the voltage across C_1 after 2 ms?

Use Figure 4–4 for problems 34–35.

34. Find the circuit current in Figure 4–4 after 5 ms.

35. Find the circuit current in Figure 4–4 after 20 ms.

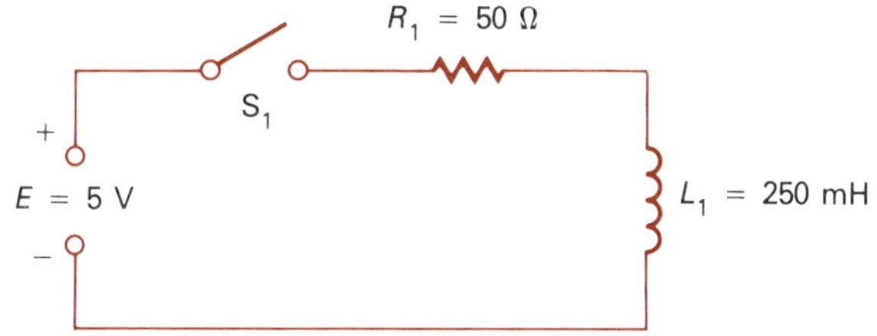

FIGURE 4–4

UNIT 4·9 PROBLEM REVIEW

Simplify each of the following problems. If necessary, round answers to three decimal places. Remove all grouping signs, radical signs, and use positive exponents only.

1. $18(3 - 2) + 4 - 6(7 \div 2)$
2. $[9(8 - 11) + 6]^3 \div 144(2 - 7)$
3. $[52(13 - 12 \div 4)]^{-4}[108 + (3 + 5)^2]^6$
4. $\dfrac{(14 - 6^2)^{-3}(2 - 10 \times 7)^2}{[(144^{0.5})^2]^{-2}}$
5. $5x^2y - (xy^2 + x^2y^2) + 3x^2(x + y)$
6. $3a[1 - 2b(1 - 5a + b) + a(2b - 5a - 4)]$
7. $mn(m + n) - [m(mn - n^2) + m^2n - n^2m]$
8. $x[x - y + 3(x + y) - 4xy - x(x + y)]$
9. $5a^4b^{-3}c - 9a^2b^4c^{-1}$ if $a = 2,\ b = -2,\ c = 5$
10. $\dfrac{16x^2y^3z^{-3} - 9x^{-3}z^5}{x^2yz^{-4}}$ if $x = -5,\ y = 6,\ z = 3$
11. $(8m^2n - 2mn^2 + 16)(-5m^{-3}n^4 + mn)^2$ if $m = 4,\ n = -4$
12. $[10x^2(x - 7)^{-2}]^2 - \left(\dfrac{3x^2y^4 - 10x}{x^5y^{-2}}\right)^{-2}$ if $x = -3,\ y = 2$
13. $(3a^4b^{-2})^2$
14. $\sqrt[4]{25^2}$
15. $\sqrt[6]{42\,798}$
16. $\left(\dfrac{4x^{-1}y^2z^3}{9x^3y^2z^{-2}}\right)^{-1}$
17. $(\sqrt[3]{183})^{0.2}$
18. $\left(\dfrac{18^{1.5}\sqrt{287}}{1.4 \times 10^3}\right)^{-2}$
19. $(\sqrt[4]{256a^{12}b^{-8}c^2})^2$

20. $\dfrac{5ab^2}{\left[\dfrac{6a^{-2}b^3}{25a^3b^{-5}}\right]^2}$

21. $[125^{1/2}(25a^4b^2c^{-3})^{-3}]^2$

22. $\left[\dfrac{(4.198 \times 10^{-5})^2(3.27 \times 10^6)^3}{(9.08 \times 10^4)^{-2}(5 \times 10^9)^{-1}}\right]^{-2}$

23. $\sqrt[n]{9a^2b^{3n}}$

24. $\dfrac{(24x^{-4}y^2z^3)(6xy^{-2})}{(3x^{-1}y^2z)^2(x^4z^{-2})^{-1}}$

25. $\dfrac{(512a^6b^{-3})^{1/3}(6a^{-2}b^2)^3}{(16ab^7)^{-2}(4a^{-1}b^6)^5}$

26. $-6x^2y - (3x^2y + 2y^2x - 32) + 12y^2x$

27. $a^2 - [(4a^2 - b^2) - (-2b^2 + 3a^2)]$

28. $(3m^2n - 5m^2n) - [(4mn^2 + 2m^2n) - (-5m^2n + 7mn^2)]$

29. $-(-12x^2y^2 + 2x^2y - 3y^2x) + [8x^2y^2 - (5x^2y + 6y^2x) - 5x^2y^2]$

30. $-x^2y^2 - (4x^2y^2 + 2x^2y^2) - x^2y^2 + 7x^2y^2$

31. $3x^2 - (4x^2y + 2y^2x - 112) + 7x^2y - (8x^2 + 5y^2x)$ if $x = -2$, $y = 4$

32. $-x^2y^2 + (6xy^2 + 4x^2y) - (5x^2y - 3x^2y^2 + 2xy^2)$ if $x = 3$, $y = -7$

33. $2z + 3x^2y^2z - (4x^2y^2z - 8xz^3 + xyz) - (2z + 4y)$ if $x = -1$, $y = 4$, $z = 8$

34. $-xyz - (5x^2yz^2 + 3xy^2z^2) + (7xy^2z^2 - 2xyz + 3x^2yz^2) - (-2x^2yz^2 + 4xy^2z^2)$ if $x = 12$, $y = -8$, $z = -2$

35. $-2(6a^2bc^{-3})(a^{-3}b^4c^5)$

36. $(4x^{-2}y^3z^{-5})^2(8x^{-2}y^5z^{-7})^{-1}$

37. $-6ab(2a - 2b)$

38. $\dfrac{(9x^2y^3z)^{-1}}{(3x^{-4}yz^{-2})^{-2}}$

39. $(x - y)^2$

40. $(x^4y)^2(3x^4y^2 - 2y + 6x^3y^4)$

41. $(a + b)(a^2 + 6ab - b^2)$

42. $(x + 4)(x - 4)$

43. $(5y^2z^3)^2(y + z)^2$

44. $\dfrac{(2a^{-2}b^3c^5)^4}{(4a^2b^{-3}c)^2}$

45. $\dfrac{(2m - n)^2}{mn}$

46. $\dfrac{(x + 2)^2(x - 2)}{x^2 - 4}$

47. $\dfrac{a^2 - 2a - 63}{a + 7}$

48. $\dfrac{a(4a + 5) - 6}{a + 2}$
49. $\dfrac{x^3 + 6x^2 - x - 40}{x + 5}$
50. $\dfrac{12a^6 - 3a^5 - 38a^4 + 39a^3 + 20a^2 - 19a + 77}{4a + 7}$
51. $(a + b)(a - b)$
52. $(m^2 - n^2)(m^2 + n^2)$
53. $(5 - x)^2$
54. $4y^2(y - 1)$
55. $(6w - 2z)^2(3wz)$
56. $(\sqrt{ab} + \sqrt{b})(\sqrt{ab} - \sqrt{b})$
57. $(x + 3)(y - 2)$
58. $x^2(x - 2)(x + 4)$
59. $(y + 2)(y - 4)^2$
60. $(2z - 4)^3$

Factor problems 61–73.

61. $6a^4bc^2 - 15a^3b^2c^2 + 9a^5bc^3$
62. $2x^2y^3 - 4x^2yz^3 + 10y^2z^4$
63. $3a^2 - 12$
64. $2x^5 - 72x$
65. $6x^4 - 6$
66. $y^2 - 12y + 27$
67. $-3a^2 - 27a - 60$
68. $2a^2 - 18$
69. $x^4 - 4$
70. $ax - 2x - 8 + 4a$
71. $-2y^2 + 14y + 36$
72. $x^3 - 12x^2 + 48x - 64$
73. $3y^4 + 9y^3 + 9y^2 + 3y$

Combine the algebraic fractions in problems 74–95.

74. $\dfrac{6x}{x - 1} - \dfrac{2x}{x + 4}$
75. $\dfrac{x(x + 2) - 8}{x^2 + x - 12} + \dfrac{2 - x}{x - 2}$
76. $1 - \dfrac{7}{y(2y + 5)} - \dfrac{1}{y}$
77. $\dfrac{2a - 6}{3a^2 - 21a + 36} + \dfrac{-4}{2a - 2}$
78. $3 + \dfrac{15}{x^2 - 6x} + \dfrac{36x}{x^2 - 12x + 36}$
79. $\dfrac{z^2 - 9}{2z^2 + 10z + 12} + \dfrac{z - 3}{2z^2 - 18}$

80. $\frac{m - 2}{m + 2} - \frac{m + 2}{m - 2}$

81. $\frac{x - 1}{2x(x - 3) + 4} - \frac{3}{x^2 - 4}$

82. $\frac{4x - 1}{x^2 - 3x - 10} - \frac{7x + 4}{x^2 - x - 6} + \frac{3x}{x^2 - 8x + 15}$

83. $\frac{6y - 2}{y^2 - 5y + 4} + \frac{7 - y}{y^2 - 16} - \frac{5y + 50}{y^2 + 3y - 4}$

84. $\frac{(x + y)^2}{x^2 + y^2} \cdot \frac{x - y}{x^2 - y^2}$

85. $\frac{(5a^3b^{-2})^2}{c^2} \cdot \frac{3a^2bc^5}{25c^{-2}ab^4}$

86. $\frac{y(y + 1) - 6}{y + 3} \div \frac{y - 2}{y - 6}$

87. $\left(1 + \frac{1}{x(x - 2)}\right) \div (x - 1)^2$

88. $\frac{x^2 - 1}{x - 2} \cdot \frac{4x - 8}{x - 1}$

89. $\frac{m^2 - m - 6}{m^2 + m - 2} \div \frac{m^2 + 2m - 15}{m^2 + 3m - 4}$

90. $\left(1 + \frac{6}{6x^2 - 13x}\right) \div (3x^2 - 2x)$

91. $\left(\frac{y}{y + 1} + \frac{1}{y - 3}\right)\left(\frac{y^2 + 2y + 1}{y^2 - 4y + 3}\right)$

92. $\left(\frac{x}{x - 2} - \frac{x}{x - 1}\right)(x^3 - 5x^2 + 6x)$

93. $\left(\frac{1}{x} - \frac{1}{x^2}\right)\left(\frac{x^3 + x^2}{x^2 - 1}\right)$

94. $\left(\frac{a^2 + a - 2}{a^2 - a - 2} \cdot \frac{a^2 + a}{a^2 - 1}\right) \div \frac{a^2 + a - 2}{2a^2 - 8a + 8}$

95. $\left(\frac{y^6 - y^2}{y + 1}\right) \div (y - 1)$

96. Find the common logarithm for each of the following. Round each mantissa to four places.
   a. 42 000
   b. 1.65×10^5
   c. $(5.7)^5$
   d. 2.3×10^{-7}

97. Find the common logarithm for each of the following. Round each mantissa to four places.
   a. $(0.37)^3$

b. 1 001
c. 876 230
d. $(9.03 \times 10^3)^4$

98. Find the decimal number corresponding to each of the following common logarithms. Round your answers to four decimal places.
    a. 1.5
    b. −3.762 4
    d. 13.82
    d. −8
99. Find the decimal number corresponding to each of the following common logarithms. Round your answers to four decimal places.
    a. 0.072 386
    b. −0.18
    c. 5.142 9
    d. 2.8
100. The input and output powers of an amplifier are 225 mW and 1.2 W respectively. Calculate the power gain in dB.
101. The input power to an amplifier is 0.04 W. The dB gain is 5. What is the output power?
102. The input power to an amplifier is 675 mW. Find the dB gain if an output current of 20 mA flows through an impedance of 1.5 kΩ.
103. The output power of a circuit is 4 times the input power. Express this gain in decibels.
104. The output power of an amplifier is 20 W. This output is to be reduced by 3 dB. What is the new output power?
105. The input and output currents of a circuit are 2.3 mA and 7.9 mA respectively. Find the power gain in decibels if the input and output impedances are equal.
106. The input voltage to an amplifier is 113 mV. The input and output impedances are 50 kΩ and 5 kΩ respectively. What is the output voltage if the power gain is 16.897 2 dB?
107. The output power of an amplifier is 0.7 W. Express the output in dBm.
108. Find the natural logarithm for each of the following numbers. Round each mantissa to four places.
    a. 427.26
    b. 1×10^{-5}
    c. $(3.2)^4$
    d. 0.087 246
109. For each of the following natural logarithms find the corresponding decimal equivalent. Round each answer to four decimal places.
    a. 4.17
    b. −1.062 4
    c. 17.984
    d. −0.624 741

110. The characteristic impedance of two-wire parallel conductors with an air dielectric is given by:

$$Z_o = 276 \log b/a$$

where a = conductor radius
b = wire spacing
Given $a = 0.032$ inches and $b = 0.25$ inches find Z_o.

Use Figure 4–5 for problems 111–113.

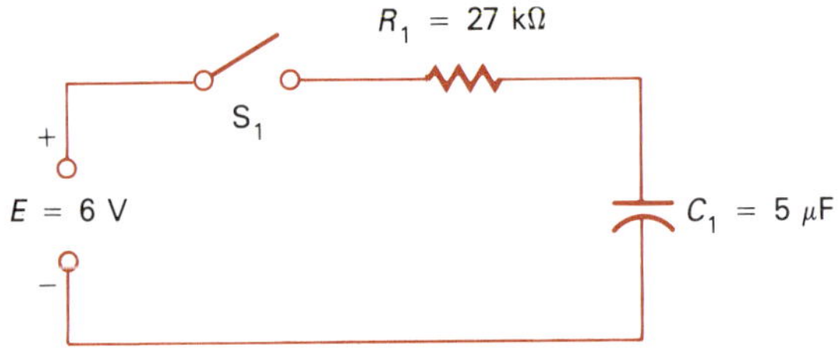

FIGURE 4–5

111. Find the voltage across C_1 in Figure 4–5 after 0.2 s.
112. At any point in time, the voltage across R_1 in Figure 4–5 is equal to the applied voltage minus the voltage across the capacitor. Given this information calculate the instantaneous current after 1 ms and 405 ms.
113. Capacitor C_1 in Figure 4–5 is changed to 0.1 μF. What is the voltage across C_1 after 100 μs?

Use Figure 4–6 for problems 114–115.

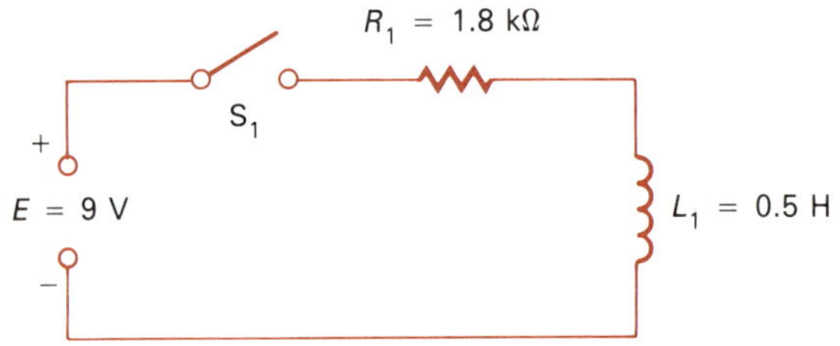

FIGURE 4–6

114. Find the circuit current in Figure 4–6 after 1 ms.
115. Calculate the voltage across R_1 in Figure 4–6 after 50 μs and 1.25 ms.

CHAPTER 5

LINEAR AND QUADRATIC EQUATIONS

An equation is a statement of equality between two expressions. Thus, $a + b = 35$ is an equation, for when the values of a and b become known, their sum will be equal to 35. Also the statement $x + y = z$ is an equation. If the values of x and y are known, then the value of z is determined. Generally, the unknown is put on the left side of the equals sign and terms whose values are known are put on the right side. Finding the value of the unknown is called *solving the equation.* If only one unknown is involved, the solution is also called the *root.*

Electronics formulas are equations. A statement such as $E = I \times R$ is an equation, for the multiplied values of I and R are equal to E. The solution of electronics problems is often equivalent to solving algebraic equations.

UNIT 5·1 SIMPLE EQUATIONS AND FORMULAS

Objectives:

After studying this unit, you should be able to

- rearrange equations to solve for a variable.
- solve equations with one unknown.
- rearrange technical formulas to solve for a variable.

Types of Equations

An equation is defined as a statement of equality between two expressions. These expressions consist of *constants* and *variables.* Constants are typically identified by letters from the beginning of the alphabet (a, b, c, etc.) and represent a fixed value. Variables are identified by letters from the end of the alphabet (x, y, z) and represent many different values.

An equation that holds true for some, but not all, of the variable values is a *conditional* equation. Two examples of conditional equations are

$$x^2 = 16 \text{ and } y + 3 = 9$$

Note that in the first equation, the variable x could be $+4$ or -4, and no other value of x would make the statement true. In the second equation only $+6$ will make the statement true.

An *identity* is an equation that is true for all values of the variables. An example of an identity equation is

$$4z = 3z + z$$

For any value of z this statement is true.

Example: Determine whether each of the following equations is conditional or an identity.

a. $3y - y = 2y$
b. $x - 8 = -4$
c. $3(a - 2) = 3a - 6$
d. $z^4 = 16$

Solution:
a. Identity
b. Conditional with only one value of x.
c. Identity
d. Conditional with two values of z

A *linear* equation is one that contains only variables raised to the first power and contains only one variable in any one term. The general form of a linear equation in one unknown is

$$ax + b = 0$$

where a and b are constants and x is a variable. Solving for the variable,

$$x = \frac{-b}{a}$$

Graphical representations of linear equations with two unknowns will be presented in a later unit.

Example: Determine whether or not these equations are linear.

a. $3x + 7 = 5$
b. $9 = 8(x + 3)$
c. $x^2 + 2x + 1 = 0$
d. $5xy + 7 = 0$

Solution:
a. $3x + 7 = 5$ or $3x - 2 = 0$ is linear.
b. $9 = 8(x + 3)$ or $8x + 15 = 0$ is linear.
c. $x^2 + 2x + 1 = 0$ is not linear due to x^2.
d. $5xy + 7 = 0$ is not linear due to xy.

Solving Equations

In order to solve, or find the root, of an equation it is necessary to ascertain the value(s) of a variable that will make the equation true. The process by which this is accomplished involves transforming the equation such that the unknown value is on one side of the equals sign and all other values are on the other. **Addition or subtraction of the same value on both sides of an equation maintains equivalent expressions.**

Example: Given the equation 4 + 3 = 7, (a) add 5 to both sides and (b) subtract 18 from both sides.

Solution: a.

$$4 + 3 = 7$$
$$4 + 3 + 5 = 7 + 5$$
$$12 = 12$$

b.

$$4 + 3 = 7$$
$$4 + 3 - 18 = 7 - 18$$
$$7 - 18 = 7 - 18$$
$$-11 = -11$$

Multiplication or division of the same value (not zero) on both sides of an equation maintains equivalent expressions.

Example: Given the equation 19 − 7 = 8 + 4, (a) multiply both sides by 3.5 and (b) divide both sides by 24.

Solution: a.

$$(3.5)(19 - 7) = (3.5)(8 + 4)$$
$$(3.5)(12) = (3.5)(12)$$
$$42 = 42$$

b.

$$(19 - 7) \div 24 = (8 + 4) \div 24$$
$$12 \div 24 = 12 \div 24$$
$$0.5 = 0.5$$

There will be many equations involving algebraic fractions. A common method employed in solving these equations is known as *cross-multiplication*. This concept was presented in previous discussions of ratio and proportion, and involves the product of the means and extremes. Follow these examples carefully.

Example: Use cross-multiplication on this equation, $\frac{36}{9} = \frac{44}{11}$.

Solution:

$$\frac{36}{9} \gtrless \frac{44}{11}$$
$$(36)(11) = (9)(44)$$
$$396 = 396$$

Example: Use cross-multiplication on this equation, $\frac{3x^2}{4} = \frac{6x^2}{8}$.

Solution:

$$\frac{3x^2}{4} \gtrless \frac{6x^2}{8}$$
$$(3x^2)(8) = (6x^2)(4)$$
$$24x^2 = 24x^2$$

Multiplication or division of the variable on both sides of an equation may result in nonequivalent expressions.

Example: Given the equation $5y = 25$, multiply both sides by y.

Solution: It should be obvious that the root of the original equation is $y = 5$. Multiplying both sides by y produces $5y^2 = 25y$ which introduces an *extraneous root* as both 5 and 0 will satisfy the equation. The point is that caution should be exercised when multiplying both sides of an equation by the variable.

Raising both sides of an equation to a power may result in nonequivalent expressions.

Example: Given the equation $z = 4$, square both sides of the equation.

Solution: Squaring both sides produces $z^2 = 16$ which again introduces an extraneous root as both 4 and -4 will satisfy the equation.

Applying these rules, in addition to the algebraic techniques presented in Chapter 4, will solve the majority of the equations you will face in your study of electronics. Once appropriate procedures have been applied, and the root(s) identified, the problem should be checked by substituting the value(s) into the original expression. Study the following examples.

Example: Solve the following equations.

a. $4y + 2 = 30$

b. $6 - 2x = 18 + x$

c. $\dfrac{9z}{4} = 3z + 2$

d. $\dfrac{1}{y + 2} = \dfrac{3}{y - 5}$

e. $x^2 - x - 6 = 0$

Solution:

a.
$$4y + 2 = 30$$
$$4y = 28 \quad \text{(Subtraction of 2 from both sides)}$$
$$y = 7 \quad \text{(Division by 4 on both sides)}$$

b.
$$6 - 2x = 18 + x$$
$$6 - 3x = 18 \quad \text{(Subtraction of } x\text{)}$$
$$-3x = 12 \quad \text{(Subtraction of 6)}$$
$$x = -4 \quad \text{(Division by } -3\text{)}$$

c.
$$\frac{9z}{4} = 3z + 2$$
$$9z = 4(3z + 2) \quad \text{(Multiplication by 4)}$$
$$9z = 12z + 8 \quad \text{(Removal of parentheses)}$$
$$-3z = 8 \quad \text{(Subtraction of } 12z\text{)}$$
$$z = -\frac{8}{3} \quad \text{(Division by } -3\text{)}$$

d.
$$\frac{1}{y + 2} = \frac{3}{y - 5}$$
$$3(y + 2) = 1(y - 5) \quad \text{(Cross-multiplication)}$$
$$3y + 6 = y - 5 \quad \text{(Removal of parentheses)}$$
$$2y + 6 = -5 \quad \text{(Subtraction of } y\text{)}$$

$2y = -11$ (Subtraction of 6)

$y = -\frac{11}{2}$ (Division by 2)

e. $x^2 - x - 6 = 0$

$(x + 2)(x - 3) = 0$ (In order for the result to be 0, one or both of the factored expressions must also be equal to 0.)

$x + 2 = 0$

$x = -2$

$x - 3 = 0$

$x = 3$ (Therefore, if $x = -2$ or 3 then the equation is true.)

Example: Solve the following equations.

a. $\frac{6(x - 2)}{x + 3} = 4$

b. $3y^2 - 3y - 126 = 0$

c. $15(z - 8) - z = 19(z - 5)$

d. $1 - \frac{x}{1 - x} = 10$

e. $\frac{x}{x - 8} + \frac{4}{x} = 0$

Solution: a.

$$\frac{6(x - 2)}{x + 3} = 4$$
$$6(x - 2) = 4(x + 3)$$
$$6x - 12 = 4x + 12$$
$$2x = 24$$
$$x = 12$$

b.

$$3y^2 - 3y - 126 = 0$$
$$3(y^2 - y - 42) = 0$$
$$3(y - 7)(y + 6) = 0$$
$$y - 7 = 0$$
$$y = 7$$
$$y + 6 = 0$$
$$y = -6$$

c.

$$15(z - 8) - z = 19(z - 5)$$
$$15z - 120 - z = 19z - 95$$
$$14z - 120 = 19z - 95$$
$$-5z = 25$$
$$z = -5$$

Check:

$$15(-5 - 8) - (-5) = 19(-5 - 5)$$
$$15(-13) + 5 = 19(-10)$$
$$-195 + 5 = -190$$
$$-190 = -190$$

d.

$$1 - \frac{x}{1 - x} = 10$$
$$1 - x\left(1 - \frac{x}{1 - x}\right) = 10(1 - x)$$
$$1 - x - x = 10 - 10x$$
$$1 - 2x = 10 - 10x$$
$$8x = 9$$
$$x = \frac{9}{8}$$

e.
$$\frac{x}{x-8} + \frac{4}{x} = 0$$
$$x(x-8)\left[\frac{x}{x-8} + \frac{4}{x}\right] = x(x-8)(0)$$
$$x^2 + 4(x-8) = 0$$
$$x^2 + 4x - 32 = 0$$
$$(x+8)(x-4) = 0$$
$$x + 8 = 0$$
$$x = -8$$
$$x - 4 = 0$$
$$x = 4$$

Formulas

Students of electronics often wonder why algebra is highly emphasized. True, a computer technician is rarely given a binomial and asked to factor it. But as students are acutely aware, there are many formulas used at all levels of electronics study. In order to understand how to rearrange a formula a basic knowledge of algebra is required.

A significant percentage of the formulas in electronics consist entirely of literal terms. In order to solve a particular problem it is necessary to transpose terms until the unknown value is on one side of the equation and all other values are on the other side of the equation.

Example: Solve $P = I^2R$ for I.

Solution:
$$P = I^2R$$
$$I^2 = \frac{P}{R}$$
$$I = \sqrt{\frac{P}{R}}$$

Example: Solve $X_C = \frac{1}{2\pi fC}$ for f.

Solution:
$$X_C = \frac{1}{2\pi fC}$$
$$fX_C = \frac{1}{2\pi C}$$
$$f = \frac{1}{2\pi CX_C}$$

Example: Solve $R_t = \frac{R_1R_2}{R_1 + R_2}$ for R_1.

Solution:

$$R_t = \frac{R_1R_2}{R_1 + R_2}$$
$$R_t(R_1 + R_2) = R_1R_2$$
$$R_tR_1 + R_tR_2 = R_1R_2$$
$$R_tR_2 = R_1R_2 - R_tR_1$$
$$R_tR_2 = R_1(R_2 - R_t)$$
$$R_1 = \frac{R_tR_2}{R_2 - R_t}$$

Once again, note that algebra was used extensively to solve the previous example. Without a knowledge of factoring it would be difficult to rearrange a formula such as this.

EXERCISE 5·1

Find the solution(s) for each of the following equations. If necessary, round the answer to three decimal places.

1. $8x + 2 = 30$
2. $\frac{y}{y - 7.2} = 10$
3. $z - \frac{z}{3} = 8$
4. $7x - (8 - x) + x = 122 - (-x)$
5. $x^2 - 25 = 0$
6. $y - 5(y + 2) = y - 10$
7. $w^2 - 8w + 10 = -5$
8. $\frac{4}{x} - \frac{1}{x - 2} = \frac{6}{x}$
9. $1 + \frac{x + 2}{x - 2} = 12$
10. $7z + 19 = 6z - 32$
11. $\frac{5y}{3} + \frac{y}{2} = y + 7$
12. $8(7x - 2) - 12x = \frac{5(x + 10) + 1}{2}$
13. $w - (3 + 3a) - 7w = -3(a - w)$
14. $z + \frac{5(z - 6)}{8} = z - 40$
15. $14 + 5y - y^2 = 0$
16. $\frac{x + 1}{x^2 - 1} = 0.1$
17. $\frac{1}{w - 1} = \frac{2}{w - 5}$
18. $6z = 0$
19. $\frac{1}{\frac{1}{x - 3}} = 4x + 3$
20. $-2[3x - (-2x + 7) - 8(2 - x)] = 5(x - 4) + 10x$
21. $\frac{x}{x + 4} = \frac{2}{x}$
22. $5(y - 1) + y - 2(y + 6) - 1 = \frac{y}{2}$
23. $\frac{x^2 - x - 20}{x^2 - 25} + 1 = 4$
24. $\frac{\frac{1}{y - 1}}{\frac{1}{y + 2}} = \frac{\frac{1}{y + 2}}{\frac{1}{y - 1}}$
25. $1 + \frac{1}{x + 1} = 1 + \frac{2}{x + 3}$

Solve for the indicated variable in each of the following electronics formulas.

26. $E = \frac{P}{I}$, solve for I.
27. $P = I^2R$, solve for I.

28. $I = \sqrt{\dfrac{P}{R}}$, solve for R.

29. $R = \dfrac{KL}{A}$, solve for L.

30. $C = \dfrac{C_1C_2}{C_1 + C_2}$, solve for C_2.

31. $G = \dfrac{1}{R}$, solve for R.

32. $R_s = \dfrac{I_m \times R_m}{I_s}$, solve for R_m.

33. $E_b = E_{cc} \times \dfrac{R_2}{R_1 + R_2}$, solve for R_2.

34. $X_L = 2\pi fL$, solve for f.

35. $X_C = \dfrac{1}{2\pi fC}$, solve for C.

36. $Z = \sqrt{R^2 + X_C^2}$, solve for R.

37. $Z = \sqrt{R^2 + (X_C - X_L)^2}$, solve for R^2.

38. $f_r = \dfrac{1}{2\pi\sqrt{LC}}$, solve for L.

39. $I_{CEO} = (\beta + 1)I_{CBO}$, solve for β.

40. $A_f = \dfrac{A}{1 - A\beta}$, solve for A.

41. $P_t = P_C\left(1 + \dfrac{M^2}{2}\right)$, solve for M.

42. $E_{R_L} = \dfrac{R_L}{R_I + R_L} \times E_s$, solve for R_I.

43. $L_t = L_1 + L_2 + 2L_m$, solve for L_m.

44. $C = 8.85 \times 10^{-12} K \dfrac{A}{d}$, solve for K.

45. $\dfrac{1}{C_t} = \dfrac{1}{C_1} + \dfrac{1}{C_2}$, solve for C_t.

46. $M = K\sqrt{L_1L_2}$, solve for L_1.

47. $I_B = I_{BQ} + \sqrt{2}\, I_s$, solve for I_s.

48. $I_s = \dfrac{V_I - V_O}{R_s}$, solve for V_I.

49. $\alpha = \dfrac{\beta}{\beta + 1}$, solve for β.

50. $\beta = \dfrac{\alpha}{1 - \alpha}$, solve for α.

UNIT 5·2 SIMULTANEOUS LINEAR EQUATIONS

Objectives:

After studying this unit, you should be able to

- solve equations with two unknowns through the elimination process.
- solve equations with two unknowns through the substitution process.
- solve equations with two unknowns through the comparison process.
- solve equations with three unknowns through the elimination process, substitution process, and through the use of determinants.

Equations with Two Unknowns

The equations presented in the previous unit contained one unknown. Through various algebraic techniques it was possible to determine the value(s) of the unknown variable. This unit will discuss methods of solving equations with more than one unknown. When two or more equations apply to a situation at the same time, they are called *simultaneous equations*.

Given values for voltage and current it is possible to use Ohm's Law to determine resistance. But consider the situation when only voltage is known. The formula might appear as

$$E = I \times R \text{ or } 12\text{ V} = I \times R$$

Obviously the value for current or resistance is unknown without additional information. When an equation has two unknowns, it requires two equations to solve for both values. This unit will focus on the elimination of values through *addition, subtraction,* and *substitution* methods for solving two equations with two unknowns.

Elimination through Addition and Subtraction

Given two equations, such as

$$4x + 2y = 24 \text{ and } x = y - 3$$

It is possible to determine the value of x if y can be eliminated. The following example will demonstrate how to solve these equations for x and y.

Example: Find the values of x and y in these equations.

$4x + 2y = 24$
$x = y - 3$

Solution: Rearrange the equations such that the unknowns are lined up under similar terms.

$4x + 2y = 24$
$x - y = -3$

Inspect the numerical coefficients to determine if two unknowns have identical coefficients. If so, then addition or subtraction may eliminate that variable. In this example it will be necessary to increase the x to $4x$ or the $-y$ to $-2y$. The latter may be the best choice due to the fact that simple addition will then eliminate y. Multiplying *both* sides of the second equation by $+2$ gives:

$$4x + 2y = 24$$
$$2x - 2y = -6$$

The next step is to add within each column. This will produce an equation with one unknown, which can be easily solved.

$$\begin{array}{rcl} 4x + 2y & = & 24 \\ +(2x - 2y) & = & +(-6) \\ \hline 6x & = & 18 \\ x & = & 3 \end{array}$$

To determine the value of y, simply substitute the value of x into *either* of the original equations. In this case we will substitute x into both to demonstrate that y can be found both ways.

$$\begin{array}{rcl} 4x + 2y & = & 24 \\ 4(3) + 2y & = & 24 \\ 12 + 2y & = & 24 \\ 2y & = & 12 \\ y & = & 6 \end{array} \qquad \text{or,} \qquad \begin{array}{rcl} x & = & y - 3 \\ 3 & = & y - 3 \\ y & = & 6 \end{array}$$

In order to check these values substitute both values into the original equations.

$$\begin{array}{rcl} 4x + 2y & = & 24 \\ 4(3) + 2(6) & = & 24 \\ 12 + 12 & = & 24 \\ 24 & = & 24 \end{array} \qquad \text{and,} \qquad \begin{array}{rcl} x & = & y - 3 \\ 3 & = & 6 - 3 \\ 3 & = & 3 \end{array}$$

Example: Find the values of x and y in these equations.

$$5x = -y - 24$$
$$x + 8 = 3y$$

Solution: Arrange the equations in the standard format.

$$5x + y = -24$$
$$x - 3y = -8$$

Multiply both sides of the second equation by 5 and subtract to solve for y.

$$\begin{array}{rcl} 5x + y & = & -24 \\ 5x - 15y & = & -40 \\ \hline 16y & = & 16 \\ y & = & 1 \end{array}$$

Substitute for y to determine x.

$$\begin{aligned} 5x &= -y - 24 \\ 5x &= -1 - 24 \\ 5x &= -25 \\ x &= -5 \end{aligned}$$

Example: Determine the values of x and y in these equations.

$$3x + y = 14$$
$$2x - 7y = -52$$

Solution: Multiply the first equation by 2 and the second by -3. Add the equations.

$$\begin{aligned} 6x + 2y &= 28 \\ -6x + 21y &= 156 \\ \hline 23y &= 184 \\ y &= 8 \end{aligned}$$

Solve for x.

$$\begin{aligned} 3x + y &= 14 \\ 3x + 8 &= 14 \\ 3x &= 6 \\ x &= 2 \end{aligned}$$

Elimination through Substitution

The substitution process involves solving one equation in terms of the unknown. This equivalent expression is then substituted in place of the unknown into the other equation. The result is an equation with one unknown, which can easily be solved. Review the following examples.

Example: Solve for x and y in these equations.

$$x + 3y = 6$$
$$-4x - y = 20$$

Solution: Select one of the equations and solve for one of the unknowns. The choice is arbitrary, but should be made so that the substitution is as simple as possible. In this case solving the first equation for x might be a logical choice.

$$x = -3y + 6$$

Substitute this equivalent expression for x into the second equation.

$$\begin{aligned} -4(-3y + 6) - y &= 20 \\ 12y - 24 - y &= 20 \\ 11y &= 44 \\ y &= 4 \end{aligned}$$

Solve for x.

$$\begin{aligned} x + 3(4) &= 6 \\ x + 12 &= 6 \\ x &= -6 \end{aligned}$$

Example: Find the values of x and y in these equations.

$$\frac{x + 1}{3} - 2y = -5$$
$$2x + 6y = 31$$

Solution: Solve for y in the second equation.

$$2x + 6y = 31$$
$$6y = 31 - 2x$$
$$y = \frac{31 - 2x}{6}$$

Substitute for y.

$$\frac{x + 1}{3} - 2\left(\frac{31 - 2x}{6}\right) = -5$$
$$\frac{x + 1}{3} - \frac{62 - 4x}{6} = -5$$
$$2x + 2 - 62 + 4x = -30$$
$$6x - 60 = -30$$
$$6x = 30$$
$$x = 5$$

Solve for y.

$$2x + 6y = 31$$
$$2(5) + 6y = 31$$
$$10 + 6y = 31$$
$$6y = 21$$
$$y = 3.5$$

Check the results.

$$\frac{x + 1}{3} - 2y = -5$$
$$\frac{5 + 1}{3} - 2(3.5) = -5$$
$$2 - 7 = -5$$
$$-5 = -5$$

Elimination through Comparison

This method of solving equations with two unknowns is essentially the same as the substitution method. The process involves solving both equations for the *same* variable. As these two variables are equal, the equivalent expressions must also be equal. Setting these two expressions equal to each other eliminates one variable, thereby leaving one unknown. Follow this example carefully.

Example: Determine the values of x and y in these equations.

$$6x + y = 2y + 36$$
$$2x - 7y = 52$$

Solution: The first step is to solve both equations for x or y. For a change, solve for y. The first equation becomes

$$\begin{aligned} 6x + y &= 2y + 36 \\ 6x - y &= 36 \\ -y &= 36 - 6x \\ y &= 6x - 36 \end{aligned}$$

The second equation is

$$\begin{aligned} 2x - 7y &= 52 \\ -7y &= 52 - 2x \\ y &= \frac{2x - 52}{7} \end{aligned}$$

Setting the two expressions for y equal results in

$$\begin{aligned} y &= y \\ 6x - 36 &= \frac{2x - 52}{7} \\ 7(6x - 36) &= 2x - 52 \\ 42x - 252 &= 2x - 52 \\ 40x &= 200 \\ x &= 5 \end{aligned}$$

Solve for y.

$$\begin{aligned} 2x - 7y &= 52 \\ 2(5) - 7y &= 52 \\ -7y &= 42 \\ y &= -6 \end{aligned}$$

Equations with Three Unknowns

When an equation contains three unknowns, it requires three equations to determine the values of each unknown. The process involves selecting two of the equations and eliminating one of the variables. Another pair of equations is selected, and the same variable is eliminated. This results in two new equations with two unknowns. These can be solved through addition, subtraction, or substitution.

Example: Solve for the unknowns in these equations.

(1) $2x + y - z = 6$
(2) $x - 3y + 4z = 28$
(3) $x - y + 2z = 16$

Solution: Subtract equation (3) from equation (2).

$$\begin{array}{rcr} x - 3y + 4z &=& 28 \\ -\ (x - \ \ y + 2z) &=& -(16) \\ \hline -\ 2y + 2z &=& 12 \end{array}$$

Multiply equation (2) by 2, and then subtract from equation (1).

$$\begin{array}{r} 2x + y - z = 6 \\ -2x + 6y - 8z = -56 \\ \hline 7y - 9z = -50 \end{array}$$

The two new equations are

(4) $-2y + 2z = 12$

(5) $7y - 9z = -50$

Multiply the new equation (4) by 7, the new equation (5) by 2, and combine the results.

$$\begin{array}{r} -14y + 14z = 84 \\ 14y - 18z = -100 \\ \hline -4z = -16 \\ z = 4 \end{array}$$

Substitute z into equation (4) to find the value of y.

$$\begin{aligned} -2y + 2z &= 12 \\ -2y + 2(4) &= 12 \\ -2y + 8 &= 12 \\ -2y &= 4 \\ y &= -2 \end{aligned}$$

Substitute y and z into equation (1) to find the value of x.

$$\begin{aligned} 2x + y - z &= 6 \\ 2x + (-2) - (4) &= 6 \\ 2x - 2 - 4 &= 6 \\ 2x - 6 &= 6 \\ 2x &= 12 \\ x &= 6 \end{aligned}$$

To check these values place x, y, and z in equation (2).

$$\begin{aligned} x - 3y + 4z &= 28 \\ 6 - 3(-2) + 4(4) &= 28 \\ 6 + 6 + 16 &= 28 \\ 28 &= 28 \end{aligned}$$

Example: Solve for the unknowns in these equations

(1) $x + 3y - 7z = -3$

(2) $-6x - 8y + z = 7$

(3) $3x + y - 6z = -18$

Solution: You should verify that $x = -5$, $y = 3$, and $z = 1$.

Determinants

Another method for solving two equations with two unknowns involves the use of *determinants*. Examine these two equations.

$$(1)\ a_1x + b_1y = c_1$$
$$(2)\ a_2x + b_2y = c_2$$

Note that there are two unknowns (x and y) and six different constants. In order to solve these two equations for x, it will be necessary to multiply equation (1) by b_2 and equation (2) by b_1. This produces

$$a_1b_2x + b_1b_2y = b_2c_1$$
$$a_2b_1x + b_1b_2y = b_1c_2$$

Subtracting gives

$$a_1b_2x - a_2b_1x = b_2c_1 - b_1c_2$$

Factoring x produces

$$x(a_1b_2 - a_2b_1) = b_2c_1 - b_1c_2$$

Solving for x gives

$$x = \frac{b_2c_1 - b_1c_2}{a_1b_2 - a_2b_1}$$

A similar process for y will give

$$y = \frac{a_1c_2 - a_2c_1}{a_1b_2 - a_2b_1}$$

Note that for both x and y, the denominators are the same. Given the numerical coefficients it is possible to substitute and quickly determine the values for both x and y. The denominator of both expressions is termed a *second-order determinant* and is written in the following format.

$$\begin{vmatrix} a_1 & b_1 \\ a_2 & b_2 \end{vmatrix} = a_1b_2 - a_2b_1$$

This determinant is second-order due to the fact that there are two *elements* in the first *row* (a_1 and b_1) and two in the second row (a_2 and b_2). The first *column* contains two elements (a_1 and a_2) and the second column contains two elements (b_1 and b_2). Two other important terms pertaining to the determinant are the *principal diagonal* and the *secondary diagonal.*

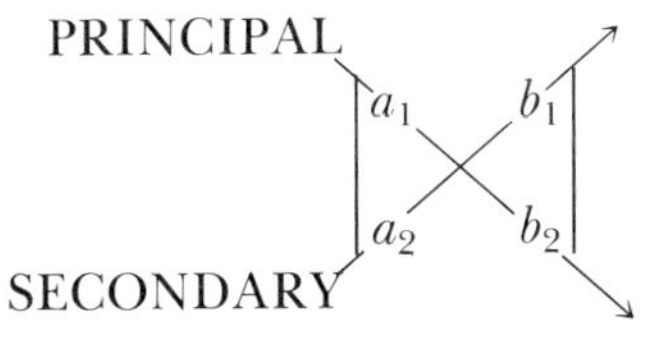

Evaluating a second-order determinant involves finding the product of the elements in the principal diagonal and then subtracting the product of the elements in the secondary diagonal.

Example: Evaluate this second-order determinant,

$$\begin{vmatrix} 5 & 1 \\ 3 & 2 \end{vmatrix}$$

Solution:

$$\begin{vmatrix} 5 & 1 \\ 3 & 2 \end{vmatrix} = (5)(2) - (3)(1) = 10 - 3 = 7$$

Example: Evaluate this second-order determinant.

$$\begin{vmatrix} 2 & -3 \\ 4 & -4 \end{vmatrix}$$

Solution:

$$\begin{vmatrix} 2 & -3 \\ 4 & -4 \end{vmatrix} = (2)(-4) - (4)(-3) = -8 + 12 = 4$$

Referring back to the expressions for x and y it is possible to write:

$$x = \frac{\begin{vmatrix} c_1 & b_1 \\ c_2 & b_2 \end{vmatrix}}{\begin{vmatrix} a_1 & b_1 \\ a_2 & b_2 \end{vmatrix}} \qquad y = \frac{\begin{vmatrix} a_1 & c_1 \\ a_2 & c_2 \end{vmatrix}}{\begin{vmatrix} a_1 & b_1 \\ a_2 & b_2 \end{vmatrix}}$$

Example: Solve these two equations using determinants.

$3x - 2y = -18$
$x + 3y = 16$

Solution:

$$x = \frac{\begin{vmatrix} -18 & -2 \\ 16 & 3 \end{vmatrix}}{\begin{vmatrix} 3 & -2 \\ 1 & 3 \end{vmatrix}} = \frac{(-18)(3) - (16)(-2)}{(3)(3) - (1)(-2)} = \frac{-22}{11} = -2$$

$$y = \frac{\begin{vmatrix} 3 & -18 \\ 1 & 16 \end{vmatrix}}{\begin{vmatrix} 3 & -2 \\ 1 & 3 \end{vmatrix}} = \frac{(3)(16) - (1)(-18)}{(3)(3) - (1)(-2)} = \frac{66}{11} = 6$$

The solution is $x = -2$ and $y = 6$

Example: Solve these two equations using determinants.

$x - 3y = 0$
$5x - 9y = 6$

Solution:

$$x = \frac{\begin{vmatrix} 0 & -3 \\ 6 & -9 \end{vmatrix}}{\begin{vmatrix} 1 & -3 \\ 5 & -9 \end{vmatrix}} = \frac{(0)(-9) - (6)(-3)}{(1)(-9) - (5)(-3)} = \frac{18}{6} = 3$$

$$y = \frac{\begin{vmatrix} 1 & 0 \\ 5 & 6 \end{vmatrix}}{\begin{vmatrix} 1 & -3 \\ 5 & -9 \end{vmatrix}} = \frac{(1)(6) - (5)(0)}{(1)(-9) - (5)(-3)} = \frac{6}{6} = 1$$

The solution is $x = 3$ and $y = 1$

Example: Use determinants to solve for x and y in these equations.

$x + 7y = -24$
$4x - 2y = 24$

Solution: You should verify that $x = 4$ and $y = -4$.

Third-Order Determinants

Third-order determinants are used to solve three equations with three unknowns. The process is the same as with second-order determinants. Given these equations:

$$\begin{aligned} &(1)\ a_1x + b_1y + c_1z = d_1 \\ &(2)\ a_2x + b_2y + c_2z = d_2 \\ &(3)\ a_3x + b_3y + c_3z = d_3 \end{aligned}$$

The following solutions for x, y, and z can be found.

$$x = \frac{b_2c_3d_1 + b_1c_2d_3 + b_3c_1d_2 - b_2c_1d_3 - b_3c_2d_1 - b_1c_3d_2}{a_1b_2c_3 + a_3b_1c_2 + a_2b_3c_1 - a_3b_2c_1 - a_1b_3c_2 - a_2b_1c_3}$$

$$y = \frac{a_1c_3d_2 + a_3c_2d_1 + a_2c_1d_3 - a_3c_1d_2 - a_1c_2d_3 - a_2c_3d_1}{a_1b_2c_3 + a_3b_1c_2 + a_2b_3c_1 - a_3b_2c_1 - a_1b_3c_2 - a_2b_1c_3}$$

$$z = \frac{a_1b_2d_3 + a_3b_1d_2 + a_2b_3d_1 - a_3b_2d_1 - a_1b_3d_2 - a_2b_1d_3}{a_1b_2c_3 + a_3b_1c_2 + a_2b_3c_1 - a_3b_2c_1 - a_1b_3c_2 - a_2b_1c_3}$$

The denominator of each expression is a third-order determinant and is written in the following format.

$$\begin{vmatrix} a_1 & b_1 & c_1 \\ a_2 & b_2 & c_2 \\ a_3 & b_3 & c_3 \end{vmatrix} = a_1b_2c_3 + a_3b_1c_2 + a_2b_3c_1 - a_3b_2c_1 - a_1b_3c_2 - a_2b_1c_3$$

Evaluating a third-order determinant involves setting up the standard format and then repeating the first and second columns immediately to the right of the determinant. As can be seen by the following diagram, there are three principal diagonals and three secondary diagonals.

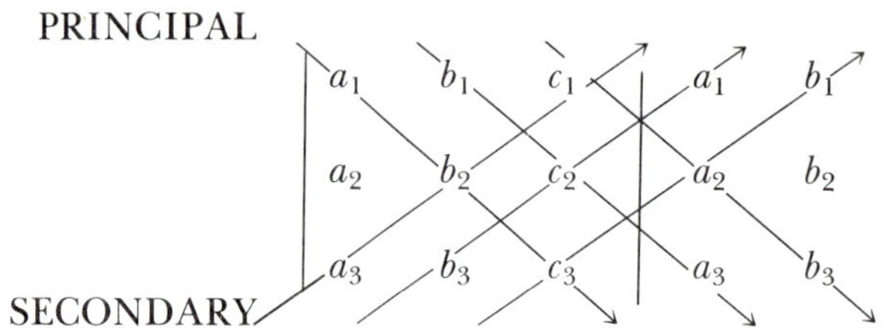

The first step is to sum the products of the three principal diagonals. From this sum the products of the three secondary diagonals are subtracted.

Example: Evaluate this third-order determinant.

$$\begin{vmatrix} 1 & 3 & 4 \\ 0 & 2 & 3 \\ -4 & 7 & -5 \end{vmatrix}$$

Solution: Repeat the first and second columns giving

$$\begin{vmatrix} 1 & -3 & 4 \\ 0 & 2 & 3 \\ -4 & 7 & -5 \end{vmatrix}\begin{matrix} 1 & -3 \\ 0 & 2 \\ -4 & 7 \end{matrix} \begin{aligned} &= (1)(2)(-5) + (-3)(3)(-4) + (4)(0)(7) \\ &\quad - (-4)(2)(4) - (7)(3)(1) - (-5)(0)(-3) \\ &= -10 + 36 + 0 + 32 - 21 + 0 \\ &= 37 \end{aligned}$$

Example: Evaluate this third-order determinant.

$$\begin{vmatrix} -2 & 4 & 3 \\ 1 & -5 & 0 \\ 6 & -2 & -1 \end{vmatrix}$$

Solution:

$$\begin{vmatrix} -2 & 4 & 3 \\ 1 & -5 & 0 \\ 6 & -2 & -1 \end{vmatrix}\begin{matrix} -2 & 4 \\ 1 & -5 \\ 6 & -2 \end{matrix} \begin{aligned} &= (-2)(-5)(-1) + (4)(0)(6) \\ &\quad + (3)(1)(-2) - (6)(-5)(3) \\ &\quad - (-2)(0)(-2) - (-1)(1)(4) \\ &= -10 + 0 - 6 + 90 + 0 + 4 \\ &= 78 \end{aligned}$$

Referring back to the three equations for x, y, and z, it is possible to write the following expressions:

$$x = \frac{\begin{vmatrix} d_1 & b_1 & c_1 \\ d_2 & b_2 & c_2 \\ d_3 & b_3 & c_3 \end{vmatrix}}{\begin{vmatrix} a_1 & b_1 & c_1 \\ a_2 & b_2 & c_2 \\ a_3 & b_3 & c_3 \end{vmatrix}} \qquad y = \frac{\begin{vmatrix} a_1 & d_1 & c_1 \\ a_2 & d_2 & c_2 \\ a_3 & d_3 & c_3 \end{vmatrix}}{\begin{vmatrix} a_1 & b_1 & c_1 \\ a_2 & b_2 & c_2 \\ a_3 & b_3 & c_3 \end{vmatrix}} \qquad z = \frac{\begin{vmatrix} a_1 & b_1 & d_1 \\ a_2 & b_2 & d_2 \\ a_3 & b_3 & d_3 \end{vmatrix}}{\begin{vmatrix} a_1 & b_1 & c_1 \\ a_2 & b_2 & c_2 \\ a_3 & b_3 & c_3 \end{vmatrix}}$$

Example: Use determinants to solve for the unknowns in these equations.

$$3x + 2y - z = 7$$
$$x - y + 5z = -15$$
$$-2x + 3y + z = 12$$

Solution:

$$x = \frac{\left|\begin{array}{ccc|cc} 7 & 2 & -1 & 7 & 2 \\ -15 & -1 & 5 & -15 & -1 \\ 12 & 3 & 1 & 12 & 3 \end{array}\right|}{\left|\begin{array}{ccc|cc} 3 & 2 & -1 & 3 & 2 \\ 1 & -1 & 5 & 1 & -1 \\ -2 & 3 & 1 & -2 & 3 \end{array}\right|}$$

$$= \frac{(-7) + (120) + (45) - (12) - (105) - (-30)}{(-3) + (-20) + (-3) - (-2) - (45) - (2)}$$

$$= \frac{71}{-71}$$

$$= -1$$

$$y = \frac{\left|\begin{array}{ccc|cc} 3 & 7 & -1 & 3 & 7 \\ 1 & -15 & 5 & 1 & -15 \\ -2 & 12 & 1 & -2 & 12 \end{array}\right|}{-71}$$

$$= \frac{(-45) + (-70) + (-12) - (-30) - (180) - (7)}{-71}$$

$$= \frac{-284}{-71}$$

$$= 4$$

$$z = \frac{\left|\begin{array}{ccc|cc} 3 & 2 & 7 & 3 & 2 \\ 1 & -1 & -15 & 1 & -1 \\ -2 & 3 & 12 & -2 & 3 \end{array}\right|}{-71}$$

$$= \frac{(-36) + (60) + (21) - (14) - (-135) - (24)}{-71}$$

$$= \frac{142}{-71}$$

$$= -2$$

The values of the unknowns are $x = -1$, $y = 4$, and $z = -2$. In order to check these values substitute the numbers into one of the original equations.

$$3x + 2y - z = 7$$
$$3(-1) + 2(4) - (-2) = 7$$
$$-3 + 8 + 2 = 7$$
$$7 = 7$$

Example: Use determinants to solve for the unknowns in these equations.

$$x - 4y + z = 20$$
$$3x - y + 5z = 37$$
$$x + 6y - 2z = -25$$

Solution: You should verify that these values are $x = 3$, $y = -3$, and $z = 5$.

EXERCISE 5·2

Determine the values of the unknown variables in each of the following problems.

1. $x + y = 18$
   $x - y = -22$
2. $3x + 51 = 7y$
   $x + \frac{y}{2} = 0$
3. $6x - 2 = y$
   $4x + 8 + y = 0$
4. $\frac{x + 3}{3} - 4y = 10$
   $4x + \frac{2y}{3} = -38$
5. $\frac{x - 1}{6} + \frac{y + 8}{12} = 1$
   $x - y = 0$
6. $3x - 7y = 35$
   $x + 4y = -20$
7. $\frac{x + 6}{2} - \frac{y + 8}{7} = 5$
   $3y - 2x = 2$
8. $18x + 5y = -3$
   $\frac{y}{x} = -3$
9. $\frac{x + 1}{y + 2} = -1$
   $x - 2y = 18$
10. $4x - y = 15$
    $x + 3y = 20$
11. $5x - \frac{y}{2} = 28 - x$
    $2x + \frac{2}{y} = 8$
12. $\frac{5x - 8}{3} - y = 0$
    $x = y$
13. $x + \frac{3 - y}{-2} = -1$
    $\frac{x}{6} - \frac{11}{2} = y$
14. $x - 2y = 18$
    $x + 3y = 10$
15. $12x - 6y = 15$
    $8x + \frac{15}{y} = -13$
16. $\frac{x - y}{2} + y = -1$
    $\frac{y - x}{5} - \frac{x + y}{2} = 3$
17. $\frac{6x}{y} - 2 = 0$
    $x + y = -12$
18. $x - \frac{4y}{8} = -1$
    $\frac{x - 5}{3} - \frac{y - 7}{2} = \frac{3}{2}$
19. $\frac{6x - 4}{5} + \frac{5y}{3} = -1$
    $\frac{6x}{y} = -8$
20. $\frac{3(x - 2)}{9} - \frac{4y}{5} = \frac{x - y}{2}$
    $2x + 3y = -5$

21. $x + y - 2z = -7$
$3x + 4y + z = 1$
$-5x - y - 3z = 0$
22. $4x + 4y + 4z = -4$
$x - 2y + z = 5$
$3x - y - 5z = 29$
23. $x - 9y + z = -1$
$4x + y - 3z = 38$
$\frac{x}{2} + 3y + 2z = -9.5$
24. $3x - y + 4z = 26$
$x + y + z = 4$
$5x - 3y - 2z = 30$
25. $x + y + z = 0$
$3x - 2y + 4z = 34$
$4x + y - 2z = -6$
26. $\frac{-2x + 3y + 2z}{3} = 7$
$4x - y - z = -32$
$5x + 12y - 7z = -82$
27. $x + y - 3z = 9$
$x + 5y - 8z = 33$
$\frac{x}{8} + \frac{5y}{2} + 6z = -22$
28. $8x + 7y - z = 9$
$x - 3y + 5z = -13$
$6x - 4y - \frac{7z}{2} = 3$
29. $3x + 4y - 5z = 18$
$x + y + z = 12$
$x - y - z = -6$
30. $\frac{4x - 2y}{3} + z = 0$
$x - \frac{y + z}{2} = -5$
$\frac{3x - 2y + 6z}{2} = \frac{3x}{-2} + 1$

UNIT 5·3 KIRCHHOFF'S LAWS AND SIMULTANEOUS EQUATIONS

Objectives:

After studying this unit, you should be able to

- use algebraic techniques and loop analysis to solve for unknown currents and voltages.
- use algebraic techniques and node analysis to solve for unknown currents and voltages.

Loop Analysis

Kirchhoff's Laws were presented in an earlier unit. The voltage law states that the sum of the voltages around a closed path or *loop* is equal to zero. The current law states that the sum of the currents entering a point or *node* is equal to the sum of the currents leaving that point. These two laws are the basis of the loop analysis and node analysis methods presented in this unit.

Loop analysis is applied to series-parallel circuits containing two or more branch currents. The circuit in Figure 5–1 has three branch currents. Current I_1 splits to form I_2 and I_3, making a total of three current values. Determining these current values through the use of Ohm's Law would involve computing total resistance, total current, and then finding the values of I_2 and I_3. Loop analysis involves writing simultaneous loop equations, and then solving for unknown values using algebraic techniques previously discussed.

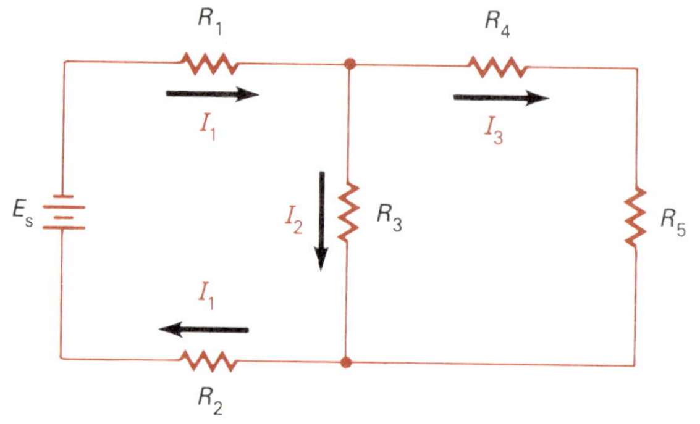

FIGURE 5–1 This series-parallel circuit contains three current values.

The current and voltage values in a circuit can be found using loop analysis. The first step is to draw loop current arrows in each closed loop. While the direction of these arrows is arbitrary, the usual convention is to draw all arrows clockwise. In the event a current is actually flowing in the opposite direction, the resultant numerical value will be negative. An independent equation must be written for *each* loop current. These equations are considered independent due to the fact that each term of the equation represents a voltage and, according to Kirchhoff, the sum of these voltages is zero.

Example: Write the loop equations for the circuit in Figure 5–2 and find the values of I_1 and I_2.

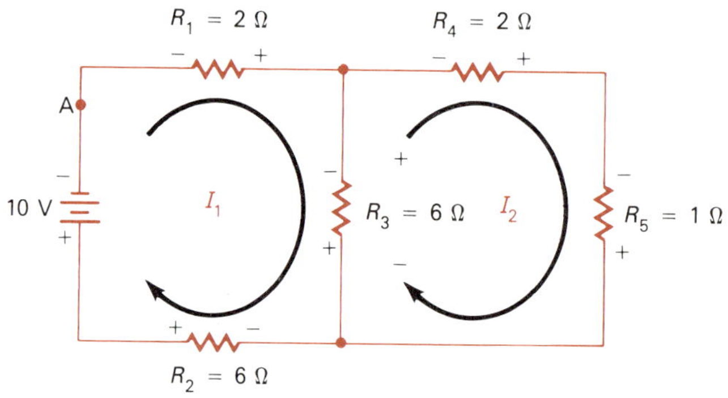

FIGURE 5–2

Solution: Pick a point in the first loop and begin moving clockwise. Each resistance will contribute a voltage expression to the equation. Those resistances with only one current passing through it will contribute positive voltages. Starting at point A the first voltage is $2I_1$. This represents the voltage developed as current I_1 passes through 2 Ω.

Continuing on around the loop brings us to R_3. The circuit indicates that two currents are passing through this resistor. It should be obvious that only one current is able to move through R_3. This current is the *difference* between the two loop currents. As we are presently concerned with I_1 this current is assumed to be larger. The voltage expression becomes $6(I_1 - I_2)$.

The third expression is $6I_1$.

The final expression in the circuit is the source voltage, E_s. Thus far in the loop an increase in potential from minus to plus has been considered a positive term. As the source voltage is passed the potential drops from positive to negative, giving a negative term. The final equation is

$$2I_1 + 6(I_1 - I_2) + 6I_1 - 10\text{ V} = 0$$
$$14I_1 - 6I_2 = 10\text{ V}$$

The equation for the right loop is

$$2I_2 + 1I_2 + 6(I_2 - I_1) = 0$$
$$-6I_1 + 9I_2 = 0$$

Note that when current I_2 passes through R_3, we assume I_2 is greater than I_1. Also, since there is no voltage source in the loop the sum of the voltage drops equals zero.

Given the two equations with two unknowns a solution may be found utilizing any of the algebraic techniques presented earlier. For this example elimination through addition will be used.

$$14I_1 - 6I_2 = 10$$
$$-6I_1 + 9I_2 = 0$$

Multiply the first equation by 1.5 and then add the equations.

$$21I_1 - 9I_2 = 15$$
$$\underline{-6I_1 + 9I_2 = 0}$$
$$15I_1 = 15$$
$$I_1 = 1\text{ A}$$

Substitute to find I_2.

$$14I_1 - 6I_2 = 10$$
$$14(1) - 6I_2 = 10$$
$$-6I_2 = -4$$
$$I_2 = \frac{2}{3}\text{ A}$$

Example: Write the loop equations for the circuit in Figure 5–3.

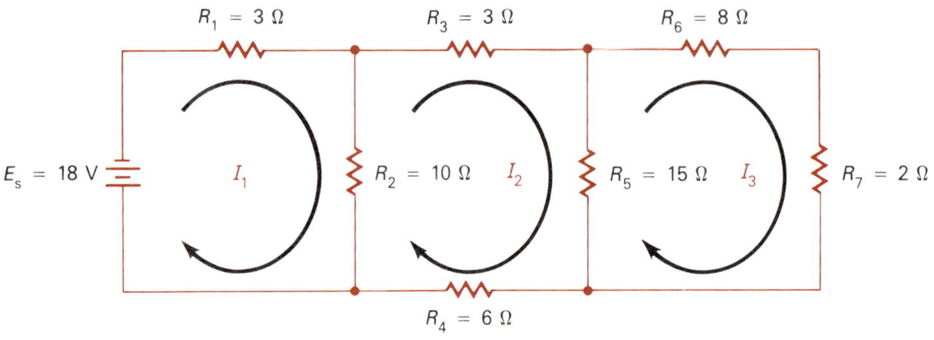

FIGURE 5–3

Solution: Loop 1: $3I_1 + 10(I_1 - I_2) - 18 = 0$

$$13I_1 - 10I_2 = 18$$

Loop 2: $3I_2 + 15(I_2 - I_3) + 6I_2 + 10(I_2 - I_1) = 0$

$$3I_2 + 15I_2 - 15I_3 + 6I_2 + 10I_2 - 10I_1 = 0$$
$$-10I_1 + 34I_2 - 15I_3 = 0$$

Loop 3: $8I_3 + 2I_3 + 15(I_3 - I_2) = 0$

$$-15I_2 + 25I_3 = 0$$

Example: Use determinants to find the values of I_1, I_2, and I_3 in Figure 5–3.

Solution: The loop equations are

$$13I_1 - 10I_2 + 0I_3 = 18$$
$$-10I_1 + 34I_2 - 15I_3 = 0$$
$$0I_1 - 15I_2 + 25I_3 = 0$$

The determinant formats are

$$I_1 = \frac{\left|\begin{array}{rrr|rr} 18 & -10 & 0 & 18 & -10 \\ 0 & 34 & -15 & 0 & 34 \\ 0 & -15 & 25 & 0 & -15 \end{array}\right|}{\left|\begin{array}{rrr|rr} 13 & -10 & 0 & 13 & -10 \\ -10 & 34 & -15 & -10 & 34 \\ 0 & -15 & 25 & 0 & -15 \end{array}\right|} = \frac{11\ 250}{5\ 625} = 2 \text{ A}$$

$$I_2 = \frac{\left|\begin{array}{rrr|rr} 13 & 18 & 0 & 13 & 18 \\ -10 & 0 & -15 & -10 & 0 \\ 0 & 0 & 25 & 0 & 0 \end{array}\right|}{5\ 625} = \frac{4\ 500}{5\ 625} = 0.8 \text{ A}$$

$$I_3 = \frac{\left|\begin{array}{rrr|rr} 13 & -10 & 18 & 13 & -10 \\ -10 & 34 & 0 & -10 & 34 \\ 0 & -15 & 0 & 0 & -15 \end{array}\right|}{5\ 625} = \frac{2\ 700}{5\ 625} = 0.48 \text{ A}$$

To check these values use the second loop equation.

$$-10I_1 + 34I_2 - 15I_3 = 0$$
$$-10(2) + 34(0.8) - 15(0.48) = 0$$
$$-20 + 27.2 - 7.2 = 0$$
$$0 = 0$$

Example: Solve for the loop currents and determine the voltage drop across R_5 in Figure 5–4.

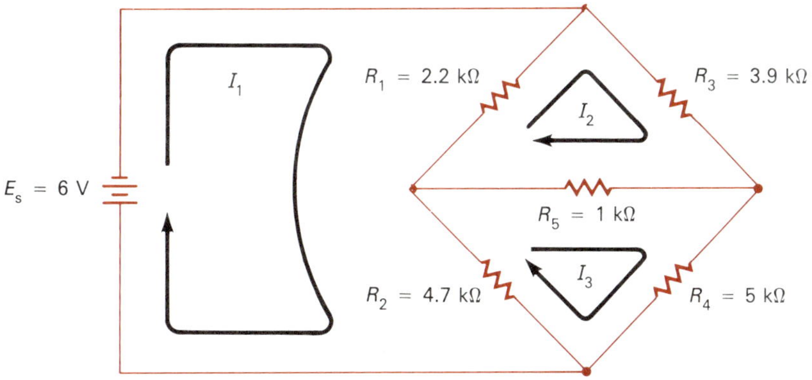

FIGURE 5–4 Loop analysis applied to a Wheatstone bridge

Solution: The three loop equations are

Loop 1: $2.2(I_1 - I_2) + 4.7(I_1 - I_3) - 6 = 0$

$$6.9I_1 - 2.2I_2 - 4.7I_3 = 6$$

Loop 2: $3.9I_2 + 1(I_2 - I_3) + 2.2(I_2 - I_1) = 0$

$$-2.2I_1 + 7.1I_2 - I_3 = 0$$

Loop 3: $5I_3 + 4.7(I_3 - I_1) + 1(I_3 - I_2) = 0$

$$-4.7I_1 - I_2 + 10.7I_3 = 0$$

You should verify that the approximate loop current values are

$I_1 = 1.57$ mA
$I_2 = 0.59$ mA
$I_3 = 0.74$ mA

The voltage drop across R_5 is found by

$$E_{R_5} = R_5(I_3 - I_2)$$
$$= 1\text{ k}\Omega\ (0.74\text{ mA} - 0.59\text{ mA})$$
$$= 0.15\text{ V}$$

Note that the actual current through R_5 is the difference between the two loop currents. The net, or resultant, current is 0.15 mA in the direction of I_3.

Node Analysis

Node analysis is similar to loop analysis in that algebraic methods are used to solve for currents and voltages in electronic circuits. A node is a reference point in a circuit and may, or may not, be the junction of two or more current paths. The circuit in Figure 5–5 has four nodes marked A through D. Node D is considered a reference node. The current through each resistor is also marked. As in loop analysis, if current is actually flowing in the opposite direction the resulting value will be negative.

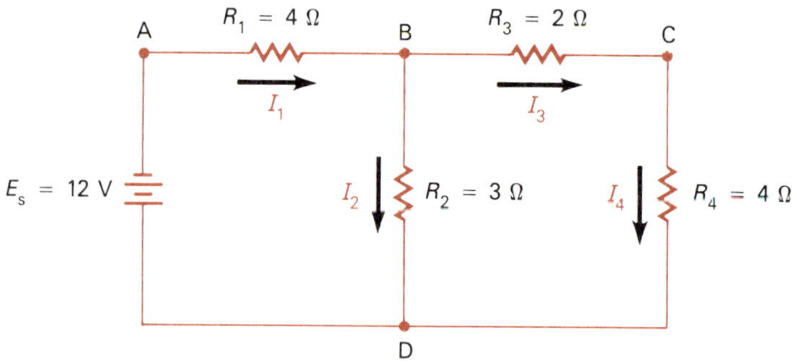

FIGURE 5–5

The first step in node analysis is to write the current equations pertaining to each node. Following this, voltage equations are written for each node. These are written with respect to the reference node. These two sets of equations are combined and solved to determine the node voltages and resistor currents. Examine each step in this example.

Example: Find the node voltages and resistor currents for the circuit in Figure 5–5.

Solution: In terms of circuit currents we know that
$I_1 = I_2 + I_3$ and $I_3 = I_4$

Also, the current through a resistor is found by dividing the voltage across the resistor by the resistor value.

$$I_1 = \frac{E_{R1}}{R_1} = \frac{E_A - E_B}{4}$$

$$I_2 = \frac{E_{R2}}{R_2} = \frac{E_B}{3}$$

$$I_3 = \frac{E_{R3}}{R_3} = \frac{E_B - E_C}{2}$$

$$I_4 = \frac{E_{R4}}{R_4} = \frac{E_C}{4}$$

By substituting the expressions for current, written in terms of the node voltages, into the equation $I_1 = I_2 + I_3$, an equation involving node voltages is obtained.

$$I_1 = I_2 + I_3$$
$$\frac{E_A - E_B}{4} = \frac{E_B}{3} + \frac{E_B - E_C}{2}$$
$$12\left(\frac{12 - E_B}{4}\right) = \left(\frac{E_B}{3} + \frac{E_B - E_C}{2}\right)12$$
$$36 - 3E_B = 4E_B + 6E_B - 6E_C$$
$$13E_B - 6E_C = 36$$

To obtain the second voltage equation, use the equation $I_3 = I_4$ and substitute the current values in terms of the node voltages.

$$\frac{E_B - E_C}{2} = \frac{E_C}{4}$$
$$2E_B - 2E_C = E_C$$
$$2E_B - 3E_C = 0$$

The two equations are

$$13E_B - 6E_C = 36$$
$$2E_B - 3E_C = 0$$

Multiply the second equation by -2, and then add the equations.

$$\begin{array}{r} 13E_B - 6E_C = 36 \\ -4E_B + 6E_C = 0 \\ \hline 9E_B + 0 = 36 \\ E_B = 4\text{ V} \end{array}$$

To find E_C substitute E_B in one of the equations.

$$2E_B - 3E_C = 0$$
$$2(4) - 3E_C = 0$$
$$8 - 3E_C = 0$$
$$E_C = 2\frac{2}{3}\text{ V}$$

You should verify that

$$I_1 = 2\text{ A}$$
$$I_2 = 1\frac{1}{3}\text{ A}$$
$$I_3 = \frac{2}{3}\text{ A}$$
$$I_4 = \frac{2}{3}\text{ A}$$

Example: Use node analysis to determine the node voltages and circuit currents in Figure 5–6.

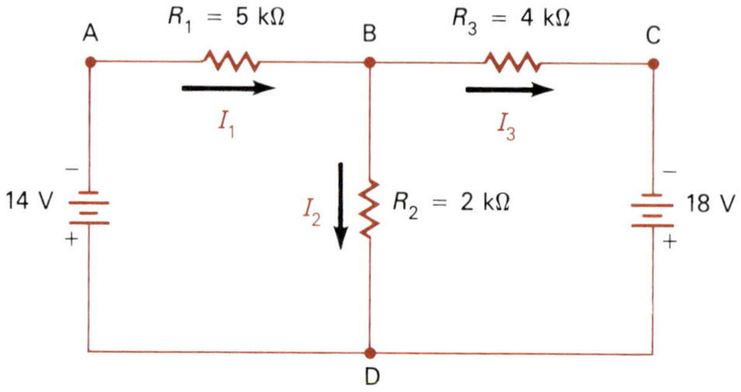

FIGURE 5–6

Solution:

$$I_1 = I_2 + I_3$$

$$\frac{E_A - E_B}{5} = \frac{E_B}{2} + \frac{E_B - E_C}{4}$$

Since $E_A = 14$ V and $E_C = 18$ V we can substitute these values into the equation.

$$\frac{14 - E_B}{5} = \frac{E_B}{2} + \frac{E_B - 18}{4}$$

$$56 - 4E_B = 10E_B + 5E_B - 90$$

$$19E_B = 146$$

$$E_B = 7.68 \text{ V}$$

The circuit currents are found as follows.

$$I_1 = \frac{E_A - E_B}{5} = \frac{14 - 7.68}{5} = 1.26 \text{ mA}$$

$$I_2 = \frac{E_B}{2} = \frac{7.68}{2} = 3.84 \text{ mA}$$

$$I_3 = \frac{E_B - E_C}{4} = \frac{7.68 - 18}{4} = -2.58 \text{ mA}$$

Note that the actual direction of I_3 is opposite of that selected.

Example: Use node analysis to find the node voltages in Figure 5–7.

Solution: Examine what we know about currents.

(1) $I_1 = I_2 + I_3$
(2) $I_3 = I_4 + I_5$

Substitute equation (2) into equation (1).

(3) $I_1 = I_2 + I_4 + I_5$

Write equation (3) in terms of node voltages.

$$\frac{10 - E_B}{6} = \frac{E_B}{12} + \frac{E_C}{6} + \frac{E_D}{4}$$

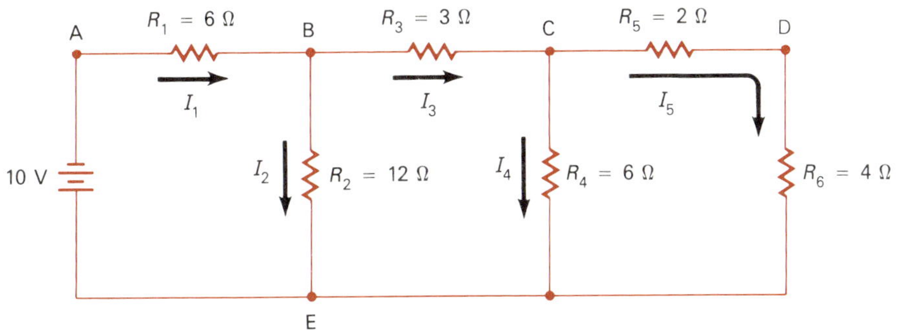

FIGURE 5–7

Multiply both sides by 12.
$20 - 2E_B = E_B + 2E_C + 3E_D$, or
(4) $3E_B + 2E_C + 3E_D = 20$

Since I_5 can be written in two forms we can derive a new expression with E_D in terms of E_C.

$$I_5 = \frac{E_D}{4} = \frac{E_C - E_D}{2}$$
$$E_D = 2E_C - 2E_D$$
$$3E_D = 2E_C$$
$$(5) \quad E_D = \frac{2E_C}{3}$$

Substitute equation (5) into equation (4).

$$3E_B + 2E_C + 3\left(\frac{2E_C}{3}\right) = 20$$
$$3E_B + 2E_C + 2E_C = 20$$
$$(6) \qquad 3E_B + 4E_C = 20$$

Equation (6) contains two unknowns. Equation (1) also contains these unknowns.

$$I_1 = I_2 + I_3$$
$$\frac{10 - E_B}{6} = \frac{E_B}{12} + \frac{E_B - E_C}{3}$$
$$20 - 2E_B = E_B + 4E_B - 4E_C$$
$$(7)\ 7E_B - 4E_C = 20$$

Add equations (6) and (7).

$$\begin{array}{r} 3E_B + 4E_C = 20 \\ 7E_B - 4E_C = 20 \\ \hline 10E_B + \quad 0 \quad = 40 \\ E_B = 4\text{ V} \end{array}$$

The remaining node voltages are

$$3E_B + 4E_C = 20$$
$$3(4) + 4E_C = 20$$
$$4E_C = 8$$
$$E_C = 2 \text{ V}$$
$$E_D = \frac{2E_C}{3}$$
$$= \frac{2(2)}{3}$$
$$= 1\frac{1}{3} \text{ V}$$

To check our results use equation (4).

$$3E_B + 2E_C + 3E_D = 20$$
$$3(4) + 2(2) + 3\left(\frac{4}{3}\right) = 20$$
$$12 + 4 + 4 = 20$$
$$20 = 20$$

EXERCISE 5·3

Use Figure 5–8 for problems 1–5.

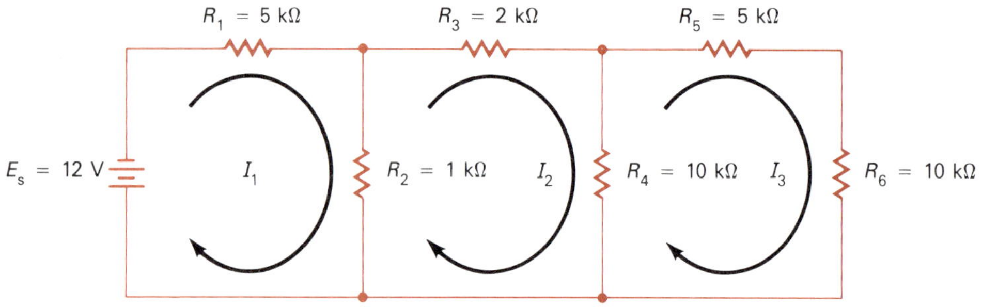

FIGURE 5–8

1. Write the loop equations for the circuit in Figure 5–8.
2. Determine the values of I_1, I_2, and I_3 in Figure 5–8.
3. What is the voltage across R_4 in Figure 5–8?
4. What voltage would be measured across R_1 in Figure 5–8?
5. If R_5 in Figure 5–8 were replaced by a short would the voltage drop across R_1 increase or decrease?

Use Figure 5–9 for problems 6–9.

6. Write the loop equations for the circuit in Figure 5–9.
7. Calculate the loop currents in Figure 5–9.
8. Determine the voltage drop across R_5 in Figure 5–9.
9. Find the current through R_2 in Figure 5–9.

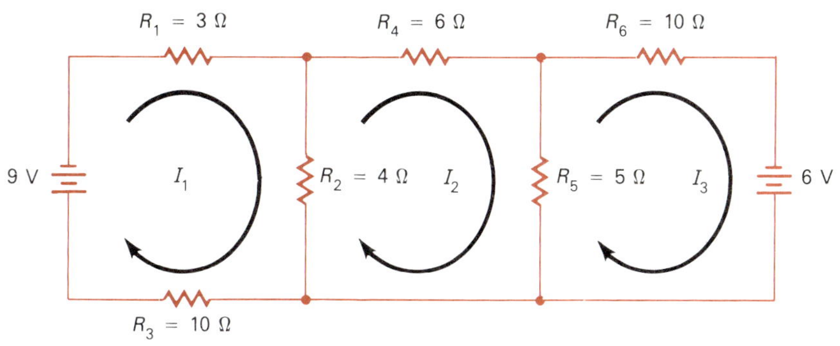

FIGURE 5–9

Use Figure 5–10 for problems 10–11.

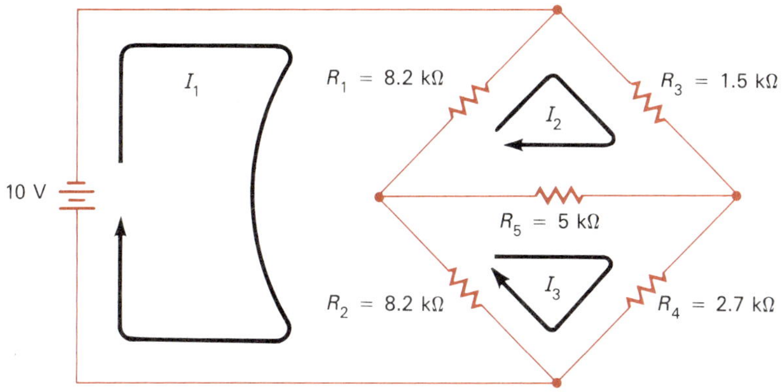

FIGURE 5–10

10. What is the magnitude and direction of the current flowing through R_5 in Figure 5–10?
11. Calculate the total current delivered by the power supply in Figure 5–10.

Use Figure 5–11 for problems 12–14.

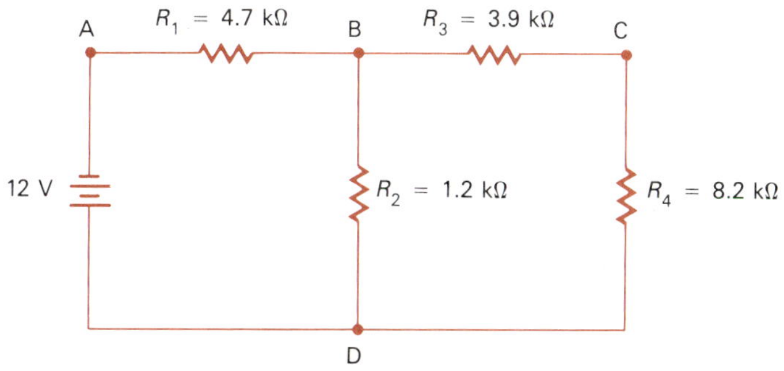

FIGURE 5–11

12. Find the three node voltages for the circuit in Figure 5–11.
13. Determine the value of the current through R_3 in Figure 5–11.
14. Calculate the voltage drop across R_1 in Figure 5–11.

Use Figure 5–12 for problems 15–17.

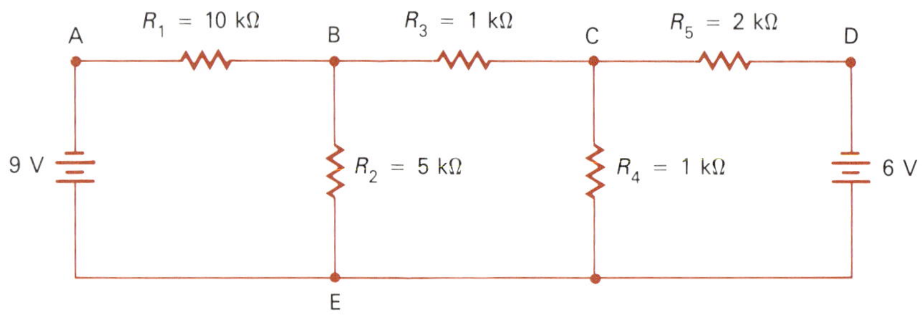

FIGURE 5–12

15. What are the four node voltages for the circuit in Figure 5–12?
16. Determine the current flowing through R_3 in Figure 5–12.
17. What current flows through R_5 in Figure 5–12?

Use Figure 5–13 for problems 18–20.

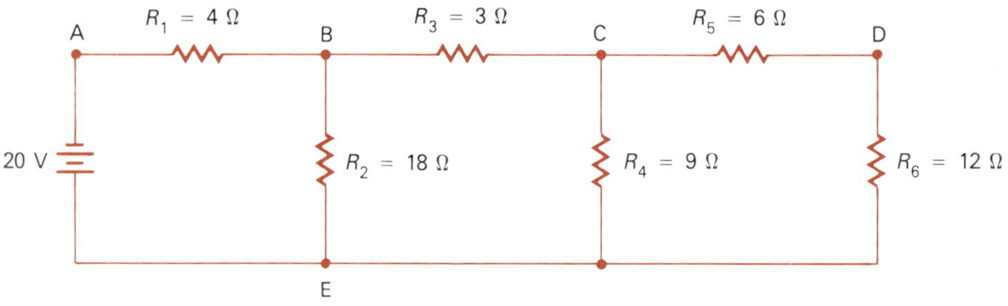

FIGURE 5–13

18. Find the four node voltages for the circuit in Figure 5–13.
19. What current flows through R_4 in Figure 5–13?
20. Find the currents flowing through R_2 and R_3 in Figure 5–13.

UNIT 5·4 QUADRATIC EQUATIONS

Objectives:

After studying this unit, you should be able to

- solve quadratic equations by factoring.
- solve quadratic equations by completing the square.
- solve quadratic equations by using the quadratic formula.

Form of a Quadratic Equation

A *quadratic equation* is one in which the unknown is raised to the second power, but not higher. By definition, then, $x^2 = 50$ is a quadratic equation, for the unknown value, x, is squared. Another quadratic equation is $4x^2 + 5x + 9 = 0$.

The *general quadratic equation* is written

$$ax^2 + bx + c = 0 \ (a \neq 0) \qquad (5\text{–}1)$$

where a, b, and c are constants. Note that coefficients b and c may be equal to zero and the equation remains quadratic. When $a = 0$ the equation is no longer quadratic as there is no x^2 term.

Example: Express each of the following equations in the general quadratic form.

a. $x^2 = 50$
b. $x^2 - 10 = 2x$
c. $4x + 3 = 7x - 3x^2$
d. $6x - x^2 = 0$
e. $3x(x + 2) = 7x - 9$
f. $6x - 12 = 6$

Solution: In order to express an equation in the general quadratic form it is often necessary to rearrange the equation.

a.
$$x^2 = 50$$
$$x^2 - 50 = 0$$
$$x^2 - 0x - 50 = 0$$
(Note the addition of $0x$ to complete the equation)

b.
$$x^2 - 10 = 2x$$
$$x^2 - 2x - 10 = 0$$

c.
$$4x + 3 = 7x - 3x^2$$
$$3x^2 + 4x - 7x + 3 = 0$$
$$3x^2 - 3x + 3 = 0$$

d.
$$6x - x^2 = 0$$
$$-x^2 + 6x + 0 = 0$$

e.
$$3x(x + 2) = 7x - 9$$
$$3x^2 + 6x = 7x - 9$$
$$3x^2 - x + 9 = 0$$

f. $6x - 12 = 6$ (Not a quadratic equation)

Solving Quadratic Equations by Finding the Square Root

For a problem such as $x^2 = 50$, the solution is to take the square root of both sides. The square root of 50 is approximately 7.1. However, there are two answers, not one, for $7.1 \times 7.1 = 50$, while -7.1×-7.1 also equals 50. Either 7.1 or -7.1 can be substituted for x in the equation $x^2 = 50$.

Example: Solve $x^2 = 81$.

Solution:
$$x^2 = 81$$
$$x = \pm\sqrt{81}$$
$$x = \pm 9$$

Example: Solve $\frac{4x^2}{144} = 1$.

Solution:
$$\frac{4x^2}{144} = 1$$
$$4x^2 = 144$$
$$x^2 = 36$$
$$x = \pm 6$$

Example: Solve $x^2 + 10 + (x + 2)(x - 5) = x^2 - 3x + 100$.

Solution:
$$x^2 + 10 + (x + 2)(x - 5) = x^2 - 3x + 100$$
$$x^2 + 10 + x^2 - 3x - 10 = x^2 - 3x + 100$$
$$2x^2 - 3x = x^2 - 3x + 100$$
$$x^2 = 100$$
$$x = \pm 10$$

To verify our answers substitute ± 10 into the original expression.

When $x = 10$
$$x^2 + 10 + (x + 2)(x - 5) = x^2 - 3x + 100$$
$$(10)^2 + 10 + (10 + 2)(10 - 5) = (10)^2 - (3)(10) + 100$$
$$100 + 10 + (12)(5) = 100 - 30 + 100$$
$$170 = 170$$

When $x = -10$
$$x^2 + 10 + (x + 2)(x - 5) = x^2 - 3x + 100$$
$$(-10)^2 + 10 + (-10 + 2)(-10 - 5) = (-10)^2 - (3)(-10) + 100$$
$$230 = 230$$

Example: Solve the quadratic equation $x^2 + 4 = 0$.

Solution:
$$x^2 + 4 = 0$$
$$x^2 = -4$$
$$x = \pm\sqrt{-4}$$

In the previous example we are asked to take the square root of -4. The square root of a number is one of two *equal* factors that, when multiplied, produce the original number. For the product of two numbers to be negative, one of the numbers must be negative and the other positive. This means that we could have $(2)(-2) = -4$, but these two numbers are not equal. The square roots of negative numbers are defined as *imaginary numbers* and are denoted by the j symbol. Imaginary numbers are used extensively in alternating current circuits addressed in a subsequent chapter. However, within this unit only quadratic equations with real values will be considered.

Solving Quadratic Equations by Factoring

An equation such as $x^2 + 5x + 6 = 0$ is a quadratic, but taking the square root of both sides will not help in the solution. However, since it is a quadratic, we may be able to factor the left side of the equation. Factoring this equation results in $(x + 3)(x + 2) = 0$. Since the product of these two factors is zero, one or both of the factors must equal zero. Set each of these factors equal to zero and solve for x.

$$x + 3 = 0 \quad \text{and} \quad x + 2 = 0$$
$$x = -3 \qquad\qquad x = -2$$

Either of the values -3 and -2 can be substituted into the original quadratic equation.

$$(-3)^2 + 5(-3) + 6 = 0 \quad \text{and} \quad (-2)^2 + 5(-2) + 6 = 0$$
$$9 - 15 + 6 = 0 \qquad\qquad 4 - 10 + 6 = 0$$
$$15 - 15 = 0 \qquad\qquad 10 - 10 = 0$$

The *roots* of $x^2 + 5x + 6 = 0$ are -3 and -2.

Some quadratics are easier to factor than others. A quadratic such as $x^2 - 3x = 0$ has an obvious common factor, x, and so the equation can be $x(x - 3) = 0$. The two factors are x and $x - 3$. When x is equated to zero, we have $x = 0$. When $x - 3$ is equated to zero, we have $x - 3 = 0$ or $x = 3$. And so the two roots of $x^2 - 3x = 0$ are 0 and 3.

Example: Find the roots of these quadratic equations.

a. $y^2 - 4 = 0$
b. $x^2 - x = 20$
c. $z^2 + 3z + 2 = 0$

Solution:

a.
$$y^2 - 4 = 0$$
$$(y + 2)(y - 2) = 0$$
$$y + 2 = 0 \quad \text{or} \quad y - 2 = 0$$
$$y = -2 \qquad\qquad y = 2$$

The roots are -2 and 2.

b.
$$x^2 - x = 20$$
$$x^2 - x - 20 = 0$$
$$(x + 4)(x - 5) = 0$$
$$x + 4 = 0 \quad \text{or} \quad x - 5 = 0$$
$$x = -4 \qquad\qquad x = 5$$

c.
$$z^2 + 3z + 2 = 0$$
$$(z + 1)(z + 2) = 0$$
$$z + 1 = 0 \quad \text{or} \quad z + 2 = 0$$
$$z = -1 \qquad\qquad z = -2$$

You must keep in mind that quadratic equations may not always be factorable. In a previous unit various methods of factoring were presented. Applying these to quadratic equations may produce roots of the equation. When factoring is not possible other methods must be employed.

Solving Quadratic Equations by Completing the Square

A process known as *completing the square* can be used to solve quadratic equations. Assume you are to find the roots of the quadratic equation

$$x^2 + 10x + 5 = 0$$

After examining this equation, you will find that it is not factorable. If the 5 was 25, then you could factor the left side of the equation into $(x + 5)^2$.

Before solving this problem a brief review of two special products will be helpful. These are

$$(x + y)^2 = x^2 + 2xy + y^2$$
$$(x - y)^2 = x^2 - 2xy + y^2$$

Note in both products that the first and last term are squares. The middle term is twice the product of the two original terms. If you were able to write an equation in this format, you could easily solve it. This is precisely what the completion of the square method allows us to do. The following example will demonstrate each step in this method.

Example: Solve $x^2 + 10x + 5 = 0$.

Solution: The first step is to move the 5 to the right side of the equation.

$$x^2 + 10x = -5$$

Next divide the coefficient of the x term by 2. This result is squared and added to *both* sides of the equation. Then solve the equation for x.

$$x^2 + 10x + 25 = -5 + 25$$
$$x^2 + 10x + 25 = 20$$
$$(x + 5)^2 = 20$$
$$x + 5 = \pm\sqrt{20}$$
$$x + 5 = \pm 4.472$$
$$x = \pm 4.472 - 5$$
$$x = -0.528$$
$$x = -9.472$$

The roots are -0.528 and -9.472. To verify the results, substitute these values into the original equation.

$$x^2 + 10x + 5 = 0$$
$$(-0.528)^2 + (10)(-0.528) + 5 = 0$$
$$0.28 - 5.28 + 5 = 0$$
$$-5 + 5 = 0$$

$$
\begin{aligned}
x^2 + 10x + 5 &= 0 \\
(-9.472)^2 + (10)(-9.472) + 5 &= 0 \\
89.72 - 94.72 + 5 &= 0 \\
-5 + 5 &= 0
\end{aligned}
$$

Example: Solve $x^2 - 3x - 7 = 0$.

Solution:

$$
\begin{aligned}
x^2 - 3x - 7 &= 0 \\
x^2 - 3x &= 7 \\
x^2 - 3x + 2.25 &= 7 + 2.25 \\
x^2 - 3x + 2.25 &= 9.25 \\
(x - 1.5)^2 &= 9.25 \\
x - 1.5 &= \pm\sqrt{9.25} \\
x &= 1.5 \pm\sqrt{9.25} \\
x &= 4.541 \\
x &= 1.541
\end{aligned}
$$

(Divide 3 by 2, square the result, and add to both sides)

Example: Solve $4x^2 + 8x - 15 = 0$

Solution:

$$
\begin{aligned}
4x^2 + 8x - 15 &= 0 \\
x^2 + 2x - 3.75 &= 0 \\
x^2 + 2x &= 3.75 \\
x^2 + 2x + 1 &= 3.75 + 1 \\
(x + 1)^2 &= 4.75 \\
x + 1 &= \pm\sqrt{4.75} \\
x &= -1 \pm\sqrt{4.75} \\
x &= 1.179 \\
x &= -3.179
\end{aligned}
$$

(Divide both sides by 4)

Solving Quadratic Equations with the Quadratic Formula

It is possible to solve the general quadratic equation for the variable x. The variable would be expressed in terms of the numerical coefficients a, b, and c.

The first step will be to solve the general equation for x.

$$ax^2 + bx + c = 0 \ (a \neq 0)$$

Moving c to the right side.

$$ax^2 + bx = -c$$

Divide both sides by a.

$$x^2 + \frac{b}{a}x = -\frac{c}{a}$$

The left side is part of a squared binomial. To complete the square we divide the coefficient by two $\left(\frac{b}{2a}\right)$ and then square it $\left(\frac{b^2}{4a^2}\right)$. This term is then added to both sides of the equation.

$$x^2 + \frac{b}{a}x + \frac{b^2}{4a^2} = -\frac{c}{a} + \frac{b^2}{4a^2}$$

Factor the left side and combine the fractions on the right side.

$$\left(x + \frac{b}{2a}\right)^2 = \frac{b^2 - 4ac}{4a^2}$$

Take the square root of both sides.

$$x + \frac{b}{2a} = \pm\sqrt{\frac{b^2 - 4ac}{4a^2}}$$
$$= \frac{\pm\sqrt{b^2 - 4ac}}{\sqrt{4a^2}}$$
$$= \frac{\pm\sqrt{b^2 - 4ac}}{2a}$$

Subtract $\frac{b}{2a}$ from both sides.

$$x = \frac{\pm\sqrt{b^2 - 4ac}}{2a} - \frac{b}{2a}$$

Combine the two fractions and you have the *quadratic formula.*

$$\mathbf{x = \frac{-b \pm\sqrt{b^2 - 4ac}}{2a}} \qquad \mathbf{(5\text{–}2)}$$

To solve a quadratic equation by using the quadratic formula, substitute the coefficients into the formula and solve for the value(s) of x.

Example: Solve $x^2 + x - 12 = 0$ using the quadratic formula.

Solution: The coefficients are $a = 1$, $b = 1$, and $c = -12$. Substitute these values into the formula.

$$x = \frac{-1 \pm\sqrt{(1)^2 - (4)(1)(-12)}}{(2)(1)}$$
$$= \frac{-1 \pm\sqrt{49}}{2}$$
$$= \frac{-1 \pm 7}{2}$$

Then the roots are

$$x = \frac{-1 + 7}{2} = 3$$

$$x = \frac{-1 - 7}{2} = -4$$

Example: Solve $2x^2 - 3x + 1 = 0$ using the quadratic formula.

Solution: The coefficients are $a = 2$, $b = -3$, and $c = 1$. The formula becomes

$$x = \frac{-(-3) \pm \sqrt{(-3)^2 - (4)(2)(1)}}{(2)(2)}$$

$$= \frac{3 \pm \sqrt{1}}{4}$$

$$= \frac{3 \pm 1}{4}$$

The roots are $x = 1$, and $x = 0.5$.

Example: Solve $x^2 - 6x + 9 = 0$ using the quadratic formula.

Solution: The coefficients are $a = 1$, $b = -6$, and $c = 9$.

$$x = \frac{-(-6) \pm \sqrt{(-6)^2 - (4)(1)(9)}}{(2)(1)}$$

$$= \frac{6 \pm \sqrt{0}}{2}$$

$$= 3$$

There is only one root for this quadratic equation.

Example: Solve $5x^2 + 13x - 17 = 0$.

Solution: See Calculator Tip for this solution.

Calculator Tip

In order to solve the quadratic equation $5x^2 + 13x - 17 = 0$ using your calculator, it will be helpful to have the capability of storing more than one number in memory.

Given that the coefficients are $a = 5$, $b = 13$, and $c = -17$ the sequence might be as follows:

$\dfrac{-13 \pm \sqrt{(13)^2 - (4)(5)(-17)}}{(2)(5)}$	13 [+/−] [STO] 1
$\dfrac{-13 \pm \sqrt{(13)^2 - (4)(5)(-17)}}{(2)(5)}$	13 [x^2] [−] 4 [×] 5 [×] 17 [+/−] [=] [√] [STO] 2
$\dfrac{-13 \pm 22.561\,028}{(2)(5)}$	2 [×] 5 [=] [STO] 3
$\dfrac{-13 \pm 22.561\,028}{10}$	[RCL] 1 [+] [RCL] 2 [=] [÷] [RCL] 3 [=] the first root,
$x = 0.956\,102\,8$	[RCL] 1 [−] [RCL] 2 [=]
$x = -3.556\,102\,8$	[÷] [RCL] 3 [=] the second root.

EXERCISE 5·4

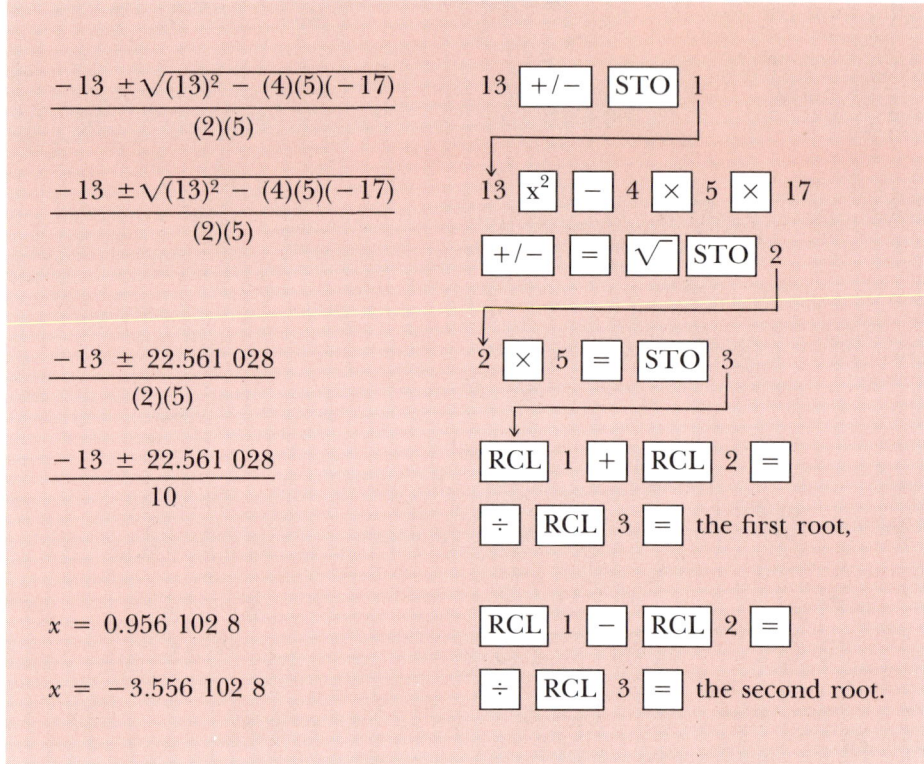

In problems 1–6 solve the quadratic equations.

1. $x^2 = 169$
2. $3x^2 = 675$
3. $x^2 + (x + 4)(x - 4) + 16 = x^2 + 64$
4. $(x^2 - 10x - 11) \div (x + 4) = -2x - 2$
5. $-(x + 2)^2 = 2x(x - 2) - 20$
6. $x(5x - 3) = 3x(x - 1) + 100$

In problems 7–15 solve the quadratic equations by factoring.

7. $x^2 - 9 = 0$
8. $x^2 + 3x - 10 = 0$
9. $x^2 + 14x + 49 = 0$
10. $x^2 - 9x + 8 = 0$
11. $2x^2 - 9x - 5 = 0$
12. $6x^2 - 5x - 6 = 0$
13. $4x^2 + 20x + 25 = 0$
14. $4x^2 - 24x + 36 = 0$
15. $4x^2 - 5x - 21 = 0$

In problems 16–25 solve the quadratic equations by completing the square.

16. $x^2 + 12x - 9 = 0$
17. $x^2 - 8x + 11 = 0$
18. $x^2 + 5x - 7 = 0$
19. $x^2 - 4x + 2 = 0$
20. $x^2 + 6x - 17 = 0$
21. $x^2 - 2x - 9 = 0$
22. $x^2 + 20x + 32 = 0$
23. $x^2 - x - 5 = 0$
24. $x^2 + 15x + 10 = 0$
25. $x^2 - 10x + 9 = 0$

In problems 26–35 solve the quadratic equations using the quadratic formula.

26. $x^2 + 7x + 3 = 0$	31. $9x^2 - 12x - 138 = 0$
27. $6x^2 + x - 13 = 0$	32. $x^2 - 9x + 7 = 0$
28. $x^2 - 9x + 12 = 0$	33. $x^2 - 14x - 20 = 0$
29. $3x^2 - 3x - 11 = 0$	34. $4x^2 + 4x + 1 = 0$
30. $4x^2 + 6x - 10 = 0$	35. $4x^2 - 13x + 10 = 0$

UNIT 5·5 GRAPHING LINEAR EQUATIONS

Objectives:

After studying this unit, you should be able to

- plot the graph of a linear equation.
- calculate the slope and x and y intercepts of a linear equation.
- develop the equation of a straight line.
- graphically solve linear equations in two variables.
- locate the intersection of two straight lines.
- determine the slope and y intercepts of parallel lines.

Coordinates

A linear equation is one whose graph is a straight line. An equation can be plotted as a graph by solving the equation. The values for x and y are regarded as coordinates or ordered pairs. A typical linear equation could be $x + y = 6$. Any number of x and y values can satisfy the equation to produce an answer of 6. If x is 1, then $y = 5$, since $1 + 5 = 6$. If $x = 2$, then $y = 4$. The following chart lists some of the values which can be substituted for the letters in the equation.

x	y
0	6
1	5
2	4
3	3
4	2
5	1
6	0

These values for x and y could be expressed as coordinates, (0, 6), (1, 5), (2, 4), etc. Other values, such as decimal and negative values, could also be included. However, with the information given above, we can plot a graph of the equation, $x + y = 6$. Figure 5–14 shows the graph.

An equation such as $x + y = 6$ is called an *indeterminate* equation because there would appear to be a limitless number of coordinates which can satisfy

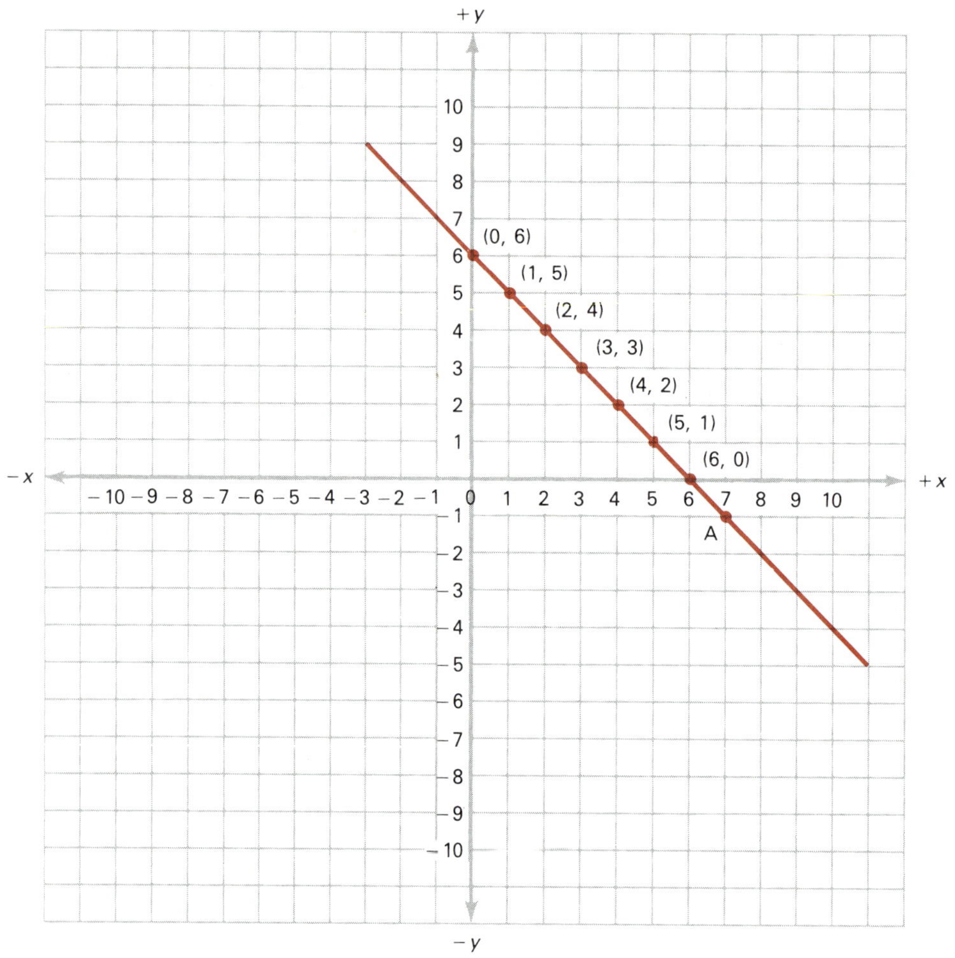

FIGURE 5–14 Graph of the equation $x + y = 6$

the equation, that is, numbers which can be substituted for x and y such that their sum is equal to 6.

In this linear equation the x value is usually identified as the *independent* variable. Given $x + y = 6$, any value of x may be selected. However, once x has been identified, the value of y is fixed. The y value is dependent upon the value of x, hence y is considered the *dependent* variable.

Note that once the linear equation has been graphed, selecting *any* point along the line will produce an ordered pair that will satisfy the equation. Point A in Figure 5–14 has an x value of 7 and a y value of -1. Substituting $(7, -1)$ into the original equation we find

$$\begin{aligned} x + y &= 6 \\ 7 + (-1) &= 6 \\ 7 - 1 &= 6 \\ 6 &= 6 \end{aligned}$$

While only two points are necessary to plot a linear graph, it is often helpful to use more points to ensure that the graph is correct. A common practice is to select both positive and negative values of x as well as zero.

Example: Identify three ordered pairs for the linear equation $2x + y = 7$.

Solution: As y is the dependent variable, it is helpful to rearrange the equation and express it in terms of y.

$$2x + y = 7$$
$$y = -2x + 7$$

Setting $x = 0$ produces

$$\begin{aligned} y &= -2x + 7 \\ &= (-2)(0) + 7 \\ &= 7 \end{aligned}$$

Setting $x = 3$ produces

$$\begin{aligned} y &= -2x + 7 \\ &= (-2)(3) + 7 \\ &= -6 + 7 \\ &= 1 \end{aligned}$$

Setting $x = -3$ produces

$$\begin{aligned} y &= -2x + 7 \\ &= (-2)(-3) + 7 \\ &= 6 + 7 \\ &= 13 \end{aligned}$$

The three ordered pairs are (0, 7), (3, 1) and (−3, 13). These points can be used to plot a graph of the equation. Note that any values of x could have been selected.

Slope of a Graph

If we have an equation such as $y = 3x - 5$, we can get the coordinates by substituting values for either x or y and solving the equation. We can, for example, choose arbitrary values for x. When we do, the corresponding values for y become fixed. Some values for x and y are shown in this chart.

x	y
1	−2
2	1
3	4
4	7
−1	−8
−2	−11

Figure 5–15 shows the graph of this equation.

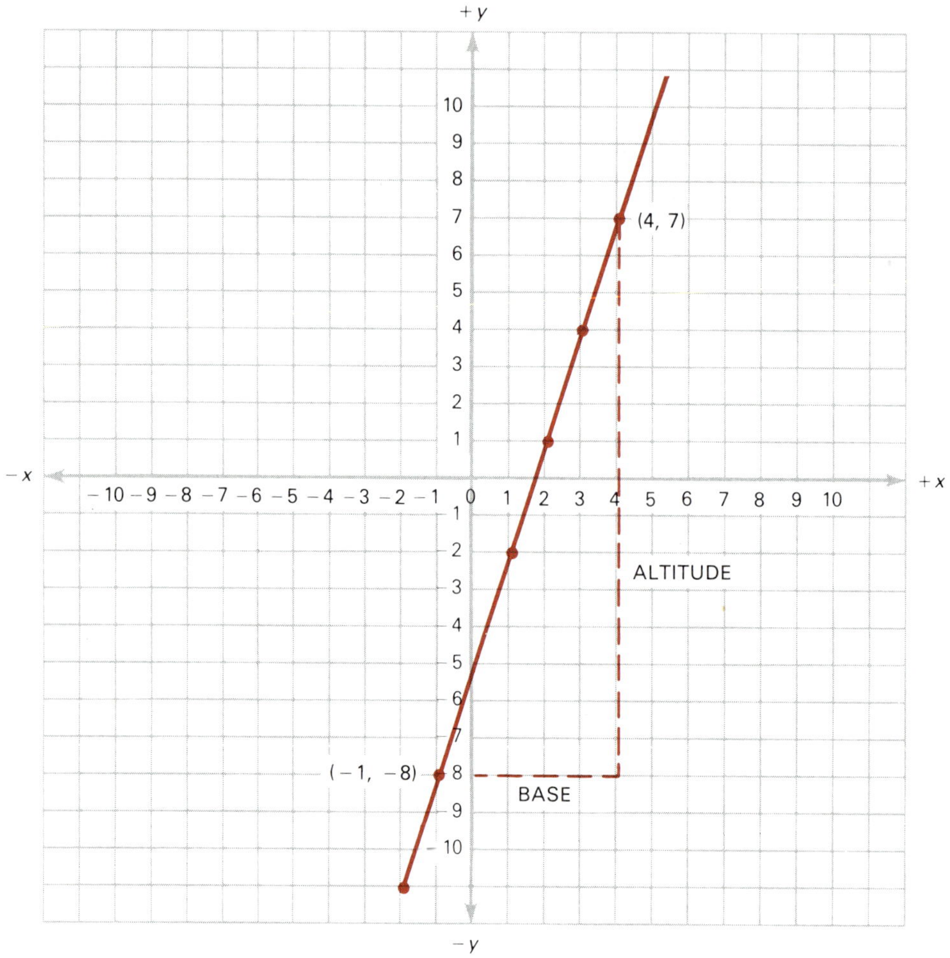

FIGURE 5–15 Graph of the equation $y = 3x - 5$

Any pair of plotted points on the graph can be connected by vertical and horizontal lines to form triangles. The ratio of the altitude to the base is the *slope* of the graph.

The slope (m) of a straight line is defined as the change (Δ) in the y value divided by the change in the x value between *any* two points. Given two points (x_1, y_1) and (x_2, y_2), the slope is written

$$m = \frac{y_2 - y_1}{x_2 - x_1} = \frac{\Delta y}{\Delta x} \qquad (5\text{–}3)$$

This equation is not valid when the line is parallel to the y axis. When this occurs both of the x values will be equal, the resulting change in x is zero, and the slope is undefined.

Example: What is the slope of the line in Figure 5–15?

Solution: Given the two points plotted on the line, let (x_1, y_1) be (4, 7) and (x_2, y_2) be (−1, −8).

$$m = \frac{y_2 - y_1}{x_2 - x_1} = \frac{(-8) - (7)}{(-1) - (4)} = \frac{-15}{-5} = 3$$

Letting $(x_1, y_1) = (-1, -8)$ and $(x_2, y_2) = (4, 7)$ the slope is

$$m = \frac{y_2 - y_1}{x_2 - x_1} = \frac{(7) - (-8)}{(4) - (-1)} = \frac{15}{5} = 3$$

The slope is constant and the selection of (x_1, y_1) and (x_2, y_2) is arbitrary.

The slope in the previous example is a positive three. Referring to Figure 5–15, note that as x increases one unit, the y value increases three units. Straight lines with positive slopes will move from lower left to upper right with high values for lines approaching a path parallel to the y axis. Lower slope values occur when the straight line moves closer to being parallel with the x axis.

Example: Determine the slope of the equation $x = 4y$.

Solution: Let $x_1 = 4$ and $x_2 = -4$.

The y values become:

$$x_1 = 4y_1 \quad \text{and} \quad x_2 = 4y_2$$
$$4 = 4y_1 \quad\quad -4 = 4y_2$$
$$y_1 = 1 \quad\quad y_2 = -1$$

Using equation 5–3 the slope is

$$m = \frac{y_2 - y_1}{x_2 - x_1} = \frac{(-1) - (1)}{(-4) - (4)} = \frac{-2}{-8} = 0.25$$

Graph the line described in the previous example to ensure that it has a positive slope and is almost parallel to the x axis.

Example: Determine the slope of a graph with two points having values of (1, 8) and (5, 3).

Solution:
$$m = \frac{y_2 - y_1}{x_2 - x_1} = \frac{(3) - (8)}{(5) - (1)} = -1.25$$

The line described in the previous example has a negative slope. This indicates that the line is moving from upper left to lower right.

Given the slope of a line, and one point on that line, it is possible to derive the linear equation. Follow this example closely.

Example: A linear graph has a slope of $m = 2$. When $x = 0$, the ordinate value is 20. What is the equation for this line?

Solution: Given $m = 2$ and a point (0, 20), equation 5–3 becomes
$$m = \frac{y_2 - y_1}{x_2 - x_1}$$
$$2 = \frac{20 - y_1}{0 - x_1}$$
$$2(0 - x_1) = 20 - y_1$$
$$y_1 = 2x_1 + 20$$
The equation is $y = 2x + 20$

Intercepts

Graphs such as the one in Figure 5–15 intersect both the x and y axes. The graph intersects the x axis when $y = 0$ and intersects the y axis when $x = 0$. When $x = 0$, the distance from the origin to the intersection of the graph with the y axis is called the y *intercept*. In Figure 5–15, the y intercept equals -5 units. In the same illustration, the x *intercept* is between $+1$ and $+2$ units.

Example: Find the x and y intercepts for the equation $y = 7x - 2$.

Solution: The x intercept occurs when $y = 0$.
$$y = 7x - 2$$
$$0 = 7x - 2$$
$$2 = 7x$$
$$x = \frac{2}{7}$$
The y intercept occurs when $x = 0$.
$$y = 7x - 2 = (7)(0) - 2 = -2$$

The line described by this equation crosses the x axis at $\frac{2}{7}$ and the y axis at -2.

The linear equation for a straight line may be determined given two points or one point and the slope. The *slope-intercept* form of a straight line equation indicates both the slope of the line and the point at which the line intercepts the y axis. Identifying the y intercept as b, the slope equation becomes

$$m = \frac{y - b}{x - 0} \qquad (5\text{–}4)$$

where m = slope of the line
(x, y) = a point on the line
$(0, b)$ = the y intercept
Solving equation 5–4 for y produces

$$\begin{aligned} m &= \frac{y - b}{x - 0} \\ y - b &= m(x - 0) \\ y - b &= mx \\ y &= mx + b \end{aligned}$$

Given the slope, m, and the y intercept, b, the equation of the line is written

$$y = mx + b \qquad (5\text{–}5)$$

When expressed in this form the numerical coefficient associated with x is the slope (m) and b is the y intercept.

Example: Determine the y intercept and slope of $3x + y = 4$.

Solution: The first step is to express the equation in the slope-intercept form.

$$\begin{aligned} 3x + y &= 4 \\ y &= -3x + 4 \end{aligned}$$

The slope is -3 and the line crosses the y axis at $(0, 4)$.

Example: Determine the slope of $3(y - 4) = 7x + 2y$.

Solution:

$$\begin{aligned} 3(y - 4) &= 7x + 2y \\ 3y - 12 &= 7x + 2y \\ y &= 7x + 12 \\ m &= 7 \end{aligned}$$

Example: What is the equation for a line with a slope of -1 and a y intercept of $(0, -5)$?

Solution:
$$y = mx + b$$
$$= (-1)(x) + (-5)$$
$$= -x - 5$$

Given two points on a straight line, it is possible to derive the linear equation. This equation will contain an unknown x and y. In addition, we will need the slope and one point to determine the numerical coefficients. Modifying equation 5–3

$$m = \frac{y - y_1}{x - x_1} \qquad \textbf{(5–6)}$$

which is equivalent to

$$y - y_1 = m(x - x_1) \qquad \textbf{(5–7)}$$

Once two points are known, the slope can be computed. The slope and either of the points is then substituted into equation 5–7 to determine the linear equation.

Example: Find the equation for a line passing through $(4, -2)$ and $(1, -8)$.

Solution: The slope is

$$m = \frac{y_2 - y_1}{x_2 - x_1}$$
$$= \frac{(-2) - (-8)}{(4) - (1)}$$
$$= \frac{6}{3}$$
$$= 2$$

Equation 5–7 becomes

$$y - y_1 = 2(x - x_1)$$

Substituting one of the points results in

$$y - (-2) = 2(x - 4)$$
$$y + 2 = 2x - 8$$
$$y = 2x - 10$$

Substituting the other point must produce the same equation.

$$y - (-8) = 2(x - 1)$$
$$y + 8 = 2x - 2$$
$$y = 2x - 10$$

The equation of a line passing through $(4, -2)$ and $(1, -8)$ is $y = 2x - 10$. It has a slope of $m = 2$, and passes through the y axis at $y = -10$.

Example: What is the equation of a line passing through (0, 6) and (4, 4)?

Solution:

$$
\begin{aligned}
m &= \frac{y_2 - y_1}{x_2 - x_1} \\
&= \frac{(6) - (4)}{(0) - (4)} \\
&= -0.5 \\
y - y_1 &= m(x - x_1) \\
y - 6 &= -0.5(x - 0) \\
y - 6 &= -0.5x \\
y &= -0.5x + 6
\end{aligned}
$$

The slope of a line parallel to the y axis is undefined. This is because the denominator of the slope equation $(x_2 - x_1)$ is zero. A straight line parallel to the x axis has a slope of zero. All values of y along this line are the same, so that the slope equation becomes

$$
\begin{aligned}
m &= \frac{y_2 - y_1}{x_2 - x_1} \\
&= \frac{0}{x_2 - x_1} \\
&= 0
\end{aligned}
$$

Example: Find the equation and slope of a line parallel to the y axis and passing through (7, −3).

Solution: As the line is parallel to the y axis, the equation is $x = 7$. For every y value the x value will be 7. The slope of this line is undefined.

Example: Find the equation and slope of a line parallel to the x axis and passing through (0, 5).

Solution: As the line is parallel to the x axis the equation is $y = 5$. For every x value the y value will be 5. The slope of this line is 0.

Simultaneous Linear Equations

It is possible to have a pair of equations, referred to as simultaneous linear equations, such that the coordinates of one of the equations will satisfy the requirements of the second equation. As an example, consider two equations:

$$
\begin{aligned}
2x + 6y &= 20 \\
3x + 4y &= 10
\end{aligned}
$$

To solve a pair of simultaneous linear equations, the algebraic techniques previously described can be used. Another method used to solve simultaneous linear equations is the graphic method. When a pair of simultaneous linear equations are plotted, the point at which they intersect supplies the coordinates which can be used in both equations. Figure 5–16 shows the graphs for $3x +$

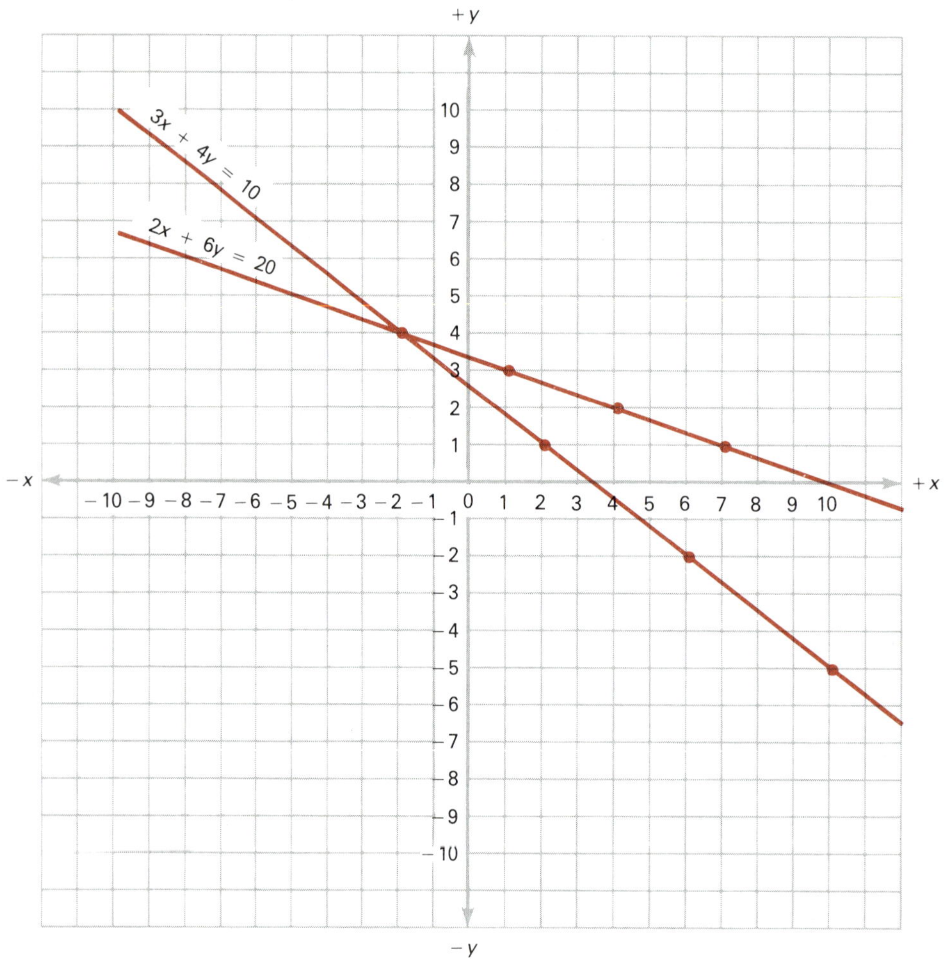

FIGURE 5–16 The intersection of both graphs supplies a set of coordinates that will satisfy both equations.

$4y = 10$ and $2x + 6y = 20$. The graph shows that intersection takes place at $x = -2$ and $y = 4$.

The equations represented by the graphs shown in Figure 5–16 are known as *independent linear equations*. While there are a large number of solutions for each equation, there is only one set of values that will satisfy both equations.

It is also possible to have a pair of linear equations that look different but which are actually identical. As an example consider the equations: $x + y = 5$ and $3x + 3y = 15$. Every value of x and y that will satisfy the first equation will satisfy the second equation. Furthermore, the second equation can be reduced by dividing through by 3 and when this is done the equation will be $x + y = 5$. Separate graphs can be plotted for each of these two equations, but will consist of one straight line superimposed on the other. Equations of this kind are called *dependent*.

A pair of equations that have no common solution is known as *inconsistent*. For the equations $x + y = 1$ and $x + y = 7$, there is no set of values which can satisfy both equations. When such equations are plotted they will appear on the graph as a pair of straight lines, and while these lines will intersect both the x and y axes, they will not cross each other.

Example: Plot the equations $x + y = 1$ and $x + y = 7$.

Solution: Expressing both in the slope-intercept form results in
$y = -x + 1$ and $y = -x + 7$
It should be evident that both lines have the same slope ($m = -1$), but cross the y axis at different points. Our graph should show that these lines are parallel. These are plotted in Figure 5–17.

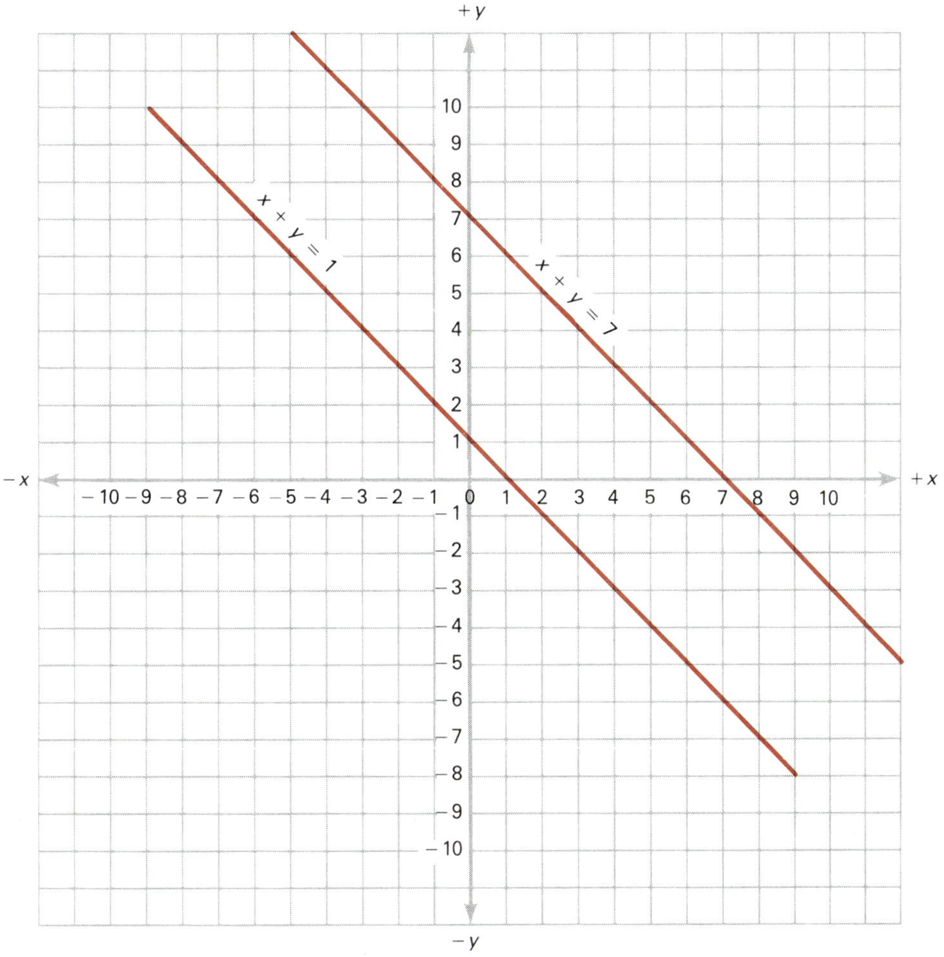

Example: Two lines are parallel. One line crosses the y axis at -6. The equation for the second line is $2x - y = -3$. Find the equation for the first line.

Solution: The first step is to express the second equation in the slope-intercept form. This produces

$y = 2x + 3$

The slope of this line is 2. Since the two lines are parallel they have the same slope. As the first line crosses the y axis at -6, the line equation is

$y = 2x - 6$

Using algebraic techniques presented earlier, it is possible to determine the point at which two straight lines will intersect. When two linear equations are expressed in the slope-intercept form the substitution method for solving simultaneous equations will produce the x and y coordinates of the intersection. While the two equations will have different slopes and y intercepts, the coordinates of the intersection will be common to both expressions.

Example: Use the substitution method to find the point where $y = -x - 2$ and $y = x + 4$ intersect. Graph the two lines to verify the answer.

Solution: The substitution method involves solving one equation for one unknown and then substituting this value into the second equation.

$$y = -x - 2 \text{ and } y = x + 4$$
$$-x - 2 = x + 4$$
$$-2x = 6$$
$$x = -3$$

Using either of the two equations, the y value can be determined as follows.

$$\begin{aligned} y &= -x - 2 \\ &= -(-3) - 2 \\ &= 1 \end{aligned} \qquad \text{and} \qquad \begin{aligned} y &= x + 4 \\ &= -3 + 4 \\ &= 1 \end{aligned}$$

The two lines intersect at $(-3, 1)$. These lines are graphed in Figure 5–18.

Example: Two straight lines are described by the equations $y = 6x + 2$ and $y = -2x + 4$. What are the coordinates of the point at which they intersect?

Solution: You should verify that the coordinates are (0.25, 3.5).

In summary, when two straight lines representing linear equations are graphed the following possibilities exist:

- The two lines may be parallel. When this is the case, the lines will have the same slopes, different y intercepts, and will not intersect.
- The two lines may intersect. When this is the case, the lines will have different slopes, one intersection point, and different y intercepts (un-

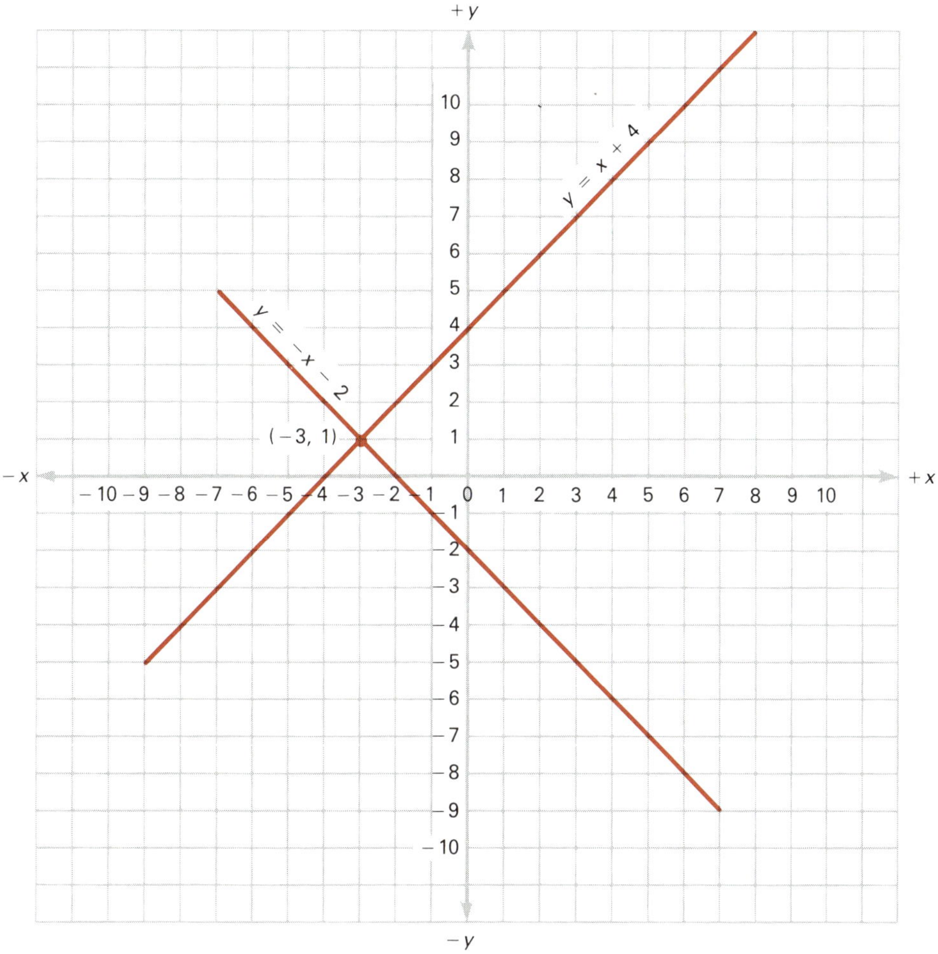

FIGURE 5–18

less they intersect on the *y* axis). The intersection point may be determined through algebraic operations or graphing.

EXERCISE 5·5

1. The abscissa refers to which axis?
2. The ordinate refers to which axis?
3. For each of the following coordinates, indicate in which quadrant the point is located.
   a. $(-5, 7)$
   b. $(0.25, 0.60)$
   c. $(-6, -1)$
   d. $(-2, 2)$

4. For each of the following coordinates, indicate in which quadrant the point is located.
   a. $(4, -0.5)$
   b. $(-4, 0.5)$
   c. $(1, 7)$
   d. $(-3.7, -4.2)$

In problems 5–14, find the linear equation for the line passing through the given coordinates. Graph each equation.

5. A line passing through $(1, 3)$ and $(4, -2)$.
6. A line passing through $(-5, 0)$ and $(-8, 3)$.
7. A line passing through $(0, 0)$ and $(6, -9)$.
8. A line passing through $(4, -6)$ and $(4, 6)$.
9. A line passing through $(2, -8)$ and having a slope of 0.5.
10. A line passing through the origin and having a slope of 4.
11. A line with a slope of 0 and passing through $(-5, -3)$.
12. A line passing through $(-4, 1)$ and having a slope of 2.
13. A line parallel to the y axis and passing through $(1, -1)$.
14. A line passing through $(3, -6)$ and having a slope of -3.

In problems 15–20, determine the slope and y intercept of each linear equation.

15. $y = 6x + 10$
16. $3x - y = 4x + 5y$
17. $6(x + y) = 5y - x - 7$
18. $\dfrac{5x + y}{2} = 3x - 7$
19. $2\left(\dfrac{4y}{3} + 5\right) = 6x + 8y$
20. $\dfrac{5(6x + 4)}{2y} = 10$
21. A straight line is described by the equation $5y - 3x = -2$. Find the slope of this line.
22. What is the slope of a line parallel to $6 - y = 4x$?
23. The y intercept of a line is -7 and is parallel to $x + y = 1$. Find the equation describing this line.
24. Two lines are parallel and have a slope of -3. Write the equations for these lines if they cross the y axis at 2 and the origin.
25. Two lines intersect at a point such that the x coordinate is -2. The first line is described by $y = x + 9$ and the slope of the second line is -3. Find the y coordinate of the intersection point.
26. What is the equation describing the second line in problem 25?
27. Find the coordinates of the intersection point for the equations $y = 14x - 19$ and $y = 7x + 12$.
28. The intersection point of two lines occurs at $(-4, -9)$. The y intercept of one of the lines is 5.2. What is the slope of this line?

29. Two lines intersect at (5, 0). The slope of the first line is 2 and the second -3. Where will each of these lines intersect the y axis?

Use Figure 5–19 for problems 30–33.

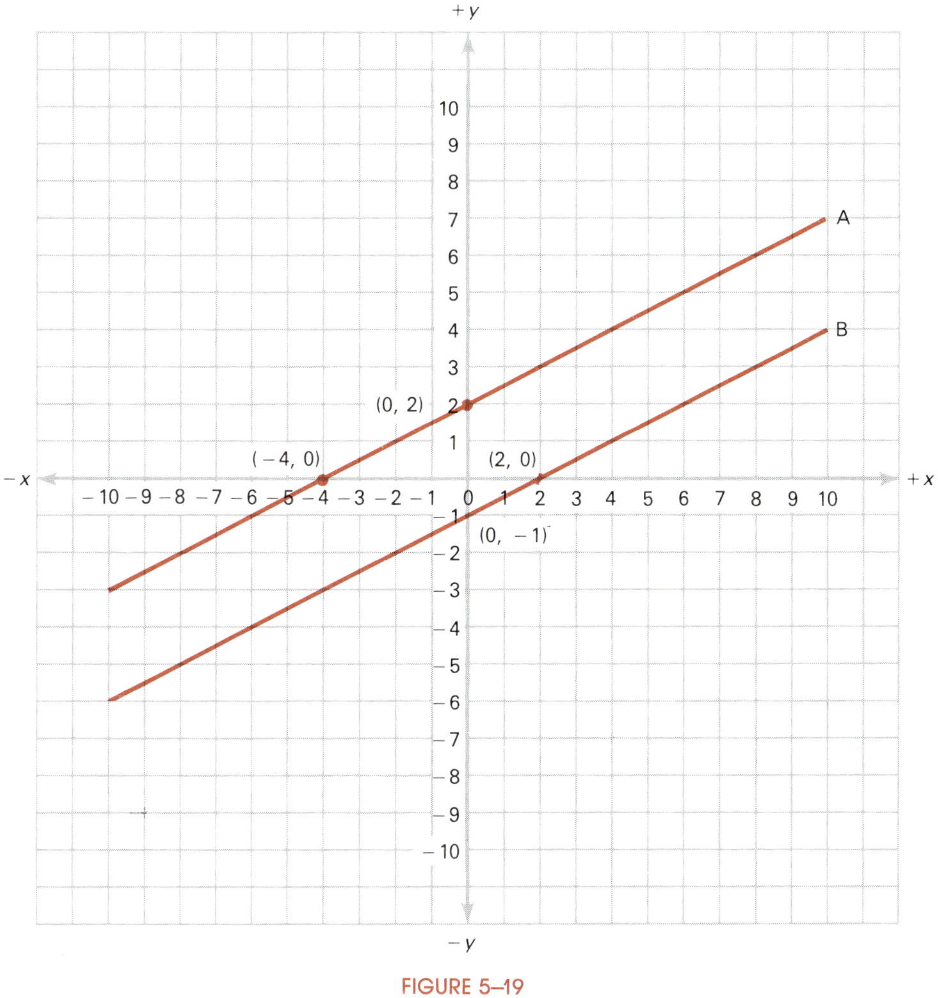

FIGURE 5–19

30. What is the equation for line A in Figure 5–19?
31. What is the slope of line B in Figure 5–19?
32. What is the equation for line B in Figure 5–19?
33. Refer to the graph in Figure 5–19. A line intersecting the y axis at $y = 6$ passes through (2, 0). What are the coordinates of the point at which this new line crosses line A?

Use Figure 5–20 for problems 34–35.

34. Find the equation which describes line A in Figure 5–20.
35. Find the equation which describes line B in Figure 5–20.

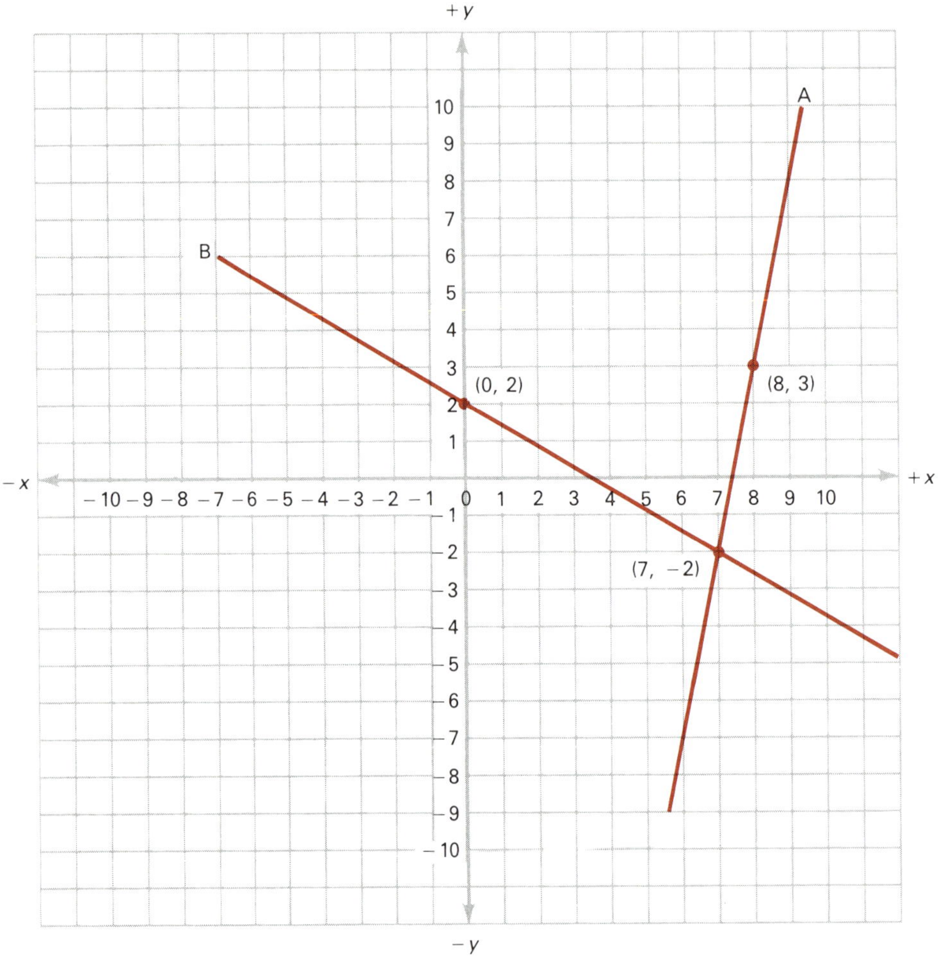

FIGURE 5–20

UNIT 5·6 GRAPHING QUADRATIC EQUATIONS

Objectives:

After studying this unit, you should be able to

- graph quadratic equations.
- determine root(s) of quadratic equations using graphs.
- determine root(s) of quadratic equations using factoring.

Graphing a Quadratic Equation

A linear equation such as $3x + 4y = 10$ is a first-degree equation since both x and y are raised to the first power. A quadratic equation such as $x^2 +$

$y^2 = 25$ is a second-degree equation, since both x and y are raised to the second power. However, both variables need not be raised to the second power. The equation, $x^2 + y = 10$, is a quadratic equation.

The roots of an equation are the numbers which can be substituted in the equation and which will still maintain the equality. A quadratic equation such as

$$x^2 = 25$$

can be solved by taking the square root of both sides. The square root of x^2 is x while the square root of 25 is $+5$ and -5. Hence, both $+5$ and -5 are the roots of this equation.

One of the methods for solving a quadratic equation is to plot it in the form of a graph. You can determine the points on the graph, that is, get the coordinates, by selecting random values for x and then solving the equation to get the corresponding value of y.

Example: Graph the quadratic equation $y = x^2 - 4x - 5$.

Solution: If random values of x are chosen and then substituted into the equation, the values for y can be found. A chart for x and y values can be set up as follows.

x	y
-2	7
-1	0
0	-5
1	-8
2	-9
3	-8
4	-5
5	0
6	7

Using these x and y values, a graph can be plotted. When the points on the graph are joined, the result is the curve shown in Figure 5–21.

Roots of Quadratic Equations

The curve, shown in Figure 5–21, is known as a *parabola*. This parabola crosses the x axis at two points, at $x = -1$ and at $x = +5$. These intersection points represent roots of the equation.

A quadratic equation can have two roots, or one root, or no roots at all, Figure 5–22. If it has two roots, it will, when graphed, cross the x axis in two places. If it has one root, it will touch the x axis at one point. And if it has no roots, it will not touch the x axis at all. A quadratic equation such as $y = x^2 + 5x + 4$ has two roots for the equation and can be factored into $(x + 4)(x + 1)$.

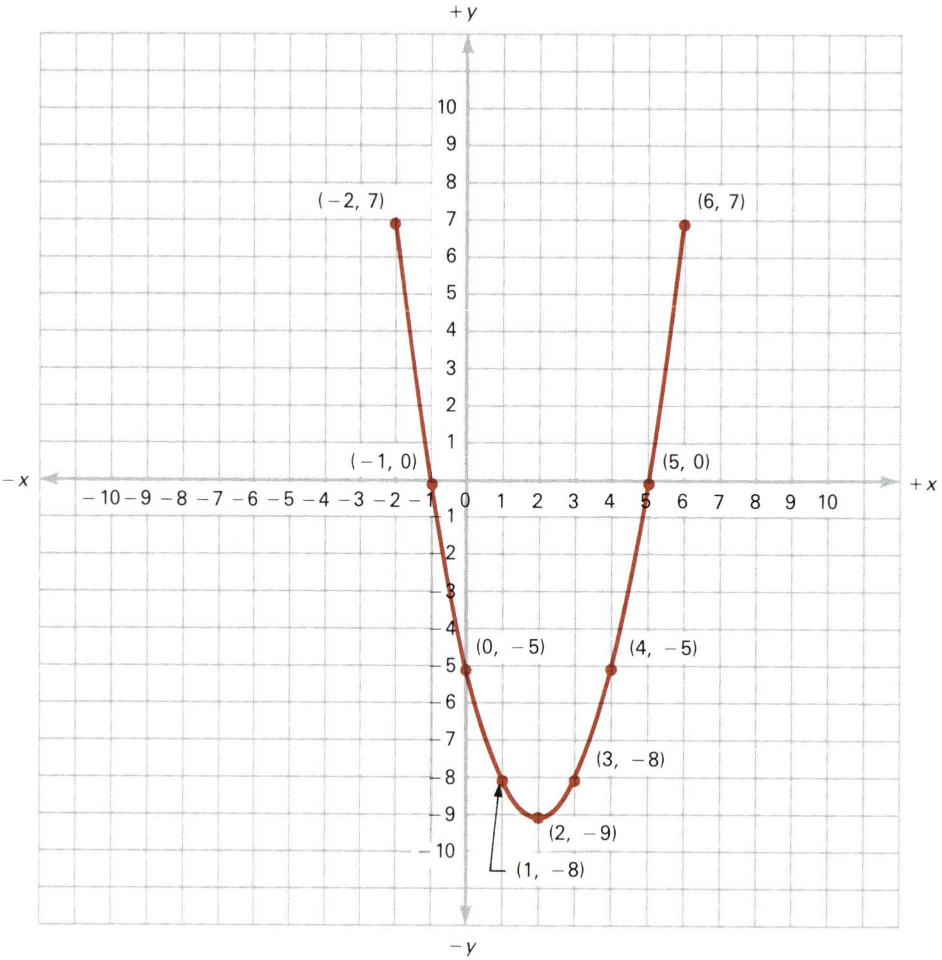

FIGURE 5–21 Graph of the quadratic equation $y = x^2 - 4x - 5$

The two roots are −4 and −1 and the graph crosses the x axis at these two points. A quadratic equation such as $y = x^2 - 6x + 9$ has just one root, for this equation factors into $(x - 3)(x - 3)$. The single root is +3. The graph of this equation touches the x axis at point 3.

An equation such as $y = x^2 + 2x + 19$ cannot be factored. When the graph of this equation is plotted, it will be found to be some distance above the x axis, but the lowest portion of the curve will not touch the x axis at any point.

If it is difficult or impossible to factor a quadratic equation, graphing it will show whether the equation has a root (or roots), or no roots at all. If the equation does have roots, graphing will show their values.

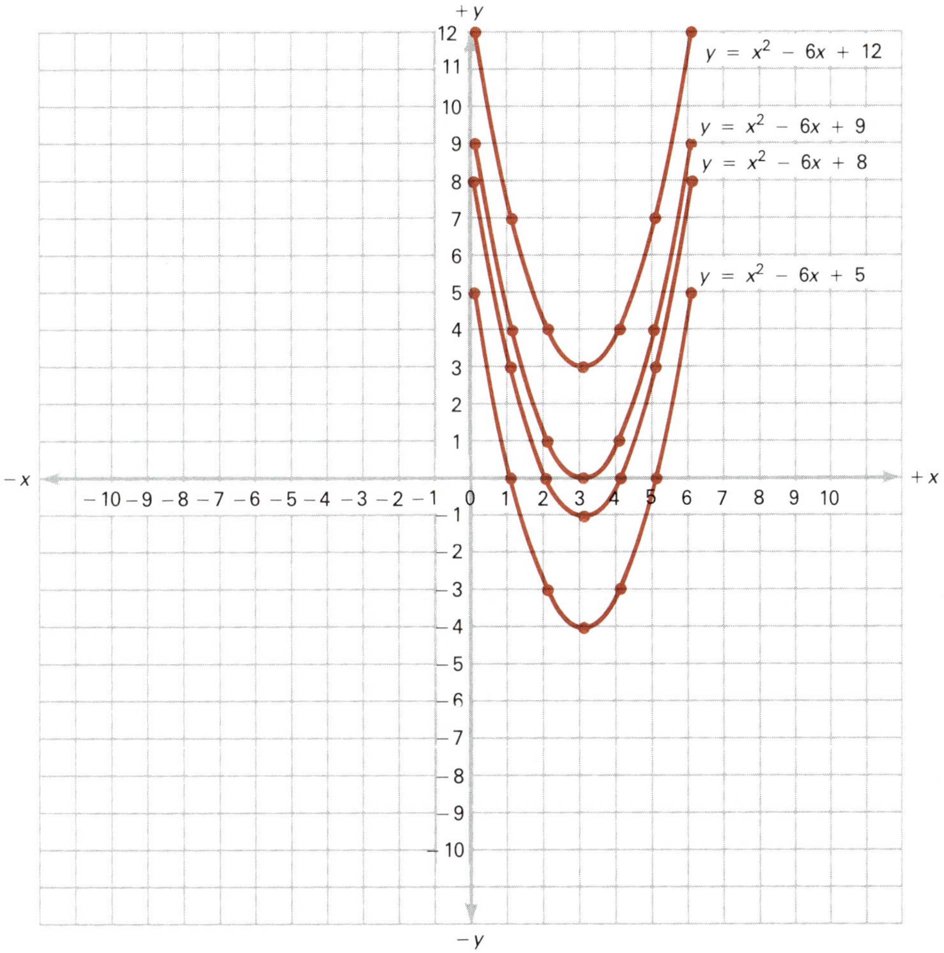

$y = x^2 - 6x + 5$

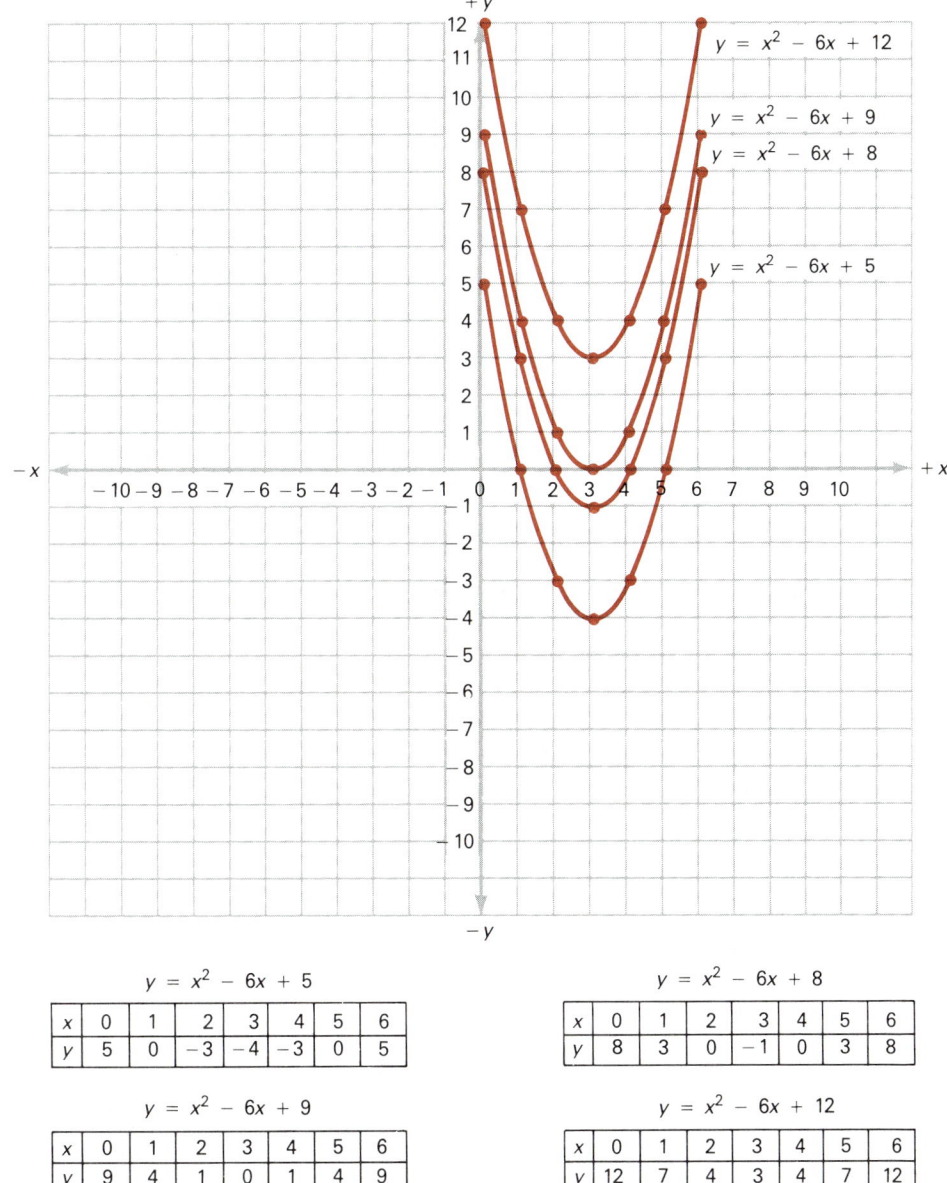

x	0	1	2	3	4	5	6
y	5	0	−3	−4	−3	0	5

$y = x^2 - 6x + 8$

x	0	1	2	3	4	5	6
y	8	3	0	−1	0	3	8

$y = x^2 - 6x + 9$

x	0	1	2	3	4	5	6
y	9	4	1	0	1	4	9

$y = x^2 - 6x + 12$

x	0	1	2	3	4	5	6
y	12	7	4	3	4	7	12

FIGURE 5–22 The graph of a quadratic equation may cross the x axis at two points, indicating two roots, touch it at one point (meaning there is just one root) or not cross the x axis at all (showing there are no roots).

EXERCISE 5·6

1. Plot and label $y = 2x^2$ and $y = -2x^2$ on the same graph.
2. Plot and label $y = x^2 + 1$ and $y = x^2 - 1$ on the same graph.
3. Plot and label $y = x^2 + 2x - 3$ and $y = 4x^2 - 3$ on the same graph.
4. Plot and label $y = x^2 + 5x + 4$ and $y = x^2 - x - 6$ on the same graph.
5. Plot and label $y = x + 3$, $y = x^2 + 3$, and $y = x^2 - 2x + 1$ on the same graph.

In problems 6–10 use factoring to determine the root(s) of the quadratic equations. Plot each of the equations to verify your results.

6. $y = x^2 - 2x + 1$
7. $y = 2x^2 + 10x + 8$
8. $y = -x^2 + 4x - 3$
9. $y = x^2 - 3x + 5$
10. $y = 6x^2 + 4x - 2$

UNIT 5·7 PROBLEM REVIEW

Solve for the indicated variable in problems 1–10.

1. $\dfrac{p_c}{p_b} = \beta_{dc}\left(\dfrac{V_c}{V_b}\right)$, solve for V_c.
2. $I_D = I_{DQ} + \dfrac{V_{DSQ}}{r_d + r_s}$, solve for r_d.
3. $V_n = \left(\dfrac{4A}{\pi}\right)\left(\dfrac{1}{4n^2 - 1}\right)$, solve for n.
4. $P_L = \dfrac{(0.707\ V_{cc})^2}{r_L}$, solve for V_{cc}.
5. $P_L = \dfrac{(0.707\ V_{CEQ})^2}{r_c + r_e}$, solve for r_c.
6. $\dfrac{X_C}{A + 1} = \dfrac{1}{2\pi fC(A + 1)}$, solve for X_C.
7. $A = \dfrac{1}{\sqrt{1 + (R^2/X_C^2)}}$, solve for R.
8. $f_c = \dfrac{1}{2\pi(R_{TH} + R_L)C}$, solve for R_{TH}.
9. $R_1 = \left(\dfrac{N_1}{N_2}\right)^2 (R_2)$, solve for N_2.
10. $f_r = \dfrac{1}{2\pi\sqrt{LC}} \cdot \sqrt{\dfrac{Q^2}{Q^2 + 1}}$, solve for Q^2.

Find the solution(s) for the equations in problems 11–35.

11. $4x + 7 = -21$
12. $3x - \frac{x}{8} = -92$
13. $\frac{x + 5}{x - 2} - \frac{1}{x - 1} = 1$
14. $49 - y^2 = 0$
15. $a^2 - 3a + 5 = 23$
16. $1 - \frac{2}{m^2 - m} = 0$
17. $\frac{x^2 + x - 12}{x^2 - 5x + 6} - 2 = 1$
18. $\frac{15}{x + 7} - (x + 5) = 10 - x$
19. $\frac{1}{x - 1} + \frac{x}{-2} = 0$
20. $4x^2 + 12x - 40 = 16x - 32$
21. $x - 2y = -16$
    $3x + 7y = 30$
22. $\frac{5x + y}{7} - 2y = x - 20$
    $\frac{6y}{x} = 12$
23. $\frac{3y - 2x}{y} + 1 = \frac{-8}{y}$
    $5x - 4y + 7 = 2y + 19$
24. $3x - 5y = 13$
    $\frac{6y - x}{8} = x - 6$
25. $4x - \frac{y}{2} = 2x + y - 18$
    $5(x + y) = 3y + 1$
26. $6\left(y - \frac{3x}{2}\right) - 2y = 8x + 5y$
    $2y - x = 2x + 25$
27. $\frac{4(5x - y)}{3} = 4(x - y)$
    $\frac{x}{y} = -1$
28. $-3[9(x - 2) - 5(2 - y)] = 12(3y - 8x) + 11$
    $\frac{9y}{x} = -18$

29. $\dfrac{5(x - 2y) + 3x}{13} = 2x$

$\dfrac{y - x}{-7} = \dfrac{1 - y}{4}$

30. $3x - y + z = 8$

$x + 2y - 8z = -27$

$4x - \dfrac{y}{2} + 2z = 11$

31. $x + y + z = 0$

$\dfrac{3x - 2z}{-2} = 10 - y$

$6x - 2y = 3 - 3z$

32. $2x - 5y + z = -20$

$2(x + y) - 3z = 22$

$4x - y = 14 - \dfrac{z}{3}$

33. $6(x - z) - 8y = 10$

$\dfrac{5(x + y)}{z} = -6$

$3x - y = 7 + z$

34. $\dfrac{10x + 3y}{2z} = -5$

$x + y + z = 0$

$\dfrac{3x - 2y + 7z}{2} = -2$

35. $9x - 2y + z - 3x - y \quad 16z$

$\dfrac{x}{3} + \dfrac{y}{z} = 7$

$x - \dfrac{y + z}{6} = -3z$

Use Figure 5–23 for problems 36–40.

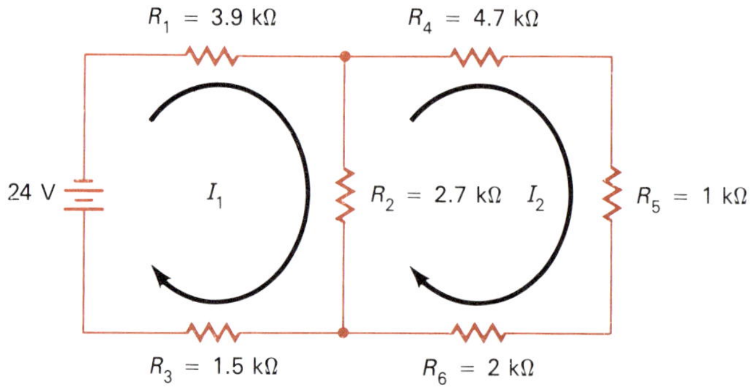

FIGURE 5–23

36. Write the loop equations for the circuit in Figure 5–23.
37. What current flows through R_2 in Figure 5–23?
38. Calculate the voltage across R_5 in Figure 5–23.
39. What current flows through R_3 in Figure 5–23?
40. Determine the total resistance in Figure 5–23.

Use Figure 5–24 for problems 41–45.

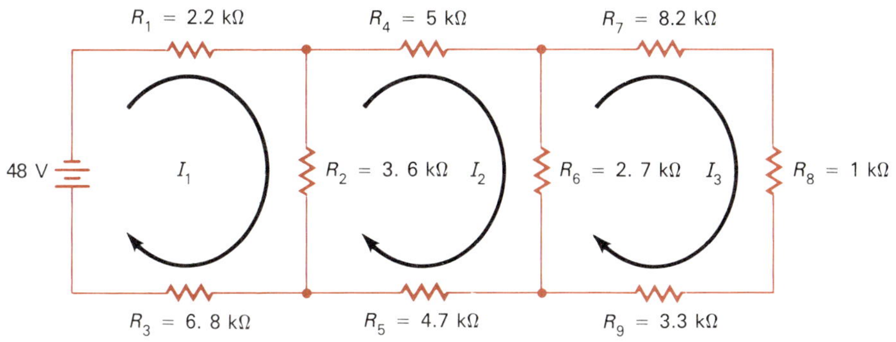

FIGURE 5–24

41. Write the loop equations for the circuit in Figure 5–24.
42. Calculate the current through R_2 in Figure 5–24.
43. Determine the voltage drop across R_5 in Figure 5–24.
44. What is the current flowing through R_6 in Figure 5–24?
45. What is the total current delivered by the power supply in Figure 5–24?

Use Figure 5–25 for problems 46–50.

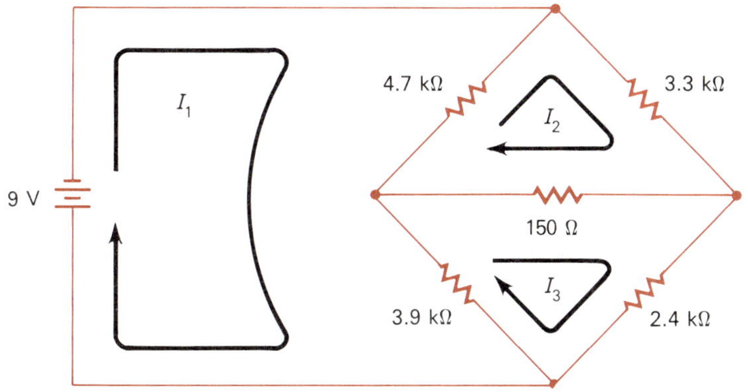

FIGURE 5–25

46. Write the loop equations for the Wheatstone bridge in Figure 5–25.
47. What is the current delivered by the power supply in Figure 5–25?
48. Calculate the voltage across the 3.3-kΩ resistor in Figure 5–25.

49. What is the direction and magnitude of the current flowing through the 150-Ω resistor in Figure 5–25?
50. Find the voltage drop across the 150-Ω resistor in Figure 5–25.

Use Figure 5–26 for problems 51–55.

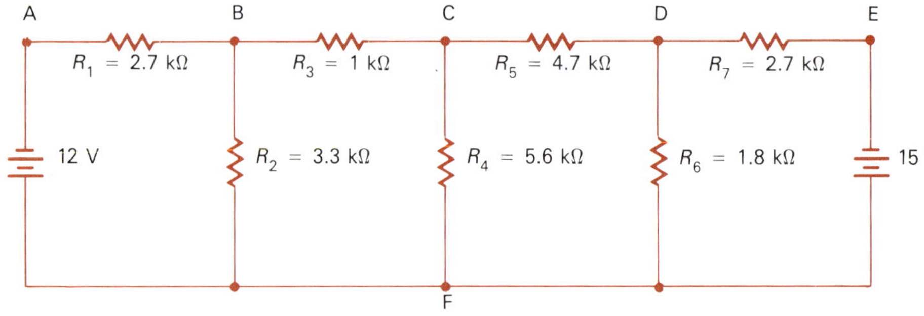

FIGURE 5–26

51. What are the five node voltages for the circuit in Figure 5–26?
52. Calculate the current value flowing through R_4 in Figure 5–26.
53. Determine the magnitude and direction of the current through R_5 in Figure 5–26.
54. What is the voltage drop across R_3 in Figure 5–26?
55. How much current flows through R_1 in Figure 5–26?

In problems 56–59, find the linear equation for the line passing through the given coordinates. Graph each equation.

56. A line passing through $(-5, 1)$ and $(5, -1)$.
57. A line passing through $(0, -8)$ and $(2, 2)$.
58. A line passing through $(-4, 6)$ and $(5, 6)$.
59. A line passing through $(1, 5)$ and $(-8, -4)$.
60. A line has a slope of -0.5 and passes through the point $(-3, 6)$. Where will this line cross the y axis?
61. Where will the line, described in problem 60, cross the x axis?
62. A line crosses the x axis at $x = -5$ and has a slope of 3. Where will this line cross the y axis?
63. Write the equation which describes the line in problem 62.
64. A line is drawn parallel to the one described in problem 62. Find the equation for this line if it passes through the origin.

Use Figure 5–27 for problems 65–68.

65. What is the slope of line A in Figure 5–27?
66. Write the equation describing line A in Figure 5–27.
67. Write the equation describing line B in Figure 5–27.
68. Write the equation describing line C in Figure 5–27. Lines B and C are parallel.

In problems 69–73 determine the slope and y intercept for each linear equation.

69. $y = 0.3x - 12$

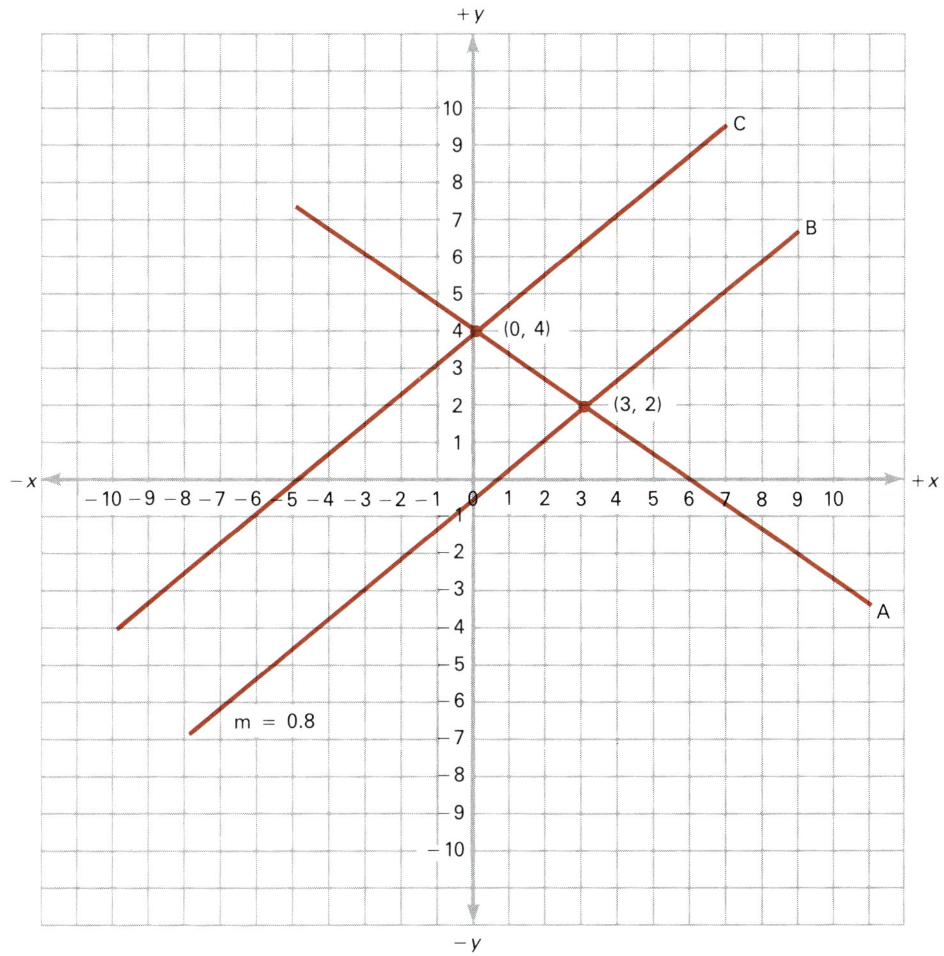

FIGURE 5–27

70. $x = \dfrac{1.5y - 3}{9}$
71. $3(y - x) = 12$
72. $x = -\dfrac{2}{9}y + \dfrac{8}{9}$
73. $x = -0.1y - 1$
74. Find the slope of a line parallel to $x = \dfrac{y}{8} + \dfrac{1}{2}$.
75. A line passes through the origin and is parallel to $y = -0.3x - 5$. Write the equation describing this line.
76. Two lines intersect at a point such that the x coordinate is -2. The first line is described by $y = -x$. The slope of the second line is 0.5. What is the y coordinate of the intersection point?

77. What is the equation describing the second line in problem 76?
78. Two lines are described by the equations $y = -2x + 3$ and $y = 4x - 6$. Find the coordinates of the intersection point.
79. The x intercept of a line is -4. This line intersects a second line at $(-1, 3)$. What is the slope of the first line?
80. What is the equation describing the first line in problem 79?
81. Three lines are plotted on a graph. The equations describing two of the lines are $y = 5$ and $x = 2$. The third line passes through the origin and the intersection of the other two lines. What is the slope of the third line?
82. What is the equation describing the third line in problem 81?
83. A fourth line is plotted on the graph described in problem 81. This line is parallel to the third line and intersects the x axis at -6. What is the equation describing the fourth line?

In problems 84–88, use factoring or graphing to determine the root(s) of the quadratic equations. Plot each of the equations to verify your results.

84. $y = x^2 - 5x + 6$
85. $y = x^2 - 3x$
86. $y = x^2 + x + 1$
87. $y = x^2 - x - 2$
88. $y = 4x^2 + 6x - 10$

In problems 89–103, solve the quadratic equations.

89. $4x^2 - x = 0$
90. $6x^2 = 150$
91. $x^2 - 16 = 0$
92. $x^2 + 7x + 12 = 0$
93. $x^2 - 8x + 7 = 0$
94. $x^2 + 5x - 15 = 0$
95. $x^2 - 6x + 2 = 0$
96. $x^2 + 24x + 80 = 0$
97. $x^2 - 7x - 3.75 = 0$
98. $x^2 + 2x - 3 = 0$
99. $3x^2 + 2x - 1 = 0$
100. $x^2 + 5x - 18 = 0$
101. $7x^2 - 3x + 9 = 0$
102. $5x^2 + 4x - 13 = 0$
103. $152x^2 + 96x + 15 = 0$

CHAPTER 6

TRIGONOMETRY FOR ELECTRONICS

Trigonometry is a specialized section of the general field of mathematics. Trigonometry is a study of the relationships that exist between the sides and the angles of triangles. As are arithmetic and algebra, trigonometry is an extremely useful tool for the solution of problems in electronics, especially those involving alternating currents.

UNIT 6·1 THE LANGUAGE OF TRIGONOMETRY

Objectives:

After studying this unit, you should be able to

- use the basic language of trigonometry.
- calculate complementary and supplementary angles.
- calculate the circumference, radius, and diameter of a circle.
- perform conversions among degree, radian, and grad units of circular measure.

Angles

An angle is formed when two straight lines meet at a point, as in Figure 6–1. The point is referred to as the *vertex* while the lines are called the *sides*. The letter *O* is often used to represent the vertex, but other letters can be used as well. The angle can be identified by a single letter, such as *A*, *O*, and *B*. If

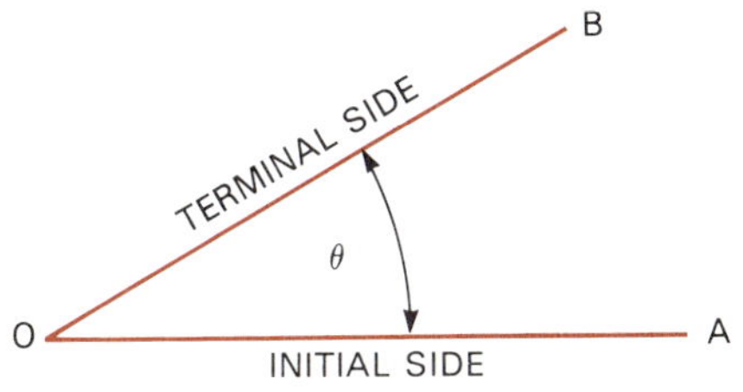

FIGURE 6–1 Formation of an angle. The junction of the two lines, O, is the vertex.

more than one angle occurs at a vertex, the angle might be identified by three letters. The vertex letter becomes the center letter of the group. Angles are also named by lowercase letters of the English alphabet placed within the angle. In electronics the angle is often named by any one of the letters of the Greek alphabet, especially the letter theta (θ). The existence of an angle is shown by an optional curved arrow as shown in Figure 6–1.

Angle Symbols

The mathematical symbol used to represent an angle is ∠. An alternative symbol has a curved line going through the symbol, ∢. When more than one angle is to be indicated, the letter *s* is incorporated within the symbol, ∠s. In some instances when a particular angle must be specified, a Greek letter such as theta, θ, will be used.

Forming an Angle

It is convenient to think of an angle as being formed by the rotation of one of its sides. One side or line of the angle is fixed while the other side moves with the vertex as a pivot point. The fixed side can be called the *base. Positive angles* are generated when the terminal side of the angle moves in a counterclockwise direction. *Negative angles* are generated when the terminal side of the angle moves in a clockwise direction away from the initial side, or base. Figure 6–2 shows both positive and negative angles.

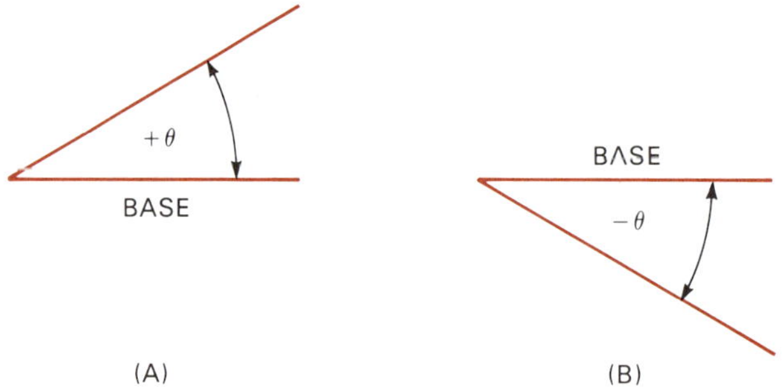

FIGURE 6–2 (A) A positive angle is represented by $+\theta$. (B) A negative angle is represented by $-\theta$.

The size of an angle is determined by the amount of rotation. Figure 6–3 shows two angles of identical size with sides of different lengths. Note that the length of the sides has no affect on the size of the angle.

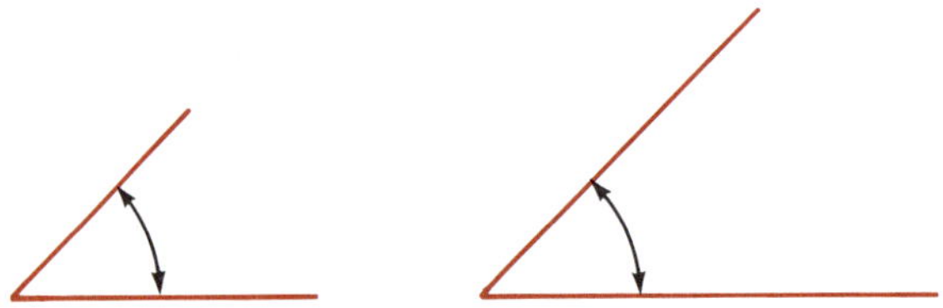

FIGURE 6–3 The lengths of the sides do not affect the size of the angles. Both angles are the same size.

The Circle

Imagine a pencil attached to the free end of the moving line, Figure 6–4. As this line continues its motion, the pencil will draw a curved line. The result will be a circle if the moving line continues completely around so that it finally comes to rest on the base line.

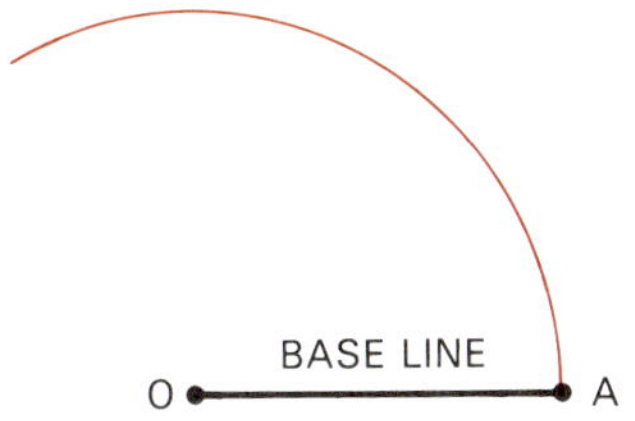

FIGURE 6–4 Base line OA is pivoted at point O and is free to rotate. If a pencil is attached to point A, its motion will draw a curved line. If continued, the result will be a circle.

The rotating line of the angle can be used to divide the circle into four equal parts, or *quadrants*. The first quadrant comes into existence when the rotating line, moving counterclockwise, is perpendicular to its original position. It produces still another quadrant when it assumes a position that is horizontal to the base line. The third quadrant occurs when the moving line is once again perpendicular to the base line. Finally, the last quadrant is formed when the moving line rotates until it is superimposed on the base line. Figure 6–5 shows the four quadrants of a circle.

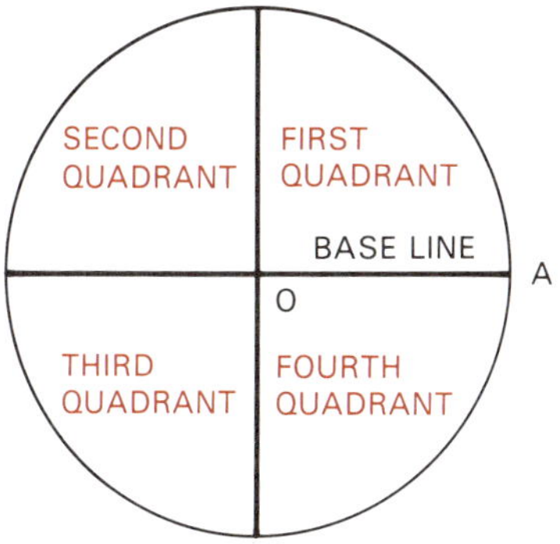

FIGURE 6–5 A circle of any size can be divided into four equal sections called quadrants. Each quadrant has a 90° angle. Quadrants are numbered counterclockwise.

The circle can be divided into 360 equal parts, each of which is called a *degree*. Since each quadrant is a fourth of a circle, then the angle of each quadrant is

$$\frac{360 \text{ degrees}}{4} = 90 \text{ degrees}$$

The symbol for a degree is a tiny superscript circle; thus 90 degrees can be more conveniently written as 90°.

Types of Angles

An angle having a value of 90° is called a *right angle*. Since a circle, any circle, contains 360°, then every circle contains four right angles. Figure 6–5 shows four right angles. A right angle is formed whenever two lines are perpendicular to each other, Figure 6–6A. The word *normal* is sometimes used as a synonym for perpendicular.

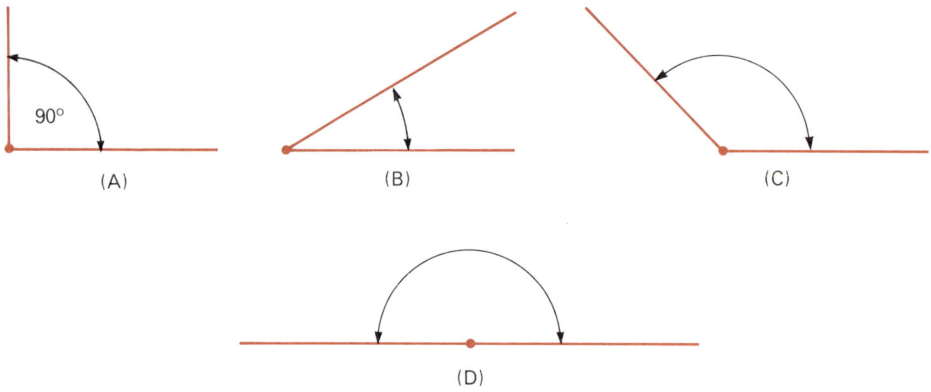

FIGURE 6–6 (A) A right angle (B) An acute angle (C) An obtuse angle (D) A straight angle

Any angle less than 90° is an *acute angle.* A 30° angle is an acute angle, Figure 6–6B. An angle greater than 90° is an *obtuse angle.* A 135° angle is an obtuse angle, Figure 6–6C. An angle having a value of 180° is called a *straight angle,* Figure 6–6D.

Complementary Angles

Figure 6–7 shows a circle divided into four quadrants, that is, four 90° angles. However, the circle can be further subdivided by including another rotating line. Thus, we now have two angles: the first angle is identified as ∠AOX while the second could be called ∠XOB. Individually, each of these angles is an acute angle, for each has a value less than 90°. The sum of the two angles is equal to 90°. These angles are called *complementary.* Any two angles whose angular sum is equal to 90° are called complementary. In Figure 6–7, ∠AOX is the complement of ∠XOB, or ∠XOB is the complement of ∠AOX.

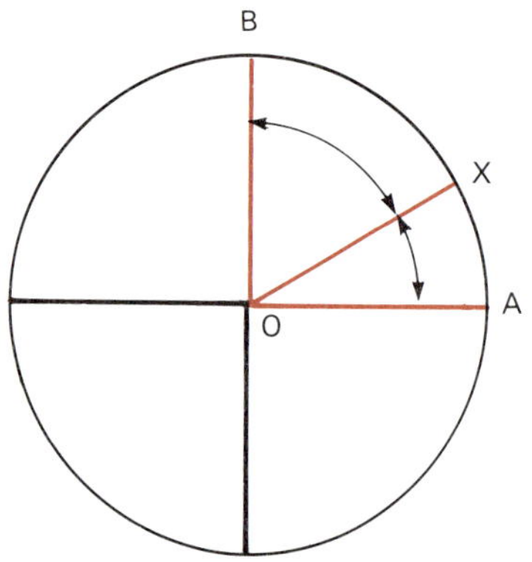

FIGURE 6–7 Two angles are complementary when their sum is equal to 90°.

Supplementary Angles

It is also possible to have a pair of angles, one of which is acute, the other obtuse. If the angular sum of two such angles is equal to 180°, as indicated in Figure 6–8, then the two angles are called *supplementary*. In Figure 6–8, ∠AOB is an acute angle, while ∠BOX is an obtuse angle. Their angular sum is 180°. ∠AOB is supplementary to ∠BOX or ∠BOX is supplementary to ∠AOB. When an angle is 180°, its two sides form a straight line and, since a pair of supplementary angles are 180°, the result is also a straight line.

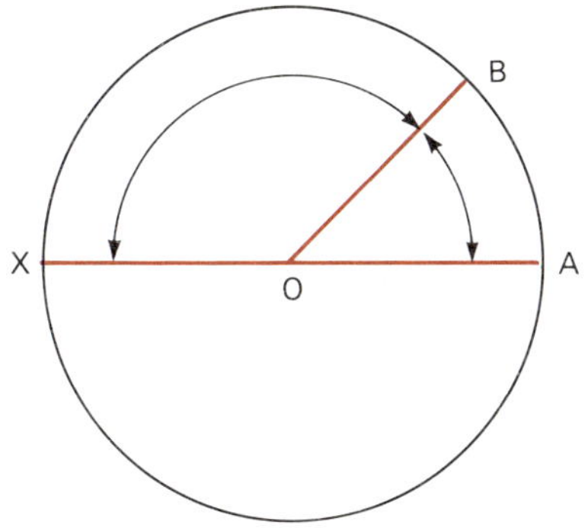

FIGURE 6–8 Two angles are supplementary when their sum is equal to 180°.

Example: Find the complementary and supplementary angles for angles of 7°, 27°, and 84°.

Solution: The complementary angles are found by subtracting each angle from 90° as follows:

$90° - 7° = 83°$
$90° - 27° = 63°$
$90° - 84° = 6°$

The supplementary angles are found by subtracting each angle from 180° as follows:

$180° - 7° = 173°$
$180° - 27° = 153°$
$180° - 84° = 96°$

Parts of a Circle

Any straight line passing through the center point of a circle and touching the circle at two points is a *diameter* of the circle, Figure 6–9. The diameter is also the longest, single straight line that can be drawn inside the circle. A *radius* is equal to one-half a diameter.

$$r = \frac{D}{2} \tag{6–1}$$

where r is the radius; D is the diameter. The radius and diameter can be measured in any convenient unit of length, inches, centimetres, etc.

There is a relationship between the diameter of a circle and the *circumference*. To calculate the length of the circumference, multiply the radius by 6.28 or 2π, assuming a value of 3.14 for π. In terms of a formula, circumference C can be calculated from

$$C = 2\pi r \tag{6–2}$$

but since the diameter D is equal to $2r$, then the formula can also be written as:

$$C = \pi D \tag{6–3}$$

From either of these formulas, knowing the radius can supply the length of the circumference, or the diameter can be calculated if the circumference is known.

Calculator Tip

There is a special relationship between the circumference and diameter of any circle. When the circumference is divided by the diameter the result is a constant. This ratio to seven decimal places is 3.141 592 7, and is represented by the Greek letter π (pi).

Depressing the π key on your calculator will display this numerical value. This value can be entered into problems just as any other number. Examine the following sequences:

a. $(2)(\pi)(5)$ = 2 [×] [π] [×] 5 [=] 31.415 927

b. $(\pi)^2$ = [π] [x^2] and 9.869 604 4 is displayed

c. $\dfrac{1}{2\pi}$ = 2 [×] [π] [=] [1/x] [=] 0.159 154 94

Example: What is the radius of a circle whose circumference measures 24 inches?

Solution: $C = 2\pi r$

$$r = \frac{C}{2\pi} = \frac{24 \text{ inches}}{2\pi} = 3.82 \text{ inches}$$

While the diameter, or radius, and circumference are the most commonly known parts of a circle, the circle can involve various other elements. These elements involve straight and curved lines, as shown in Figure 6–9.

A *chord* is a straight line drawn anywhere inside the circle and touching the circumference at two points. Unlike a diameter, a chord need not pass through the center of the circle. A diameter can, however, be regarded as a special kind of chord, although it is not generally thought of in that way. A

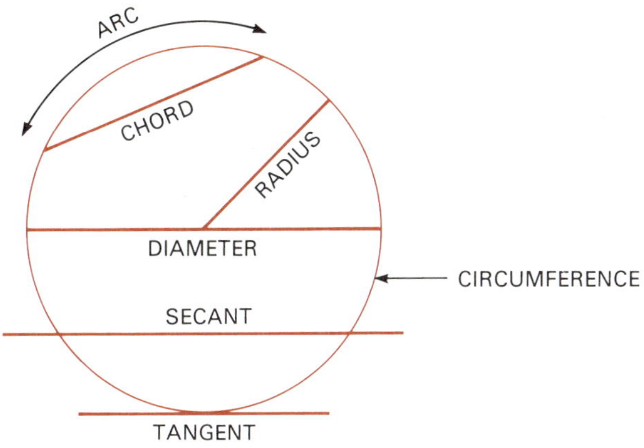

FIGURE 6–9 Lines of a circle

secant is similar to a chord, except that it continues through the circle at two points. A *tangent* is a straight line that touches the circumference at only one point. For this reason, any tangent must be outside the circle, since if it were inside it would cut the circle at two points and would then be regarded as a secant. The diameter, radius, chord, secant and tangent, then, are the straight lines that can be somehow involved with the circle.

Any part of the circumference is called an *arc*. Diameters can be drawn as a pair of lines perpendicular to each other, producing four quadrants, each of which has an arc of equal length. An area bounded by two radii and the intercepted arc is a *sector* while the area between a chord and its arc is known as a *segment*, as shown in Figure 6–10.

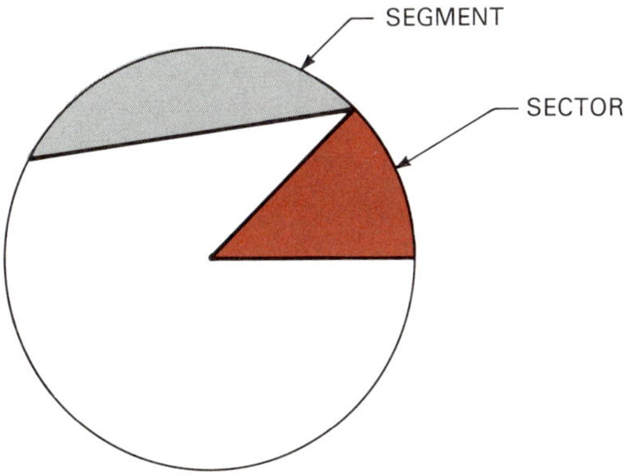

FIGURE 6–10 An area bounded by two radii and the intercepted arc is called a *sector*. The area between a chord and the arc it produces is known as a *segment*.

Circular Measure

The *degree* is a commonly used circular measure and can be subdivided into 60 equal parts, each of which is known as a *minute*. The minute can also be subdivided into 60 equal parts, with each part called a *second*. A table of circular measure would then be:

60 seconds = 1 minute
60 minutes = 1 degree
90 degrees = 1 quadrant
4 quadrants = 1 circle

The symbol for degrees, as indicated earlier, is °, as in 65°. The symbol for seconds is the double quotation mark ("), as in 35". The symbol for the minute is the single quotation mark: 45′. A measurement of 38°27′47" is equivalent to 38 degrees, 27 minutes, 47 seconds. Unless a high order of accuracy is required, electronics problems involving angles are limited to measurements in degrees.

Example: Complete the following problems:

a. 67°24′56" + 12°39′18"
b. 90° − 5°2′42"
c. 18°34" − 6°40′19"

Solution:

a.
$$\begin{array}{r} 67°24'56'' \\ +\,12°39'18'' \\ \hline 80°\ \ 4'14'' \end{array}$$

b.
$$90° = \begin{array}{r} 89°59'60'' \\ -\ \ 5°\ \ 2'42'' \\ \hline 84°57'18'' \end{array}$$

c.
$$18°0'34'' = \begin{array}{r} 17°60'34'' \\ -\ \ 6°40'19'' \\ \hline 11°20'15'' \end{array}$$

Degrees may also be expressed in decimal form. This occurs often, especially when calculators are used. An angle of 15.5° is obviously equal to 15°30′ in that 0.5 × 60′ = 30′. In order to convert minutes to decimal parts of a degree simply multiply by

$$\frac{1°}{60'}$$

This factor is equal to 1, and results in dividing the minute value by 60.

Example: Convert the following angles to their decimal equivalents.

a. 15′
b. 48′
c. 53′
d. 12°36′

Solution:

a. $15' = 15' \times \dfrac{1°}{60'} = \dfrac{15°}{60} = 0.25°$

b. $48' = 48' \times \dfrac{1°}{60'} = \dfrac{48°}{60} = 0.8°$

c. $53' = 53' \times \dfrac{1^\circ}{60'} = \dfrac{53^\circ}{60} = 0.883^\circ$

d. $12^\circ 36' = 12^\circ + \left(36' \times \dfrac{1^\circ}{60'}\right) = 12^\circ + \dfrac{36^\circ}{60} = 12.6^\circ$

The process to convert seconds to decimal parts of a degree is the same as for minutes, except that the conversion factor is

$$\frac{1^\circ}{3\ 600''}$$

This is due to the fact that there are 3 600″ in 1°.

Example: Convert 5°18′45″ to a decimal.

Solution:

$$5^\circ = 5^\circ$$
$$18' = 18' \times \frac{1^\circ}{60'} = 0.3^\circ$$
$$45'' = 45'' \times \frac{1^\circ}{3\ 600''} = 0.012\ 5^\circ$$
$$5^\circ 18' 45'' = 5^\circ + 0.3^\circ + 0.012\ 5^\circ = 5.312\ 5^\circ$$

To convert a degree expressed as a decimal back to minutes and seconds involves multiplying by

$$\frac{60'}{1^\circ} \text{ and then } \frac{60''}{1'}$$

Follow this example carefully.

Example: Convert 75.36° to degrees, minutes, and seconds.

Solution:

$$\begin{aligned} 75.36^\circ &= 75^\circ + \left(0.36^\circ \times \frac{60'}{1^\circ}\right) \\ &= 75^\circ + 21.6' \\ &= 75^\circ + 21' + \left(0.6' \times \frac{60''}{1'}\right) \\ &= 75^\circ + 21' + 36'' \\ &= 75^\circ 21' 36'' \end{aligned}$$

Due to the expanded use of computers and electronic calculators the majority of calculations involving degrees utilize the decimal form. With this in mind this text will use the decimal degree exclusively.

Radian Measure

Another unit for circular measure is the radian. A *radian* is an angle having an arc whose length is equal to that of a radius of the circle. Figure 6–11 shows how a radian is calculated. The circle has a radius whose length is the

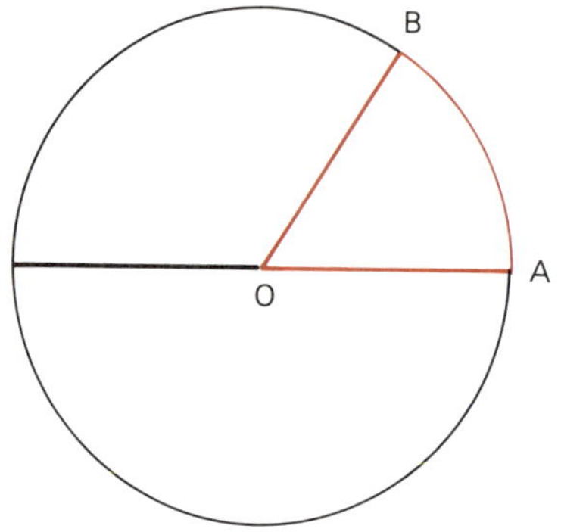

FIGURE 6–11 Line OA is a radius of this circle. When arc AB has the same length as radius OA, ∠AOB is equal to 1 radian.

distance from the center of the circle to a point on the circumference, identified here by the letter A. Line OA is a radius of the circle and so is line OB. Angle AOB intercepts an arc on the circumference. The length of this arc is the distance along the circumference from point A to point B. When this distance is equal in length to radius OA, or OB, or any other radius of this circle, then ∠AOB is equal to 1 radian.

A little earlier, the circumference of any circle was given as

$$C = 2\pi r$$

If both sides are divided by r, the formula becomes

$$\frac{C}{r} = 2\pi$$

This means the circumference of any circle, divided by its radius, is equal to a constant, and this constant has a numerical value of approximately 6.28. Another way of considering this is to regard the circumference as having a length that is always 6.28 times greater than the radius. Since an arc can be made equal in length to the radius, then a circumference contains 6.28 radians or 2π radians. However, when we measure off 6.28 radians along the circumference of a circle, we will be back at the starting point, or will have completed 360°. Hence:

$$6.28 \text{ radians} = 2\pi \text{ radians} = 360°$$

Dividing each of these by 2, we will then have

$$3.14 \text{ radians} = \pi \text{ radians} = 180°$$

And dividing by 3.141 59 (a more accurate value for π) we have

$$1 \text{ radian} = \frac{180°}{3.141\ 59} = 57.295\ 827° \text{ or } 57.3° \text{ (rounded)}$$

Radian-Degree Conversions

In electronics problems it may be necessary or convenient to change from radians to degrees or degrees to radians. The relationship between radians and degrees can be obtained by using ordinary algebra. **To change radians to degrees,**

$$\text{degrees} = \text{radians} \times \frac{180°}{\pi}$$

This equation can be simplified by dividing 180° by π.

$$\textbf{degrees} = \textbf{radians} \times \textbf{57.295 78°/radian} \qquad (6–4)$$

To change degrees to radians,

$$\text{radians} = \text{degrees} \times \frac{\pi}{180°}$$

This equation can be simplified by dividing π by 180°.

$$\textbf{radians} = \textbf{degrees} \times \textbf{0.017 453 radian/degree} \qquad (6–5)$$

Thus, we now have two equations, one for finding the value in radians when the number of degrees is known and the other for finding the number of degrees when information in terms of radians is supplied.

Example: Change 3.4 radians to degrees.

Solution:

$$\begin{aligned} \text{degrees} &= \text{radians} \times \frac{180°}{\pi} \\ &= 3.4 \text{ radians} \times 57.295\ 78°/\text{radian} \\ &= 194.8° \end{aligned}$$

Example: Convert 37°30′18″ to radians.

Solution:

$$\begin{aligned} 37°30'18'' &= 37° + \left(30' \times \frac{1°}{60'}\right) + \left(18'' \times \frac{1°}{3\ 600''}\right) \\ &= 37° + 0.5° + 0.005° \\ &= 37.505° \end{aligned}$$

$$\begin{aligned} \text{radians} &= 37.505° \times 0.017\ 453 \text{ radian/degree} \\ &= 0.65 \text{ radian} \end{aligned}$$

Example: Convert 8.5 radians to degrees

Solution:

$$\begin{aligned} \text{degrees} &= 8.5 \text{ radians} \times 57.295\ 78°/\text{radian} \\ &= 487.0° \\ &= 360° + 127.0° \\ &= 1 \text{ revolution} + 127.0° \end{aligned}$$

Grad Measure

A third unit of circular measure is defined as grad measure. There are 400 grads in a circle, or 100 grads in a 90° quadrant. One *grad* is equal to $\frac{1}{400}$ of a circle. You should verify that the following conversion factors are correct.

$$\text{grads} = \text{degrees} \times 1.11\bar{1} \text{ grads/degree} \quad (6\text{–}6)$$

$$\text{grads} = \text{radians} \times 63.662 \text{ grads/radian} \quad (6\text{–}7)$$

$$\text{degrees} = \text{grads} \times 0.9 \text{ degrees/grad} \quad (6\text{–}8)$$

$$\text{radians} = \text{grads} \times 0.015\ 708 \text{ radians/grad} \quad (6\text{–}9)$$

Example: Convert 275 grads to (a) degrees and (b) radians.

Solution: a. degrees = 275 grads × 0.9 degrees/grad = 247.5°
b. radians = 275 grads × 0.015 708 radians/grad = 4.319 7 radians

Calculator Tip

Electronic calculators, using an algebraic entry system, will often provide for conversions among degrees, radians, and grads. Consult the owner's manual for your calculator for the sequence to perform these conversions.

EXERCISE 6·1

1. Find the complementary angle for each of these angles.
   a. 35°
   b. 72°37′14″
   c. 0°82′46″
2. Find the complementary angle for each of these angles.
   a. 20°0′27″
   b. 62°1′3″
   c. 45°
3. Find the supplementary angle for each of these angles.
   a. 3°8′
   b. 175°0′40″
   c. 147.28°

4. Find the supplementary angle for each of these angles.
   a. 100°13′30″
   b. 36.08°
   c. 179.87°
5. What is the diameter of a circle with a radius of 20 centimetres?
6. Calculate the radius of a circle with a circumference of 113.1 inches.
7. What is the circumference of a circle with a diameter of 4.5 inches?
8. The circumference of a circle is $2\pi^2$. What is the diameter of this circle?
9. Complete the following problems and express the result as a degree decimal accurate to three decimal places.
   a. 84° − 17°12′32″
   b. 107.2° + 35°8′15″
   c. 342°49′19″ − 46.38°
   d. 20°0′52″ + 8.75° + 138°4′16″
10. Complete the following problems and express the result as a degree decimal accurate to three decimal places.
    a. 240° − 172.5° + 14°0′37″
    b. 62° + 31°17′2″ − 0.9°
    c. 135° − 29°12′ − 15°48′
    d. 327.8° − (113°12′50″ − 264.36°)
11. Express the following angles in radians accurate to two decimal places.
    a. 62°
    b. 137°
    c. 243.8°
    d. 172°12′35″
    e. 400°
    f. 27°12′13″
    g. 36.29°
    h. 114.592°
12. Express the following angles in grads accurate to two decimal places.
    a. 225°
    b. 137.28°
    c. 14°12′18″
    d. 62.93°
    e. 300°57′52″
    f. 625°
    g. 8.1°
    h. 310.5°
13. Convert the following measurements as indicated. Express answers to two decimal places.
    a. 4.7 radians to grads
    b. 372 grads to degrees
    c. 295° to radians
    d. 62.5 radians to degrees
    e. 97.3 grads to radians
    f. 183°19′12″ to grads
14. Convert the following measurements as indicated. Express answers to two decimal places.
    a. 427 grads to degrees
    b. 1.5 radians to degrees
    c. 180° to grads
    d. 582° to radians
    e. 3.27 radians to grads
    f. 14 grads to radians

UNIT 6·2 THE TRIGONOMETRIC FUNCTIONS

Objectives:

After studying this unit, you should be able to

- use a trigonometric table to determine the function of any angle.
- use an electronic calculator to determine the function of any angle.
- determine the two angles of a right triangle given two sides of the triangle.
- determine the signs of the functions in all four quadrants.

Angles and Sides of a Right Triangle

Trigonometry, as defined earlier, is not only concerned with angles but with the sides of triangles. A triangle is a closed, three-sided geometric figure. Every triangle, no matter what its shape, has six parts: three sides and three angles.

Figure 6–12 shows a right triangle, so called since one of its angles is a right angle or a 90° angle. The horizontal line, *b,* is the *base:* the vertical line meeting the base, *a,* is the *altitude.* The slant line joining the altitude and the base, *c,* is the *hypotenuse.*

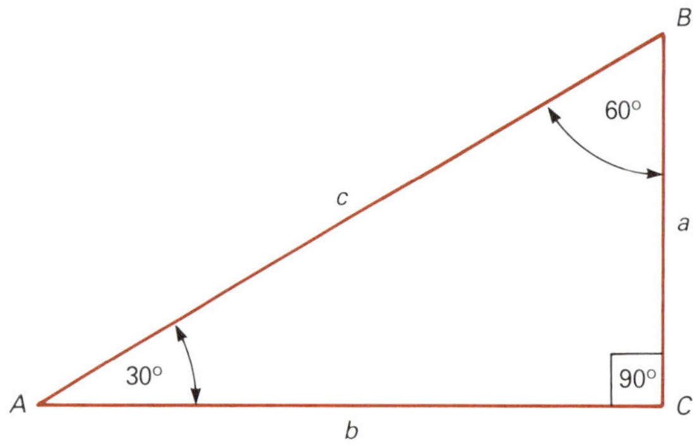

FIGURE 6–12

These three lines, the base, the altitude and the hypotenuse, enclose three angles, the sum of whose angular measure is always equal to 180°. This total angular measure of 180° is applicable to all triangles, regardless of shape. In the case of the right triangle, since one of the angles is 90°, then the sum of the two remaining angles must also be 90°, since 90° + 90° = 180°. There are an infinite number of right triangles, but in each of them, one of the angles is 90°.

In solving triangle problems in electronics, the sides of a right triangle are identified as *opposite* (opp) or *adjacent* (adj) to some reference angle. The side opposite the right angle is the *hypotenuse* (hyp). In Figure 6–12, the vertex angles are marked with capital letters *A, B,* and *C.* The sides across from these angles are labeled with lowercase letters *a, b,* and *c*. Any letters may be used to identify the sides and angles. Using $\angle A$ as the reference angle,

a is opposite $\angle A$
b is adjacent to $\angle A$
c is the hypotenuse

Using $\angle B$ as the reference angle,

a is adjacent to $\angle B$
b is opposite $\angle B$
c is the hypotenuse

Figure 6–13 shows a right triangle *ABC* with its base and hypotenuse extended to points C' and B'. In the right triangles, $\angle C$ is equal to $\angle C'$ since they are both right angles; $\angle A$ and $\angle A$ are the same angle; $\angle B$ and $\angle B'$ must also be equal since their values result from subtracting equal values from 180°. Increasing the length of the base and the length of the hypotenuse does not change the sizes of the 3 angles of the right triangle.

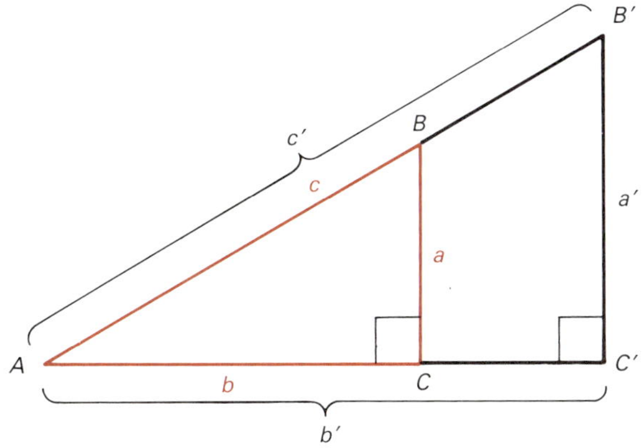

FIGURE 6–13 Increasing the length of the base and the hypotenuse does not change the size of the three angles of the triangle.

Sine of an Angle

The sine (sin) of an angle is the ratio of the opposite side to the hypotenuse. This ratio is a constant for a given angle regardless of the size of the right triangle. Referring to $\angle A$ in Figure 6–14, the opposite side is side a. The sine of $\angle A$ is

$$\sin \angle A = \frac{\text{opp}}{\text{hyp}} = \frac{a}{c}$$

Referring to $\angle B$ in Figure 6–14, the opposite side is side b. The sine of $\angle B$ is

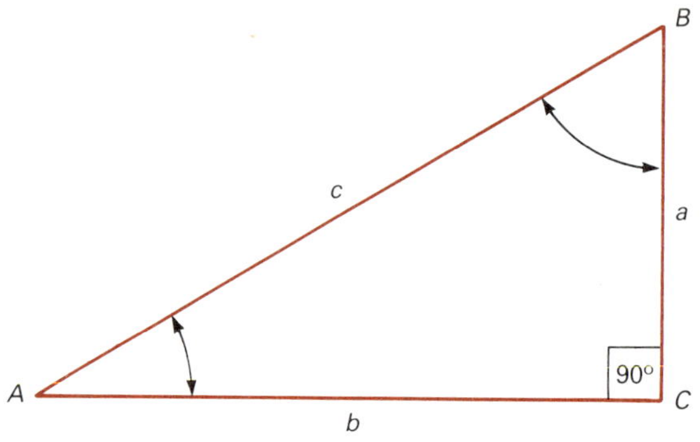

FIGURE 6–14

$$\sin \angle B = \frac{\text{opp}}{\text{hyp}} = \frac{b}{c}$$

Example: Use the triangle shown in Figure 6–14, let the length of line a equal 2 inches and the length of line c equal 4 inches. Find the sine of $\angle A$.

Solution: $\sin \angle A = \dfrac{\text{opp}}{\text{hyp}} = \dfrac{2 \text{ inches}}{4 \text{ inches}} = 0.5$

Example: A right triangle has an altitude of 15 centimetres and a hypotenuse of 22 centimetres. What is the sine of the angle between the hypotenuse and base?

Solution: $\sin \angle A = \dfrac{\text{opp}}{\text{hyp}} = \dfrac{15 \text{ cm}}{22 \text{ cm}} = 0.681\ 8$

This sine value of 0.681 8 represents a specific angle. When this angle is present in a right triangle the ratio of the angle's opposite side to the hypotenuse will always be 0.681 8. Note that the sine of an angle is a numerical value with no units of measurement.

Cosine of an Angle

The cosine (cos) of an angle is the ratio of the adjacent side to the hypotenuse. This ratio is a constant for a given angle regardless of the size of the right triangle. Referring to $\angle A$ in Figure 6–14, the adjacent side is side b. The cosine of $\angle A$ is

$$\cos \angle A = \frac{\text{adj}}{\text{hyp}} = \frac{b}{c}$$

Referring to $\angle B$ in Figure 6–14, the adjacent side is side a. The cosine of $\angle B$ is

$$\cos \angle B = \frac{\text{adj}}{\text{hyp}} = \frac{a}{c}$$

As in the sine of an angle, the cosine is a constant for a specific angle.

Tangent of an Angle

The tangent (tan) of an angle is the ratio of the opposite side to the adjacent side. This ratio is a constant for a given angle regardless of the size of the right triangle. Referring to $\angle A$ in Figure 6–14, the opposite side is side a and the adjacent side is side b. The tangent of $\angle A$ is

$$\tan \angle A = \frac{\text{opp}}{\text{adj}} = \frac{a}{b}$$

Referring to $\angle B$ in Figure 6–14, the opposite side is side b and the adjacent side is side a. The tangent of $\angle B$ is

$$\tan \angle B = \frac{\text{opp}}{\text{adj}} = \frac{b}{a}$$

As in the sine and cosine, the tangent is a constant for a specific angle.

Values of Trigonometric Functions

The sine, cosine, and tangent are referred to as *trigonometric functions.* With the help of a table of trigonometric functions, such as Table 6–1, any side of a right triangle can be calculated provided the value of one angle and one side are known. Table 6–1 shows the value of the sine, cosine, and tangent for all values of the angle ranging from 0° to 90°. To use the table, locate the value of the angle in degrees in the column marked "Angle." For example, to determine sine of 38°, locate 38° in the "Angle" column. Move one column to the right to see that the sine of this angle is 0.615 7. As another example, the cosine of 73° can be found by locating 73° in the "Angle" column. Move two columns to the right and the cosine of 73° is 0.292 4.

Calculator Tip

To find the sine of 74.83° in Table 6–1 would require *interpolation* or calculation of the sine value 83% of the way between the values of 0.961 3 and 0.965 9, or approximately 0.965 1. This can be a time-consuming process.

The use of trigonometric tables has declined rapidly with the expanded use of electronic calculators. Most calculators will not only display the function of an angle, but given a function will determine the angle. A typical sequence to determine the sine of 74.83° would be

74.83 [sin] and 0.965 153 6 is displayed

The sequence to find the angle with a cosine of 0.874 6 would be

0.874 6 [INV] [cos] and 29° is displayed

Example: Using Figure 6–15, find (a) $\angle A$, (b) $\angle B$.

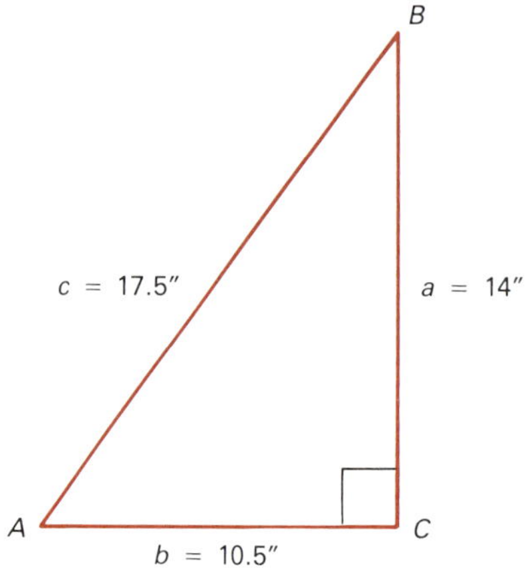

FIGURE 6–15

Solution: a. Since all three of the sides of the right triangle are known, it is possible to use any of the three functions presented thus far.

$$\sin \angle A = \frac{\text{opp}}{\text{hyp}} = \frac{14''}{17.5''} = 0.8, \text{ therefore } \angle A = 53.13°$$

$$\cos \angle A = \frac{\text{adj}}{\text{hyp}} = \frac{10.5''}{17.5''} = 0.6, \text{ therefore } \angle A = 53.13°$$

$$\tan \angle A = \frac{\text{opp}}{\text{adj}} = \frac{14''}{10.5''} = 1.33\overline{3}, \text{ therefore } \angle A = 53.13°$$

b. $\angle B$ may be found using any of four methods. The first three involve using the sin, cos, and tan functions. The fourth involves finding the complementary angle for $\angle A$.

$$\sin \angle B = \frac{\text{opp}}{\text{hyp}} = \frac{10.5''}{17.5''} = 0.6, \text{ therefore } \angle B = 36.87°$$

TABLE 6–1

Trigonometric Functions

Angle	sin	cos	tan	Angle	sin	cos	tan
1°	0.017 5	0.999 8	0.017 5	46°	0.719 3	0.694 7	1.035 5
2°	0.034 9	0.999 4	0.034 9	47°	0.731 4	0.682 0	1.072 4
3°	0.052 3	0.998 6	0.052 4	48°	0.743 1	0.669 1	1.110 6
4°	0.069 8	0.997 6	0.069 9	49°	0.754 7	0.656 1	1.150 4
5°	0.087 2	0.996 2	0.087 5	50°	0.766 0	0.642 8	1.191 8
6°	0.104 5	0.994 5	0.105 1	51°	0.777 1	0.629 3	1.234 9
7°	0.121 9	0.992 5	0.122 8	52°	0.788 0	0.615 7	1.279 9
8°	0.139 2	0.990 3	0.140 5	53°	0.798 6	0.601 8	1.327 0
9°	0.156 4	0.987 7	0.158 4	54°	0.809 0	0.587 8	1.376 4
10°	0.173 6	0.984 8	0.176 3	55°	0.819 2	0.573 6	1.428 1
11°	0.190 8	0.981 6	0.194 4	56°	0.829 0	0.559 2	1.482 6
12°	0.207 9	0.978 1	0.212 6	57°	0.838 7	0.544 6	1.539 9
13°	0.225 0	0.974 4	0.230 9	58°	0.848 0	0.529 9	1.600 3
14°	0.241 9	0.970 3	0.249 3	59°	0.857 2	0.515 0	1.664 3
15°	0.258 8	0.965 9	0.267 9	60°	0.866 0	0.500 0	1.732 1
16°	0.275 6	0.961 3	0.286 7	61°	0.874 6	0.484 8	1.804 0
17°	0.292 4	0.956 3	0.305 7	62°	0.882 9	0.469 5	1.880 7
18°	0.309 0	0.951 1	0.324 9	63°	0.891 0	0.454 0	1.962 6
19°	0.325 6	0.945 5	0.344 3	64°	0.898 8	0.438 4	2.050 3
20°	0.342 0	0.939 7	0.364 0	65°	0.906 3	0.422 6	2.144 5
21°	0.358 4	0.933 6	0.383 9	66°	0.913 5	0.406 7	2.246 0
22°	0.374 6	0.927 2	0.404 0	67°	0.920 5	0.390 7	2.355 9
23°	0.390 7	0.920 5	0.424 5	68°	0.927 2	0.374 6	2.475 1
24°	0.406 7	0.913 5	0.445 2	69°	0.933 6	0.358 4	2.605 1
25°	0.422 6	0.906 3	0.466 3	70°	0.939 7	0.342 0	2.747 5
26°	0.438 4	0.898 8	0.487 7	71°	0.945 5	0.325 6	2.904 2
27°	0.454 0	0.891 0	0.509 5	72°	0.951 1	0.309 0	3.077 7
28°	0.469 5	0.882 9	0.531 7	73°	0.956 3	0.292 4	3.270 9
29°	0.484 8	0.874 6	0.554 3	74°	0.961 3	0.275 6	3.487 4
30°	0.500 0	0.866 0	0.577 4	75°	0.965 9	0.258 8	3.732 1
31°	0.515 0	0.857 2	0.600 9	76°	0.970 3	0.241 9	4.010 8
32°	0.529 9	0.848 0	0.624 9	77°	0.974 4	0.225 0	4.331 5
33°	0.544 6	0.838 7	0.649 4	78°	0.978 1	0.207 9	4.704 6
34°	0.559 2	0.829 0	0.674 5	79°	0.981 6	0.190 8	5.144 6
35°	0.573 6	0.819 2	0.700 2	80°	0.984 8	0.173 6	5.671 3
36°	0.587 8	0.809 0	0.726 5	81°	0.987 7	0.156 4	6.313 8
37°	0.601 8	0.798 6	0.753 6	82°	0.990 3	0.139 2	7.115 4
38°	0.615 7	0.788 0	0.781 3	83°	0.992 5	0.121 9	8.144 3
39°	0.629 3	0.777 1	0.809 8	84°	0.994 5	0.104 5	9.514 4
40°	0.642 8	0.766 0	0.839 1	85°	0.996 2	0.087 2	11.430 1
41°	0.656 1	0.754 7	0.869 3	86°	0.997 6	0.069 8	14.300 7
42°	0.669 1	0.743 1	0.900 4	87°	0.998 6	0.052 3	19.081 1
43°	0.682 0	0.731 4	0.932 5	88°	0.999 4	0.034 9	28.636 3
44°	0.694 7	0.719 3	0.965 7	89°	0.999 8	0.017 5	57.290 0
45°	0.707 1	0.707 1	1.000 0	90°	1.000 0	0.000 0	

$$\cos \angle B = \frac{\text{adj}}{\text{hyp}} = \frac{14''}{17.5''} = 0.8, \text{ therefore } \angle B = 36.87^\circ$$

$$\tan \angle B = \frac{\text{opp}}{\text{adj}} = \frac{10.5''}{14''} = 0.75, \text{ therefore } \angle B = 36.87^\circ$$

$$\angle B = 90^\circ - 53.13^\circ = 36.87^\circ$$

Interpolation

The increasing use of electronic calculators has greatly diminished the use of trigonometric tables such as Table 6–1. However, as there will be times when a table such as this will be used, it is important that an understanding of the interpolation process be acquired. This process is demonstrated in the following examples.

Example: Find the sine of 36.58°.

Solution: Referring to Table 6–1 the value will lie between the sine of 36° and 37°.

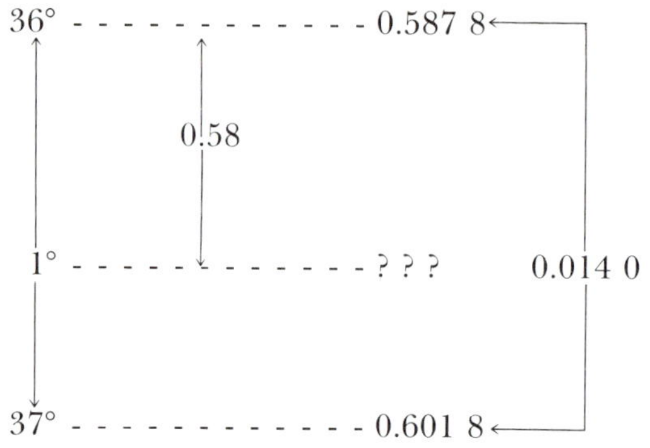

It is obvious from the above diagram that the value lies 58% of the way between 0.587 8 and 0.601 8. This would be

$0.58 \times 0.014\,0 = 0.008\,12$

Adding this value to the sine of 36° gives the sine of 36.58°, or

$0.587\,8 + 0.008\,12 = 0.595\,92$

Example: What is the cosine of 78°18′45″?

Solution:

$$78^\circ 18' 45'' = 78^\circ + \left(18' \times \frac{1^\circ}{60'}\right) + \left(45'' \times \frac{1^\circ}{3\,600''}\right)$$

$$= 78.\,312\,5^\circ$$

cos 78° = 0.207 9 and cos 79° = 0.190 8. The difference between cos 78° and cos 79° is 0.017 7.

Diagramming these values gives

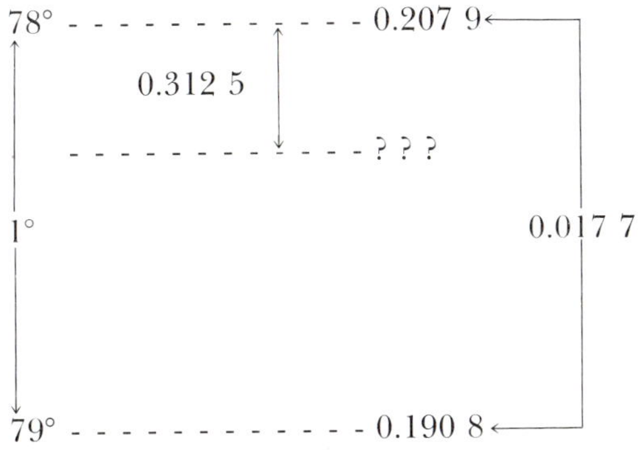

The unknown distance is
$0.312\ 5 \times 0.017\ 7 = 0.005\ 53$

Since the cosine values are decreasing, this value is subtracted from the cosine of 78°, giving
$\cos 78.312\ 5° = 0.207\ 9 - 0.005\ 53 = 0.202\ 4$

Example: Determine what angle has a tangent of 0.926 0?

Solution: Examining Table 6–1 tells us that this angle falls between 42° and 43°. A diagram will help in visualizing this problem.

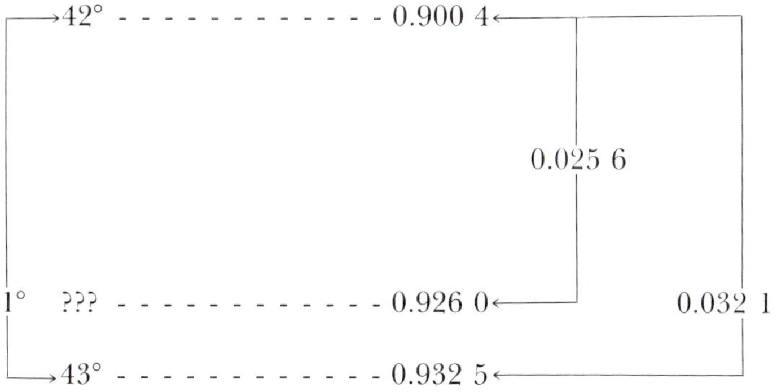

The unknown degree value falls at a point that is equivalent to
$$\left(\frac{0.025\ 6}{0.032\ 1}\right) \times 1° = 0.797\ 5° \text{ or } 0.8°$$

The unknown angle is
$42° + 0.8° = 42.8°$

Reciprocal Functions

There are relationships between the sides of a right triangle other than the sine, cosine, and tangent. The inverse of the sine is called the cosecant (csc). The *cosecant* of an angle is the ratio of the hypotenuse to the opposite side. The inverse of the cosine is called the secant (sec). The *secant* of an angle is the ratio of the hypotenuse to the adjacent side. The inverse of the tangent is called the cotangent (cot). The *cotangent* is the ratio of the adjacent side to the opposite side.

Basic Functions		Reciprocal Functions
sin	⟷	csc
cos	⟷	sec
tan	⟷	cot

Example: Using Figure 6–15, find the cotangent of (a) $\angle A$, and (b) $\angle B$.

Solution: a. $\cot \angle A = \frac{\text{adj}}{\text{opp}} = \frac{10.5''}{14''} = 0.75$

b. $\cot \angle B = \frac{\text{adj}}{\text{opp}} = \frac{14''}{10.5''} = 1.3\overline{3}$

Example: What angle has a cotangent of 2.5?

Solution: Since the cotangent is the inverse of the tangent, simply invert the 2.5 and find the angle corresponding to this value.

$\frac{1}{2.5} = 0.4$

tan = 0.4, therefore the angle is 21.8°.

Calculator Tip

Electronic calculators are able to determine the cotangent, cosecant, and secant of angles through the use of the reciprocal function [1/x]. The cosecant, for example, is the reciprocal of the sine. With this in mind, review these examples:

1. csc of 34.5°: 34.5 [sin] [1/x] and 1.765 5 is displayed
2. cot of 60°: 60 [tan] [1/x] and 0.577 4 is displayed
3. sec of 12°8′48″: 12 [+] 8 [÷] 60 [+] 48 [÷] 3 600 [=] [cos] [1/x] and 1.022 9 is displayed
4. What angle has a cotangent of 0.208 9?

   0.208 9 [1/x] [INV] [tan] and 78.2° is displayed

Cofunctions

In the right triangle shown in Figure 6–16, there is a relationship between $\angle A$ and $\angle B$. These two angles are complementary and their sum is therefore 90°. It can also be shown that the same ratio will appear for the cofunctions of these angles. The sine and cosine are cofunctions; the tangent and the cotangent are cofunctions; and the secant and the cosecant are cofunctions.

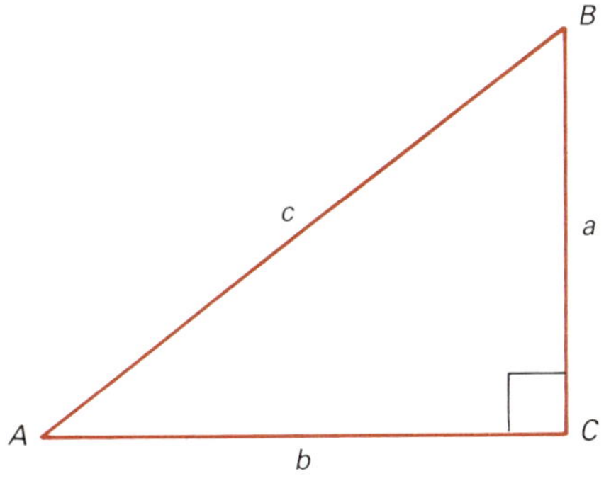

$$\sin \angle A = \frac{a}{c} = \cos \angle B$$

$$\cos \angle A = \frac{b}{c} = \sin \angle B$$

$$\tan \angle A = \frac{a}{b} = \cot \angle B$$

$$\cot \angle A = \frac{b}{a} = \tan \angle A$$

$$\sec \angle A = \frac{c}{b} = \csc \angle B$$

$$\csc \angle A = \frac{c}{a} = \sec \angle B$$

FIGURE 6–16

Example: The sine of 30° equals 0.5. Find the cosine of 60°.

Solution: The cofunctions of complementary angles are equal. The cosine of 60° equals 0.5.

Ranges of the Trigonometric Functions in the First Quadrant

When working with the three basic trigonometric functions, it is helpful to acquire an awareness of the numerical range of each function. In Figure 6–17, triangle *ABC* is drawn in the first quadrant. To examine the range of the sine of $\angle A$, begin with the basic ratio for the sine of an angle.

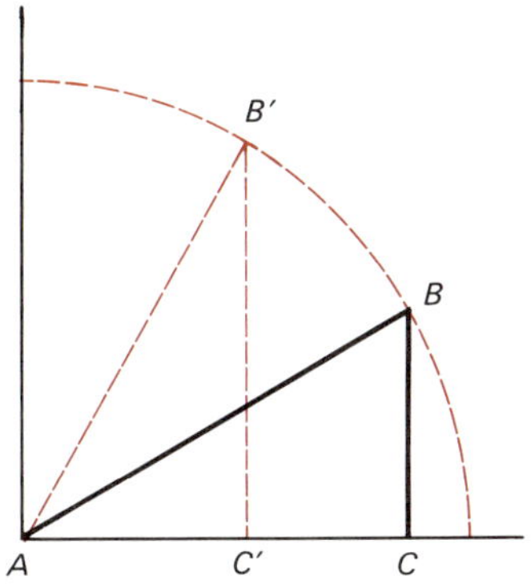

	0°	30°	45°	60°	90°
sin *A*	0	0.5	0.707	0.866	1
cos *A*	1	0.866	0.707	0.5	0
tan *A*	0	0.577	1	1.732	undefined

FIGURE 6–17 The ranges of the sine, cosine, and tangent functions in the first quadrant.

$$\sin \angle A = \frac{\text{opp}}{\text{hyp}}$$

Keeping the hypotenuse a constant value, increase and decrease the size of the angle and compare the ratios. As the size of $\angle A$ approaches zero, the length of the opposite side approaches zero.

$$\sin 0° = \frac{0}{\text{hyp}} = 0$$

As $\angle A$ increases towards 90°, the length of the opposite side approaches the same length as the hypotenuse

$$\sin 90° = 1$$

The table in Figure 6–17 gives the values of the three functions at 0°, 30°, 45°, 60°, and 90°. In the first quadrant, the ranges of the three functions are

The sine values range from 0 to 1.
The cosine values range from 1 to 0.
The tangent values range from 0 to undefined.

Trigonometric Functions in the Second, Third, and Fourth Quadrants

An angle of 130° rotates the hypotenuse to a point in the second quadrant as shown in Figure 6–18. Dropping a perpendicular to the x axis forms a triangle with an altitude a, hypotenuse c, and a base $-b$. Note that the base is a negative value due to the direction on the x axis. The two triangles in Figure

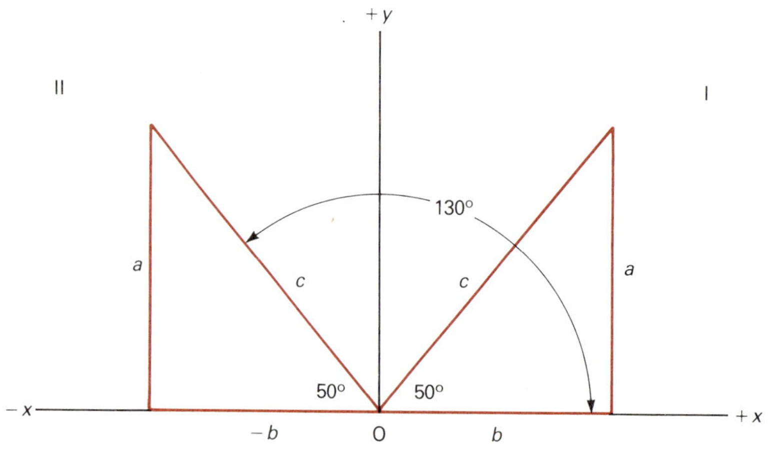

FIGURE 6–18

6–18 are identical except that the one in the first quadrant has been rotated about the y axis to the second quadrant.

The sine of 50° in both triangles would be the same. The hypotenuse c is positive in all four quadrants while a is positive only in quadrants I and II.

In examining Figure 6–18 it should be evident that an angle of 130° also describes the line 50° from the negative x axis. The sine of 130°, therefore, is the same as that of 50°. With a and c both positive the sign is also positive.

The cosine of an angle in the second quadrant would be negative due to $-b$. The numerical value of the cosine of 130° would again be the same as the cosine of 50°. The tangent of 130° would also be negative.

Determining the signs of the functions in the third and fourth quadrants involves the same logic as was followed in the second quadrant. The signs for the three functions in all four quadrants are given in Figure 6–19.

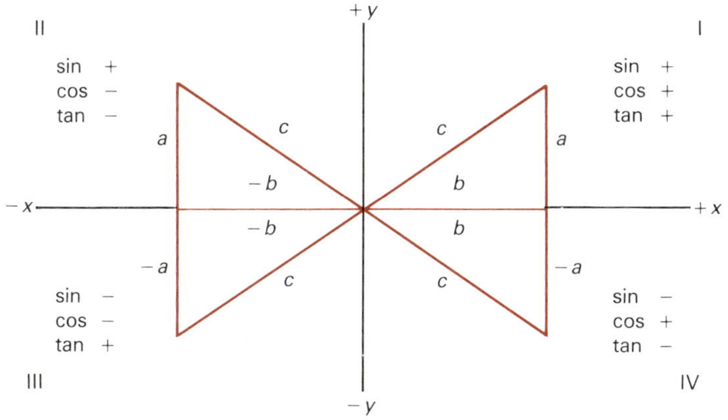

FIGURE 6–19

The process to determine the numerical value of a trigonometric function of any angle is

1. Extend a perpendicular to the x axis.
2. Determine the angle between the hypotenuse and the x axis.
3. Determine the value of the trigonometric function for this angle.
4. Indicate the appropriate sign depending upon the quadrant and the specific function.

Example: What is the sine of 237°?

Solution: An angle of 237° is in the third quadrant so that when a perpendicular is extended to the x axis, the angle is

$237° - 180° = 57°$

Table 6–1 lists the sine of 57° as 0.838 7. As the angle is in the third quadrant the value is

$\sin 237° = -0.838\ 7$

Example: Find the cosine of 315°.

Solution: $360° - 315° = 45°$
$\cos 45° = 0.707\ 1$
or, $\cos 315° = 0.707\ 1$

Example: What is the tangent of 265°?

Solution: $265° - 180° = 85°$
$\tan 85° = 11.43$
or, $\tan 265° = 11.43$

Example: What is the cotangent of 162°?

Solution: $180° - 162° = 18°$
$\tan 18° = 0.324\ 9$
$\cot 18° = 3.077\ 9$
in the second quadrant, $\cot 162° = -3.077\ 9$

Calculator Tip

Electronic calculators will automatically affix the correct sign to functions for angles greater than 90°. Also, it is not necessary to first subtract a given angle from 180° or 360° to determine the first quadrant equivalent. The sequences to solve the previous four examples would be:

237 [sin] and −0.838 7 is displayed

315 [cos] and 0.707 1 is displayed

265 [tan] and 11.430 1 is displayed

162 [tan] [1/x] and −3.077 7 is displayed.

EXERCISE 6·2

1. One of the angles in a right triangle is 14°. What are the other two angles?
2. Two angles in a right triangle total 114.8°. What is the value of the third angle?
3. The altitude of a right triangle is 11 inches. If the base is 4.5 inches, find the three angles within the triangle.
4. The hypotenuse of a right triangle is 18 centimetres in length. The altitude is 9 centimetres. What are the three angles?

Given two of three sides of a triangle (hypotenuse c, and legs a and b) in problems 5–15, determine the triangle's two acute angles.

5. $a = 3.1''$, $b = 7.5''$
6. $a = 12$ cm, $c = 19.3$ cm
7. $b = 2''$, $c = 17''$
8. $b = 1.7''$, $a = 4.62''$
9. $c = 11.3''$, $a = 9.7''$
10. $c = 25''$, $b = 20''$
11. $a = 29$ cm, $b = 11.08$ cm
12. $a = 8.95$ cm, $c = 30.19$ cm
13. $b = 12''$, $a = 12''$
14. $b = 25''$, $c = 50''$
15. $c = 4.5''$, $a = 3.9''$
16. Find the value for each of these functions accurate to four decimal places.
    a. sin 83.2°　　e. sec 82°41′7″
    b. cot 12°　　f. csc 0°
    c. tan 47°13′51″　　g. cot 45°
    d. cos 34.07°　　h. sin 0°57′19″
17. Find the value for each of these functions accurate to four decimal places.
    a. csc 60°　　e. cos 31°
    b. sin 75°0′13″　　f. cot 1°
    c. tan 89.9°　　g. tan 50°39′15″
    d. sec 11.19°　　h. cos 83°27′41″
18. Find the angle in the first quadrant associated with each of these functions. Express angles to two decimal places.
    a. sin = 0.125 7　　e. csc = 45.840 3
    b. cos = 0.350 2　　f. cot = 0.827 3
    c. tan = 7.115 4　　g. cos = 0.747 5
    d. sec = 1.024 5　　h. sin = 0.422 6
19. Find the angle in the first quadrant associated with each of these functions. Express angles to two decimal places.
    a. cot = 1.247 6　　e. csc = 1
    b. sin = 0.969 4　　f. sec = 1.007 1
    c. cos = 0.999 3　　g. cos = 0.406 7
    d. tan = 0.665 7　　h. sin = 0.345 1
20. Find the value of each of these functions accurate to four decimal places.
    a. cos 341°　　c. cot 271.24°
    b. sin 184°18′36″　　d. csc 100°
21. Find the value of each of these functions accurate to four decimal places.
    a. tan 194°59′47″　　c. sin 135°
    b. sec 286.13°　　d. cos 315°
22. Find the angles in the four quadrants associated with these functions. Express angles to two decimal places.
    a. sin = −0.707 1　　c. sec = −1.087 9
    b. tan = −0.052 4　　d. cos = 0.290 5

23. Find the angles in the four quadrants associated with these functions. Express angles to two decimal places.
    a. cos = −0.707 1
    b. cot = −0.463 1
    c. csc = 1.555 7
    d. sin = 0.999 8

UNIT 6·3 RIGHT TRIANGLES

Objectives:

After studying this unit, you should be able to

- solve right triangles using the Pythagorean Theorem.
- solve right triangles given one side and one angle.

Pythagorean Theorem

The triangle shown in Figure 6–20 is a right triangle whose base is 3 inches long, altitude is 4 inches long, and hypotenuse is 5 inches long. Attached to each of the sides is a square. The base square is 3 × 3 = 9 *square* inches.

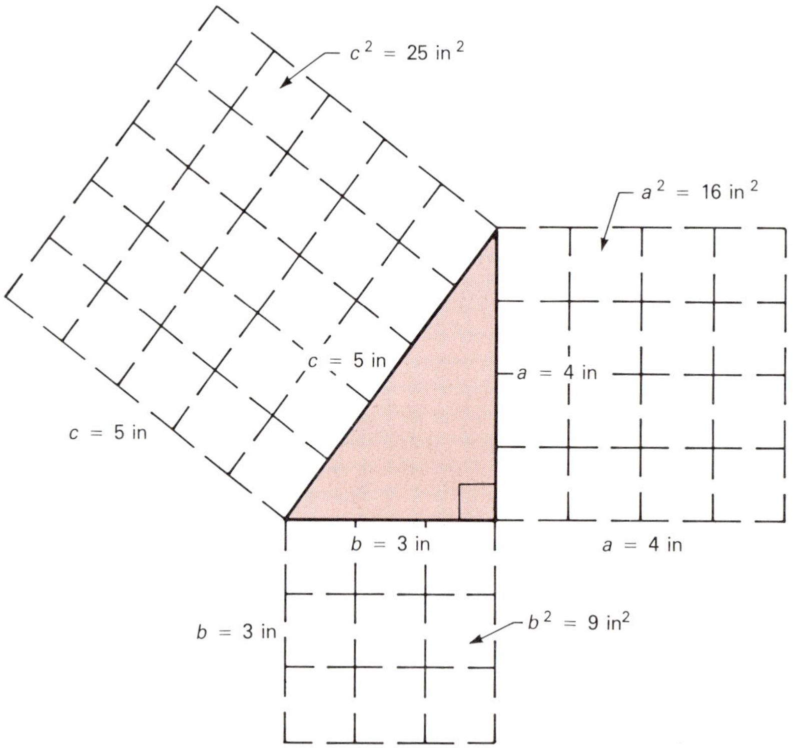

FIGURE 6–20 The area of the square connected to the base plus the area of the square connected to the altitude is equal to the area of the square on the hypotenuse.

The square using the altitude as one of its sides is 4 × 4 = 16 *square* inches. The square on the hypotenuse is 5 × 5 = 25 *square* inches. Interestingly, the sum of the base and altitude areas, 9 square inches + 16 square inches is equal to the area of the square on the hypotenuse, or 25 square inches. If the altitude is a, then any other side of the square attached to the altitude is also a, and the area is $a \times a = a^1 \times a^1 = a^2$. Similarly, for the base, b, $b \times b = b^1 \times b^1 = b^2$. And for the hypotenuse, $c \times c = c^1 \times c^1 = c^2$. Based on the previous statement, then, $a^2 + b^2 = c^2$, simply a mathematically shorthand way of writing that **the sum of the squares on the base and the altitude is equal to the area of the square on the hypotenuse.** This rule is known as the *Pythagorean Theorem.* Taking the square root of both sides of the equation $a^2 + b^2 = c^2$ will produce

$$c = \sqrt{a^2 + b^2} \tag{6–10}$$

Solving for a gives

$$a = \sqrt{c^2 - b^2} \tag{6–11}$$

Similarily, solving for b produces

$$b = \sqrt{c^2 - a^2} \tag{6–12}$$

The importance of these formulas can never be over emphasized. Given any two sides of a right triangle, the third can be calculated.

Example: A right triangle has a base 4 inches in length and an altitude 7.25 inches in length. Find the length of the hypotenuse.

Solution:

$$\begin{aligned} c &= \sqrt{a^2 + b^2} \\ &= \sqrt{(7.25)^2 + (4)^2} \\ &= \sqrt{52.562\,5 + 16} \\ &= \sqrt{68.562\,5} \\ &= 8.28'' \end{aligned}$$

Example: Find the missing side and angles in Figure 6–21.

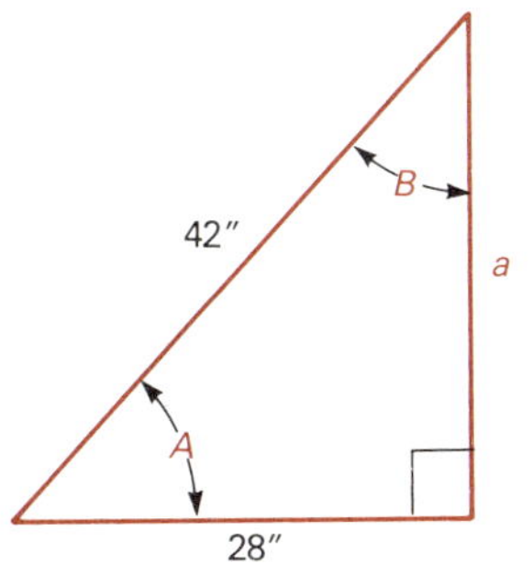

FIGURE 6–21

Solution:
$$a = \sqrt{c^2 - b^2}$$
$$= \sqrt{(42)^2 - (28)^2}$$
$$= 31.3''$$
$$\cos \angle A = \frac{28}{42} = 0.666, \angle A = 48.19°$$
$$\angle B = 90° - 48.19° = 41.81°$$

Example: Find the missing side and angles in Figure 6–22.

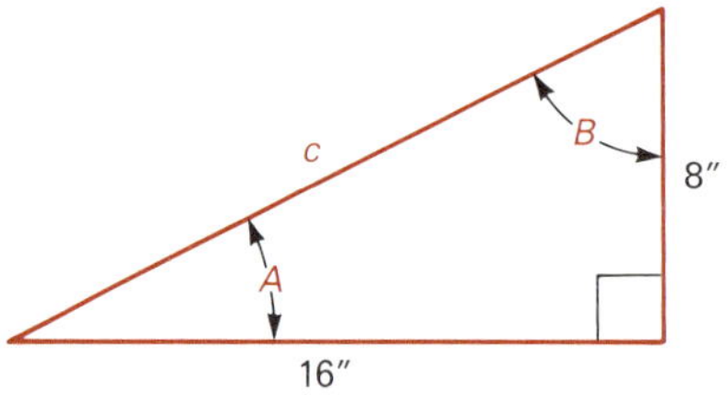

FIGURE 6–22

Solution:
$$c = \sqrt{a^2 + b^2}$$
$$= \sqrt{(16)^2 + (8)^2}$$
$$= 17.89''$$
$$\tan \angle A = \frac{8}{16} = 0.5, \angle A = 26.57°$$
$$\angle B = 90° - 26.57° = 63.43°$$

In reviewing the three previous examples there are some observations to be made about right triangles. One is that if $\angle A$ is greater than $\angle B$, side a will be greater than side b. Note in the last example that $\angle B$ was slightly more than twice $\angle A$, and that b was twice a. A second observation relates to the length of the hypotenuse compared to the altitude and base. The length of c will *always* be greater than a or b, but will always be less than $a + b$. When possible, utilize the values originally given in the problem. This prevents using numbers that have been rounded to two or three places and improves the accuracy of the final results.

Given One Side and One Angle

The information given for the right triangle in Figure 6–23 indicates that one side (a) and one angle are known. The following example will demonstrate a number of alternative methods for solving this problem.

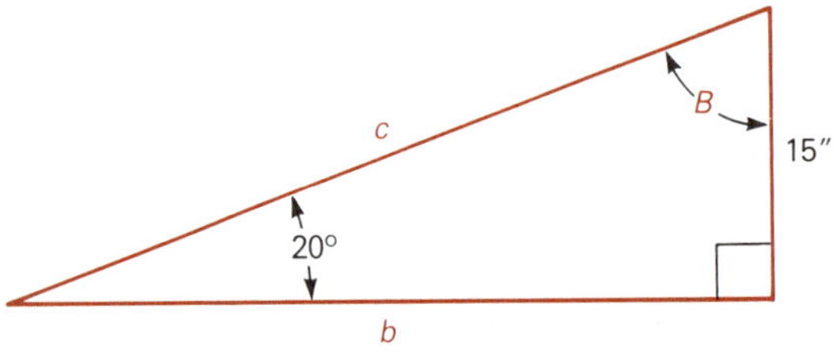

FIGURE 6–23

Example: Find the missing values in Figure 6–23.

Solution: The first step is to find $\angle B$.

$\angle B = 90° - 20° = 70°$

To find the remaining unknowns, starting with either unknown is appropriate. For the purpose of demonstration both approaches will be given. In order to find c, the formula for sine will be used.

$$\sin 20° = \frac{15''}{c}$$

$$0.342\ 0 = \frac{15''}{c}$$

$$c = \frac{15''}{0.342\ 0}$$

$$c = 43.86''$$

To find the base (b) the Pythagorean Theorem or the tangent could be used. As the value of c is a rounded value it will be interesting to compare the results of both. Using the Pythagorean Theorem we find

$$b = \sqrt{c^2 - a^2}$$

$$= \sqrt{(43.86'')^2 - (15'')^2}$$

$$= 41.22''$$

Using the formula for the tangent produces

$$\tan 20° = \frac{15''}{b}$$

$$0.364\ 0 = \frac{15''}{b}$$

$$b = \frac{15''}{0.364\ 0}$$

$$= 41.21''$$

Example: Find the missing values in Figure 6–24.

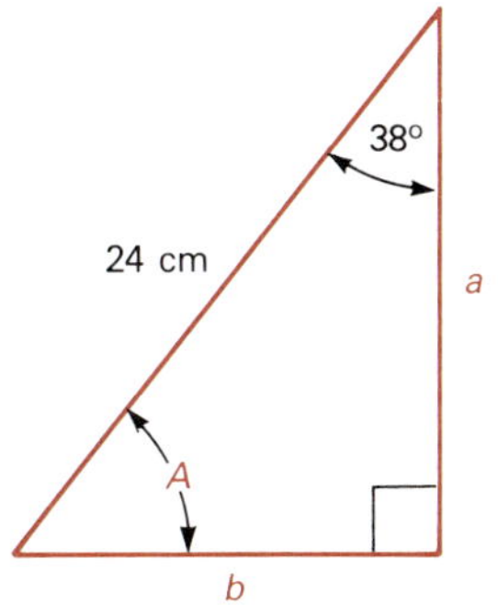

FIGURE 6–24

Solution: $\angle A = 90° - 38° = 52°$

$$\sin 38° = \frac{b}{24 \text{ cm}}$$
$$b = (24 \text{ cm})(\sin 38°)$$
$$= 14.78 \text{ cm}$$

$$\cos 38° = \frac{a}{24 \text{ cm}}$$
$$a = (24 \text{ cm})(\cos 38°)$$
$$= 18.91 \text{ cm}$$

Using the Pythagorean Theorem as a check gives

$$c = \sqrt{a^2 + b^2}$$
$$24 \text{ cm} = \sqrt{(18.91 \text{ cm})^2 + (14.78 \text{ cm})^2}$$
$$24 \text{ cm} = 24 \text{ cm}$$

Example: Find the missing values in Figure 6–25.

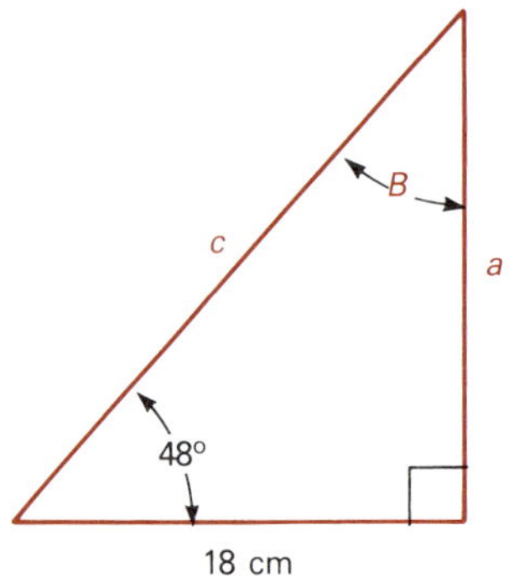

FIGURE 6–25

Solution: $\angle B = 90° - 48° = 42°$

$$\cos 48° = \frac{18 \text{ cm}}{c}$$
$$c = \frac{18 \text{ cm}}{\cos 48°}$$
$$= 26.9 \text{ cm}$$

$$\tan 48° = \frac{a}{18}$$
$$a = (18 \text{ cm})(\tan 48°)$$
$$= 19.99 \text{ cm}$$

As a check,

$$b = \sqrt{c^2 - a^2}$$

$$18 \text{ cm} = \sqrt{(26.9 \text{ cm})^2 - (19.99 \text{ cm})^2}$$

$$18 \text{ cm} = 18 \text{ cm}$$

EXERCISE 6·3

In problems 1–10, find the length of the unknown side and both of the triangle's acute angles.

1. $a = 11''$, $b = 19''$
2. $a = 16''$, $c = 42''$
3. $b = 3''$, $c = 25''$
4. $a = 6.2$ cm, $b = 6.2$ cm
5. $a = 13.8$ cm, $c = 25$ cm
6. $b = 17.4$ cm, $c = 34.8$ cm
7. $a = 1''$, $b = 5''$
8. $a = 8$ cm, $c = 30.5$ cm
9. $b = 8$ cm, $c = 30.5$ cm
10. $a = 19.7$ cm, $b = 7.1$ cm

In problems 11–20, find the length of the unknown sides and the value of the unknown angles.

11. $a = 17''$, $A = 41°$
12. $b = 23.8$ cm, $A = 19.8°$
13. $c = 37.1$ cm, $A = 36°$
14. $a = 8''$, $B = 24.7°$
15. $b = 11.9$ cm, $B = 71.2°$
16. $c = 62.5$ cm, $B = 3.6°$
17. $a = 38''$, $A = 19.3°$
18. $b = 20''$, $A = 30°$
19. $c = 40.4$ cm, $A = 61°37'28''$
20. $a = 16''$, $B = 44°57'50''$

UNIT 6·4 IMAGINARY AND COMPLEX NUMBERS

Objectives:

After studying this unit, you should be able to

- perform basic operations with imaginary numbers.
- perform basic operations with complex numbers.
- identify points in a complex plane using rectangular notation.
- identify points in a complex plane using polar notation.
- perform conversions between rectangular and polar expressions.

The *j*-Operator

Up to now, we have been concerned only with the roots of positive quantities. However, the root of a positive term is both positive and negative. Thus the square root of 4 results in two roots: $+2$ and -2, since $+2 \times +2 = 4$ and -2×-2 also equals 4.

Any negative number can be regarded as the product of -1 and a positive number. Thus, -9 can be thought of as 9×-1. The square root of -9 can then be written as $\sqrt{9 \times -1}$. Since we can take the square root of either term separately, $\sqrt{9 \times -1} = 3\sqrt{-1}$ or $\sqrt{-1} \times \sqrt{9}$. The square root of -1 is sometimes called an *imaginary number,* simply to distinguish it from real numbers. Quite often the letter i is used in place of $\sqrt{-1}$ but since this letter is also used as a symbol for current, it is commonly replaced by the letter j for electronics purposes. To emphasize that j is equivalent to $\sqrt{-1}$ it is often called the *operator-j* or the *j-operator*. The j-operator is used extensively in ac electronics.

When you are confronted with problems involving the square roots of negative numbers, each should be expressed as a j-operator before proceeding. As a general equation we have

$$\sqrt{-x} = \sqrt{x}\, j \qquad \textbf{(6–13)}$$

where x is a positive number greater than zero.

Example: What is the square root of -36?

Solution: $\sqrt{-36} = \sqrt{36 \times -1} = \sqrt{-1} \times 6 = j6$

Example: Simplify this expression: $-\sqrt{-16x^2}$

Solution:

$$\begin{aligned} -\sqrt{-16x^2} &= -\sqrt{(-1)16x^2} \\ &= -\sqrt{-1} \cdot \sqrt{16x^2} \\ &= -\sqrt{-1} \cdot 4x \\ &= -j4x \end{aligned}$$

Powers of the *j*-Operator

All of the fundamental operations of arithmetic (addition, subtraction, multiplication and division) are applicable to imaginary numbers. Imaginary numbers can also be raised to a power. Thus:

$$\begin{aligned} j^1 &= \sqrt{-1} = j \\ j^2 &= j \times j = \sqrt{-1} \times \sqrt{-1} = -1 \\ j^3 &= j \times j \times j = j^2 \times j = -1 \times j = -j \\ j^4 &= j^2 \times j^2 = -1 \times -1 = 1 \\ j^5 &= j^3 \times j^2 = -j \times -1 = j \end{aligned}$$

However, if we continue with powers of j, the answers will begin to repeat. Thus:

$$j^6 = j^4 j^2 = 1 \times -1 = -1$$

Note that the powers of j begin to repeat after j is raised to the fourth power. The sequence is j, -1, $-j$, 1 and the cycle repeats itself. This makes it possible to separate a power of j into multiples of 4.

Example: Evaluate the following powers of j:

a. j^{14} b. j^{24} c. j^{49} d. j^{103}

Solution:

a. $j^{14} = j^{12}j^2 = (1)(-1) = -1$

b. $j^{24} = 1$

c. $j^{49} = j^{48}j^1 = (1)(j) = j$

d. $j^{103} = j^{100}j^3 = (1)(-j) = -j$

Addition of Imaginary Numbers

Imaginary numbers can be added by converting to the j form.

Example: Add $\sqrt{-64}$, $\sqrt{-81}$, and $\sqrt{-49}$.

Solution:

$$\sqrt{-64} + \sqrt{-81} + \sqrt{-49}$$
$$= j8 + j9 + j7$$
$$= j24$$

Coefficients preceding the radical sign are handled in the usual manner.

Example: Add. $4\sqrt{-2} + 3\sqrt{-8} + 6\sqrt{-18}$

Solution:

$$4\sqrt{-2} + 3\sqrt{-8} + 6\sqrt{-18}$$
$$= j4\sqrt{2} + j3\sqrt{8} + j6\sqrt{18}$$
$$= j4\sqrt{2} + j3\sqrt{4 \times 2} + j6\sqrt{9 \times 2}$$
$$= j4\sqrt{2} + j3 \times 2\sqrt{2} + j6 \times 3\sqrt{2}$$
$$= j4\sqrt{2} + j6\sqrt{2} + j18\sqrt{2}$$
$$= j28\sqrt{2}$$

Subtraction of Imaginary Numbers

Subtraction of imaginary numbers is the inverse of the addition process.

Example: Subtract $\sqrt{-25}$ from $\sqrt{-100}$.

Solution:

$$\sqrt{-100} - \sqrt{-25}$$
$$= j10 - j5$$
$$= j5$$

Multiplication of Imaginary Numbers

Multiplication of imaginary numbers follows the usual steps in ordinary multiplication.

Example: Multiply $\sqrt{-36}$ by $\sqrt{-25}$.

Solution:
$$\begin{aligned} & \sqrt{-36}\,(\sqrt{-25}) \\ &= j6(j5) \\ &= j^2 30 \\ &= -1(30) \\ &= -30 \end{aligned}$$

Divison of Imaginary Numbers

Division of imaginary numbers is the inverse of multiplication.

Example: Divide $\sqrt{-81}$ by $\sqrt{-9}$.

Solution:
$$\begin{aligned} & \frac{\sqrt{-81}}{\sqrt{-9}} \\ &= \frac{j9}{j3} \\ &= 3 \end{aligned}$$

Note that j divided by j is equal to 1.

Complex Numbers

A *complex number* is a real number and an imaginary number united by a plus or a minus sign. For example, $5 - j7$ is a complex number. So is $a + jx$. However, since two terms are involved, complex numbers in the form of $a + jx$ can be considered as binomials.

Addition of Complex Numbers

Complex numbers can be subjected to the usual arithmetic operations by first combining the real parts and then combining the imaginary parts.

Example: Add $4 + j6$ and $2 + j7$.

Solution: $(4 + j6) + (2 + j7) = 6 + j13$

If the complex number is supplied in the form of the square root of a negative number, simply change the j-operator form.

Example: Add $3 + \sqrt{-49}$ and $5\sqrt{-25}$.

Solution:
$$\begin{aligned} & (3 + \sqrt{-49}) + (5\sqrt{-25}) \\ &= (3 + j7) + (5 \times j5) \\ &= 3 + j7 + j25 \\ &= 3 + j32 \end{aligned}$$

Subtraction of Complex Numbers

To subtract complex numbers, change the sign of the subtrahend and proceed with the rules for addition. This rule is applicable whether the subtraction problem is arranged in vertical or horizontal form.

Example: Subtract $7 - j6$ from $3 - j2$.

Solution:

$$\begin{aligned} & (3 - j2) - (7 - j6) \\ &= 3 - j2 - 7 + j6 \\ &= -4 + j4 \end{aligned}$$

Multiplication of Complex Numbers

Multiplication of complex numbers follows the same procedure used previously in connection with binomials.

Example: Multiply $3 - j6$ by $4 + j2$.

Solution:

$$\begin{aligned} & (3 - j6)(4 + j2) \\ &= 12 - j24 + j6 - 12j^2 \\ &= 12 - j18 - 12j^2 \\ &= 12 - j18 + 12 \\ &= 24 - j18 \end{aligned}$$

Division of Complex Numbers

As a first step in division, the denominator must be rationalized. To get a real number as the divisor, multiply numerator and denominator by the *conjugate* of the denominator. The conjugate is the denominator with its connector sign changed.

Example: Divide 8 by $4 + \sqrt{-2}$.

Solution: The first step is to multiply numerator and denominator by the conjugate of the denominator.

$$\begin{aligned} & \frac{8}{4 + \sqrt{-2}} \times \frac{4 - \sqrt{-2}}{4 - \sqrt{-2}} \\ &= \frac{8(4 - \sqrt{-2})}{(4 + \sqrt{-2})(4 - \sqrt{-2})} \\ &= \frac{32 - j8\sqrt{2}}{16 - j^2 2} \\ &= \frac{32 - j8\sqrt{2}}{18} \end{aligned}$$

Summarizing the basic operations as they apply to complex numbers results in:

- Addition and subtraction

$$(a + jb) + (c + jd) = (a + c) + j(b + d) \qquad (6\text{–}14)$$

- Multiplication

$$\begin{aligned} (a + jb)(c + jd) &= ac + jad + jcb + j^2bd \\ &= ac + jad + jcb - bd \\ &= (ac - bd) + j(ad + cb) \end{aligned}$$

therefore,

$$(a + jb)(c + jd) = (ac - bd) + j(ad + cb) \tag{6–15}$$

- Division

$$\begin{aligned}\frac{a + jb}{c + jd} &= \frac{a + jb}{c + jd} \times \frac{c - jd}{c - jd} \\ &= \frac{(a + jb)(c - jd)}{(c + jd)(c - jd)} \\ &= \frac{ac - jad + jbc - j^2bd}{c^2 - jcd + jcd - j^2d^2} \\ &= \frac{(ac + bd) + j(bc - ad)}{c^2 + d^2}\end{aligned}$$

therefore,

$$\frac{a + jb}{c + jd} = \frac{(ac + bd) + j(bc - ad)}{c^2 + d^2} \tag{6–16}$$

Rectangular and Polar Notation

Complex numbers can be expressed in *rectangular* form or *polar* form. Each of these forms can represent the location of a point in a plane. As a complex number contains an imaginary part, the corresponding point exists in a *complex* plane as shown in Figure 6–26. The horizontal axis is the real number axis and the vertical is the imaginary axis. In a later unit these will be referred to as resistance axis and reactance axis respectively.

The rectangular form of a complex number is written

$$a + jb$$

where a is the real number and jb is the imaginary number.

The point in the first quadrant of Figure 6–26 represents the complex number $3 + j4$. To locate this point find 3 on the $+x$ axis, 4 on the $+j$ axis, and then locate the intersection of the lines extended from these points. Rectangular notation identifies both the point and the *vector* drawn from the origin to that point.

Example: Locate the points and draw the vectors for these complex numbers: $-2 + j$, $-4 - j3$, and $1 - j$.

Solution: Each of these complex numbers has been graphed in Figure 6–27. Note the location of each vector and the corresponding signs of the real and imaginary parts of each complex number.

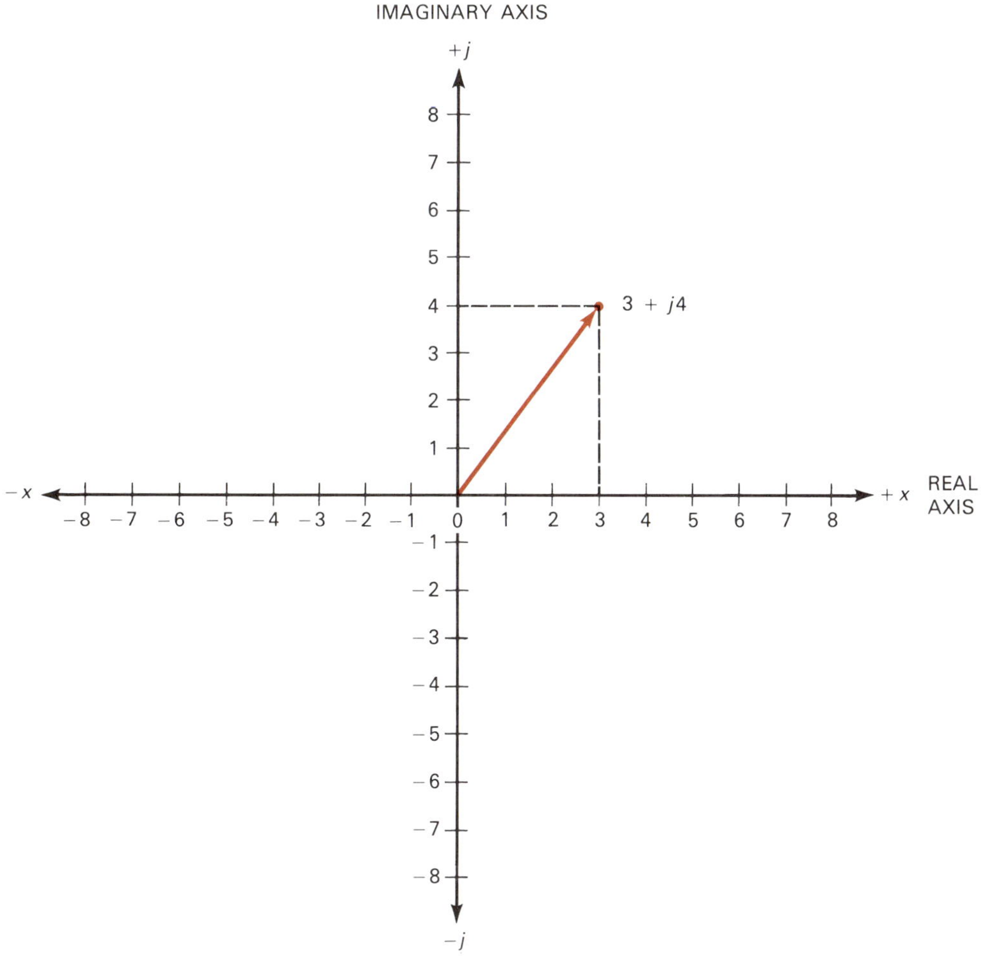

FIGURE 6–26 Points in the complex plane

The rectangular form of complex numbers is one method of describing the location of a vector. Another method of describing the vector is to indicate the length, or magnitude, and the angle measured with respect to the positive real axis. The vector in Figure 6–28 is described by the rectangular number $a + jb$. The polar form of this vector is $r \angle \theta$, where r is the length of the vector and θ is the angle. Given the distance from the origin to a point, and the angle between that point and the positive real axis, it is possible to locate a point in the complex plane.

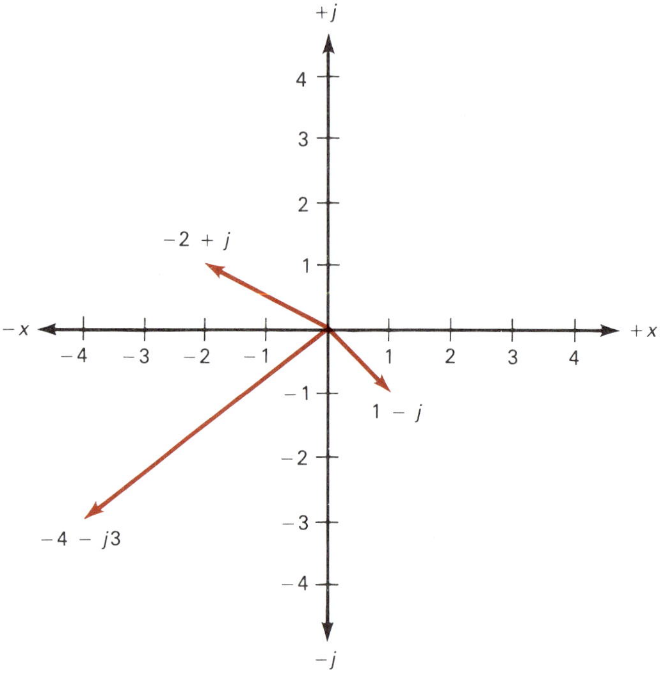

FIGURE 6–27

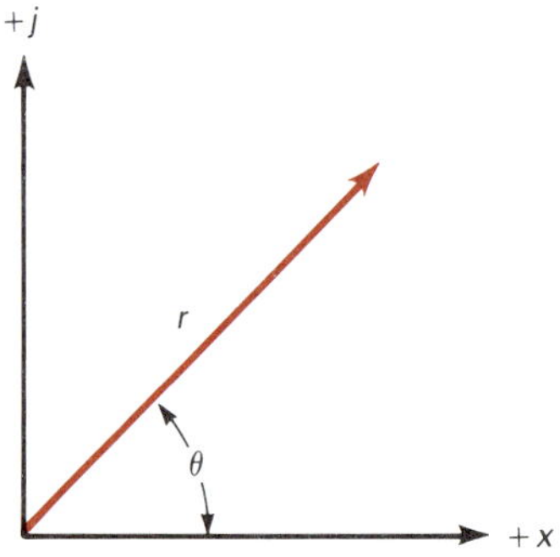

FIGURE 6–28

Example: Locate these points in the complex plane: $3\angle 30°$, $2\angle 165°$, $3\angle 210°$, and $4\angle 360°$.

Solution: Each of these points has been graphed in Figure 6–29. Note that r is *always* positive.

Mathematical Operations with Polar Numbers

Basic operations using rectangular notation and imaginary numbers were presented earlier in this unit. Similar operations are also used with polar notation.

Polar addition and subtraction may only be performed when each com-

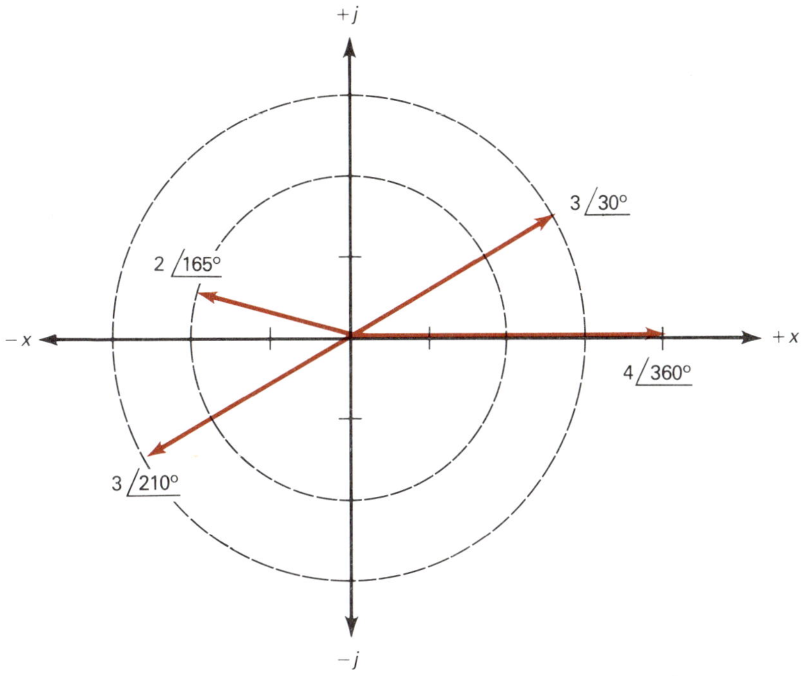

FIGURE 6–29

plex number contains the same phase angle, or 180° multiples of that angle. When vectors are in phase, and in the same direction, the lengths are summed and the angle remains the same. When two vectors are 180° out of phase, therefore moving in opposite directions, the resultant length is the difference between the two vectors and the resulting angle is that of the larger vector.

Example: Combine the following polar numbers as indicated.

a. $5\angle 30^\circ + 10\angle 30^\circ$
b. $8\angle 0^\circ + 12\angle 180^\circ$
c. $20\angle 70^\circ + 30\angle 250^\circ$
d. $9\angle 10^\circ + 6\angle 370^\circ$

Solution: a. $5\angle 30^\circ + 10\angle 30^\circ = 15\angle 30^\circ$
b. $8\angle 0^\circ + 12\angle 180^\circ = 4\angle 180^\circ$
c. $20\angle 70^\circ + 30\angle 250^\circ = 10\angle 250^\circ$
d. $9\angle 10^\circ + 6\angle 370^\circ = 15\angle 10^\circ = 15\angle 370^\circ$

These polar problems appear in Figure 6–30.

Multiplication of complex numbers in polar form involves multiplying the vector length and adding the phase angles.

$$(r_1 \angle \theta_1)(r_2 \angle \theta_2) = (r_1)(r_2) \angle \theta_1 + \theta_2 \qquad (6\text{–}17)$$

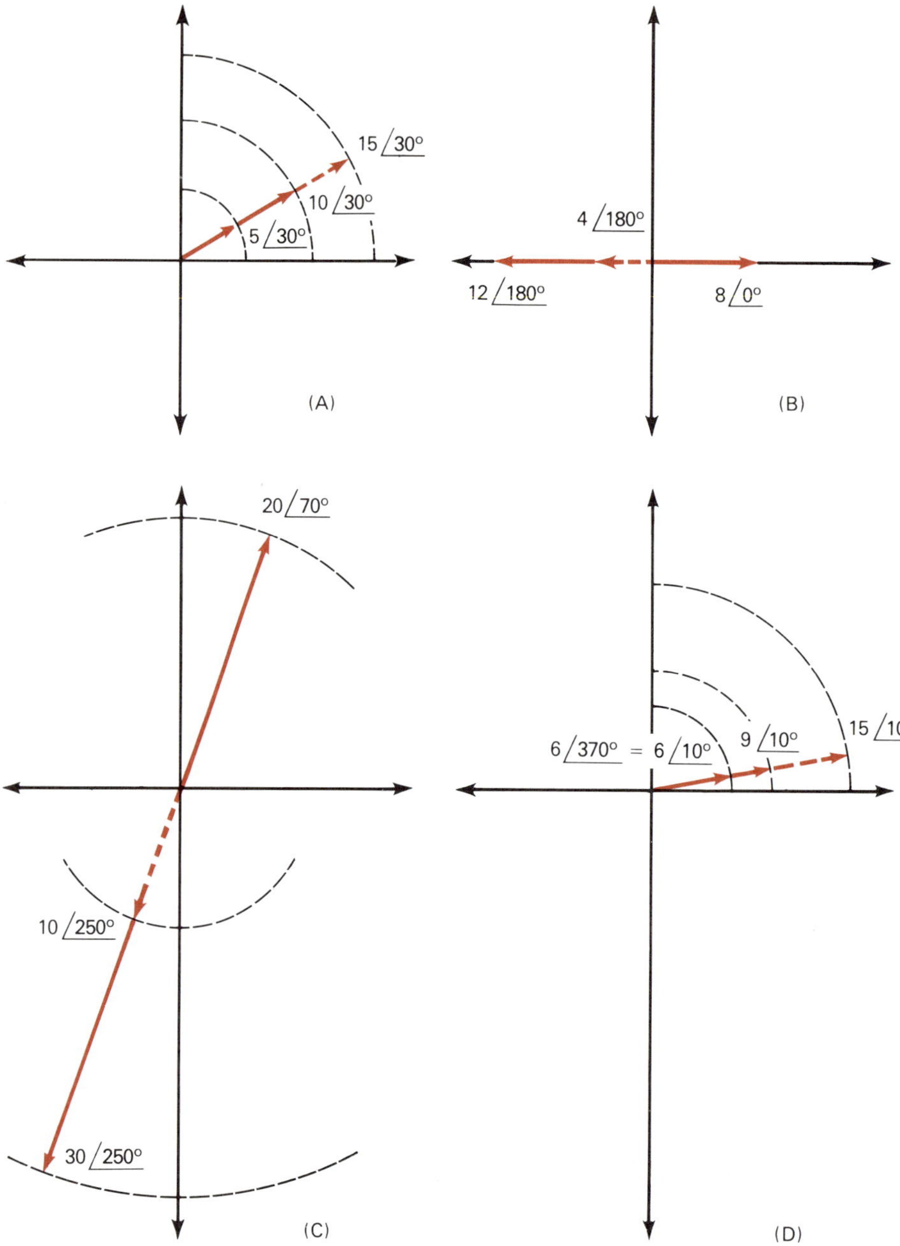

15 /30°
10 /30°
5 /30°
(A)
4 /180°
12 /180°
8 /0°
(B)
20 /70°
10 /250°
30 /250°
(C)
6 /370° = 6 /10°
9 /10°
15 /10°
(D)

FIGURE 6–30

Example: Multiply each of these problems.

a. $(4\angle 15^\circ)(7\angle 30^\circ)$
b. $(10\angle -60^\circ)(3\angle 20^\circ)$
c. $(-12\angle 90^\circ)(5\angle 0^\circ)$
d. $(1\angle 57^\circ)(4\angle -60^\circ)$

Solution:

a. $(4\angle 15^\circ)(7\angle 30^\circ) = (7)(4)\angle 15^\circ + 30^\circ = 28\angle 45^\circ$
b. $(10\angle -60^\circ)(3\angle 20^\circ) = (10)(3)\angle -60^\circ + 20^\circ = 30\angle -40^\circ$
c. $(-12\angle 90^\circ)(5\angle 0^\circ) = (-12)(5)\angle 90^\circ + 0^\circ = -60\angle 90^\circ$
d. $(1\angle 57^\circ)(4\angle -60^\circ) = (1)(4)\angle 57^\circ - 60^\circ = 4\angle -3^\circ$

Division of polar numbers involves dividing the vector length in the numerator by the length in the denominator. The resultant phase angle is found by subtracting the denominator's phase angle from that found in the numerator.

$$\frac{r_1\angle \theta_1}{r_2\angle \theta_2} = \frac{r_1}{r_2}\angle \theta_1 - \theta_2 \qquad \textbf{(6–18)}$$

Example: Divide each of the following as indicated.

a. $10\angle 30^\circ \div 5\angle 18^\circ$
b. $60\angle 0^\circ \div 15\angle 90^\circ$
c. $28\angle 25^\circ \div 7\angle -5^\circ$
d. $17\angle 0^\circ \div 34\angle -20^\circ$

Solution:

a. $10\angle 30^\circ \div 5\angle 18^\circ = \frac{10}{5}\angle 30^\circ - 18^\circ = 2\angle 12^\circ$

b. $60\angle 0^\circ \div 15\angle 90^\circ = \frac{60}{15}\angle 0^\circ - 90^\circ = 4\angle -90^\circ$

c. $28\angle 25^\circ \div 7\angle -5^\circ = \frac{28}{7}\angle 25^\circ - (-5^\circ) = 4\angle 30^\circ$

d. $17\angle 0^\circ \div 34\angle -20^\circ = \frac{17}{34}\angle 0^\circ - (-20^\circ) = 0.5\angle 20^\circ$

Rectangular-Polar Conversions

It should be obvious that these two methods of locating vectors are related. The vector in Figure 6–31 has been located by both rectangular and polar coordinates.

Conversion from rectangular to polar notation is accomplished by use of these equations:

$$r = \sqrt{a^2 + b^2} \qquad \textbf{(6–19)}$$

$$\tan \theta = \frac{b}{a} \qquad \textbf{(6–20)}$$

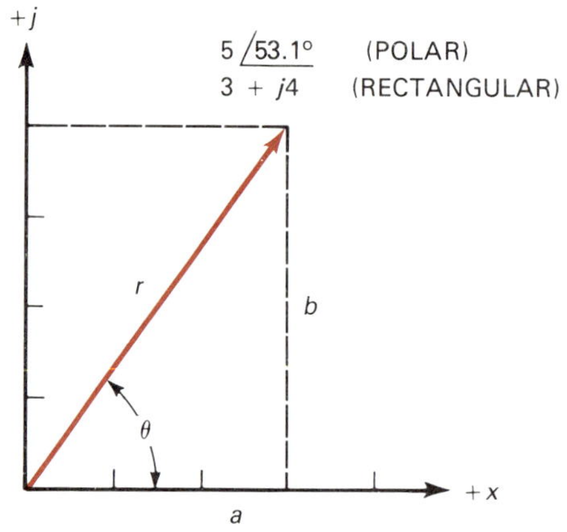

FIGURE 6–31

Conversions from polar to rectangular notation utilize these equations:

$$a = r(\cos \theta) \quad (6\text{–}21)$$

$$b = r(\sin \theta) \quad (6\text{–}22)$$

Example: Convert $3 + j4$ to polar notation.

Solution: $r = \sqrt{3^2 + 4^2} = \sqrt{9 + 16} = \sqrt{25} = 5$

$\tan \theta = \dfrac{4}{3} = 1.333, \ \theta = 53.13°$

$3 + j4 = 5\underline{/53.13°}$

Example: Convert $12\underline{/137°}$ to rectangular notation.

Solution:

$$\begin{aligned} a &= r(\cos \theta) \\ &= 12(\cos 137°) \\ &= 12(-0.731\ 4) \\ &= -8.78 \\ b &= r(\sin \theta) \\ &= 12(\sin 137°) \\ &= 12(0.682) \\ &= 8.18 \end{aligned}$$

$12\underline{/137°} = -8.78 + j8.18$

Calculator Tip

Scientific calculators often provide the capability for direct conversions between rectangular and polar notation. Consult the owner's manual for your calculator for specific sequences to perform these conversions.

Example: Convert $-6 - j6$ to polar notation.

Solution: $r = \sqrt{(-6)^2 + (-6)^2} = \sqrt{72} = 8.49$

$\tan\theta = \dfrac{-6}{-6} = 1$, therefore $\theta = 45°$. Since this point is in the third quadrant, the angle is $180° + 45° = 225°$.

$-6 - j6 = 8.49\angle 225°$

Example: Convert $10\angle 290°$ to rectangular notation.

Solution:
$$a = 10(\cos 290°) = 10(0.342) = 3.42$$
$$b = 10(\sin 290°) = 10(-0.939\,7) = -9.4$$
$$10\angle 290° = 3.42 - j9.4$$

EXERCISE 6·4

In problems 1–10, express each number in terms of j.

1. $\sqrt{-9}$
2. $\sqrt{-49}$
3. $-\sqrt{-\dfrac{9}{16}}$
4. $-\sqrt{-144}$
5. $-\sqrt{-8x^2}$
6. $\sqrt{-81a^4b^2}$
7. $-\sqrt{-0.01x^4}$
8. $\sqrt{-32}$
9. $-\sqrt{-147}$
10. $-\sqrt{-48x^4y^2}$

In problems 11–15, simplify each in terms of j.

11. j^{16}
12. j^{31}
13. j^7
14. j^{93}
15. j^{54}

Perform the indicated operations in problems 16–30.

16. $\sqrt{-49} - \sqrt{-16}$
17. $\sqrt{-36} + \sqrt{-64}$
18. $(-\sqrt{-100})(\sqrt{-25})$
19. $(\sqrt{-24})(\sqrt{-54})$
20. $\sqrt{-225} \div \sqrt{-25}$
21. $(7 + j9) + (9 + j3)$
22. $(6 - j11) - (-2 - j7)$
23. $(-6 + j)(2 - j3)$
24. $(5 + j5)(5 - j5)$
25. $(10 - j15) \div (3 + j2)$
26. $(-9 + j9) \div (6 - j)$
27. $(14 - j2) + (7 + j8)$
28. $(4 + j7)(2 - j5)$
29. $(-12 - j9) - (-11 - j6)$
30. $(1 + j2) \div (1 - j7)$

In problems 31–40, indicate what quadrant the point is in.

31. $-5 + j7$
32. $-5 - j7$
33. $3 + j12$
34. $6 - j4$
35. $-8 + j$
36. $1 + j$
37. $-2 + j2$
38. $-4 - j6$
39. $6 - j4$
40. $-12 + j11$

Perform the indicated operations in problems 41–50.

41. $7\angle 80° + 17\angle 80°$
42. $6\angle 15° + 4\angle 195°$
43. $20\angle 40° - 20\angle 220°$
44. $(3\angle 18°)(5\angle 20°)$
45. $(6\angle 67°)(3\angle 13°)$
46. $(4\angle 45°)(4\angle -45°)$
47. $12\angle 15° \div 4\angle 30°$
48. $16\angle 30° \div 5\angle -30°$
49. $25\angle -63° \div 9\angle -12°$
50. $14\angle 50° \div 4\angle -10°$

In problems 51–55, convert the rectangular expressions to polar expressions.

51. $-8 + j16$
52. $12 - j7$
53. $-6 - j10$
54. $1 + j$
55. $14 + j5$

In problems 56–60, convert the polar expressions to rectangular expressions.

56. $15\angle 100°$
57. $9\angle 317°$
58. $6\angle 90°$
59. $18.5\angle 3°$
60. $24\angle 265.8°$

UNIT 6·5 PROBLEM REVIEW

1. Find the complementary angle for each of these angles.
   a. 78.2°
   b. 14°
   c. 44°59′12″
   d. 0°4′47″
2. Find the complementary angle for each of these angles.
   a. 11.08°
   b. 84°0′19″
   c. 20°
   d. 0.89°
3. Find the supplementary angle for each of these angles.
   a. 78.2°
   b. 101.9°
   c. 41°0′58″
   d. 174°
4. Find the supplementary angle for each of these angles.
   a. 16°12′38″
   b. 138.65°
   c. 167°
   d. 90°
5. What is the circumference of a circle with a diameter of 7.5 inches?
6. Find the radius of a circle with a circumference of 31.42 cm.
7. The circumference of a circle is π^2. What is the radius of the circle?
8. Complete the following problems and express the result as a degree decimal:
   a. 19.7° − 11°12′45″

b. $157°46' + 47° - 65.7°$
c. 1.5 radians $- 64.3°$
d. $270° - (112°18' - 0.4$ radians)

9. Complete the following problems and express the result as a degree decimal.
   a. 1 radian $+ 45°$
   b. $187°19'6'' - 57.5° + 0.5$ radians
   c. $485° - (2.8$ radians $+ 312°)$
   d. 150 grads $+ 1.9$ radians $+ 63.2°$
10. Convert the following as indicated.
    a. 1.2 radians to degrees
    b. 327° to grads
    c. 162.8 grads to radians
    d. 93°52′37″ to radians
11. Convert the following as indicated.
    a. 3.5 radians to grads
    b. 428 grads to degrees
    c. 217.92° to radians
    d. 10 radians to degrees
12. One of the angles in a right triangle is 68.5°. What are the two other angles?
13. Two of the angles in a right triangle total 90°. What is the value of the third angle?
14. The altitude and base of a right triangle are 12 cm and 15 cm respectively. Find the three angles within the triangle.

Given two of three sides of a triangle (hypotenuse c, sides a and b). In problems 15–20, find the triangle's two acute angles.

15. $a = 10''$, $b = 20''$
16. $c = 18.8$ cm, $a = 11.6$ cm
17. $b = 1.5$ cm, $c = 3.7$ cm
18. $b = 16.1$ cm, $a = 37.5$ cm
19. $c = 100$ cm, $a = 46.2$ cm
20. $c = 57.6$ cm, $b = 24.9$ cm
21. Find the value for each of these functions.
    a. sin 112.7°
    b. tan 11°18′12″
    c. cos 62°
    d. tan 1.5 radians
    e. csc 328.73 grads
    f. sec 119°
22. Find the value for each of these functions.
    a. sin 5 radians
    b. tan 225°
    c. cos 62 grads
    d. sin 726 grads
    e. cot 180°
    f. sec 12°13′57″
23. Find the angles in all four quadrants associated with these functions.

a. tan −1.271 6
b. sin 0.872 6
c. cos −0.377 6
d. cot 1.076 5

24. Find the angles in all four quadrants associated with these functions.
a. sin −0.127 9
b. cos 0.473 8
c. sec −1.264 1
d. tan −1.264 1

In problems 25–35, find the unknown side(s) and the unknown angle(s) as shown in Figure 6–32.

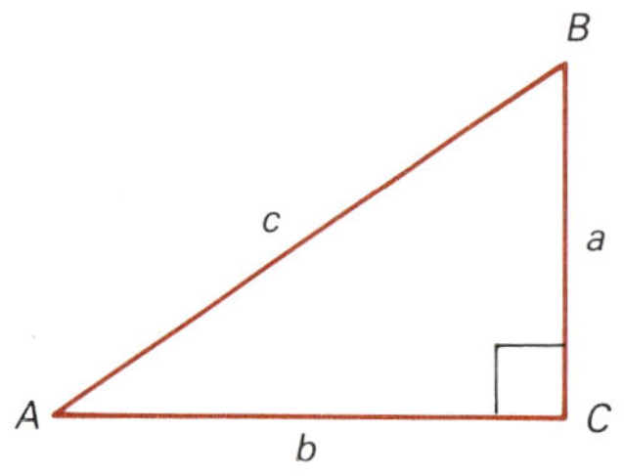

FIGURE 6–32

25. $a = 15''$, $b = 13''$
26. $a = 10''$, $A = 37.2°$
27. $b = 42$ cm, $c = 69.8$ cm
28. $b = 16''$, $B = 21°$
29. $c = 25''$, $B = 54.8°$
30. $c = 30.2$ cm, $a = 24.5$ cm
31. $a = 19.8$ cm, $B = 11.7°$
32. $a = 54''$, $c = 71.9''$
33. $c = 8.24$ cm, $A = 13.5°$
34. $a = 37.6$ cm, $b = 28.9$ cm
35. $b = 12.36$ cm, $A = 45°$

Perform the indicated operations in problems 36–55.

36. $(3 + j3) + (4 + j4)$
37. $(7 - j6) + (3 + j2)$
38. $(9 - j5) - (-4 - j6)$
39. $(-2 + j5) + (-5 - j3)$
40. $(9 + j) - (j5)$
41. $(3 + j7)(2 - j6)$
42. $(10 - j6)(5)$
43. $(-8 + j8)(8 + j8)$
44. $(12 + j4) \div (3 + j)$
45. $(-15 - j3) \div (3 + j5)$
46. $5\angle 100° + 10\angle 100°$
47. $15\angle 0° + 10\angle 180°$
48. $(7\angle 20°)(4\angle -20°)$
49. $(10\angle 90°)(10\angle 0°)$
50. $(5.8\angle -10°)(6\angle 100°)$
51. $(1.5\angle 180°)(2\angle 45°)$
52. $16\angle 15° \div 4\angle 38°$
53. $37\angle 112° \div 8\angle 220°$
54. $24\angle 20° \div 16\angle -20°$
55. $9\angle 45° \div 5\angle 135°$

In problems 56–60, convert the rectangular form to polar notation.

56. $12 + j6$
57. $-8 - j3$
58. $9 - j5$
59. $j4$
60. $1 - j6$

In problems 61–65, convert the polar form to rectangular notation.

61. $21\angle 38°$
62. $62\angle -19°$
63. $20\angle 45°$
64. $5\angle 127°$
65. $10\angle 283°$

CHAPTER 7

TRIGONOMETRY FOR ALTERNATING CURRENT CIRCUITS IN SERIES

The electronics technician, working with alternating current circuits, will find that a complete understanding of these circuits is based upon a solid foundation in trigonometry. Through trigonometry, you will be able to predict instantaneous current and voltage values. Trigonometry is also used to diagram circuit waveforms and to explain the complexities of alternating current circuits. We will begin our study of ac series circuits by looking closely at the sine wave.

UNIT 7·1 THE SINE WAVE

Objectives:

After studying this unit, you should be able to

- calculate the wavelength, frequency, and period of a sine wave.
- calculate the average, effective, peak, and peak-to-peak values of a sine wave.
- calculate the instantaneous values of a sine wave.

Forming the Sine Wave

In Figure 7–1, the circle shown at the left in the illustration, between the North (N) and South (S) poles of the magnet, represents a coil in cross section. The rotation of the coil, known as an *armature,* is represented in the drawing

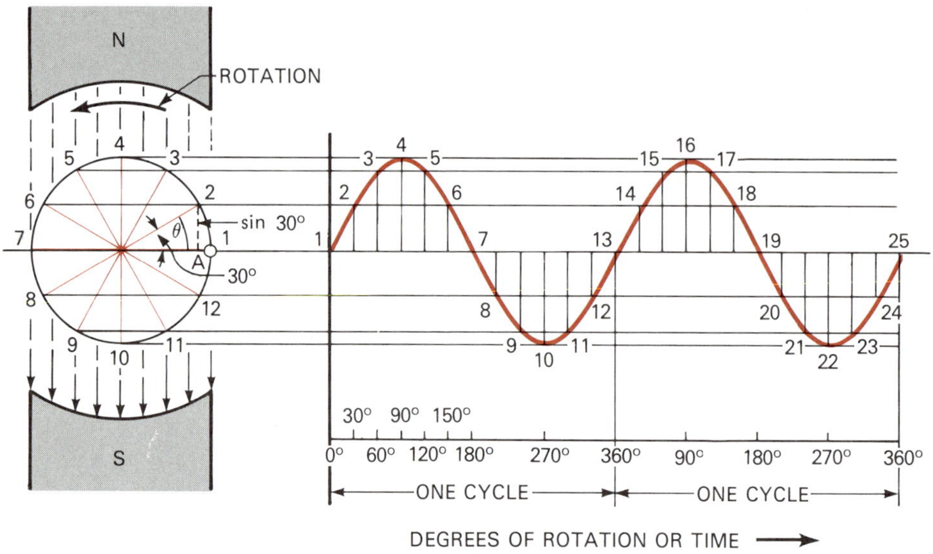

FIGURE 7–1 A sine wave of voltage and current is produced when a coil, known as an armature, rotates in a magnetic field.

by a numbered circle. Point 1, on the outside right-hand side of the circle, is the starting point of the armature coil. At this moment, the output of the generator is zero. When the coil moves up to point 2, the voltage gradually increases from 1 to 2, as shown on the graph at the right. This is the beginning of the graph or waveform of the output voltage of the generator.

Beneath the waveform is a line measured off in degrees. The line starts with 0°, the next division is 30°, and so on. Vertical lines could be drawn from each division on this line.

When the armature reaches point 2, a horizontal line is drawn from this point and intersects a vertical line projected from 30°. Similarly, when the armature reaches point 3 on the circle, a horizontal line projected from this point will intersect a vertical line drawn from the 60° point. These intersection points can be connected and when this is done the graph is that of a sine wave.

The output voltage of the generator actually moves through each degree value between 0° and 360°. When the armature reaches 90° it is moving perpendicular to the magnetic field. The voltage output is maximum at this point. As the armature moves past 90° the output voltage begins to decrease towards zero. At 180° (point 7) the armature is moving parallel to the magnetic field and, as no lines of force are being cut, the generator output is zero.

As the armature moves past point 7 it is cutting through the lines of magnetic force in a direction *opposite* of that during the first 180°. At 270° (point 10) the armature is again moving perpendicular to the field and a maximum voltage *peak* is produced. The voltage output decreases towards zero as the armature approaches 360° of rotation.

The waveform in Figure 7–1 displays the output of two complete rotations of the armature. This results in two *cycles,* each of which encompasses 360°. As the armature speed increases the time required for one cycle will decrease, but the shape of the waveform remains essentially the same.

Wavelength and Frequency

The wave shown in Figure 7–2 is a sine wave and is a *periodic* type, meaning that the current (or voltage) it represents changes its direction of flow (or polarity) at regular time intervals. The portion of the wave above a horizontal line drawn through its center is ordinarily regarded as positive; the lower portion, negative. The plus and minus signs, though, shown in Figure 7–2, are simply used to indicate a change of polarity (in the case of voltage) or change of direction of flow (in the case of current). The horizontal axis for the sine wave is marked both in degrees (0°, 90°, 180°, 270°, and 360°) and the corresponding radian values (0, $\frac{\pi}{2}$, π, $\frac{3\pi}{2}$, and 2π).

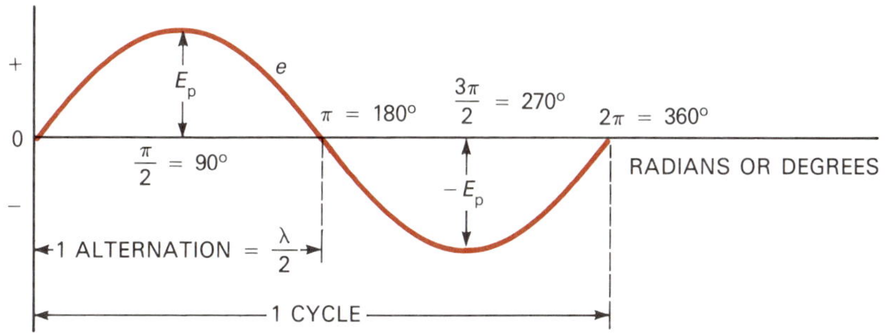

FIGURE 7–2 Sine wave of the voltage in an alternating current circuit

The *wavelength* is the distance from the start of a single cycle to the end of that single cycle. The wavelength is also the distance between successive positive peaks or successive negative peaks. Wavelength is represented by the Greek letter lambda (λ).

The *frequency* of a periodic wave is the number of complete wavelengths per second. As the number of waves per second increases, the wavelength decreases. This is an inverse relationship and can be expressed by the formula:

$$f = \frac{v}{\lambda} \qquad \textbf{(7–1)}$$

where, f = the frequency in hertz
v = the wave velocity, nominally the speed of light at 300 000 000 metres per second
λ = the wavelength in metres

Wavelength is often given in metres, although a submultiple such as the centimetre is also used. Frequency is in hertz, comparable to the cycle per second. The hertz is the basic unit of frequency, but multiples such as the kilohertz (kHz), 1 000 hertz, and the megahertz (MHz), 1 000 000 hertz, are also used.

Wavelength to Frequency Conversions

The velocity of a radio wave in free space is 300 000 000 metres per second, which is also, for practical purposes, the velocity of light. **To convert wavelength (in metres) to frequency (in hertz):**

$$f = \frac{300\ 000\ 000 \text{ m/s}}{\lambda} = \frac{3 \times 10^8 \text{ m/s}}{\lambda} \qquad (7\text{–}2)$$

Example: What is the frequency, in hertz, of a wave whose length is 10 000 metres?

Solution: $f = \dfrac{300\ 000\ 000 \text{ m/s}}{\lambda} = \dfrac{300\ 000\ 000 \text{ m/s}}{10\ 000 \text{ m}} = 30\ 000 \text{ Hz} = 30 \text{ kHz}$

Example: What is the frequency, in megahertz, of a high-frequency wave whose wavelength is 20 cm?

Solution:

$$\begin{aligned} f &= \frac{3 \times 10^8 \text{ m/s}}{\lambda} \\ &= \frac{3 \times 10^8 \text{ m/s}}{0.2 \text{ m}} \\ &= 1.5 \times 10^9 \text{ Hz} \\ &= 1\ 500 \text{ MHz} \end{aligned}$$

Calculator Tip

Electronic calculators often provide the capability to display all numbers as a number times ten raised to a power which is a multiple of three. When operating in this mode all numbers entered and displayed are converted to 10^6, 10^{-3}, 10^0, 10^3, 10^6, etc.

As these powers of ten are used extensively in electronics, this mode can be very useful. When in this *engineering* mode, a problem such as the previous example might be completed with this sequence:

Step	*Enter*	*Displayed*
1	3 [EE] 8	3×10^8
2	[÷]	300×10^6
3	20 [EE] [+/−] 2	20×10^{-2}
4	[=]	1.5×10^9 or 1 500 MHz

Frequency to Wavelength Conversions

Since wavelength and frequency are reciprocals, the formula previously supplied for wavelength to frequency conversions can also be used for frequency to wavelength conversions. **To convert frequency (in hertz) to wavelength (in metres):**

$$\lambda = \frac{300\ 000\ 000\ \text{m/s}}{f} = \frac{3 \times 10^8\ \text{m/s}}{f} \qquad (7\text{–}3)$$

Example: What is the wavelength of a wave whose frequency is 1 000 MHz?

Solution:
$$\begin{aligned}\lambda &= \frac{3 \times 10^8\ \text{m/s}}{f}\\ &= \frac{3 \times 10^8\ \text{m/s}}{1 \times 10^9\ \text{Hz}}\\ &= 3 \times 10^{-1}\ \text{m}\\ &= 30\ \text{cm}\end{aligned}$$

Period of a Wave

Frequency is the number of complete cycles per second, or it can be regarded as the time duration of a number of cycles. A wave, with a frequency of 60 Hz, has 60 complete cycles per second. The time duration of a single cycle would be 1/60 second, and this is known as the *period* (T) of the wave. As a formula, the period of a wave is:

$$T = \frac{t}{N} \qquad (7\text{–}4)$$

where: T = the period in seconds of one cycle
t = time in seconds during which cycles are measured
N = number of cycles measured during time t

Example: Find the period for each of the following.
a. 60 cycles in 1 second
b. 1×10^6 cycles in 7.8 milliseconds
c. 400 cycles in 25 μs

Solution:
a. $T = \frac{t}{N} = \frac{1\ \text{s}}{60} = 1.67 \times 10^{-2}\text{s} = 16.7\ \text{ms}$

b. $T = \frac{7.8 \times 10^{-3}\ \text{s}}{1 \times 10^6} = 7.8 \times 10^{-9}\ \text{s} = 7.8\ \text{ns}$

c. $T = \frac{25 \times 10^{-6}\ \text{s}}{400} = 6.25 \times 10^{-8}\ \text{s} = 62.5\ \text{ns}$

Frequency (f) was previously defined as the number of cycles per second. Assuming the period (T) of one cycle is 0.1 seconds, then it is logical that 10 of these would occur in 1 second. The reciprocal relationship between frequency and period is expressed as

$$f = \frac{1}{T} \tag{7–5}$$

or,

$$T = \frac{1}{f} \tag{7–6}$$

Example: What is the period of a wave with a frequency of 180 kHz?

Solution:
$$\begin{aligned} T &= \frac{1}{f} \\ &= \frac{1}{180 \times 10^3 \text{ Hz}} \\ &= 5.6 \times 10^{-6} \text{ s} \\ &= 5.6\ \mu\text{s} \end{aligned}$$

Example: A technician reads the period of a signal on an oscilloscope. Find the frequency if the period is 4 nanoseconds (4×10^{-9} s).

Solution:
$$\begin{aligned} f &= \frac{1}{T} \\ &= \frac{1}{4 \times 10^{-9} \text{ s}} \\ &= 250 \times 10^6 \text{ Hz} \\ &= 250 \text{ MHz} \end{aligned}$$

Sine Wave Voltage and Current Reference Points

Voltages and currents in dc circuits are easily measured since they are ordinarily fixed amounts, or can be made so for the purpose of measurement. But a sine wave voltage or current is constantly changing, with its range extending from zero, when the voltage or current waveform crosses the x axis, to some maximum or peak amount. For this reason, various references are used in connection with such voltages and currents.

The *peak* value of a sine wave of voltage or current is its maximum amount. A complete cycle consists of two alternations or half waves, and each cycle has a positive peak and a negative peak. Since the upper and lower halves of a perfect sine wave are symmetrical, the positive peak is equal to the negative peak.

Averaging is a method of adding numbers and then dividing by the total numbers involved. To average a sine wave, a number of vertical lines are drawn from the x axis as shown in Figure 7–3. The height of each of these lines, from the x axis to the sine curve, is measured. The total of all the heights of these vertical lines, also known as ordinates, is divided by the number of lines used. The larger the number of vertical lines involved, the more accurate the measurement. The average value of a sine wave of voltage or current, using

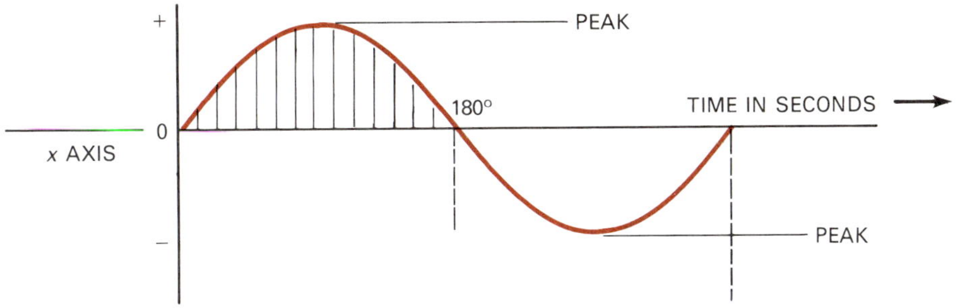

FIGURE 7–3 Technique for finding the average value of a sine wave of voltage or current.

this technique, results in a formula which can be used for any sine wave of voltage or current.

$$E_{av} = 0.637E_p \quad (7\text{–}7)$$

The number 0.637 is equal to 63.7 percent. The average value is 63.7 percent of the peak, whether that peak is positive or negative. Note that we are averaging only half of the sine wave. In a symmetrical sine wave the average of the complete cycle is zero.

Example: A sine wave has a negative peak of 134.5 V. What is the average value of this wave?

Solution: $E_{av} = 0.637E_p = 0.637 \times 134.5\text{ V} = 85.676\ 5\text{ V}$

The formula for average voltage can be rearranged in terms of peak voltage. This expression is

$$E_p = \frac{E_{av}}{0.637} = 1.57E_{av} \quad (7\text{–}8)$$

In terms of current, this formula becomes

$$I_{av} = 0.637I_p \quad (7\text{–}9)$$

or

$$I_p = 1.57I_{av} \quad (7\text{–}10)$$

If the peak value in the previous example was 134.5 A, the average value would be 85.676 5 A. If the peak had been in milliamperes, then the calculated average value would also have been in milliamperes.

Example: A resistive circuit shunted across an ac generator has an average current of 345 μA. What is the peak current in amperes?

Solution: $I_p = 1.57I_{av} = 1.57 \times 345\ \mu\text{A} = 541.65\ \mu\text{A}$

Example: A resistor carrying an average sine current of 45 mA is connected across a 10 V source. What is the peak current in amperes?

Solution: $I_p = 1.57I_{av} = 1.57 \times 45 \text{ mA} = 70.65 \text{ mA}$

Example: A sine wave voltage on an oscilloscope has a 15-volt peak. What is the average value of this signal?

Solution: $E_{av} = 0.637E_p = 0.637 \times 15 \text{ V} = 9.56 \text{ V}$

Peak-to-Peak Values

The peak value of a sine wave of current or voltage is the maximum amplitude of a half cycle. The *peak-to-peak* value (abbreviated as p-p) is the strength of the sine wave of voltage (or current) measured from the positive peak to the negative peak. But since the positive and negative peaks have the same value, then:

$$E_{p\text{-}p} = 2E_p \tag{7–11}$$

and

$$I_{p\text{-}p} = 2I_p \tag{7–12}$$

These formulas can be transposed so that:

$$E_p = \frac{E_{p\text{-}p}}{2} = 0.5E_{p\text{-}p} \tag{7–13}$$

and

$$I_p = \frac{I_{p\text{-}p}}{2} = 0.5I_{p\text{-}p} \tag{7–14}$$

Example: A peak-to-peak current of 640 mA flows in a circuit when 85 V is applied. What is the average current in milliamperes?

Solution: $I_p = 0.5I_{p\text{-}p} = 0.5 \times 640 \text{ mA} = 320 \text{ mA}$
$I_{av} = 0.637I_p = 0.637 \times 320 \text{ mA} = 203.84 \text{ mA}$

When peak-to-peak values are given and the answer is required in terms of average voltage or current, either the peak-to-peak value can be divided by 2 or 0.637 can be divided by 2.

$$E_{av} = 0.637 \times \frac{E_{p\text{-}p}}{2} = 0.318\,5E_{p\text{-}p} \tag{7–15}$$

and

$$I_{av} = 0.637 \times \frac{I_{p\text{-}p}}{2} = 0.318\,5I_{p\text{-}p} \tag{7–16}$$

Effective or Root-Mean-Square Values

The peak value of a sine wave of voltage or current, or the average value, represented two attempts to find a meaningful way in which to measure changing waveforms. However, neither the peak value, peak-to-peak value, or the average value, have any relationship to direct current measurements. To get a relative value between dc and sine wave ac, electrical energy is changed to heat energy which then becomes the common denominator between dc and ac.

When a current, whether dc or ac, flows through a resistor, heat is produced. The *effective* value of an alternating sine current is that value which will produce the same amount of heat in a resistor as would a direct current.

However, the effective value of a sine wave can also be determined mathematically, as shown in Figure 7–4. As in the case of finding the average value

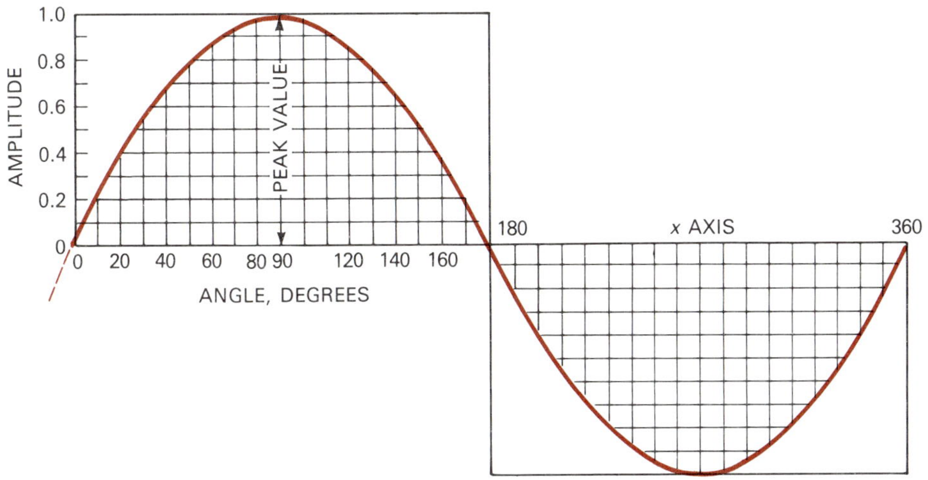

FIGURE 7–4 A first step toward determining the rms or effective value of a sine wave of voltage or current is to obtain a large number of instantaneous values. This can be done over a half cycle of the wave.

of a sine wave, many ordinates are erected from the base line or *x* axis until they intersect the graph of the sine wave. Each of these values is then squared—that is, multiplied by itself. These squared values are added and their average is found. The procedure here is the same as finding the average value, except that the value represented by each ordinate is squared. The square root of the average squared values is then obtained. This square root value is known as the *effective* value or the *root-mean-square* value, abbreviated as *rms,* and is approximately 70 percent of the peak value:

$$E_{rms} = 0.707E_p \qquad (7\text{–}17)$$

and

$$I_{rms} = 0.707I_p \qquad (7\text{–}18)$$

Example: A sine wave current has a peak value of 83 mA. What is its effective (or rms) value?

Solution: $I_{rms} = 0.707 I_p = 0.707 \times 83 \text{ mA} = 58.681 \text{ mA}$

Equations 7–17 and 7–18 can be rearranged in terms of peak current and voltage:

$$E_p = \frac{E_{rms}}{0.707} = 1.414 E_{rms} \tag{7–19}$$

and

$$I_p = \frac{I_{rms}}{0.707} = 1.414 I_{rms} \tag{7–20}$$

Example: (a) What is the peak value of a sine wave voltage when its effective value is 83.6 V? (b) What is the peak-to-peak value?

Solution: a. $E_p = 1.414 E_{rms} = 1.414 \times 83.6 \text{ V} = 118.21 \text{ V}$
b. $E_{p\text{-}p} = 2E_p = 2 \times 118.21 \text{ V} = 236.42 \text{ V}$

Sine Wave Relationships

The four ways of defining or expressing values of voltage or current sine waves are *average, effective, peak,* and *peak-to-peak.* It is possible to move back and forth between these values. Figure 7–5 shows the relationships and how to manipulate them to get any desired value.

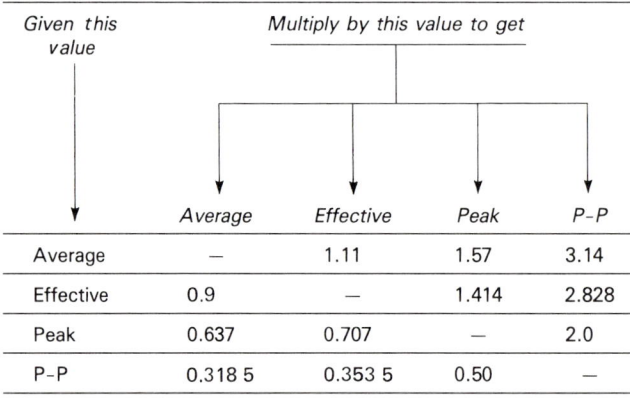

Given this value	*Multiply by this value to get:* Average	Effective	Peak	P-P
Average	—	1.11	1.57	3.14
Effective	0.9	—	1.414	2.828
Peak	0.637	0.707	—	2.0
P-P	0.318 5	0.353 5	0.50	—

FIGURE 7–5 Relationships between the various values of a sine wave of voltage or current.

Example: The average value of a sine wave of voltage is 83 V. What are the (a) effective, (b) peak, and (c) peak-to-peak equivalents?

Solution: a. $E_{rms} = 1.111 \times E_{av} = 1.111 \times 83\ V = 92.21\ V$
b. $E_p = 1.57 \times E_{av} = 1.57 \times 83\ V = 130.31\ V$
c. $E_{p\text{-}p} = 3.14 \times E_{av} = 3.14 \times 83\ V = 260.62\ V$

Instantaneous Values of Sine Wave Voltage

The maximum voltage of a sine wave as described earlier occurs twice during any single cycle of the wave; once at the 90° point and again at the 270° point, assuming the wave starts at 0°. These two conditions are *instantaneous* values, but we can also have instantaneous values at other points along the *x* axis, such as 27°, 45°, 195°, etc. Since the instantaneous values of the sine wave voltage are maximum or peak at 90° and 270°, these other instantaneous values will all be smaller.

If a sine wave has a peak or maximum of 200 V at 90°, for example, its voltage at 0° will be zero. The voltage will also be zero at 180° and again at 360°. This zero-voltage condition can be indicated in two ways: (1) by illustrating the wave graphically as shown earlier or by means of a counterclockwise rotating vector as in Figure 7–6. As the vector starts to rotate, it develops an

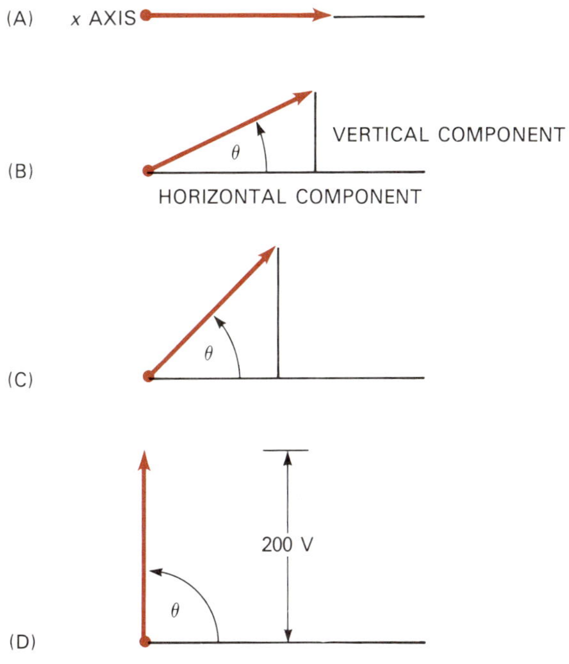

FIGURE 7–6 The voltage in an ac circuit, having a sine waveform, can be represented by a counterclockwise rotating vector. In (A) the voltage is 0 and the phase angle is 0 degrees. In (B) the vector has turned a number of degrees, represented by θ. The vector has a horizontal and a vertical component. The vertical component represents the instantaneous voltage at that phase angle. As the vector continues to rotate (C) it finally reaches its peak value (D).

angle between its horizontal starting line and its position at any moment. The vector has a horizontal and vertical component. The horizontal component, the distance along the x axis from the origin, 0, represents a condition of zero voltage. The vertical component, the distance from the arrowhead of the vector to the x axis, is the voltage at any particular instant. Figure 7–6 also shows successive positions for the vector. When the vector reaches the 90° point, the voltage is maximum and, in this particular example, is 200 V. When the vector is at 90°, there is no horizontal component, and the distance from the arrowhead of the vector to the x axis is maximum.

The value of the voltage at any particular instant, then, depends on the size of the angle θ. As the angle becomes larger, the instantaneous voltage also becomes larger. To calculate the voltage at any instant:

$$e = E_{max} \sin \theta \qquad (7\text{–}21)$$

where θ is the angle between the rotating vector and the x axis. The lowercase letter e is used to indicate an instantaneous value.

Example: A sine wave has a peak value of 200 V. What is the voltage when the angle opposite the vertical component of the vector is 43°?

Solution:

$$\begin{aligned} e &= E_{max} \sin \theta \\ &= 200 \text{ V} \times \sin 43° \\ &= 200 \text{ V} \times 0.682\,0 \\ &= 136.4 \text{ V} \end{aligned}$$

Conventionally, voltages in the range from 0° to 180° are considered positive; those from 180° to 360° are regarded as negative. In the example just supplied, the voltage becomes 136.40 V once again at 137° and two more times after the sine wave passes the 180° point. At 180° + 43° or 223° the voltage is 136.40 V, but at this time it is −136.40 V. The voltage at 317° is also −136.40 V. To find the instantaneous voltage, then, at any point between 180° and 360°, subtract 180°, find the answer using the formula, and then put a minus sign in front of the answer.

Calculator Tip

When using an electronic calculator to determine instantaneous voltage and current values, the calculator will automatically assign negative signs to values occurring between 180° and 360°. This is due to the fact that sine values are negative in the third and fourth quadrants. Examine the following problems and the sequence to solve each. The peak voltage of the sine wave is 100 V_p.

$$e = E_{max} \sin \theta$$

1. Instantaneous voltage at 118°

   100 [×] 118 [sin] [=] 88.29 V

2. Instantaneous voltage at 300°

   100 [×] 300 [sin] [=] −86.6 V

Example: A sine wave has a peak value of 176.5 V. What is its value at 235°?

Solution:

$$\begin{aligned} e &= E_{max} \sin \theta \\ &= 176.5\ \text{V} \times \sin 235° \\ &= 176.5\ \text{V} \times (-0.819\,2) \\ &= -144.58\ \text{V} \end{aligned}$$

Instantaneous Values of Sine Wave Current

Although the previous discussion was solely in terms of sine wave voltage, the same thinking applies to sine wave current. The same formula can be used, with the substitution of I for current, in place of E for voltage.

$$i = I_{max} \sin \theta \qquad (7\text{–}22)$$

Example: A sine wave current reaches a peak of 345.7 mA at the end of $\frac{1}{4}$ cycle. What is the instantaneous value of this current at 220°?

Solution:

$$\begin{aligned} i &= I_{max} \sin \theta \\ &= 345.7\ \text{mA} \times \sin 220° \\ &= 345.7\ \text{mA} \times (-0.642\,8) \\ &= -222.21\ \text{mA} \end{aligned}$$

A minus sign is put in front of this answer to indicate that the current flow at 220° is opposite in direction to the movement of current for the first half cycle of the wave, from 0° to 180°. In this formula, conversion of current from a submultiple to the basic current unit, amperes, is not required.

Current and Voltage in Terms of Radians

In the previous examples, the vertical component of the vector is expressed in terms of its relationship to its opposite angle. A complete cycle consists of 360° or, as explained previously, 2π radians. A frequency of 1 Hz (or one cycle per second) would be equivalent to 6.28 radians per second. The instantaneous value of a sine wave of voltage or current can then be found by using these formulas:

$$e = E_{max} \sin \omega t \qquad (7\text{–}23)$$

and

$$i = I_{max} \sin \omega t \qquad (7\text{–}24)$$

The Greek lowercase omega (ω) is used instead of $6.28 \times f$, or as a substitute for $2\pi f$. Omega (ω) is the symbol for angular velocity. This velocity refers to the rate at which the rotating vector in Figure 7–6 moves about the x axis. Omega is defined by

$$\omega = \frac{\text{angular distance in radians}}{\text{time in seconds}}$$

If this vector rotates through 360°, or 2π radians, the amount of time involved is the period (T). Therefore,

$$\omega = \frac{2\pi}{T}$$

Since $T = \frac{1}{f}$, then

$$\omega = 2\pi \div \frac{1}{f} = 2\pi f \qquad (7\text{–}25)$$

Also, since velocity × time = distance then

$$\omega t = \theta \qquad (7\text{–}26)$$

This formula states that the angular velocity multiplied by the time of rotation produces the angle in radians the vector travels during that time period.

The two equations for instantaneous voltage and current in angular velocity time notation are

$$e = E_{max} \sin \omega t$$
$$i = I_{max} \sin \omega t$$

Example: Express a voltage of 220 V at a frequency of 60 Hz in angular velocity time notation.

Solution:

$$\begin{aligned} E_{max} &= E_{rms} \times 1.414 \\ &= 220 \text{ V} \times 1.414 \\ &= 311 \text{ V} \end{aligned}$$

$$\begin{aligned} \omega &= 2\pi f \\ &= (2)(\pi)(60) \\ &= 377 \text{ radians per second} \end{aligned}$$

$$e = E_{\max} \sin \omega t$$
$$= 311 \text{ V} \sin 377t$$

Example: What is the instantaneous voltage in the previous example after 5 milliseconds has elapsed?

Solution:
$$\begin{aligned} e &= 311 \text{ V} \sin 377t \\ &= 311 \text{ V} \sin (377 \text{ radians/s} \times 5 \text{ ms}) \\ &= 311 \text{ V} \sin 1.885 \text{ radians} \\ &= 311 \text{ V} \times 0.951 \\ &= 295.77 \text{ V} \end{aligned}$$

In order to provide a better understanding of the relationship between angular velocity and instantaneous values visualize the armature of a generator rotating at a specific velocity. Through the use of angular velocity time notation it is possible to determine the voltage output at any instant in time. To clarify this process, repeat the previous example without the use of time notation.

Example: What is the voltage level after 5 ms for a 220-V, 60-Hz signal?

Solution:
$$\begin{aligned} E_{\max} &= E_{\text{rms}} \times 1.414 \\ &= 220 \text{ V} \times 1.414 \\ &= 311 \text{ V} \end{aligned}$$

$$\begin{aligned} t &= \frac{1}{f} \\ &= \frac{1}{60 \text{ Hz}} \\ &= 16.67 \text{ ms} \end{aligned}$$

After 5 ms the vector has rotated

$$\left(\frac{5 \text{ ms}}{16.67 \text{ ms}}\right)(360°) = 108°$$

The instantaneous voltage is found by

$$\begin{aligned} e &= E_{\max} \sin \theta \\ &= 311 \text{ V} \sin 108° \\ &= (311 \text{ V})(0.951\ 1) \\ &= 295.78 \text{ V} \end{aligned}$$

Example: Assume a voltage sine wave starts at time $t = 0$. The voltage has a maximum value of 183 V. The frequency is 60 Hz. What is the instantaneous value of the wave two thousandths of a second (0.002 s) after the wave crosses the x axis?

Solution:
$$\begin{aligned} e &= E_{\max} \sin \omega t \\ &= 183 \text{ V} \sin [2\pi(60 \text{ Hz})(0.002 \text{ s})] \\ &= 183 \text{ V} \sin 0.754 \text{ radians} \\ &= 183 \text{ V} \times 0.684\ 5 \\ &= 125.27 \text{ V} \end{aligned}$$

In the example just supplied, the frequency was given as 60 Hz. This means that 60 complete cycles are finished in 1 second. One cycle, then, would need only $\frac{1}{60}$ of a second or 0.016 66 second. In the problem, though, the instantaneous value was required at 0.002 second. Dividing this number by the time it takes to complete a full cycle, we would have:

$$\frac{0.002 \text{ s}}{0.016\ 666 \text{ s}} = 0.12 \text{ or } 12\%$$

This means that when 12% of the single cycle is completed, the instantaneous voltage at that moment is 125.27 V. However, at $\frac{1}{4}$ cycle, or 90°, the voltage is at its peak, or 183 V. Since 12% is almost half of one-quarter (0.25), it might have been logical to expect that the voltage would have been approximately half of 183 V or about 91 V. It is not 91 V, because a sine wave does not "grow" in linear fashion. The voltage grows rapidly at the start, but its rate of increase slows as it reaches its peak.

As a check on the work:

$$\frac{0.002 \text{ s}}{0.016\ 66 \text{ s}} \times 360° = 43°$$

This 43° agrees with the previously obtained value of the phase angle. The problem can then be solved by using the original formula:

$$e = E_{\text{max}} \sin \theta \ (\theta = 43°)$$

This check also supplies an alternative method for handling problems of this kind.

There is one caution. The instantaneous value was calculated as 125.27 V. But this is not the peak value of 183 V, which exists only at 90° and 270°. Thus, for problems involving rms or average values of a sine wave, it is the instantaneous peak value that must be used, and not the instantaneous value at any other phase angle.

Problems involving current, when the frequency and time are supplied, are handled in exactly the same way as the sample problem supplied. In either case, voltage or current, no conversion is required to basic units.

EXERCISE 7·1

In problems 1–10, find the wavelength for the given frequency.

1. 60 Hz
2. 150 MHz
3. 300 kHz
4. 2.7 MHz
5. 45×10^8 Hz
6. 1.7×10^{12} Hz
7. 300 MHz
8. 94 MHz
9. 3.5×10^9 Hz
10. 600 kHz

In problems 11–20, find the frequency for the given wavelength.

11. 7.5×10^5 m
12. 36.585 m
13. 1.5×10^3 m
14. 10 cm
15. 60 cm
16. 3×10^5 m
17. 5 m
18. 555.56 m
19. 200 m
20. 1 m

In problems 21–25, find the period of the given frequency.

21. 1.8 MHz
22. 400 Hz
23. 108 MHz
24. 5×10^{10} Hz
25. 120 kHz

In problems 26–30, find the frequency of the given period.

26. 16.667 ms
27. 5×10^{-9} s
28. 25 μs
29. 0.4 μs
30. 2 μs

In problems 31–40, convert the given value to the indicated value.

31. 100 V_{av} to V
32. 55.8 V to V_p
33. 48 mA_p to mA_{av}
34. 400 V_{pp} to V
35. 6.3 V to V_{pp}
36. 60 A_{av} to A_p
37. 120 V to V_{av}
38. 18.9 V_p to V
39. 35.7 V_{av} to V_{pp}
40. 12.6 mA to mA_p

In problems 41–50, find the instantaneous value at the indicated angle. The voltage given is the peak value.

41. 150 V_p, $\theta = 45°$
42. 100 V_p, $\theta = 317°$
43. 12 V_p, $\theta = 180°$
44. 60 V_p, $\theta = 197°$
45. 90 V_p, $\theta = 265°$
46. 25 V_p, $\theta = 387°$
47. 220 V_p, $\theta = 62°38'56''$
48. 64 V_p, $\theta = 1.877$ radians
49. 50 V_p, $\theta = 2\pi$ radians
50. 1 000 V_p, $\theta = \pi/2$ radians

In problems 51–55, express the given voltage and frequency in angular velocity time notation.

51. 100 V, 60 Hz
52. 80 V_{pp}, 120 Hz
53. 15 V_{av}, 1 000 Hz
54. 1.8 V_p, 10 kHz
55. 5 V_p, 150 MHz
56. What is the instantaneous voltage in problem 51 after 1.2 ms?
57. What is the instantaneous voltage in problem 52 after 0.5 ms?
58. What is the instantaneous voltage in problem 53 after 2.3 radians?
59. What is the instantaneous voltage in problem 54 after 8 μs?
60. What is the instantaneous voltage in problem 55 after 1 μs?

UNIT 7·2 VECTORS AND PHASORS

Objectives:

After studying this unit, you should be able to

- determine the phase angle between sinusoidal waveforms.
- calculate instantaneous values of waveforms with phase angles.

- convert waveform equations between the phasor and time domains.
- use phasor notation to combine sinusoidal waveforms.
- calculate instantaneous values of composite waveforms

Phase Angle

A circle or a straight line can be measured off in degrees. Just as either a small circle or a large one has 360°, any length of straight line can have 360°. If, as shown in Figure 7–7A a selected straight line is 360° then its halfway

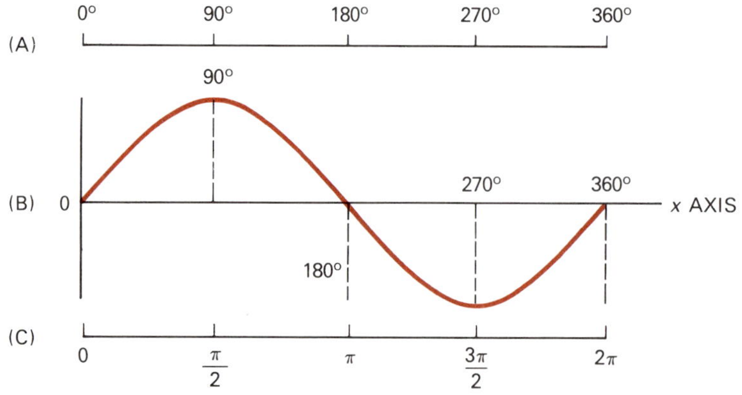

FIGURE 7–7 A straight line having a length of 360° can be used to measure a sine wave.

point will be 180°. The line can be further subdivided into equal quarter sections, so selected points on the line can be 90°, 180°, 270°, and 360°. A voltage (or current) sine wave drawn on this line, referred to as an *x axis,* can start at 0°, reach its maximum value at the 90° point and gradually decrease to zero at the 180° mark. At this point, the polarity reverses; the voltage reaches a peak at 270° and then gradually decreases once again to zero at the 360° point.

While Figure 7–7B represents a single sine wave of voltage (or current), there is no reason why a second voltage (or current) cannot be measured along the same *x* axis, nor is there any reason why this second voltage (or current) should start at the same time as the first.

The circumference of any circle can be divided into 360°. If this circle is cut at any point and formed into a straight line, then this straight line will also be divided into 360°. The 360° though, as mentioned earlier, is the same as 2π radians. And so, as indicated in Figure 7–7C, the length of a line is 2π, equivalent to 360°. Half the length would be π or 180°, 90° is equivalent to $\pi/2$ and 270° can also be written as $3\pi/2$. In electronics problems, a line may be divided in either way: in *degrees,* or in *radians.* Figure 7–8A shows two currents, I_1 and I_2, with both currents starting at the same time and crossing the *x* axis at the same time. These two currents are then said to be in step or *in phase.* Note that the two currents need not have the same amplitude.

Figure 7–8B shows that current I_1 *leads* current I_2 by some angle, *theta.*

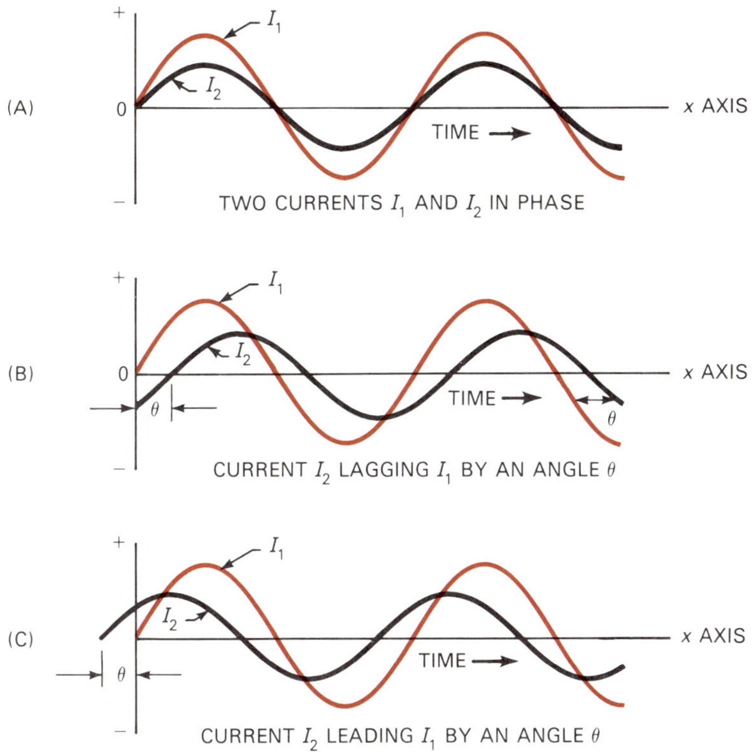

FIGURE 7–8 A pair of currents (or voltages) can be in phase, with both starting at the same time, or one of the currents (or voltages) can lag or lead the other.

Or, we could describe these currents by saying that current I_2 *lags* current I_1. The angle existing between these two currents, measured along the *x* axis, is known as the *phase angle.* Although the *x* axis is not marked as such, it could be divided into 360° in a manner similar to the straight line shown earlier in Figure 7–7A. Note that currents I_1 and I_2 in Figure 7–8B do not reach their maximum positive or negative peaks simultaneously, nor do they cross the *x* axis at the same point.

The situation shown in Figure 7–7B is reversed in Figure 7–7C such that current I_2 *leads* current I_1, or, conversely, current I_1 *lags* current I_2.

While Figure 7–8 graphs a pair of currents, it is also possible to have two voltages in an ac circuit in phase or out of phase, or to compare the phase relationship between a voltage and a current.

Vectors

A vector is a straight line used to indicate the strength and direction of a quantity, such as a force, a voltage, or current. However, it is also sometimes used to represent values such as resistance, reactance (both inductive and capacitive) and impedance. The vector is terminated in an arrow to indicate direction.

Figure 7–9 shows a single arrow representing 10 V ac. If this vector is made 1 inch long, then 20 V could be represented by doubling the length of the vector. This can be done, as indicated, simply by adding another 1 inch vector to the original.

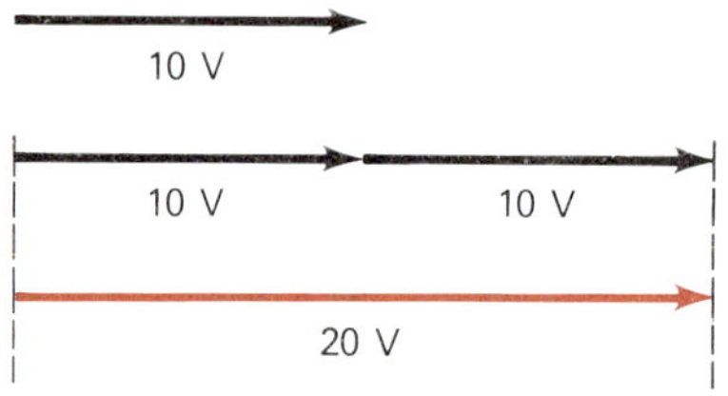

FIGURE 7–9 A vector can be made to represent 20 V, in this example, by doubling the length of the original vector. This is equivalent to adding two vectors, each having a value of 10 V.

A vector quantity involves two concepts: magnitude and direction. In an ac circuit, we are not only concerned with quantity, but with phase as well. Phase separation is indicated in a vector circuit by having vector lines drawn in different directions. But not all quantities are represented by vectors, nor need they be. Numbers are often used in electronics without reference to direction; such numbers are called *scalars*. If a resistance of 20 Ω is measured in a circuit, 20 is a scalar quantity, or simply a scalar. An inductance of 5 mH, a capacitance of 2 μF, a signal strength of 5 V are all scalars. Since scalars and vectors are different, it is necessary to emphasize which is which. Vectors are sometimes written in boldface type. Another method is to use a dot above or below the letter representing the vector. There is no standardization but there is no difficulty as long as a vector quantity is clearly identified.

Addition of Vectors

Vectors can be added directly only if they represent in-phase quantities. The example in Figure 7–9, showing a pair of voltages, represents the direct addition of two in-phase vector quantities. The problem is handled as one of simple arithmetic addition. However, this technique cannot be used where there is a phase difference.

It is helpful to compare vectors with the hands of a clock. Vectors, however, rotate in a counterclockwise direction, as shown in Figure 7–10. The in-

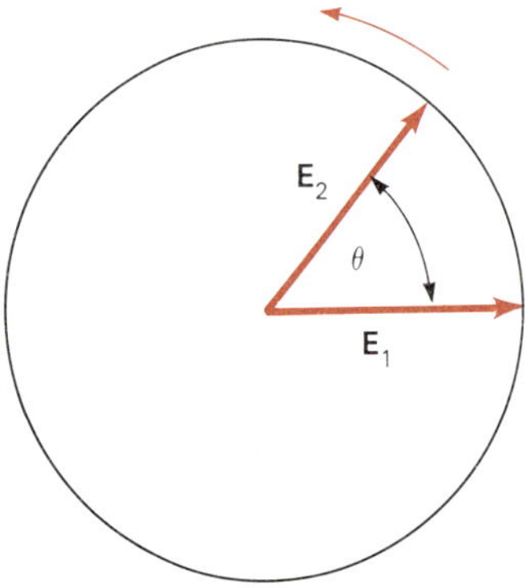

FIGURE 7–10 Two out-of-phase voltages can be represented by a pair of vectors. The angle separating them is their *phase difference* or *phase angle*. Although $\mathbf{E_1}$ and $\mathbf{E_2}$ are shown here as equal, they can have different values.

cluded angle between the two vectors, marked θ in the drawing, represents the *phase difference* or *phase angle*. The circle in Figure 7–10, drawn here to emphasize the clocklike nature of a pair of vectors, is ordinarily omitted.

Parallelogram Method of Adding Vectors

There are various ways of adding vectors and the technique known as the parallelogram method is shown in Figure 7–11. Here we have two vectors hav-

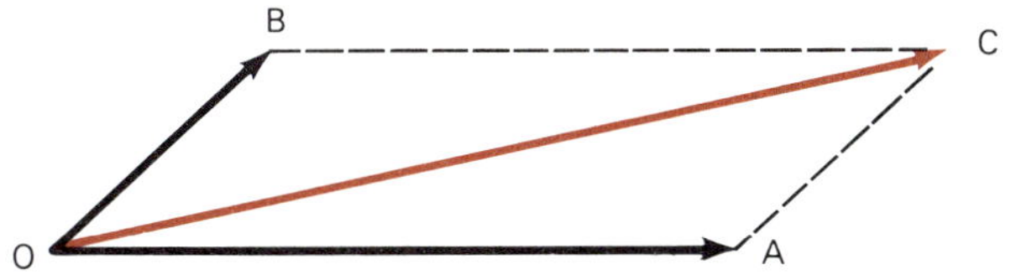

FIGURE 7–11 When two vectors, such as **OA** and **OB,** are out of phase, they cannot be added directly. In the parallelogram method of adding such vectors, vector **OC** is the vector sum of **OA** and **OB.**

ing different magnitudes and a phase angle of less than 90°. The two vectors are **OA** and **OB** with the letter O representing the point of origin of both vectors. Vectors **OA** and **OB** could be a pair of out-of-phase voltages, with **OA** the drop across one component and **OB** the voltage drop across another component. To determine the total voltage, line AC is drawn parallel to line OB and also equal in length to line OB. Line BC is constructed parallel to line OA. Both constructed lines are shown in dash form. A vector is drawn from the origin O to point C. This vector is the diagonal of the parallelogram and represents the vector sum of the two voltages.

Vector Sum of Right-Angle Components

The parallelogram method of finding the vector sum can also be used when a pair of vectors are at right angles to each other, as shown in Figure 7–12. Here we have two voltages, E_1 and E_2, with E_1 smaller than E_2. Since E_1 and E_2 are out of phase, they cannot be added arithmetically. As in the previous example, the sum of the two voltages can be found by forming a parallelogram and drawing a diagonal. The diagonal is line OC. Note, however, that an equal diagonal can also be drawn from point A to point B. When this is done, we have the formation of a right-angle triangle. For this special case where the two values are 90° out of phase, the vector sum can be determined algebraically. In this illustration, the length of the hypotenuse is

$$AB = \sqrt{(OA)^2 + (OB)^2}$$

According to the vector diagram we have three voltages. One of these is $\mathbf{E_1}$, the other is $\mathbf{E_2}$. The third voltage, the vector sum of $\mathbf{E_1}$ and $\mathbf{E_2}$, is now represented by vector **AB.** But since it is the sum of $\mathbf{E_1}$ and $\mathbf{E_2}$, it must be the applied voltage. Voltages E_1 and E_2 are out-of-phase drops across two elec-

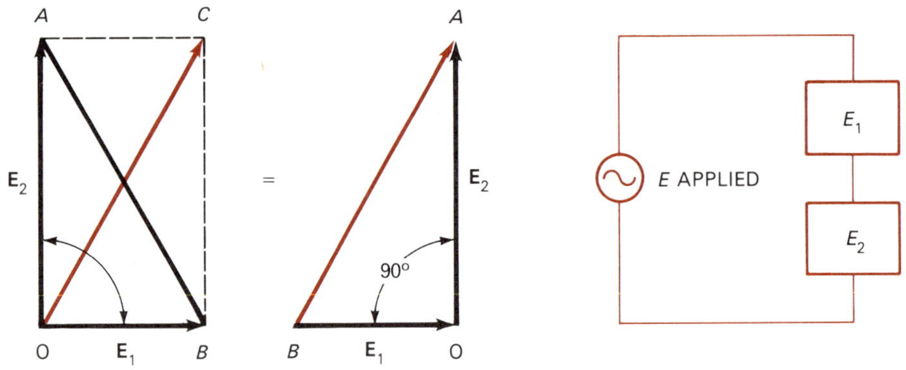

FIGURE 7–12 A pair of voltages that are 90° out of phase can be represented in vector form by a right triangle.

tronic parts. In a dc circuit, the applied voltage is the arithmetic sum of the voltage drops. **In ac circuits, where the voltages across parts can be out of phase, the applied voltage is the vector sum.** The vector sum can be determined graphically by using the parallelogram method in which the vectors are scaled to represent the different quantities, or it can be calculated by trigonometric functions, using the formula for finding the length of the hypotenuse of a right triangle. In Figure 7–12, line AB is the hypotenuse, but it also represents the applied voltage; so, instead of writing AB we can call it $E_{applied}$. Similarly, E_1 is the same as OB and E_2 is equivalent to OA. The equation can now be written as:

$$E_{applied} = \sqrt{E_1^2 + E_2^2}$$

Phase Relations

Earlier in this unit the concept of a phase angle was introduced. The diagram in Figure 7–8A presents a pair of waveforms that are in phase. Both of these waveforms pass through the origin so that the formula for either of these would be

$$i = I_{max} \sin (\omega t)$$

In Figure 7–13 a waveform has been shifted to the left of the origin. The basic formula now becomes

$$e = E_{max} \sin (\omega t + \theta) \qquad (7\text{–}27)$$

where, e = instantaneous voltage at time t

E_{max} = peak voltage (E_p)

ω = angular velocity in radians per second

t = a specific point in time

θ = the angle in degrees the waveform has been shifted to the *left* of the origin

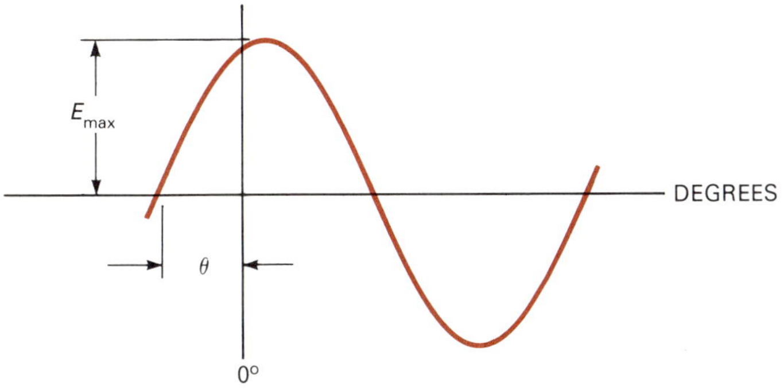

FIGURE 7–13

The waveform in Figure 7–13 crosses the horizontal axis in a positive direction prior to crossing the vertical axis. The sine wave in Figure 7–14

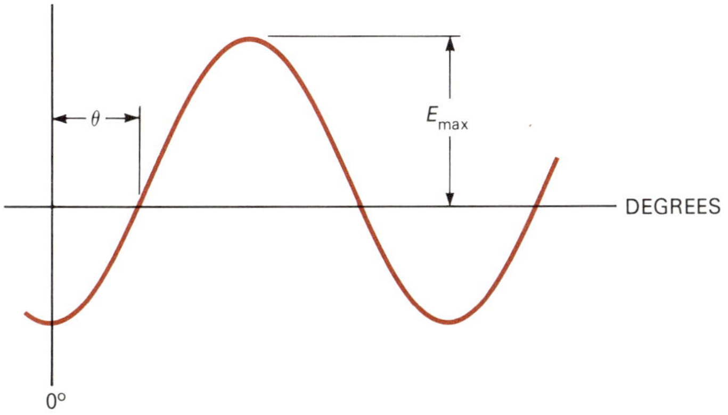

FIGURE 7–14

crosses the horizontal axis to the right of the origin at an angle equal to θ. The basic formula now becomes

$$e = E_{max} \sin(\omega t - \theta) \qquad (7\text{–}28)$$

Example: Both of the waveforms in Figure 7–15 are 120 V, 60 Hz. (a) Find the instantaneous voltage, for waveform A, 10 ms after the start of the cycle. (b) Find the same value for waveform B.

Solution:

$$E_p = E \times 1.414$$
$$= 120\text{ V} \times 1.414$$
$$= 169.68\text{ V}$$
$$\omega = 2\pi f$$
$$= (2)(\pi)(60)$$
$$= 377\text{ radians per second}$$

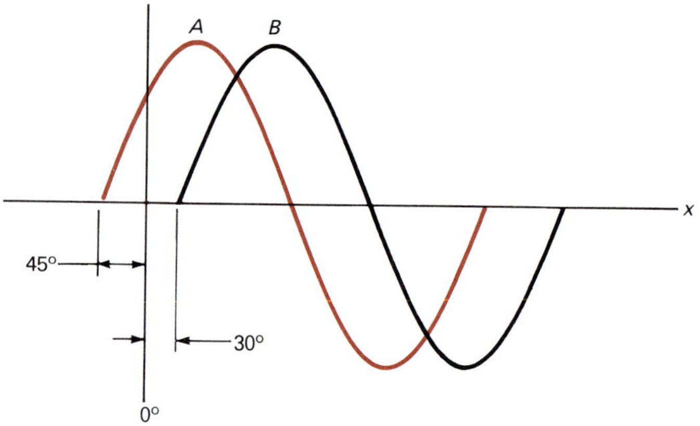

FIGURE 7–15

a. $e_A = E_p \sin(\omega t + \theta)$
$= (169.68 \text{ V}) \sin [(377)(0.01) + 45°]$
$= (169.68 \text{ V}) \sin (3.77 + 45° \times 0.017\,453 \text{ radians/degree})$
$= (169.68 \text{ V}) \sin (3.77 + 0.785)$
$= (169.68 \text{ V}) \sin 4.555 \text{ radians}$
$= (169.68 \text{ V})(-0.987\,6)$
$= -167.58 \text{ V}$

b. $e_B = E_p \sin(\omega t - \theta)$
$= (169.68 \text{ V}) \sin [(377)(0.01) - 30°]$
$= (169.68 \text{ V}) \sin (3.77 - 30° \times 0.017\,453 \text{ radians/degree})$
$= (169.68 \text{ V}) \sin (3.77 - 0.524)$
$= (169.68 \text{ V}) \sin 3.246 \text{ radians}$
$= (169.68 \text{ V})(-0.104\,2)$
$= -17.68 \text{ V}$

Example: What is the instantaneous voltage at the point where waveform A crosses the vertical axis in Figure 7–15?

Solution: As the waveform has traveled through 45° of rotation the value is
$e = E_p \sin 45°$
$= (169.68 \text{ V}) \sin 45°$
$= (169.68 \text{ V})(0.707)$
$= 119.96 \text{ V}$ (or approximately 120 V)

Example: What is the phase relationship between the waveforms in Figure 7–15?

Solution: Waveform A leads waveform B by 75°, *or* B lags A by 75°

Example: Given the following waveforms for current and voltage determine the phase relationship.

$i = I_p \sin(\omega t + 15°)$
$e = E_p \sin(\omega t - 15°)$

Solution: i leads e by 30°

Phasors

A rotating vector that generates a sinusoidal waveform is called a *phasor*. One end of the phasor is connected to the origin. A phasor represents a sine wave at the instant where time equals zero ($t = 0$).

As with vectors, the standard direction of rotation is counterclockwise. Two waveforms with the same frequency, but different peak values, that cross the origin together at $t = 0$ would appear as in Figure 7–16.

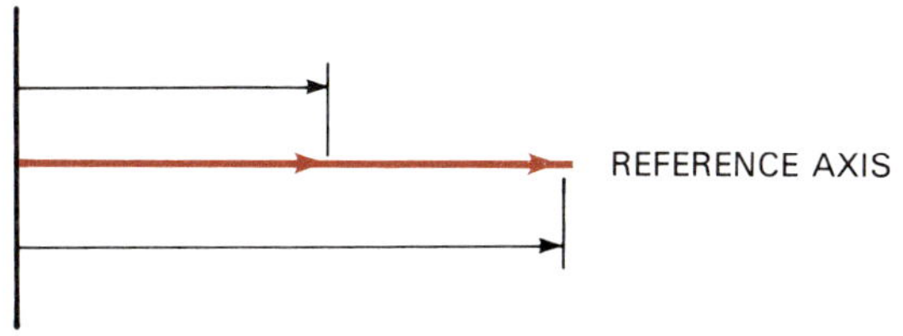

FIGURE 7–16 Phasor representation of two waveforms in phase with each other and the reference axis.

The position of a phasor indicates its relationship to other phasors and the reference axis. Within a specific problem all waveforms *must* be of the same frequency. If this is true then the rotating phasors representing the sine waves will maintain their same phase angle relationships. In terms of the magnitude of these phasors, each is the effective or rms value of the voltage or current.

The general form for expressing the phasor form of a sine wave is

$$\mathbf{E} = E \angle \theta \qquad (7\text{–}29)$$

and

$$\mathbf{I} = I \angle \theta \qquad (7\text{–}30)$$

where, E = effective voltage
I = effective current
θ = the phase angle

Note that the phasor form of a sine wave is the same as the polar notation of a complex number.

Example: Convert the following time domain expressions to the phasor domain.

a. $e = 100\text{ V} \sin \omega t$
b. $i = 25\text{ mA} \sin(\omega t + 35°)$
c. $e = 15\text{ V} \sin(\omega t - 10°)$

Solution: a. As there is no phase angle this waveform passes through the

origin at $t = 0$, resulting in a magnitude equal to the rms value, and a phase angle of zero.
$\mathbf{E} = 70.7\text{ V }\angle 0^\circ$

b. At $t = 0$ this waveform leads by 35°, resulting in
$\mathbf{I} = 17.675\text{ mA }\angle 35^\circ$

c. At $t = 0$ this waveform lags by 10°, resulting in
$\mathbf{E} = 10.605\text{ V }\angle -10^\circ$

The diagram in Figure 7–17 displays the phasor and time domain representations for these waveforms.

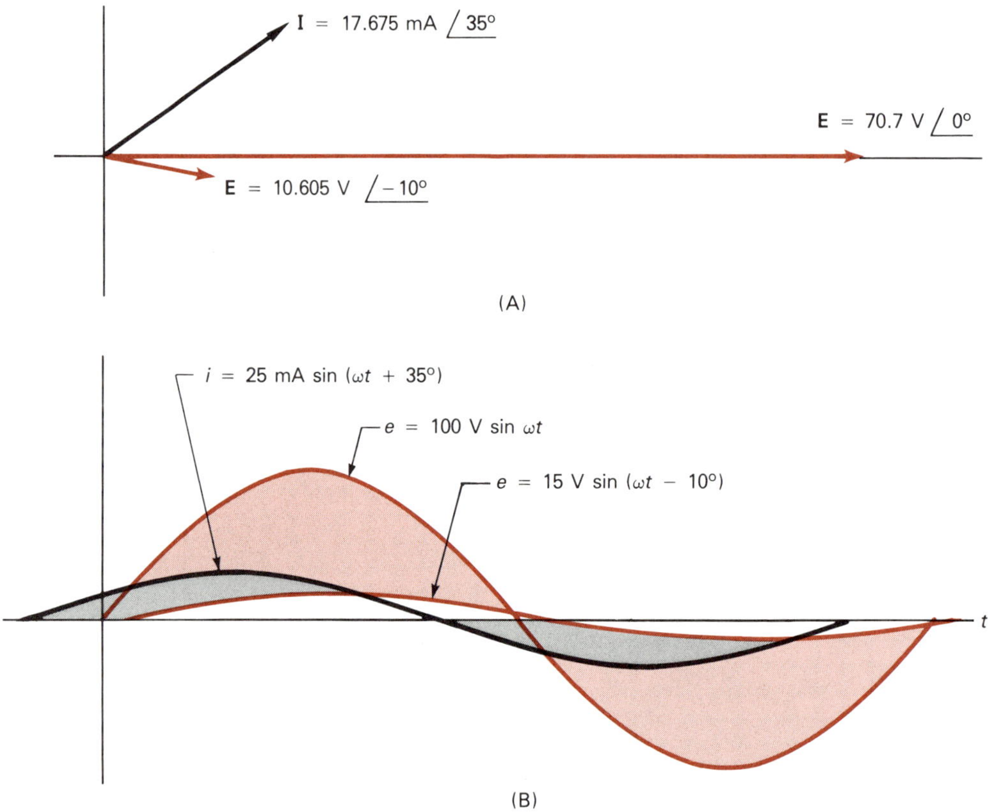

FIGURE 7–17 Phasor (A) and time (B) representations of three waveforms

Example: Convert the following phasor domain expressions to the time domain.

a. $\mathbf{E} = 50\text{ V }\angle 0^\circ$
b. $\mathbf{E} = 120\text{ V }\angle -42^\circ$
c. $\mathbf{I} = 30\text{ mA }\angle 20^\circ$

Solution: a. As there is no phase angle this waveform passes through the origin at $t = 0$, resulting in
$e = 70.7\text{ V}\sin\omega t$

b. $e = 169.68\text{ V}\sin(\omega t - 42°)$
c. $i = 42.42\text{ mA}\sin(\omega t + 20°)$

Addition and Subtraction of Phasors

As in the case of dc values, it is often necessary to add and subtract ac values. One method of accomplishing this is to add each point along the x axis as shown in Figure 7–18A. Note that in Figure 7–18B, the height to point a on

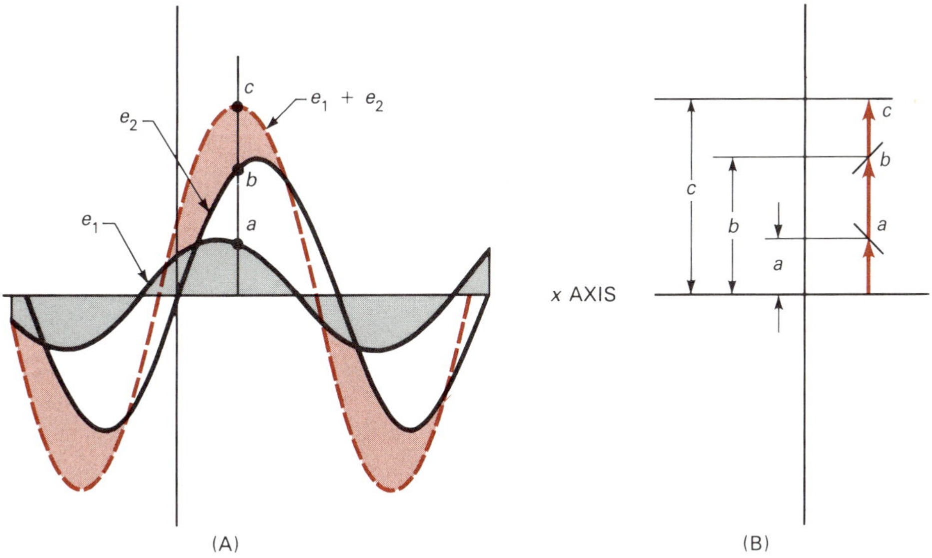

FIGURE 7–18 The addition of two ac waveforms

wave e_1 is added to the distance to point b on e_2, resulting in point c on the *composite* waveform. As one of the waveforms crosses the x axis in a negative direction the points are subtracted until both cross the axis. At this point the composite also moves across the axis in a negative direction.

This is obviously a lengthy process to follow in order to add two waveforms. A simpler, more expedient, method is to convert the time domain expressions to the phasor domain. These expressions can be easily converted to rectangular notation for addition and subtraction. The resultant can then be converted back to phasor and time domain expressions. Again, all waveforms must be of the same frequency to utilize this process.

Example: Add the two voltage waveforms in Figure 7–19.

Solution: Given: $e_1 = 100\text{ V}\sin\omega t$
$e_2 = 80\text{ V}\sin(\omega t - 30°)$

Converting to the phasor domain
$\mathbf{E_1} = 70.7\text{ V}\angle 0°$
$\mathbf{E_2} = 56.56\text{ V}\angle -30°$

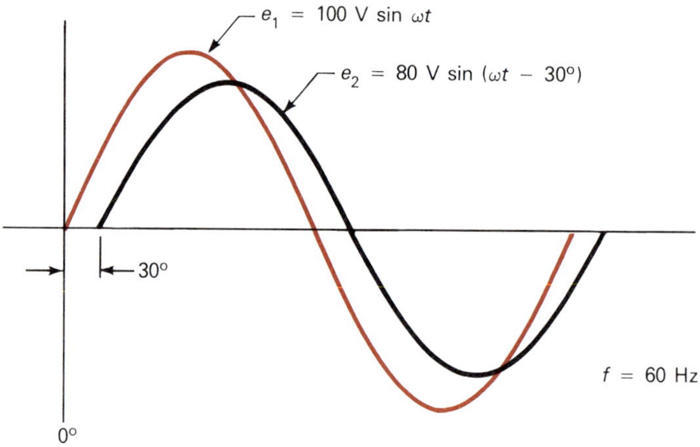

FIGURE 7–19

Converting to rectangular notation
$70.7 + j0$
$48.98 - j28.28$

Adding the two expressions

$$\begin{array}{r} 70.7 + j0 \\ + \; 48.98 - j28.28 \\ \hline 119.68 - j28.28 \end{array}$$

Converting back to the phasor domain
$\mathbf{E}_t = 122.98 \text{ V} \angle -13.29^\circ$

Converting back to the time domain
$e_t = 173.89 \text{ V} \sin(\omega t - 13.29^\circ)$

Note that the peak value of the composite wave is less than the sum of the peak voltages. This is obviously due to the fact that the peaks occur 30° apart. Also observe that the phase angle is less than the 30° found in one of the original waveforms.

Example: Add the two current waveforms in Figure 7–20.

Solution: Given: $i_1 = 5 \text{ A} \sin(\omega t + 35^\circ)$
$i_2 = 5 \text{ A} \sin(\omega t - 45^\circ)$

Converting to the phasor domain
$\mathbf{I}_1 = 3.535 \text{ A} \angle 35^\circ$
$\mathbf{I}_2 = 3.535 \text{ A} \angle -45^\circ$

Converting to rectangular notation
$2.9 + j2.03$
$2.5 - j2.5$

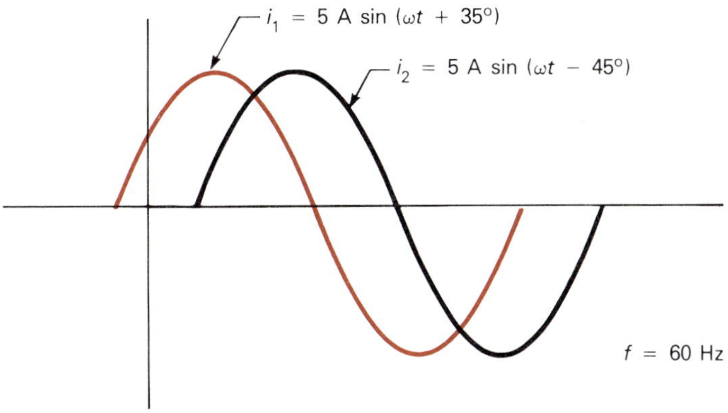

FIGURE 7–20

Adding the two expressions

$$\begin{array}{r} 2.9 + j2.03 \\ + \; 2.5 - j2.5 \\ \hline 5.4 - j0.47 \end{array}$$

Converting back to the phasor domain

$\mathbf{I}_t = 5.42 \text{ A } \angle -4.97°$

Converting back to the time domain

$i_t = 7.66 \text{ A} \sin (\omega t - 4.97°)$

The phasor equivalents of the waveforms in Figure 7–20 are shown in Figure 7–21.

Example: What is the instantaneous current after 450 μs for the composite waveform in Figure 7–20?

Solution: The composite expression is

$i_t = 7.66 \text{ A} \sin (\omega t - 4.97°)$

$$\begin{aligned} \omega &= 2\pi f \\ &= (2)(\pi)(60) \\ &= 377 \text{ radians per second} \end{aligned}$$

$$\begin{aligned} \omega t &= 377 \text{ radians per second} \times 450\ \mu\text{s} \\ &= 0.17 \text{ radians} \end{aligned}$$

$4.97° = 0.087 \text{ radians}$

$$\begin{aligned} i_t &= (7.66 \text{ A}) \sin (0.17 \text{ radians} - 0.087 \text{ radians}) \\ &= (7.66 \text{ A}) \sin 0.083 \text{ radians} \\ &= 0.64 \text{ A} \end{aligned}$$

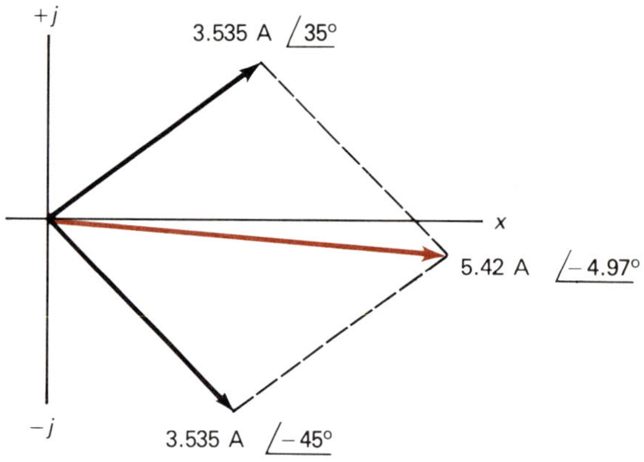

FIGURE 7–21 The phasor representations of the waveforms in Figure 7–20

EXERCISE 7·2

In problems 1–10, determine the phase angle relationship between the waveform expressions.

1. $i_1 = I_p \sin \omega t$
   $i_2 = I_p \sin (\omega t + 30°)$
2. $e_1 = E_p \sin (\omega t + 90°)$
   $e_2 = E_p \sin (\omega t - 20°)$
3. $e = E_p \sin (\omega t - 24°)$
   $i = I_p \sin \omega t$
4. $e = E_p \sin (\omega t + 15°)$
   $i = I_p \sin (\omega t + 15°)$
5. $\mathbf{E} = 20 \text{ V} \angle -20°$
   $i = I_p \sin (\omega t + 15°)$
6. $\mathbf{I_1} = 3 \text{ A} \angle 45°$
   $\mathbf{I_2} = 2 \text{ A} \angle -10°$
7. $\mathbf{E_1} = 100 \text{ V} \angle 0°$
   $\mathbf{E_2} = 100 \text{ V} \angle -15°$
8. a: $20 \text{ V} + j12$
   b: $45 \text{ V} - j17$
9. $\mathbf{I} = 15 \text{ mA} \angle 38°$
   $18 \text{ V} - j10$
10. $e = E_p \sin (\omega t - 11°)$
    $i = I_p \sin (\omega t - 8°)$
11. Find e after 1 ms for the waveform $e = E_p \sin (\omega t - 10°)$ if f = 400 Hz and E = 12 V.
12. What is i after 1 ms for the waveform $i = I_p \sin (\omega t + 60°)$ if f = 1 000 Hz and I = 10 mA?
13. The equation for a waveform is $e = E_p \sin (\omega t - 45°)$. What is e after 1.51 radians if E_p = 50 V and f = 60 Hz?
14. Find i after 50 μs for the waveform $i = I_p \sin (\omega t - 40°)$ if f = 10 kHz and I = 100 μA.

In problems 15–20, convert the phasor expressions to time domain expressions.

15. $\mathbf{E} = 115 \text{ V} \angle -18°$
16. $\mathbf{I} = 3.2 \text{ mA} \angle 0°$
17. $\mathbf{E} = 6 \text{ V} \angle 48°$
18. $\mathbf{I} = 7.07 \text{ A} \angle 60°$
19. $\mathbf{E} = 50 \text{ V} \angle -27°$
20. $\mathbf{I} = 65 \text{ mA} \angle 1.5°$

In problems 21–25, convert the time domain expressions to phasor domain expressions.

21. $i = 10 \text{ A} \sin (\omega t + 15°)$
22. $e = 1.5 \text{ V} \sin \omega t$

23. $i = 20 \text{ mA} \sin(\omega t - 72°)$
24. $e = 80 \text{ V} \sin(\omega t + 37°)$
25. $i = 9 \text{ mA} \sin wt$

In problems 26–34, combine the two waveforms and express the composite waveform in phasor and time domain notation.

26. $i_1 = 15 \text{ mA} \sin \omega t$
    $i_2 = 20 \text{ mA} \sin(\omega t + 40°)$
27. $e_1 = 48 \text{ V} \sin(\omega t - 20°)$
    $e_2 = 65 \text{ V} \sin(\omega t + 15°)$
28. $e_1 = 120 \text{ V} \sin(\omega t + 90°)$
    $e_2 = 120 \text{ V} \sin \omega t$
29. $i_1 = 1 \text{ A} \sin(\omega t + 10°)$
    $i_2 = 1 \text{ A} \sin(\omega t + 18°)$
30. $e_1 = 80 \text{ V} \sin(\omega t - 60°)$
    $e_2 = 35 \text{ V} \sin(\omega t - 30°)$
31. $i_1 = 120 \ \mu\text{A} \sin \omega t$
    $i_2 = 100 \ \mu\text{A} \sin(\omega t - 8°)$
32. $i_1 = 2.5 \text{ A} \sin(\omega t + 25°)$
    $i_2 = 2.5 \text{ A} \sin(\omega t - 25°)$
33. $e_1 = 160 \text{ V} \sin(\omega t + 50°)$
    $e_2 = 120 \text{ V} \sin(\omega t + 45°)$
34. $e_1 = 160 \text{ V} \sin(\omega t + 50°)$
    $e_2 = 120 \text{ V} \sin(\omega t - 45°)$

UNIT 7·3 Alternating Current Circuits in Series

Objectives:

After studying this unit, you should be able to

- calculate voltage, current, reactance, and impedance values in series ac circuits.
- express circuit values in both the time and phasor domains.
- interpret time domain graphs of voltage, current, and power.
- calculate phase angles.
- calculate circuit values in *RL, RC,* and *RCL* circuits.
- calculate the resonant frequency for an *RCL* circuit.
- calculate the real, apparent, and reactive power in series ac circuits.

Resistive Alternating Current Circuits

The circuit in Figure 7–22 shows an ac power source connected to a resistor. The characteristics of a resistor are such that peak current occurs at the same point that peak voltage occurs. The formulas for the circuit voltage and

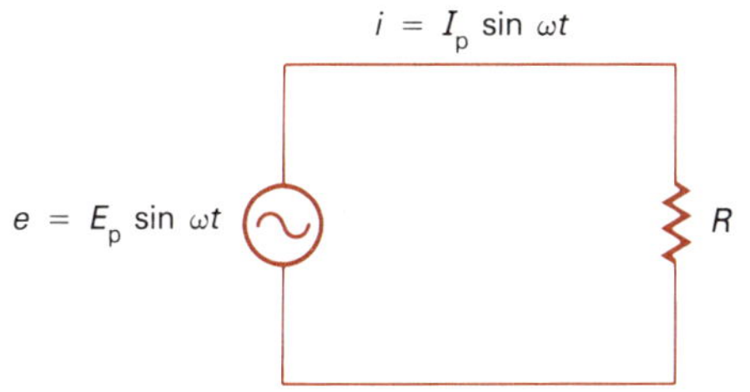

FIGURE 7–22 Resistive ac circuit

current were presented earlier and are

$$e = E_p \sin \omega t$$
$$i = I_p \sin \omega t$$

As can be seen by the sine waves of voltage and current in Figure 7–23A, the two are in phase. As voltage increases across R the current also increases. At any point in time the value of R is given by

$$R = \frac{e}{i}$$

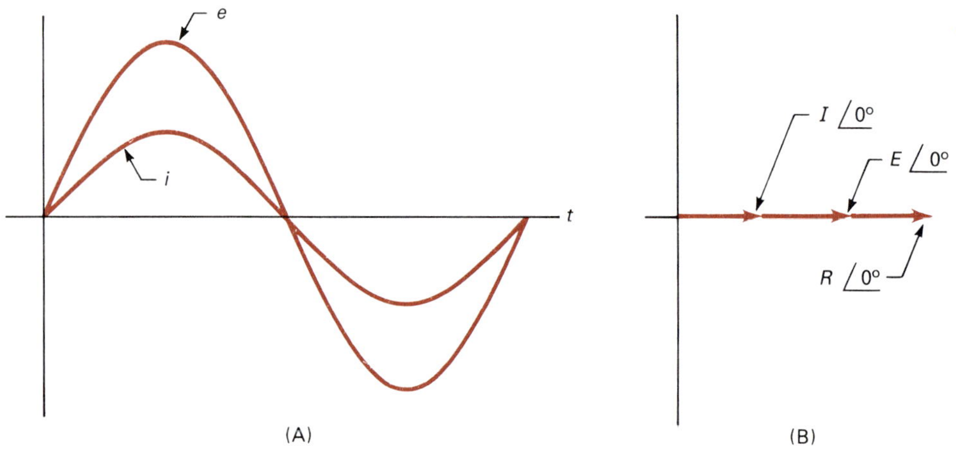

FIGURE 7–23 Time (A) and phasor (B) representations of a resistive ac circuit

Since e and i are in phase, R can also be found using average, rms, or peak-to-peak values.

$$R = \frac{e}{i} = \frac{E}{I} = \frac{E_{av}}{I_{av}} = \frac{E_{p\text{-}p}}{I_{p\text{-}p}}$$

In Figure 7–23B, the current and voltage phasors have been drawn along the x axis. This is due to the waveforms being in phase. Logically, then, a phasor representing R would appear as

$$\mathbf{R} = \frac{E \angle 0°}{I \angle 0°} = R \angle 0°$$

Example: Find the rms current for the circuit in Figure 7–24. What is the instantaneous current after 10 ms?

Solution: As this is a purely resistive circuit the standard form of Ohm's Law may be used.

$$I = \frac{E}{R} = \frac{120 \text{ V}}{10 \text{ k}\Omega} = 12 \text{ mA}$$

FIGURE 7–24

To determine the instantaneous current write the expression for current

$$
\begin{aligned}
i &= I_p \sin \omega t \\
&= (12 \text{ mA} \times 1.414) \sin [(2)(\pi)(60)(0.01)] \\
&= (16.968 \text{ mA})(-0.587\,8) \\
&= -9.97 \text{ mA}
\end{aligned}
$$

Example: A sinusoidal voltage is expressed as $e = 60 \text{ V} \sin (\omega t - 60°)$. This voltage is applied to a 2-kΩ resistor. Write the expression for the current waveform.

Solution: $I_p = \dfrac{E_p}{R} = \dfrac{60 \text{ V}}{2 \text{ k}\Omega} = 30 \text{ mA}$

$i = 30 \text{ mA} \sin (\omega t - 60°)$

Note that in the two previous resistive circuit examples, voltage and current are in phase and the usual form of Ohm's Law applies. This will not be the case for inductive and capacitive circuits.

Capacitive Alternating Current Circuits

The circuit in Fig. 7–25 contains an ac source and a capacitor. Due to the charging and discharging action of the capacitor the voltage and current waveforms are 90° out of phase in a purely capacitive circuit. The time domain expressions for the circuit voltage and current are

$$e = E_p \sin \omega t \qquad (7\text{–}31)$$

$$i = I_p \sin (\omega t + 90°) \qquad (7\text{–}32)$$

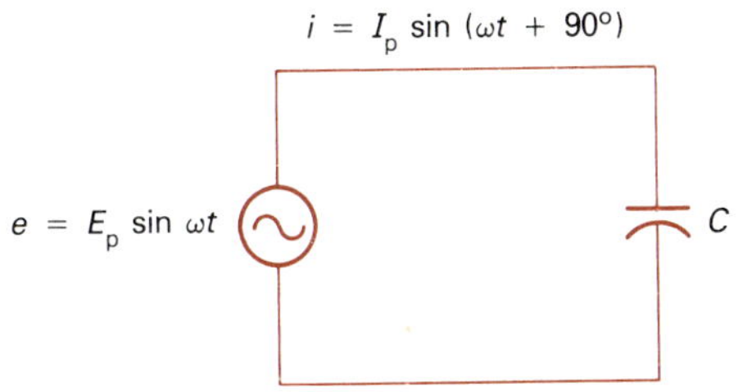

FIGURE 7–25 Capacitive ac circuit

The circuit waveforms in Figure 7–26A show that current leads voltage by 90° in a purely capacitive circuit. Converting these expressions to the phasor domain gives

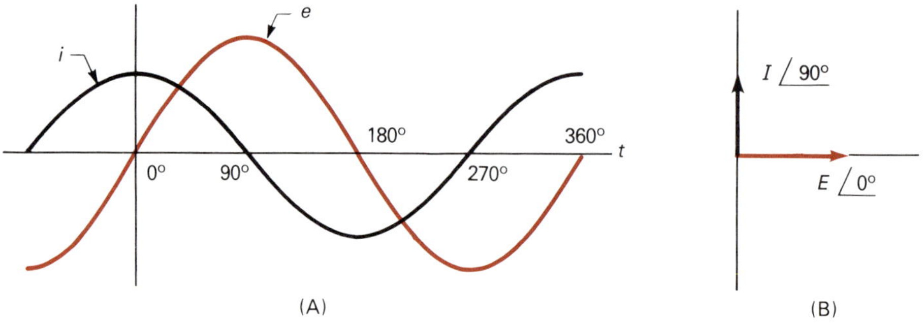

FIGURE 7–26 Time (A) and phasor (B) representations of a purely capacitive circuit

$$\mathbf{E} = E\,\angle 0^\circ$$
$$\mathbf{I} = I\,\angle 90^\circ$$

These phasors are diagrammed in Figure 7–26B.

The opposition to alternating current flow due to a capacitor is termed *capacitive reactance* (X_C). As in the case of resistance the unit of measurement for reactance is the ohm. The reactance of a capacitor is given by

$$X_C = \frac{1}{\omega C} \qquad (7\text{–}33)$$

where, ω = angular velocity in radians per second
C = capacitor value in farads

Since $\omega = 2\pi f$, the formula for capacitive reactance can be written

$$X_C = \frac{1}{\omega C} = \frac{1}{2\pi f C} \qquad (7\text{–}34)$$

It is extremely important to note that, for a given frequency, as the value of C increases the opposition to current flow decreases. Also, for a specific capacitor value, as the frequency increases the reactance in ohms decreases.

Example: Calculate the reactance each of these capacitors would have in a circuit operating at 10 kHz.
a. 100 pF b. 1 μF c. 50 μF

Solution: a. $X_C = \dfrac{1}{(2)(\pi)(10 \times 10^3\text{ Hz})(100 \times 10^{-12}\text{ F})} = 159.15\text{ k}\Omega$

b. $X_C = \dfrac{1}{(2)(\pi)(10 \times 10^3\text{ Hz})(1 \times 10^{-6}\text{ F})} = 15.92\ \Omega$

c. $X_C = \dfrac{1}{(2)(\pi)(10 \times 10^3\text{ Hz})(50 \times 10^{-6}\text{ F})} = 0.32\ \Omega$

Example: Calculate the reactance a 0.05-μF capacitor would have at (a) 60 Hz, (b) 1 kHz, and (c) 5 MHz.

Solution: a. $X_C = \dfrac{1}{(2)(\pi)(60\text{ Hz})(0.05 \times 10^{-6}\text{ F})} = 53.05\text{ k}\Omega$

b. $X_C = \dfrac{1}{(2)(\pi)(1 \times 10^{3}\text{ Hz})(0.05 \times 10^{-6}\text{ F})} = 3.18\text{ k}\Omega$

c. $X_C = \dfrac{1}{(2)(\pi)(5 \times 10^{6}\text{ Hz})(0.05 \times 10^{-6}\text{ F})} = 0.637\ \Omega$

Ohm's Law assumes a slightly different format in a capacitive ac circuit. In terms of the basic values of voltage, current, and resistance the formulas are

$$E = IX_C$$

$$I = \frac{E}{X_C}$$

$$X_C = \frac{E}{I}$$

In terms of polar notation the formulas are

$$\mathbf{X_C} = \frac{\mathbf{E}}{\mathbf{I}} = \frac{E\angle 0^\circ}{I\angle 90^\circ} = X_C\angle -90^\circ$$

Accepting the expression $X_C\angle -90^\circ$ may take a little thought. It is simple to imagine a rotating vector of current or voltage. A rotating vector of reactance is a bit more abstract. Remember that instantaneous opposition to current flow is e/i. But as these values are out of phase it is logical to think of inductive reactance as a phasor.

Example: The voltage applied to a 10-μF capacitor is $e = 50\text{ V}\sin 377t$. Find the capacitive reactance and write the expression for the circuit current in time domain notation.

Solution: Capacitive reactance is

$$X_C = \frac{1}{\omega C} = \frac{1}{(377)(10 \times 10^{-6}\text{ F})} = 265.25\ \Omega$$

$$I_p = \frac{E_p}{X_C} = \frac{50\text{ V}}{265.25\ \Omega} = 0.19\text{ A}$$

The current expressed in time domain notation is

$$i = 0.19\text{ A}\sin(377t + 90^\circ)$$

Example: A 0.02-μF capacitor is connected to a 15-V, 1 000-Hz circuit. What is the current level expressed as a phasor?

Solution: $$X_C = \frac{1}{2\pi fC} = \frac{1}{(2)(\pi)(1 \times 10^3\ \text{Hz})(0.02 \times 10^{-6}\ \text{F})} = 7.96\ \text{k}\Omega$$

$$\mathbf{I} = \frac{E \angle 0^\circ}{X_C \angle -90^\circ} = \frac{15\ \text{V} \angle 0^\circ}{7.96\ \text{k}\Omega \angle -90^\circ} = 1.88\ \text{mA} \angle 90^\circ$$

Example: A 1-μF capacitor has a reactance of 100 Ω. If the applied voltage is 20 V find (a) the circuit frequency, (b) the circuit current, and (c) the time domain expressions.

Solution: a. Solving the reactance formula for frequency gives

$$f = \frac{1}{2\pi X_C C} = \frac{1}{(2)(\pi)(100\ \Omega)(1 \times 10^{-6}\ \text{F})} = 1.59\ \text{kHz}$$

b. $$\mathbf{I} = \frac{E \angle 0^\circ}{X_C \angle -90^\circ} = \frac{20\ \text{V} \angle 0^\circ}{100\ \Omega \angle -90^\circ} = 200\ \text{mA} \angle 90^\circ$$

c. Converting to the time domain

$i = 282.8\ \text{mA} \sin(\omega t + 90^\circ)$

$e = 28.28\ \text{V} \sin \omega t$

Example: The current in a purely capacitive circuit containing a 0.5-μF capacitor is 10 mA. (a) What is the circuit frequency if the applied voltage is 10 V? Write (b) the phasor and (c) time domain expressions for voltage and current.

Solution: a. $$\mathbf{X_C} = \frac{\mathbf{E}}{\mathbf{I}} = \frac{10\ \text{V} \angle 0^\circ}{10\ \text{mA} \angle 90^\circ} = 1\ \text{k}\Omega \angle -90^\circ$$

$$f = \frac{1}{2\pi X_C C} = \frac{1}{(2)(\pi)(1 \times 10^3\ \Omega)(0.5 \times 10^{-6}\ \text{F})} = 318.3\ \text{Hz}$$

b. The phasor expressions are

$\mathbf{E} = 10\ \text{V} \angle 0^\circ$

$\mathbf{I} = 10\ \text{mA} \angle 90^\circ$

c. The time domain expressions are

$e = 14.14\ \text{V} \sin \omega t$

$i = 14.14\ \text{mA} \sin(\omega t + 90^\circ)$

Inductive Alternating Current Circuits

The circuit in Figure 7–27 contains an ac power source and an inductor. Due to the expanding and collapsing of magnetic lines of force around the inductor, the voltage and current waveforms are 90° out of phase. This rela-

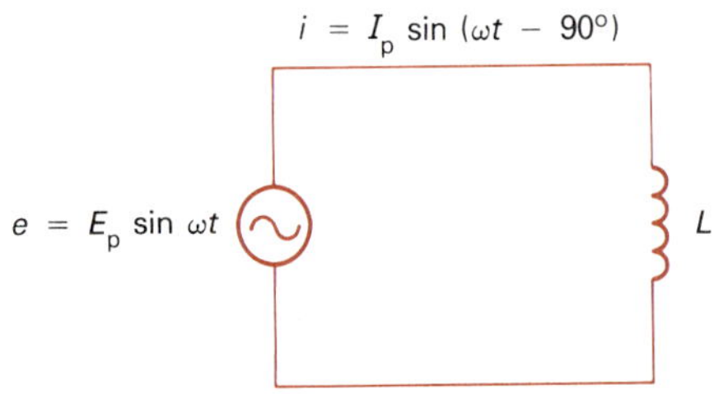

FIGURE 7–27 Inductive ac circuit

tionship is similar to that found in capacitive circuits, although in a purely inductive circuit current *lags* voltage by 90°. The time domain expressions for voltage and current are

$$e = E_p \sin \omega t \tag{7–35}$$

$$i = I_p \sin (\omega t - 90°) \tag{7–36}$$

The circuit waveforms in Figure 7–28A indicate that current lags voltage by 90° in an inductive circuit. Converting these expressions to the phasor domain gives

$$\mathbf{E} = E \angle 0°$$
$$\mathbf{I} = I \angle -90°$$

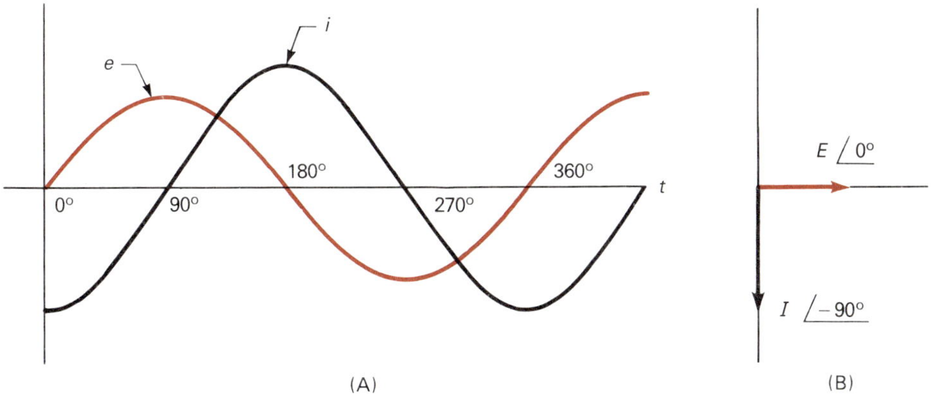

FIGURE 7–28 Time (A) and phasor (B) representations of a purely inductive circuit

These phasors are diagrammed in Figure 7–28B.

The opposition to alternating current flow due to an inductor is termed *inductive reactance* (X_L). The unit of measurement is the ohm and the basic formula is

$$X_L = \omega L \tag{7–37}$$

where, ω = angular velocity in radians per second
L = inductor value in henrys

Since $\omega = 2\pi f$, the formula can be written

$$X_L = 2\pi f L \tag{7–38}$$

For a given frequency, as the value of the inductor increases, the value of the inductive reactance increases. Also, for a fixed inductor value the reactance will increase as frequency increases.

Example: Calculate the reactance that each of these inductors would have in a circuit operating at 1 kHz.
a. 100 μH b. 50 mH c. 2 H

Solution: a. $X_L = (2)(\pi)(1 \times 10^3 \text{ Hz})(100 \times 10^{-6} \text{ H}) = 0.63\ \Omega$
b. $X_L = (2)(\pi)(1 \times 10^3 \text{ Hz})(50 \times 10^{-3} \text{ H}) = 314.16\ \Omega$
c. $X_L = (2)(\pi)(1 \times 10^3 \text{ Hz})(2 \text{ H}) = 12.57 \text{ k}\Omega$

Example: Calculate the reactance a 25-mH inductor would have at
a. 60 Hz b. 10 kHz c. 5 MHz

Solution: a. $X_L = (2)(\pi)(60 \text{ Hz})(25 \times 10^{-3} \text{ H}) = 9.42\ \Omega$
b. $X_L = (2)(\pi)(10 \times 10^3 \text{ Hz})(25 \times 10^{-3} \text{ H}) = 1.57 \text{ k}\Omega$
c. $X_L = (2)(\pi)(5 \times 10^6 \text{ Hz})(25 \times 10^{-3} \text{ H}) = 785.4 \text{ k}\Omega$

The use of Ohm's Law in inductive circuits follows the same logic as was used in capacitive circuits. Given a purely inductive circuit the formulas for circuit values are

$$E = IX_L$$

$$I = \frac{E}{X_L}$$

$$X_L = \frac{E}{I}$$

When phase angles are included the formulas are expressed in polar form

$$\mathbf{X_L} = \frac{\mathbf{E}}{\mathbf{I}} = \frac{E \angle 0^\circ}{I \angle -90^\circ} = X_L \angle 90^\circ$$

Example: A-200 mH inductor is connected to a 60-V, 60-Hz supply. What is (a) the inductive reactance, (b) the circuit current, (c) the time domain, and (d) the phasor domain expressions for voltage and current?

Solution: a. $X_L = (2)(\pi)(60 \text{ Hz})(200 \times 10^{-3} \text{ H}) = 75.4\ \Omega$
b. $I_p = \frac{E_P}{X_L} = \frac{(60 \text{ V})(1.414)}{75.4\ \Omega} = 1.13 \text{ A}$
c. The expression for circuit voltage is
$e = 84.84 \text{ V} \sin 377t$
and since current lags voltage by 90°
$i = 1.13 \text{ A} \sin (377t - 90^\circ)$
d. The phasor expressions then become
$\mathbf{E} = 60 \text{ V} \angle 0^\circ$
$\mathbf{I} = 0.8 \text{ A} \angle -90^\circ$

Example: Given $\mathbf{E} = 12 \text{ V} \angle 0^\circ$ and $\mathbf{I} = 2 \text{ mA} \angle -90^\circ$ in a 60-Hz circuit, find the value of (a) the inductor and (b) the time domain equations for voltage and current. (c) Graph the waveforms and phasors.

Solution: a. $\mathbf{X_L} = \dfrac{\mathbf{E}}{\mathbf{I}} = \dfrac{12\text{ V }\angle 0°}{2\text{ mA }\angle -90°} = 6\text{ k}\Omega\ \angle 90°$

Solving the reactance formula for the inductor gives

$$L = \frac{X_L}{2\pi f} = \frac{6\text{ k}\Omega}{(2)(\pi)(60\text{ Hz})} = 15.92\text{ H}$$

b. Converting to the time domain gives

$e = 16.97\text{ V} \sin 377t$

$i = 2.83\text{ mA} \sin (377t - 90°)$

c. Figure 7–29 displays these waveforms.

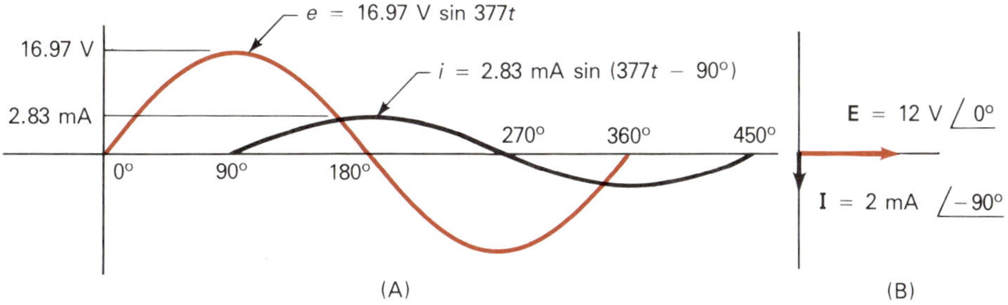

FIGURE 7–29

Example: The voltage applied to a 0.5-H inductor is $e = 24\text{ V} \sin (377t - 15°)$. (a) Find the expression for circuit current and (b) graph the waveforms and phasors.

Solution: a. $X_L = \omega L = (377)(5 \times 10^{-1}\text{ H}) = 188.5\ \Omega$

$$I_{\text{P}} = \frac{E_{\text{P}}}{X_L} = \frac{24\text{ V}}{188.5\ \Omega} = 0.13\text{ A} = 130\text{ mA}$$

As the current lags by 90° the expression is

$i = 130\text{ mA} \sin (377t - 105°)$

The phasor expressions are

$\mathbf{E} = 16.97\text{ V }\angle -15°$

$\mathbf{I} = 91.91\text{ mA }\angle -105°$

b. These waveforms are graphed in Figure 7–30.

Example: The current passing through a 1.5-H coil is $i = 20\text{ mA} \sin (754t - 10°)$. Find (a) f, (b) X_L, (c) E_{p}, and (d) the time domain expression for voltage.

Solution: a. $\omega = 2\pi f$, so that the expression for frequency is

$$f = \frac{\omega}{2\pi} = \frac{754}{2\pi} = 120\text{ Hz}$$

b. $X_L = \omega L = (754)(1.5\text{ H}) = 1\ 131\ \Omega$

c. $E_{\text{p}} = (I_{\text{p}})(X_L) = (20 \times 10^{-3}\text{ A})(1\ 131\ \Omega) = 22.62\text{ V}$

d. $e = 22.62\text{ V} \sin (754t + 80°)$

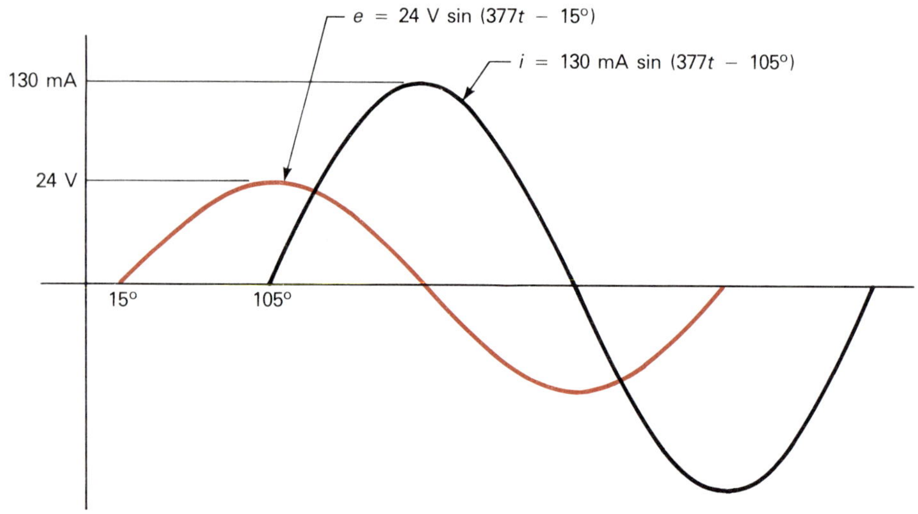

(A)

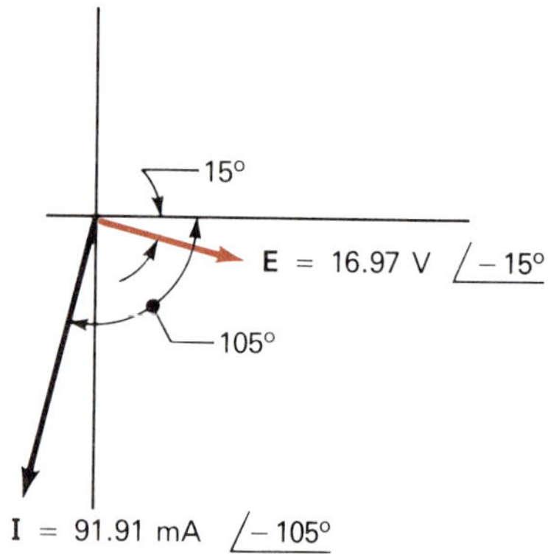

(B)

FIGURE 7–30

Impedance

The opposition to ac in a purely resistive circuit is resistance. In a purely inductive or capacitive circuit the opposition is reactance. The opposition due to a combination of any two, or all three of these in one circuit, is termed *impedance* (Z). The unit of measurement for impedance is the ohm.

Earlier in this unit the phasor expressions for resistance, inductive reactance, and capacitive reactance were developed. These are:

$$\mathbf{R} = R \angle 0^\circ$$
$$\mathbf{X_L} = X_L \angle 90^\circ$$
$$\mathbf{X_C} = X_C \angle -90^\circ$$

These phasors are diagrammed in Figure 7–31.

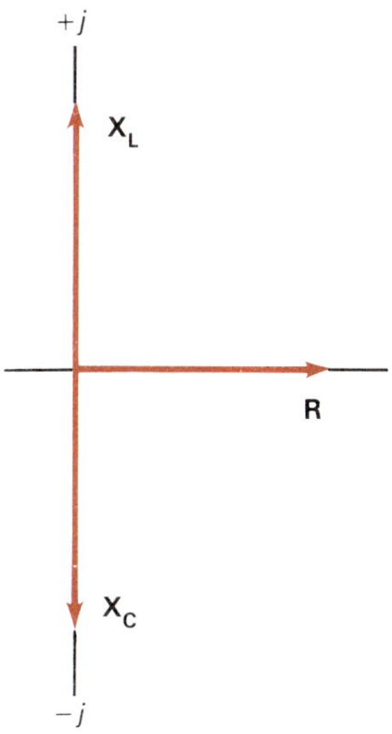

FIGURE 7–31

The formula for impedance is

$$\mathbf{Z} = \frac{\mathbf{E}}{\mathbf{I}} = \mathbf{Z}\,\underline{/\theta} \qquad (7\text{–}39)$$

where **Z** = circuit impedance in ohms
E = voltage in volts
I = current in amperes

For circuits containing only resistance, inductance, or capacitance the impedance is equal to the circuit opposition. These would be expressed as

$$\text{Resistive: } \mathbf{Z} = \mathbf{R} = R\,\underline{/0^\circ}$$
$$\text{Inductive: } \mathbf{Z} = \mathbf{X_L} = X_L\,\underline{/90^\circ}$$
$$\text{Capacitive: } \mathbf{Z} = \mathbf{X_C} = X_C\,\underline{/-90^\circ}$$

In those circuits containing only resistance and capacitance (*RC*), resistance and inductance (*RL*), or all three (*RCL*), the process to determine impedance is a bit more involved.

RC Circuits

The circuit in Figure 7–32 contains a resistor and a capacitor, hence the expression *RC* circuit. The first step in the mathematical analysis is to determine the capacitive reactance. This is

$$X_C = \frac{1}{2\pi fC}$$
$$= \frac{1}{(2)(\pi)(1 \times 10^3\ \text{Hz})(15 \times 10^{-6}\ \text{F})}$$
$$= 10.61\ \Omega$$

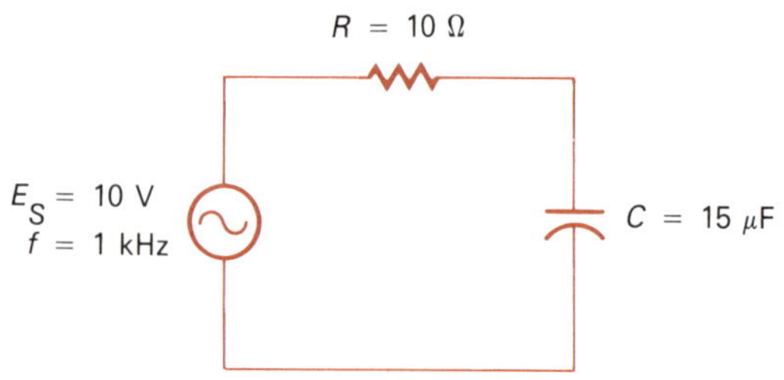

FIGURE 7–32

or, expressed as a phasor,

$$\mathbf{X_C} = 10.61\ \Omega \angle -90°$$

The diagram in Figure 7–33 shows the phasors for resistance and capacitive reactance. The addition of these vectors results in the circuit impedance.

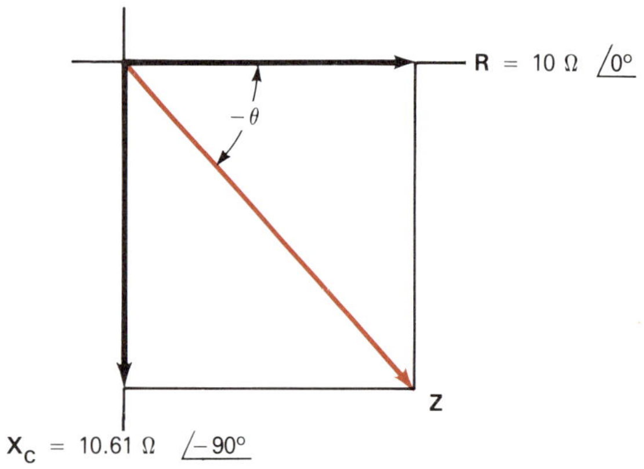

FIGURE 7–33

The impedance may be found by expressing the resultant vector in rectangular notation, and then converting to the polar form. Examine the following sequence

$$\begin{aligned} \mathbf{Z} &= R - jX_C \\ &= (10 - j10.61)\ \Omega && \text{(Rectangular notation)} \\ &= 14.58\ \Omega \angle -46.67° && \text{(Polar notation)} \end{aligned}$$

A related method is to use the Pythagorean Theorem.

$$\begin{aligned} \mathbf{Z} &= \sqrt{R^2 + X_C^2} \\ &= \sqrt{10^2 + 10.61^2} \\ &= 14.58\ \Omega \end{aligned}$$

$$\tan\theta = \frac{-X_C}{R} = \frac{-10.61\ \Omega}{10\ \Omega} = -1.061,\ \theta = -46.67°$$

Resulting in

$$\mathbf{Z} = 14.58\ \Omega\ \underline{/-46.67^\circ}$$

Once the circuit impedance has been determined circuit current may be computed.

$$\begin{aligned}\mathbf{I} &= \frac{\mathbf{E}}{\mathbf{Z}}\\ &= \frac{10\text{ V }\underline{/0^\circ}}{14.58\ \Omega\ \underline{/-46.67^\circ}}\\ &= 0.69\text{ A }\underline{/46.67^\circ}\end{aligned}$$

Given the current, it is possible to determine the voltage drop across the resistor. Keep in mind that in a series circuit, there is only one current. In a purely capacitive circuit this current lags voltage by 90°. With the addition of resistance this angle has been reduced to 46.67°. Note that with a single current, and with R and X_C out of phase, it is logical to expect the voltages across these components to also be out of phase. The current must also be in phase with the voltage across the resistor. The current must, therefore, lead the voltage across the capacitor.

Computing the voltages across the resistor and capacitor in this example produces

$$\mathbf{E_R} = \mathbf{IR} = (0.69\text{ A }\underline{/46.67^\circ})(10\ \Omega\ \underline{/0^\circ}) = 6.9\text{ V }\underline{/46.67^\circ}$$
$$\mathbf{E_C} = \mathbf{IX_C} = (0.69\text{ A }\underline{/46.47^\circ})(10.61\ \Omega\ \underline{/-90^\circ}) = 7.32\text{ V }\underline{/-43.33^\circ}$$

The phasor diagram in Figure 7–34 shows the relationship among the

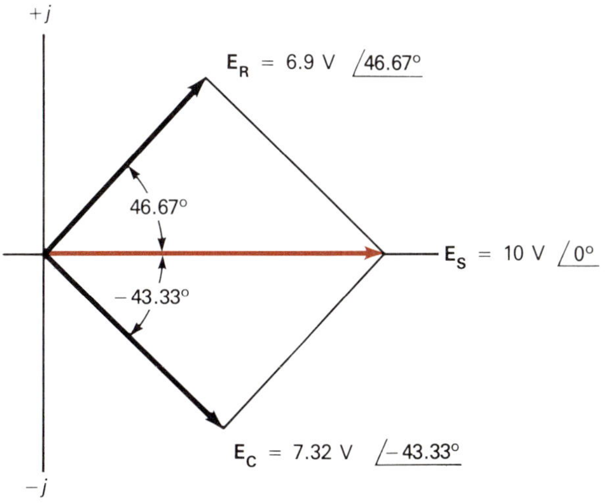

FIGURE 7–34

voltages in the *RC* circuit. Note that E_R and E_C are 90° out of phase and when these two are added as vectors the resultant is the source voltage.

In order to check the values of E_R and E_C the polar values can be converted to rectangular notation and summed. This sum should equal the applied voltage.

$$\mathbf{E_R} = 6.9 \text{ V} \angle 46.67^\circ = (4.73 + j5.02) \text{ V}$$
$$\mathbf{E_C} = 7.32 \text{ V} \angle -43.33^\circ = (5.32 - j5.02) \text{ V}$$
$$\mathbf{E_S} = \mathbf{E_R} + \mathbf{E_C} = (4.73 + j5.02) \text{ V} + (5.32 - j5.02) \text{ V} = 10 \text{ V}$$

The current and voltages in this circuit may also be expressed in time domain notation. The reader should verify that these are

$$e_S = 14.14 \text{ V} \sin \omega t$$
$$e_R = 9.76 \text{ V} \sin (\omega t + 46.67^\circ)$$
$$e_C = 10.35 \text{ V} \sin (\omega t - 43.33^\circ)$$
$$i = 0.98 \text{ A} \sin (\omega t + 46.67^\circ)$$

These waveforms are displayed in Figure 7–35.

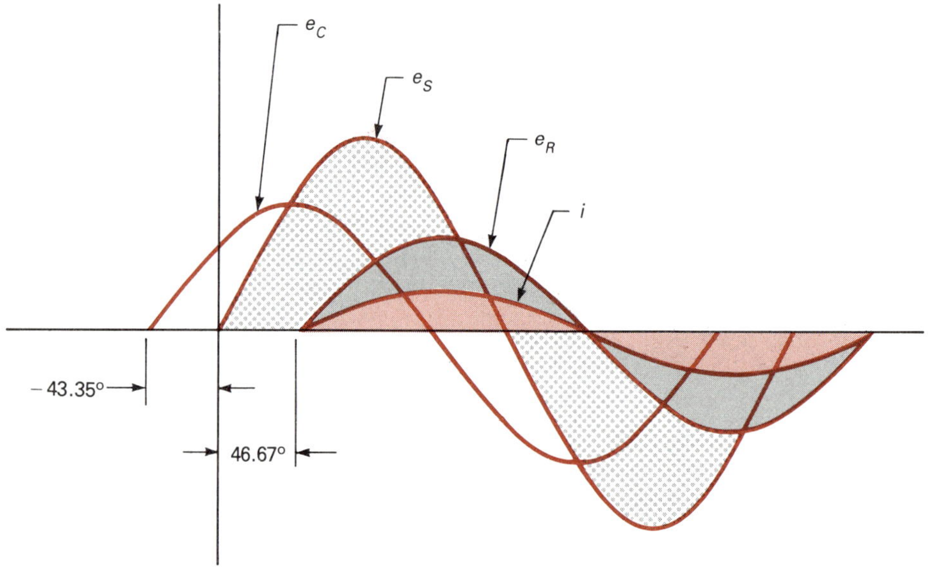

FIGURE 7–35

Example: For the circuit in Figure 7–36 find each of the following:

a. X_C	d. $\mathbf{E_R}$	g. e_R
b. $\mathbf{Z}$	e. $\mathbf{E_C}$	h. e_C
c. $\mathbf{I}$	f. e_S	i. i

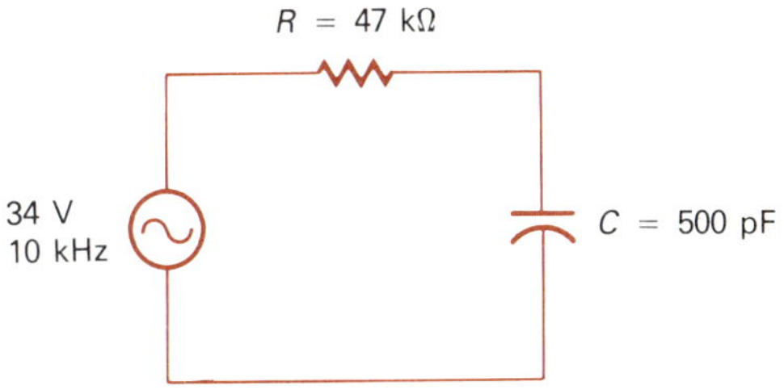

FIGURE 7–36

Solution: a. $X_C = \dfrac{1}{2\pi fC}$

$= \dfrac{1}{(2)(\pi)(10 \times 10^3\ \text{Hz})(500 \times 10^{-12}\ \text{F})}$

$= 31.83\ \text{k}\Omega$

b. $\mathbf{Z} = R - jX_C$

$= (47 - j31.83)\ \text{k}\Omega$

$= 56.76\ \text{k}\Omega\ \angle -34.11°$

c. $\mathbf{I} = \dfrac{\mathbf{E}}{\mathbf{Z}}$

$= \dfrac{34\ \text{V}\ \angle 0°}{56.76\ \text{k}\Omega\ \angle -34.11°}$

$= 0.6\ \text{mA}\ \angle 34.11°$

d. $\mathbf{E_R} = \mathbf{IR}$

$= (0.6\ \text{mA}\ \angle 34.11°)(47\ \text{k}\Omega\ \angle 0°)$

$= 28.2\ \text{V}\ \angle 34.11°$

e. $\mathbf{E_C} = \mathbf{IX_C}$

$= (0.6\ \text{mA}\ \angle 34.11°)(31.83\ \text{k}\Omega\ \angle -90°)$

$= 19.1\ \text{V}\ \angle -55.89°$

f. $e_S = 48.08\ \text{V} \sin \omega t$

g. $e_R = 39.87\ \text{V} \sin (\omega t + 34.11°)$

h. $e_C = 27.01\ \text{V} \sin (\omega t - 55.89°)$

i. $i = 0.85\ \text{mA} \sin (\omega t + 34.11°)$

Example: The capacitor in Figure 7–36 is replaced with a 0.001 μF capacitor. Express the voltage across the capacitor in both phasor and time domain notation.

Solution: $X_C = \dfrac{1}{2\pi fC} = 15.92\ \text{k}\Omega$

$\mathbf{Z} = R - jX_C$

$= (47 - j15.92)\ \text{k}\Omega$

$= 49.62\ \text{k}\Omega\ \angle -18.71°$

$$\mathbf{I} = \frac{\mathbf{E}}{\mathbf{Z}}$$
$$= \frac{34\text{ V}\angle 0^\circ}{49.62\text{ k}\Omega\angle -18.71^\circ}$$
$$= 0.69\text{ mA}\angle 18.71^\circ$$

$$\mathbf{E_C} = \mathbf{IX_C}$$
$$= (0.69\text{ mA}\angle 18.71^\circ)(15.92\text{ k}\Omega\angle -90^\circ)$$
$$= 10.98\text{ V}\angle -71.29^\circ$$

$$e_C = 15.53\text{ V}\sin(\omega t - 71.29^\circ)$$

Example: What is the value of C in Figure 7–37?

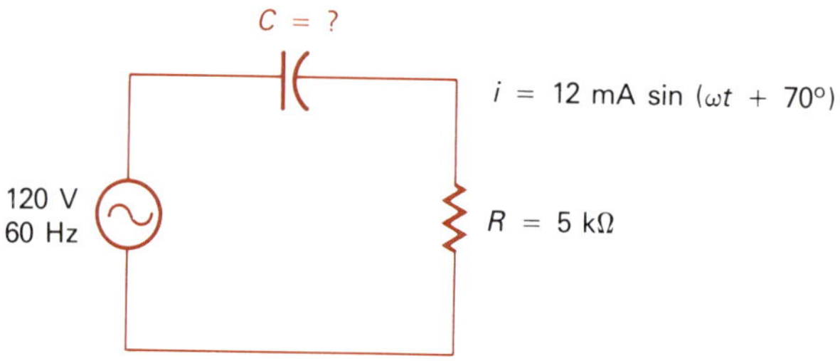

FIGURE 7–37

Solution: Converting the time domain current to polar notation
$\mathbf{I} = 8.48\text{ mA}\angle 70^\circ$

Circuit impedance is then found as follows
$$\mathbf{Z} = \frac{\mathbf{E}}{\mathbf{I}} = \frac{120\text{ V}\angle 0^\circ}{8.48\text{ mA}\angle 70^\circ} = 14.15\text{ k}\Omega\angle -70^\circ$$

Reactance is found using the impedance diagram in Figure 7–38.
$X_C = \sqrt{Z^2 - R^2} = \sqrt{(14.15)^2 - (5)^2} = 13.24\text{ k}\Omega$

A related method for determining the reactance is to convert impedance to rectangular notation. Remember that $Z = R - jX_C$. This should give us the value of the resistor *and* the inductive reactance.

$\mathbf{Z} = 14.15\text{ k}\Omega\angle -70^\circ$ (Polar notation)
$= (4.8 - j13.29)\text{ k}\Omega$ (Rectangular notation)

Note that there is a slight difference in the two methods of computing X_C. This is due to rounding of previous answers. Solving the reactance formula for capacitance:

$$C = \frac{1}{2\pi f X_C} = \frac{1}{(2)(\pi)(60\text{ Hz})(13.24\times 10^3\ \Omega)} = 0.2\ \mu\text{F}$$

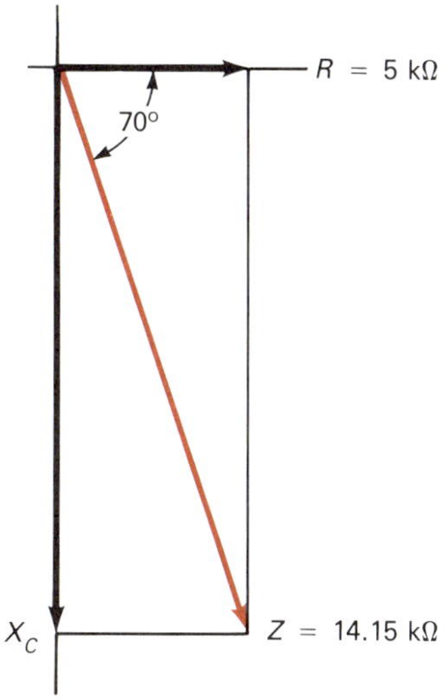

FIGURE 7–38

RL Circuits

The circuit in Figure 7–39 contains a resistor and an inductor. The first step in the analysis of this *RL* circuit is to determine the inductive reactance.

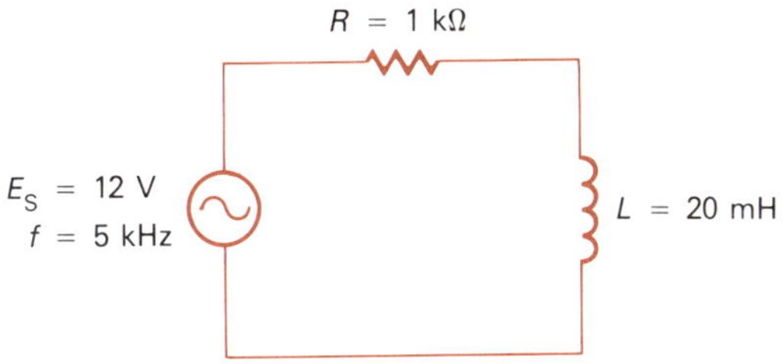

FIGURE 7–39

This is

$$\begin{aligned} X_L &= 2\pi f L \\ &= (2)(\pi)(5 \times 10^3 \text{ Hz})(20 \times 10^{-3} \text{ H}) \\ &= 628.32\ \Omega \end{aligned}$$

or, expressed as a phasor

$$\mathbf{X_L} = 628.32\ \Omega \angle 90^\circ$$

The diagram in Figure 7–40 shows the phasors for resistance and inductive reactance. The addition of these vectors results in the circuit impedance. As in *RC* circuits, the impedance may be found through a rectangular to polar conversion, or by utilizing the Pythagorean Theorem.

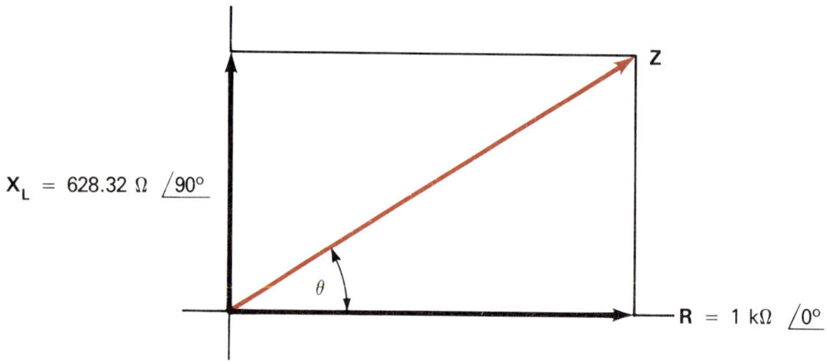

FIGURE 7–40

$$\mathbf{Z} = R + jX_L$$
$$= (1\,000 + j628.32)\ \Omega$$
$$= 1.18\ \text{k}\Omega \angle 32.14°$$

or,

$$Z = \sqrt{R^2 + X_L^2}$$
$$= \sqrt{(1\,000)^2 + (628.32)^2}$$
$$= 1.18\ \text{k}\Omega$$
$$\tan 0 = \frac{X_L}{R} - \frac{628.32\ \Omega}{1\,000\ \Omega} = 0.628\,32,\ \theta = 32.14°$$

resulting in

$$\mathbf{Z} = 1.18\ \text{k}\Omega \angle 32.14°$$

The circuit current is found by

$$\mathbf{I} = \frac{\mathbf{E}}{\mathbf{Z}} = \frac{12\ \text{V} \angle 0°}{1.18\ \text{k}\Omega \angle 32.14°} = 10.17\ \text{mA} \angle -32.14°$$

Remember in a purely inductive circuit that voltage leads current by 90°. With the addition of resistance to this circuit the phase angle has been reduced to 32.14°. The individual voltages due to the circuit current are

$$\mathbf{E_R} = \mathbf{IR}$$
$$= (10.17\ \text{mA} \angle -32.14°)(1\ \text{k}\Omega \angle 0°)$$
$$= 10.17\ \text{V} \angle -32.14°$$

$$\mathbf{E_L} = \mathbf{IX_L}$$
$$= (10.17\ \text{mA} \angle -32.14°)(628.32\ \Omega \angle 90°)$$
$$= 6.39\ \text{V} \angle 57.86°$$

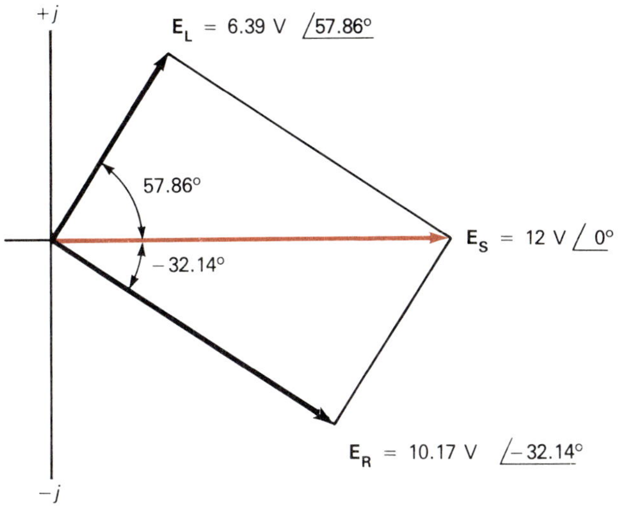

FIGURE 7–41

The phasor diagram in Figure 7–41 shows the relationship among the voltages. Note that E_R and E_L are 90° out of phase. The addition of these two voltages will result in the source voltage. This can be verified as follows

$$\mathbf{E_R} = 10.17\text{ V}\,\angle{-32.14^\circ} = (8.6 - j5.41)\text{ V}$$
$$\mathbf{E_L} = 6.39\text{ V}\,\angle{57.86^\circ} = (3.4 + j5.41)\text{ V}$$
$$\mathbf{E_S} = \mathbf{E_R} + \mathbf{E_L} = (8.6 - j5.41)\text{ V} + (3.4 + j5.41)\text{ V} = 12\text{ V}$$

The time domain expressions for the voltages and current in this circuit are

$$e_S = 16.97\text{ V}\sin\omega t$$
$$e_R = 14.38\text{ V}\sin(\omega t - 32.41^\circ)$$
$$e_L = 9.04\text{ V}\sin(\omega t + 57.86^\circ)$$
$$i = 16.97\text{ mA}\sin(\omega t - 32.41^\circ)$$

Example: For the circuit shown in Figure 7–42, find each of these values:

a. $\mathbf{X_L}$ d. $\mathbf{E_R}$ g. e_R
b. $\mathbf{Z}$ e. $\mathbf{E_L}$ h. e_L
c. $\mathbf{I}$ f. e_S i. i

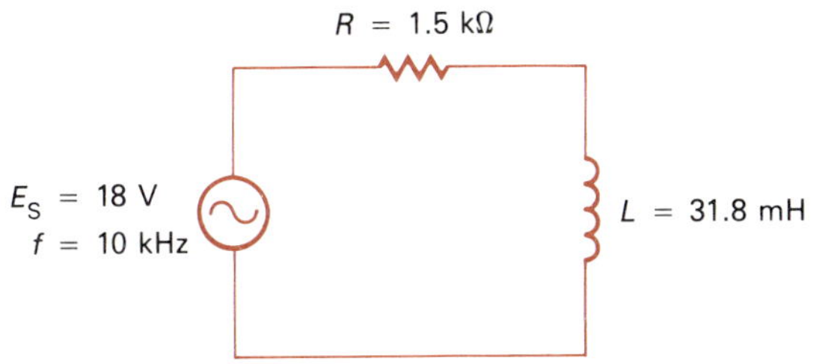

FIGURE 7–42

Solution: a. $\mathbf{X_L} = 2\pi f L \angle 90°$
$= (2)(\pi)(10 \times 10^3 \text{ Hz})(31.8 \times 10^{-3} \text{ H})$
$= 2 \text{ k}\Omega \angle 90°$

b. $\mathbf{Z} = R + jX_L = (1.5 + j2) \text{ k}\Omega = 2.5 \text{ k}\Omega \angle 53.1°$

c. $\mathbf{I} = \dfrac{\mathbf{E}}{\mathbf{Z}} = \dfrac{18 \text{ V} \angle 0°}{2.5 \text{ k}\Omega \angle 53.1°} = 7.2 \text{ mA} \angle -53.1°$

d. $\mathbf{E_R} = \mathbf{IR} = (7.2 \text{ mA} \angle -53.1°)(1.5 \text{ k}\Omega \angle 0°) = 10.8 \text{ V} \angle -53.1°$

e. $\mathbf{E_L} = \mathbf{IX_L} = (7.2 \text{ mA} \angle -53.1°)(2 \text{ k}\Omega \angle 90°) = 14.4 \text{ V} \angle 36.9°$

f. $e_S = 24.45 \text{ V} \sin \omega t$

g. $e_R = 15.2 \text{ V} \sin (\omega t - 53.1°)$

h. $e_L = 20.36 \text{ V} \sin (\omega t + 36.9°)$

i. $i = 10.18 \text{ mA} \sin (\omega t - 53.1°)$

These waveforms are graphed in Figure 7–43.

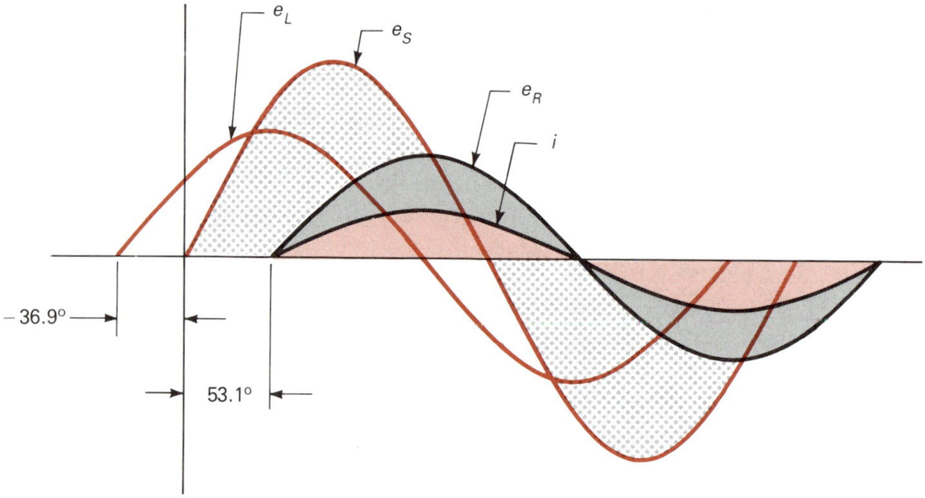

FIGURE 7–43

Example: What are the values of R and L in Figure 7–44?

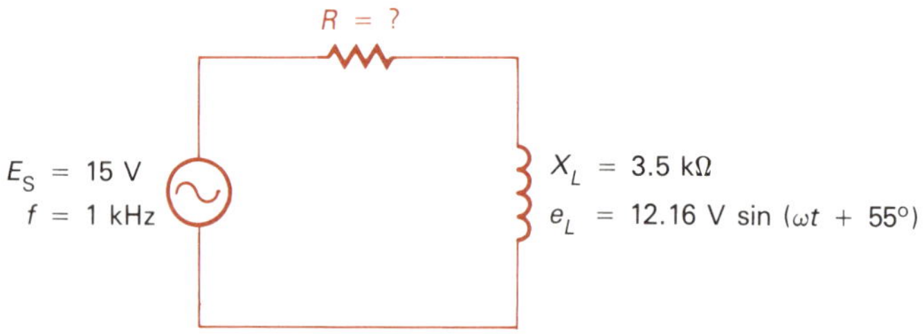

FIGURE 7–44

Solution: Convert e_L to a phasor

$\mathbf{E_L} = 8.6 \text{ V} \angle 55° = (4.93 + j7.04) \text{ V}$

The value of $\mathbf{E_R}$ is

$$\begin{aligned}\mathbf{E_R} &= \mathbf{E_S} - \mathbf{E_L}\\ &= (15 + j0)\ \text{V} - (4.93 + j7.04)\ \text{V}\\ &= (10.07 - j7.04)\ \text{V}\\ &= 12.29\ \text{V} \angle -34.96^\circ\end{aligned}$$

Current is found by

$$\mathbf{I} = \frac{\mathbf{E_L}}{\mathbf{X_L}} = \frac{8.6\ \text{V} \angle 55^\circ}{3.5\ \text{k}\Omega \angle 90^\circ} = 2.46\ \text{mA} \angle -35^\circ$$

The phasor notation for R is

$$\mathbf{R} = \frac{\mathbf{E_R}}{\mathbf{I}} = \frac{12.29\ \text{V} \angle -34.96^\circ}{2.46\ \text{mA} \angle -35^\circ} = 5\ \text{k}\Omega \angle 0^\circ$$

The value of L is found by

$$L = \frac{X_L}{2\pi f} = \frac{3.5\ \text{k}\Omega}{(2)(\pi)(1 \times 10^3\ \text{Hz})} = 0.56\ \text{H}$$

Example: What is the circuit current in Figure 7–45?

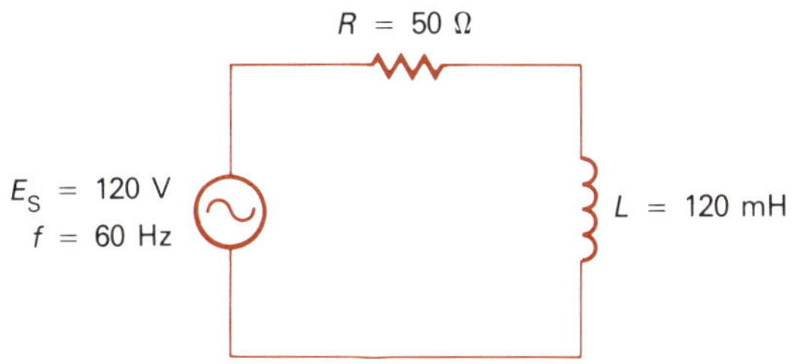

FIGURE 7–45

Solution:

$$\begin{aligned}\mathbf{X_L} &= 2\pi f L \angle 90^\circ\\ &= (2)(\pi)(60\ \text{Hz})(0.12\ \text{H}) \angle 90^\circ\\ &= 45.24\ \Omega \angle 90^\circ\end{aligned}$$

$$\begin{aligned}\mathbf{Z} &= R + jX_L\\ &= (50 + j45.24)\ \Omega\\ &= 67.43\ \Omega \angle 42.14^\circ\end{aligned}$$

$$\mathbf{I} = \frac{\mathbf{E}}{\mathbf{Z}} = \frac{120\ \text{V} \angle 0^\circ}{67.43\ \Omega \angle 42.14^\circ} = 1.78\ \text{A} \angle -42.14^\circ$$

RCL Circuits

The circuit in Figure 7–46 is a series *RCL* circuit. The first step in the mathematical analysis of this configuration is to determine the values of X_C and X_L.

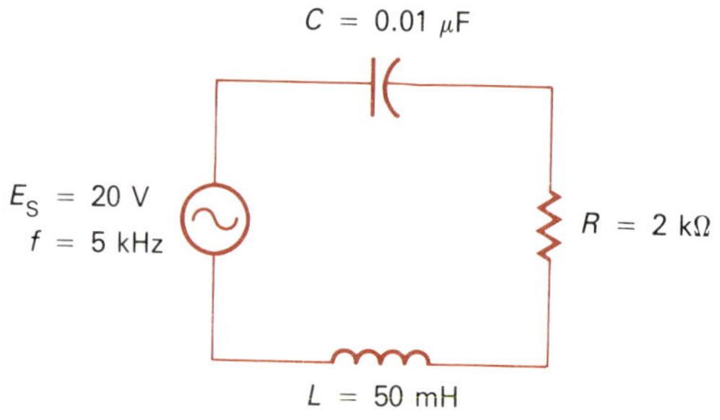

FIGURE 7–46

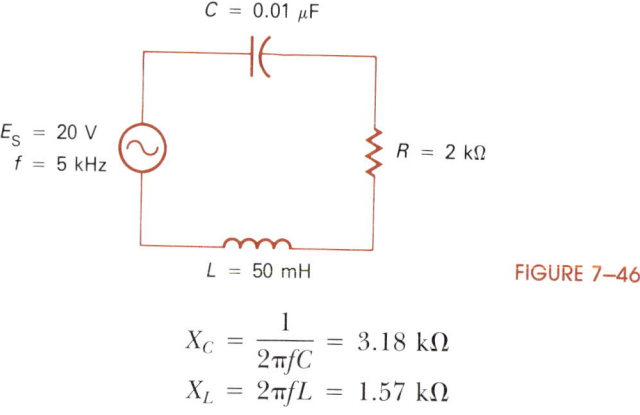

$$X_C = \frac{1}{2\pi f C} = 3.18 \text{ k}\Omega$$
$$X_L = 2\pi f L = 1.57 \text{ k}\Omega$$

The phasor notations for these values are

$$\mathbf{X_C} = 3.18 \text{ k}\Omega \angle -90^\circ$$
$$\mathbf{X_L} = 1.57 \text{ k}\Omega \angle 90^\circ$$

The diagram in Figure 7–47A displays the phasors for resistance, capacitive reactance, and inductive reactance. As X_L and X_C are 180° out of phase they may be summed in both the polar and the rectangular forms. The vector sum of X_C and X_L in a series *RCL* circuit is the *net reactance,* or *X*. For this example the net reactance would be

$$\begin{aligned}\mathbf{X} &= \mathbf{X_C} + \mathbf{X_L}\\ &= 3.18 \text{ k}\Omega \angle -90^\circ + 1.57 \text{ k}\Omega \angle 90^\circ\\ &= 1.61 \text{ k}\Omega \angle -90^\circ\end{aligned}$$

or, in rectangular notation,

$$\begin{aligned}\mathbf{X} &= \mathbf{X_C} + \mathbf{X_L}\\ &= (0 - j3.18) \text{ k}\Omega + (0 + j1.57) \text{ k}\Omega\\ &= 0 - j1.61 \text{ k}\Omega\\ &= 1.61 \text{ k}\Omega \angle -90^\circ\end{aligned}$$

Circuit impedance for a series *RCL* circuit is the phasor sum of reactance and resistance as shown in Figure 7–47B. In this example the capacitive reactance is greater than the inductive reactance. The net reactance, therefore, is capacitive in nature, resulting in a negative phase angle for circuit impedance. Impedance for this circuit is calculated as follows

$$\begin{aligned}\mathbf{Z} &= R + jX_L - jX_C\\ &= R + j(X_L - X_C)\\ &= R + jX\\ &= [2 + j(1.57 - 3.18)] \text{ k}\Omega\\ &= (2 - j1.61) \text{ k}\Omega\end{aligned}$$

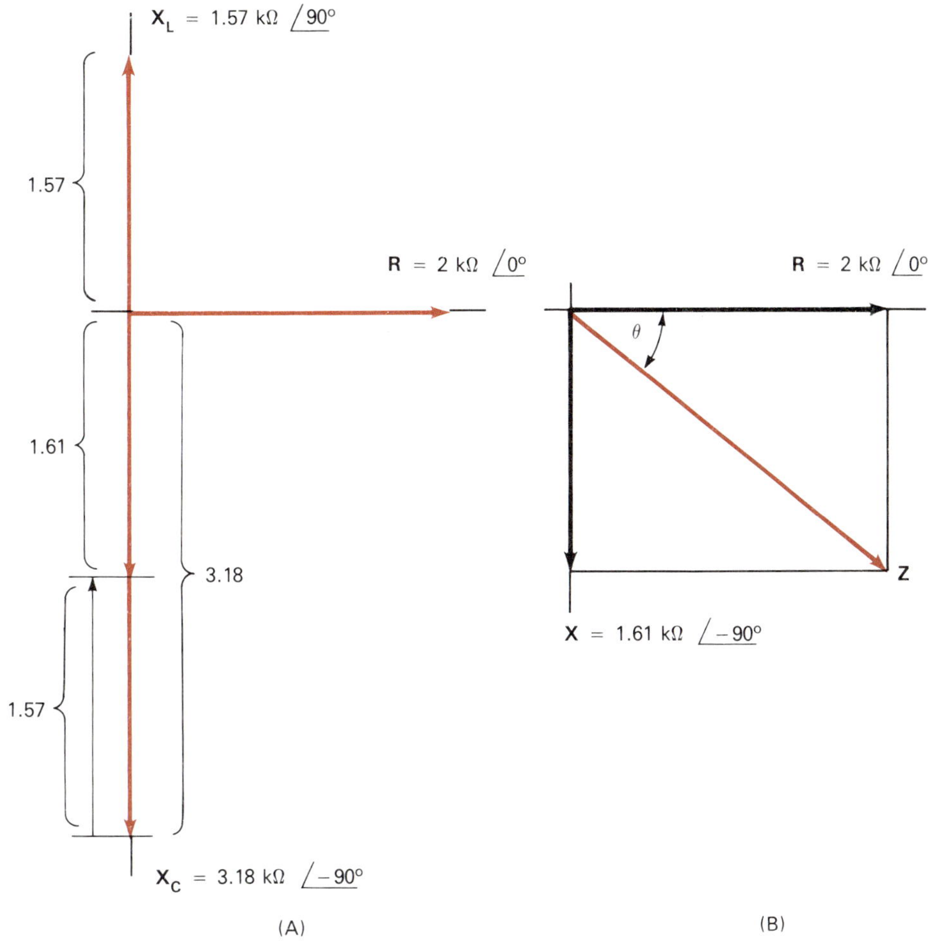

FIGURE 7–47

which, when converted to polar notation, gives

$$\mathbf{Z} = 2.57 \text{ k}\Omega \angle -38.83^\circ$$

Note that when X_C is greater than X_L, the expression for impedance is $R - jX$. When X_L is greater than X_C the expression becomes $R + jX$.

Another method for determining the value of Z is

$$\begin{aligned} Z &= \sqrt{R^2 + (X_L - X_C)^2} \\ &= \sqrt{(2)^2 + (1.57 - 3.18)^2} \\ &= \sqrt{6.592\ 1} \\ &= 2.57 \text{ k}\Omega \end{aligned}$$

$$\tan\theta = \frac{X}{R} = \frac{-1.61 \text{ k}\Omega}{2 \text{ k}\Omega} = -0.805,\ \theta = -38.83^\circ$$

therefore,

$$\mathbf{Z} = 2.57\ \text{k}\Omega\ \underline{/-38.83^\circ}$$

The current, in the series *RCL* circuit shown in Figure 7–46, is found by

$$\mathbf{I} = \frac{\mathbf{E}}{\mathbf{Z}} = \frac{20\ \text{V}\ \underline{/0^\circ}}{2.57\ \text{k}\Omega\ \underline{/-38.83^\circ}} = 7.78\ \text{mA}\ \underline{/38.83^\circ}$$

As this current flows through each reactance and resistance a voltage will be developed across each component. These are

$$\mathbf{E_R} = \mathbf{IR} = (7.78\ \text{mA}\ \underline{/38.83^\circ})(2\ \text{k}\Omega\ \underline{/0^\circ}) = 15.56\ \text{V}\ \underline{/38.83^\circ}$$
$$\mathbf{E_C} = \mathbf{IX_C} = (7.78\ \text{mA}\ \underline{/38.83^\circ})(3.18\ \text{k}\Omega\ \underline{/-90^\circ}) = 24.74\ \text{V}\ \underline{/-51.17^\circ}$$
$$\mathbf{E_L} = \mathbf{IX_L} = (7.78\ \text{mA}\ \underline{/38.83^\circ})(1.57\ \text{k}\Omega\ \underline{/90^\circ}) = 12.21\ \text{V}\ \underline{/128.83^\circ}$$

The phasor diagram in Figure 7–48 shows the relationship among these three voltages. Kirchhoff's Voltage Law states that the sum of the voltages around a closed loop will equal zero. Examining the magnitudes in Figure 7–48, the sum appears to exceed the applied voltage. This is not actually true

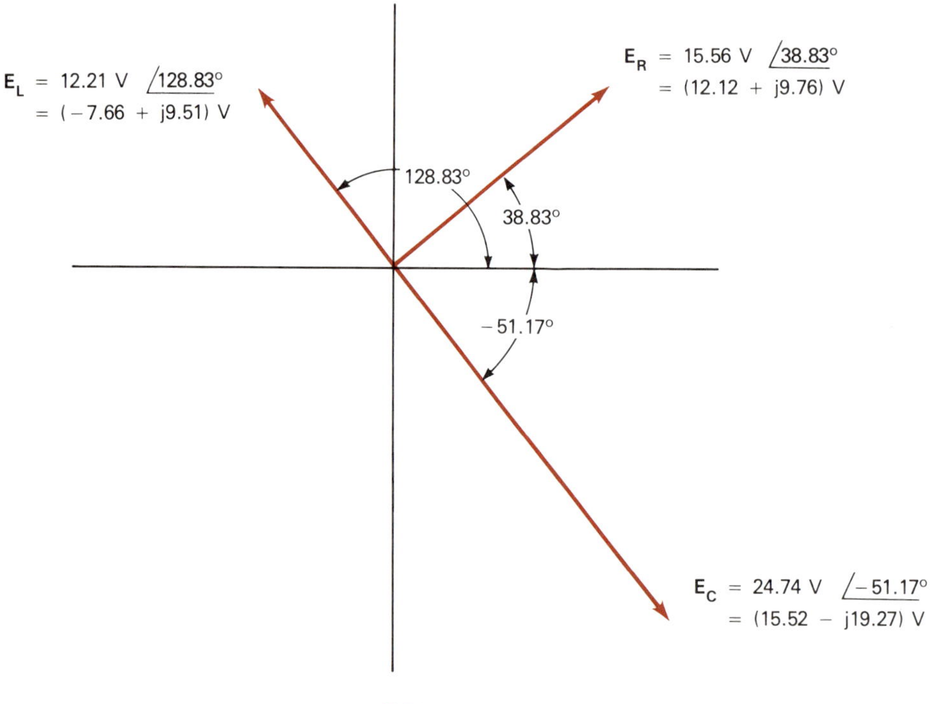

FIGURE 7–48

in that these are phasors, and when summed as phasors the total will equal the applied voltage. The applied voltage is

$$\mathbf{E_S} = \mathbf{E_R} + \mathbf{E_L} + \mathbf{E_C}$$
$$20\text{ V }\angle 0^\circ = (15.56\text{ V }\angle 38.83^\circ) + (12.21\text{ V }\angle 128.83^\circ) + (24.74\text{ V }\angle -51.17^\circ)$$
$$(20 + j0)\text{ V} = (12.12 + j9.76)\text{ V} + (-7.66 + j9.51)\text{ V} + (15.52 - j19.27)\text{ V}$$
$$(20 + j0)\text{ V} = (20 + j0)\text{ V}$$
$$20\text{ V} = 20\text{ V}$$

The time domain voltages and current in this circuit are graphed in Figure 7–49. The corresponding expressions are

$$e_S = 28.28\text{ V}\sin \omega t$$
$$e_R = 22\text{ V}\sin(\omega t + 38.83^\circ)$$
$$e_C = 34.98\text{ V}\sin(\omega t - 51.17^\circ)$$
$$e_L = 17.26\text{ V}\sin(\omega t + 128.83^\circ)$$
$$i = 11\text{ mA}\sin(\omega t + 38.83^\circ)$$

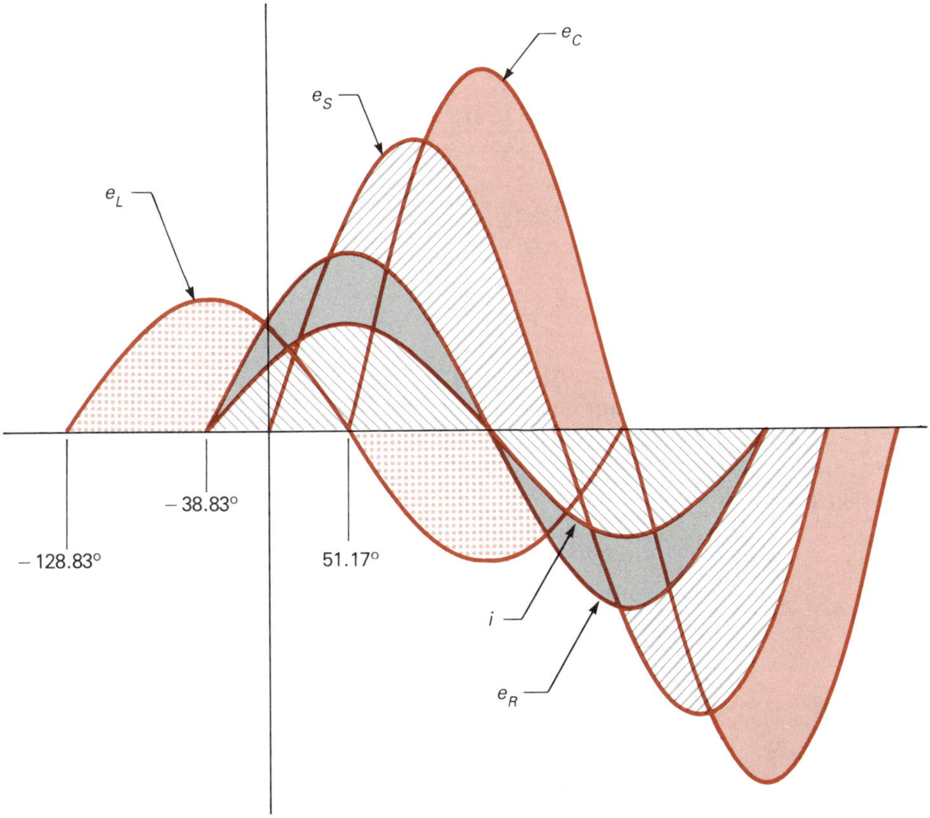

FIGURE 7–49

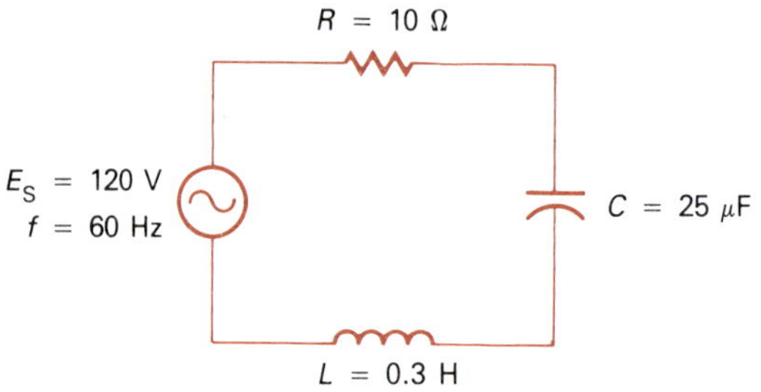

FIGURE 7–50

Example: For the circuit in Figure 7–50 find

a. X_L c. X e. $\mathbf{I}$ g. $\mathbf{E_L}$

b. X_C d. $\mathbf{Z}$ f. $\mathbf{E_C}$ h. $\mathbf{E_X}$

Solution:

a. $X_L = 2\pi fL = 113.1\ \Omega$

b. $X_C = \dfrac{1}{2\pi fC} = 106.1\ \Omega$

c. $X = X_L - X_C = 113.1\ \Omega - 106.1\ \Omega = 7\ \Omega$

d. $\mathbf{Z} = R + jX = (10 + j7)\ \Omega = 12.21\ \Omega\ \angle 34.99°$

e. $\mathbf{I} = \dfrac{\mathbf{E_S}}{\mathbf{Z}} = \dfrac{120\text{ V}\ \angle 0°}{12.21\ \Omega\ \angle 34.99°} = 9.83\text{ A}\ \angle -34.99°$

f. $\mathbf{E_C} = \mathbf{IX_C}$
$= (9.83\text{ A}\ \angle -34.99°)(106.1\ \Omega\ \angle -90°)$
$= 1\ 042.96\text{ V}\ \angle -124.99°$

g. $\mathbf{E_L} = \mathbf{IX_L}$
$= (9.83\text{ A}\ \angle -34.99°)(113.1\ \Omega\ \angle 90°)$
$= 1\ 111.78\text{ V}\ \angle 55.01°$

h. $\mathbf{E_X} = \mathbf{IX}$
$= (9.83\text{ A}\ \angle -34.99°)(7\ \Omega\ \angle 90°)$
$= 68.81\text{ V}\ \angle 55.01°$

Another method to determine the voltage drop across the net reactance is to sum the voltages across C and L.

$\mathbf{E_X} = \mathbf{E_C} + \mathbf{E_L}$
$= (1\ 042.96\text{ V}\ \angle -124.99°) + (1\ 111.78\text{ V}\ \angle 55.01°)$
$= 68.81\text{ V}\ \angle 55.01°$

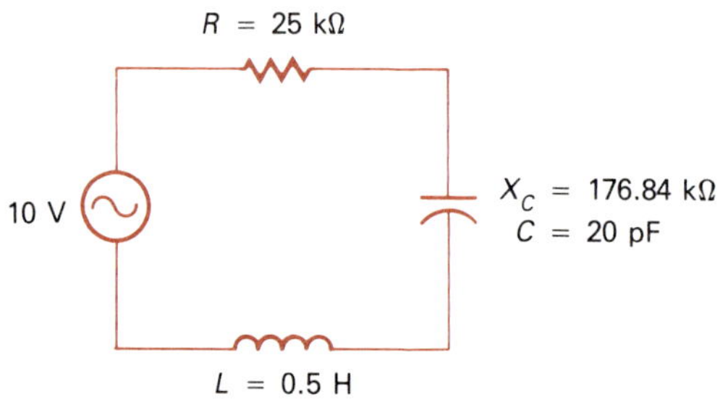

FIGURE 7–51

Example: In the circuit in Figure 7–51 find the magnitudes only for these values:
a. f b. X_L c. X d. Z e. I

Solution: a. Solving the reactance formula for frequency

$$f = \frac{1}{2\pi X_C C}$$
$$= \frac{1}{(2)(\pi)(176.84 \times 10^3\ \Omega)(20 \times 10^{-12}\ \text{F})}$$
$$= 45\ \text{kHz}$$

b. $X_L = 2\pi f L$
$$= (2)(\pi)(45 \times 10^3\ \text{Hz})(0.5\ \text{H})$$
$$= 141.37\ \text{k}\Omega$$

c. $X = X_C - X_L$
$$= 176.84\ \text{k}\Omega - 141.37\ \text{k}\Omega$$
$$= 35.47\ \text{k}\Omega$$

d. $Z = \sqrt{R^2 + X^2}$
$$= \sqrt{(25 \times 10^3\ \Omega)^2 + (35.47 \times 10^3\ \Omega)^2}$$
$$= 43.39\ \text{k}\Omega$$

e. $I = \dfrac{E_S}{Z} = \dfrac{10\ \text{V}}{43.39\ \text{k}\Omega} = 0.23\ \text{mA}$

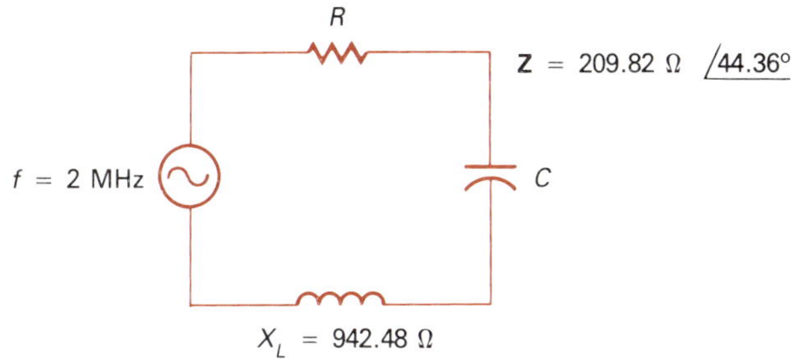

FIGURE 7–52

Example: For the circuit in Figure 7–52, determine the value of
a. R b. X_C c. C d. L

Solution: a. Converting impedance to rectangular notation will give the values of R and X.

$$\mathbf{Z} = 209.82\ \Omega\ \angle 44.36^\circ$$
$$= R + jX$$
$$= (150 + j146.7)\ \Omega$$
$$R = 150\ \Omega$$

b. Since j is positive, $X_L > X_C$,

$$X_C = X_L - X$$
$$= 942.48\ \Omega - 146.7\ \Omega$$
$$= 795.78\ \Omega$$

c. $C = \frac{1}{2\pi f X_C} = \frac{1}{(2)(\pi)(2 \times 10^6 \text{ Hz})(795.78 \ \Omega)} = 100 \text{ pF}$

d. $L = \frac{X_L}{2\pi f} = \frac{942.48 \ \Omega}{(2)(\pi)(2 \times 10^6 \text{ Hz})} = 75 \ \mu\text{H}$

Series Resonance

At this point it should be obvious that a change in frequency affects inductive and capacitive reactance differently. Given a series *RCL* circuit with a very low applied frequency the capacitor will have an extremely high reactance. As frequency increases X_C will decrease and X_L will increase. When the frequency is extremely high, X_C will be very low and X_L will be very high. For a series *RCL* circuit, there will be one frequency where $X_L = X_C$. When this occurs the circuit is at *resonance,* and the frequency is the *resonant frequency* (f_r).

The formula for resonant frequency can be derived by setting the expressions for X_L and X_C equal and solving for f.

$$X_L = X_C$$

$$2\pi f L = \frac{1}{2\pi f C}$$

$$(2\pi f)^2 = \frac{1}{LC}$$

$$2\pi f = \frac{1}{\sqrt{LC}}$$

$$f = \frac{1}{2\pi\sqrt{LC}}$$

The formula for resonant frequency is

$$f_r = \frac{1}{2\pi\sqrt{LC}} \qquad \textbf{(7–40)}$$

where, f_r = resonant frequency in hertz
L = inductance in henrys
C = capacitance in farads

When series resonance occurs $X_L = X_C$. At this frequency the reactance values are 180° out of phase and of equal magnitude. The circuit impedance at this frequency then is equal to the resistance.

Example: Calculate the resonant frequency for the circuit in Figure 7–53. Also, find the values of $\mathbf{X_L}$, $\mathbf{X_C}$, **Z, I,** and the voltages across each component.

Solution: $f_r = \frac{1}{2\pi\sqrt{LC}}$

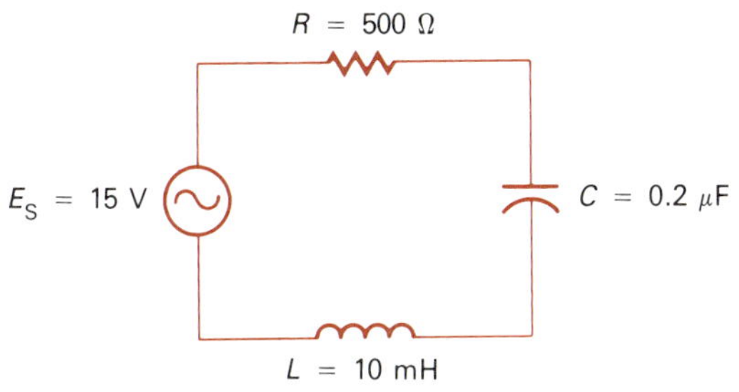

FIGURE 7–53

$$= \frac{1}{(2)(\pi)\sqrt{(10 \times 10^{-3}\text{ H})(0.2 \times 10^{-6}\text{ F})}}$$

$$= 3\,558.81\text{ Hz}$$

$$\mathbf{X_L} = 2\pi fL \angle 90^\circ$$

$$= (2)(\pi)(3\,558.81\text{ Hz})(10 \times 10^{-3}\text{ H}) \angle 90^\circ$$

$$= 223.61\ \Omega \angle 90^\circ$$

$$\mathbf{X_C} = \frac{1}{2\pi fC} \angle -90^\circ$$

$$= \frac{1}{(2)(\pi)(3\,558.81\text{ Hz})(0.2 \times 10^{-6}\text{ F})} \angle -90^\circ$$

$$= 223.61\ \Omega \angle -90^\circ$$

$$\mathbf{Z} = R \angle 0^\circ = 500\ \Omega \angle 0^\circ$$

$$\mathbf{I} = \frac{\mathbf{E_S}}{\mathbf{Z}} = \frac{15\text{ V} \angle 0^\circ}{500\ \Omega \angle 0^\circ} = 30\text{ mA} \angle 0^\circ$$

$$\mathbf{E_R} = \mathbf{IR} = (30\text{ mA} \angle 0^\circ)(5\text{ k}\Omega \angle 0^\circ) = 15\text{ V} \angle 0^\circ$$

$$\mathbf{E_C} = \mathbf{IX_C} = (30\text{ mA} \angle 0^\circ)(223.61\ \Omega \angle -90^\circ) = 6.71\text{ V} \angle -90^\circ$$

$$\mathbf{E_L} = \mathbf{IX_L} = (30\text{ mA} \angle 0^\circ)(223.61\ \Omega \angle 90^\circ) = 6.71\text{ V} \angle 90^\circ$$

Example: The circuit in Figure 7–54 is at resonance. Find the value of $\mathbf{X_L}$, L, **Z, I,** and the voltage across each component.

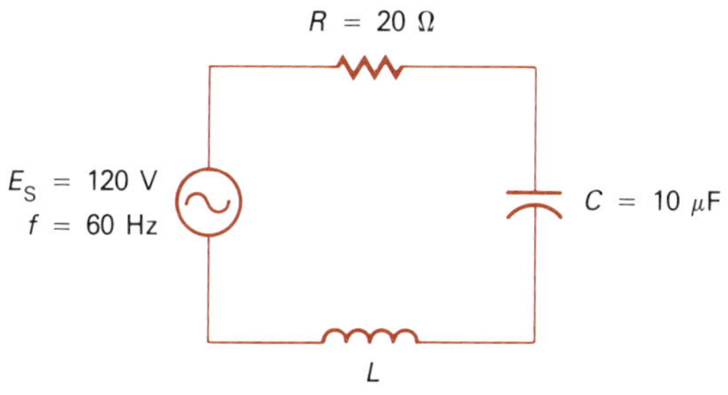

FIGURE 7–54

Solution: At resonance $X_C = X_L$, therefore

$$X_L = X_C = \frac{1}{2\pi fC} = \frac{1}{(2)(\pi)(60\text{ Hz})(10 \times 10^{-6}\text{ F})} = 265.26\ \Omega$$

In terms of phasor notation
$\mathbf{X_L} = 265.26\ \Omega\ \underline{/90^\circ}$

The value of L is found by

$$L = \frac{X_L}{2\pi f} = \frac{265.26\ \Omega}{(2)(\pi)(60\text{ Hz})} = 703.6\text{ mH}$$

At resonance $Z = R$, therefore
$\mathbf{Z} = 500\ \Omega\ \underline{/0^\circ}$

Circuit current is

$$\mathbf{I} = \frac{\mathbf{E_S}}{\mathbf{Z}} = \frac{120\text{ V}\ \underline{/0^\circ}}{500\ \Omega\ \underline{/0^\circ}} = 240\text{ mA}\ \underline{/0^\circ}$$

Component voltages are
$\mathbf{E_R} = \mathbf{IR} = (240\text{ mA}\ \underline{/0^\circ})(500\ \Omega\ \underline{/0^\circ}) = 120\text{ V}\ \underline{/0^\circ}$
$\mathbf{E_C} = \mathbf{IX_C} = (240\text{ mA}\ \underline{/0^\circ})(265.26\ \Omega\ \underline{/-90^\circ}) = 63.66\text{ V}\ \underline{/-90^\circ}$
$\mathbf{E_L} = \mathbf{IX_L} = (240\text{ mA}\ \underline{/0^\circ})(265.26\ \Omega\ \underline{/90^\circ}) = 63.66\text{ V}\ \underline{/90^\circ}$

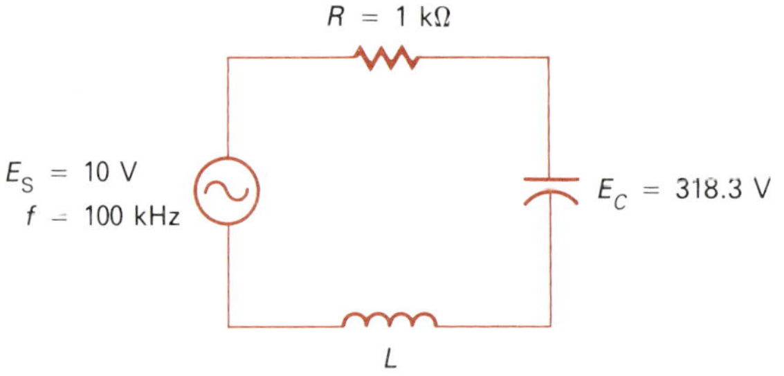

FIGURE 7–55

Example: For the resonant circuit in Figure 7–55 find the values of
a. I b. X_C c. C d. X_L e. L

Solution: a. As the magnitude values only are required, we need not be concerned about phase angles. At resonance $Z = R$, so that current is

$$I = \frac{E_S}{R} = \frac{10\text{ V}}{1\text{ k}\Omega} = 10\text{ mA}$$

b. Capacitive reactance can be found by

$$X_C = \frac{E_C}{I} = \frac{318.3\text{ V}}{10\text{ mA}} = 31.83\text{ k}\Omega$$

c. The value of C is

$$C = \frac{1}{2\pi f X_C} = \frac{1}{(2)(\pi)(100 \times 10^3\text{ Hz})(31.83 \times 10^3\ \Omega)} = 50\text{ pF}$$

d. $X_L = X_C = 31.83\text{ k}\Omega$

e. $$L = \frac{X_L}{2\pi f} = \frac{31.83 \times 10^3\ \Omega}{(2)(\pi)(100 \times 10^3\text{ Hz})} = 50.66\text{ mH}$$

Power in Alternating Current Circuits

The concept of power was first presented in dc circuits. In a purely resistive dc circuit power, in watts, was the product of current and voltage. This is also the case for *instantaneous* power in ac circuits. This may be calculated as follows

$$p = ei = \frac{e^2}{R} = i^2R$$

where, p = instantaneous power in watts
e = instantaneous voltage in volts
i = instantaneous current in amperes
R = resistance in ohms

In a purely resistive circuit the waveforms of current and voltage are in phase. These waveforms are displayed in Figure 7–56 along with the waveform

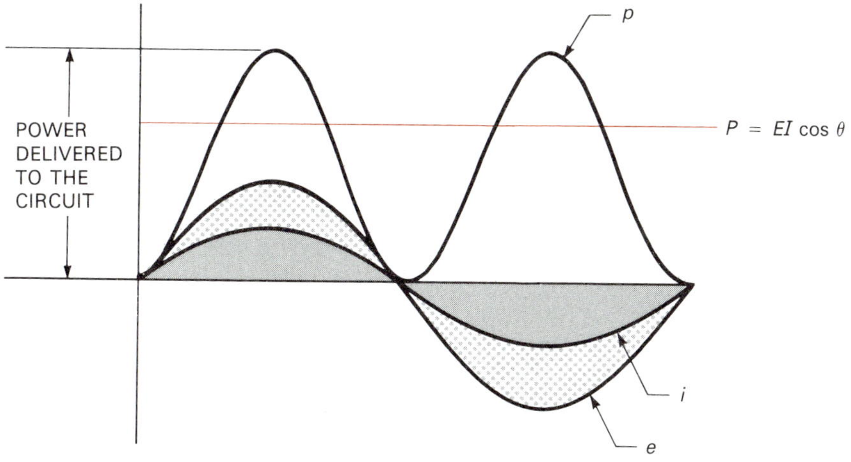

FIGURE 7–56

of power. Note that during the first 180° power is the product of positive values of voltage and current. As the values of voltage and current both go negative the resulting product remains positive. All power is therefore delivered to the circuit and none is returned to the source by the circuit. Also, note that the frequency of the power waveform is twice that of the applied voltage.

The *average* or *real* power in an ac circuit is

$$P = EI \cos \theta \qquad (7\text{–}41)$$

where, P = real power in watts
E = voltage in volts
I = current in amperes
θ = phase angle between voltage and current

Example: What is the power dissipated by the resistor in Figure 7–57?

FIGURE 7–57

Solution:

$$I = \frac{E}{R} = \frac{120\text{ V}}{100\ \Omega} = 1.2\text{ A}$$

$$P = EI \cos\theta = (120\text{ V})(1.2\text{ A})(\cos 0°) = 144\text{ W}$$

In a purely resistive circuit the power delivered to the load is the product of voltage and current. As reactance is introduced into a circuit the *apparent* power will be greater than the actual power. The formula for apparent power is

$$P_A = EI \qquad (7\text{–}42)$$

where, P_A = apparent power in *volt-amperes* (VA)
E = voltage in volts
I = current in amperes

Note that this expression does not take into account the phase angle. In a resistive circuit $P = P_A$. For phase angles greater than 0° the apparent power is greater than the real power. The *power factor* of a circuit is real power divided by apparent power. Follow this derivation

$$P = EI \cos\theta$$
$$P_A = EI$$

Substituting P_A for EI in the first formula results in

$$P = P_A \cos\theta$$

with the expression for the power factor being

$$F_p = \frac{P}{P_A} = \cos\theta \qquad (7\text{–}43)$$

Example: Voltage leads current by 15° in a circuit. Compute the real power and power factor if the apparent power is 25 VA.

Solution:

$$P = P_A \cos\theta$$
$$= (25\text{ VA})(\cos 15°)$$
$$= 24.15\text{ W}$$
$$F_p = \cos 15° = 0.965\,9$$

In a purely inductive circuit voltage leads current by 90°. Instantaneous power is again the product of voltage and current. The current, voltage, and power waveforms in an inductive circuit are shown in Figure 7–58. Note that

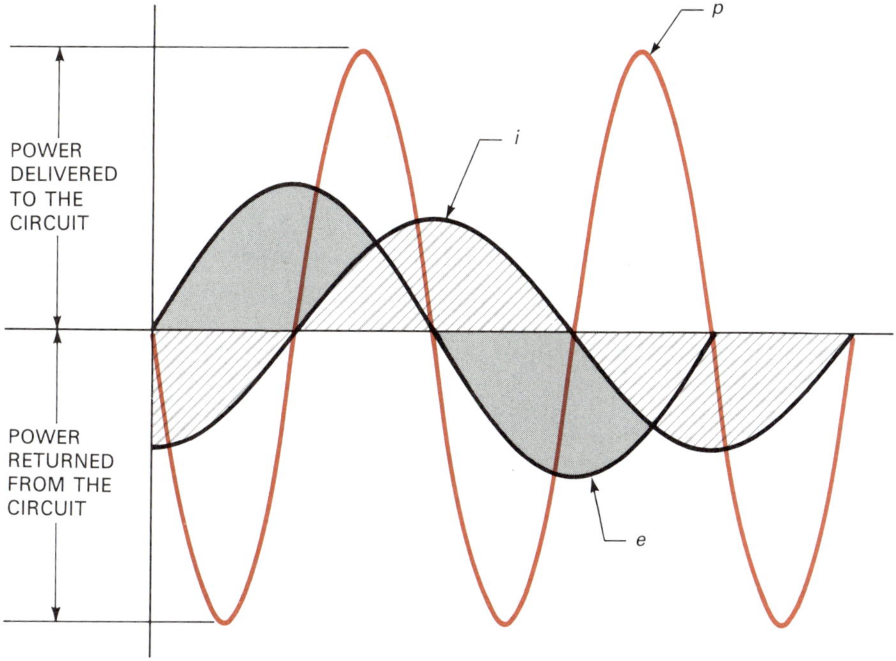

FIGURE 7–58

during one half of the power wave, power is delivered to the inductor and stored in the coil's magnetic field. During the other half of the cycle, power is returned to the source as the field collapses.

The average power for an inductive circuit is

$$P = EI \cos \theta = EI \cos 90° = 0$$

This is logical in that the average value of this sine wave is the horizontal axis. The peak value of the curve is termed the *reactive power* (P_Q). The general formula for reactive power in an inductive circuit is

$$\mathbf{P_{Q_L} = EI \sin \theta} \qquad \textbf{(7–44)}$$

where, P_{Q_L} = reactive power in *vars* (volt-ampere reactive)

E = voltage in volts

I = current in amperes

θ = phase angle between voltage and current

Power, whether real, apparent, or reactive, my also be expressed in terms of current and resistance or voltage and resistance. These include

$$P = EI = I^2R = \frac{E^2}{R}$$

Example: A 50-mH inductor conducts 100 mA from a 400-Hz source. Calculate the real, apparent, and reactive power.

Solution: Assuming the coil has negligible resistance compared to the reactance, the first step is to determine X_L.

$X_L = 2\pi fL = (2)(\pi)(400 \text{ Hz})(50 \times 10^{-3} \text{ H}) = 125.66\ \Omega$

As this is a purely reactive circuit, the power of primary interest is reactive. For the purpose of demonstration, all three will be computed.

$P = I^2X_L \cos\theta = (0.1 \text{ A})^2(125.66\ \Omega)(\cos 90°) = 0$

$P_A = I^2X_L = (0.1 \text{ A})^2(125.66\ \Omega) = 1.26 \text{ VA}$

$P_{Q_L} = I^2X_L \sin\theta = (0.1 \text{ A})^2(125.66\ \Omega)(\sin 90°) = 1.26 \text{ vars}$

Note that for a purely inductive circuit the apparent and reactive power are the same. This will not be the case in *RL* circuits.

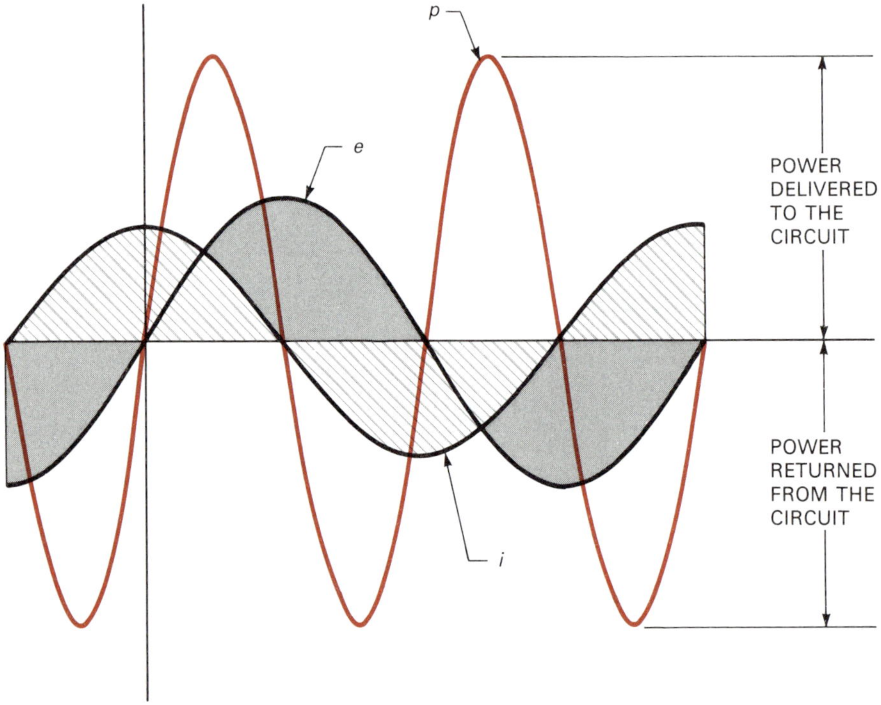

FIGURE 7–59

In a purely capacitive circuit current leads voltage by 90°. The current, voltage, and power waveforms in a capacitive circuit are shown in Figure 7–59. Note that power is delivered to the capacitor and stored in the electrostatic field during the first half of the power cycle. During the second half of the cycle power is returned to the source as the capacitor discharges.

As in the case of the inductor, the average power for a capacitive circuit is zero. The expression for *reactive* power is

$$P_{Q_C} = EI \sin \theta \qquad (7\text{–}45)$$

where P_{Q_C} is reactive power in vars.

Power In Series *RCL* Circuits

The diagram in Figure 7–60 shows the power triangles relating real, apparent, and reactive power. The angle θ is the phase angle between voltage and current. For a purely resistive circuit, reactive power is zero and apparent

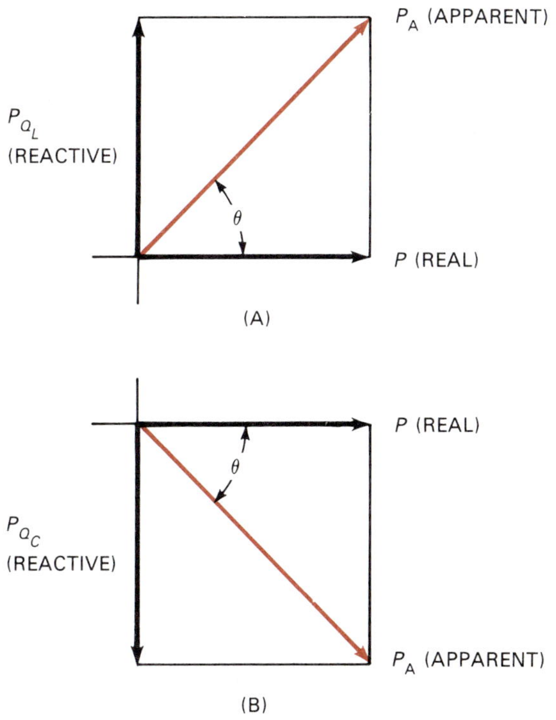

FIGURE 7–60

and real power are equal. In a purely reactive circuit, the real power is zero and reactive and apparent power are equal. In *RL*, *RC*, and *RCL* circuits, each of these values will be different.

Example: For the series *RL* circuit in Figure 7–61, calculate total power in watts, vars, and volt-amperes. Also find the power factor.

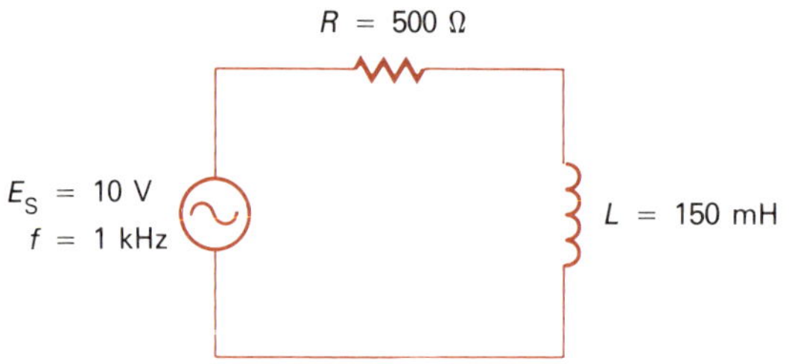

FIGURE 7–61

Solution:

$$X_L = 2\pi fL = 942.48\ \Omega$$

$$Z = \sqrt{R^2 + X_L{}^2} = \sqrt{(500\ \Omega)^2 + (942.48\ \Omega)^2} = 1\ 066.9\ \Omega$$

$$I = \frac{E_S}{Z} = \frac{10\ \text{V}}{1\ 066.9\ \Omega} = 9.37\ \text{mA}$$

$$\tan\theta = \frac{X_L}{R} = \frac{942.48\ \Omega}{500\ \Omega} = 1.885,\ \theta = 62.1°$$

$$P_t = EI\cos\theta = (10\ \text{V})(9.37\ \text{mA})(\cos 62.1°) = 0.044\ \text{W}$$

$$P_{Q(t)} = EI\sin\theta = (10\ \text{V})(9.37\ \text{mA})(\sin 62.1°) = 0.083\ \text{vars}$$

$$P_{A(t)} = EI = (10\ \text{V})(9.37\ \text{mA}) = 0.094\ \text{VA}$$

$$F_p = \cos\theta = \frac{P}{P_A} = \frac{P_t}{P_{A(t)}} = \frac{0.044}{0.094} = 0.47$$

The values in the previous example have been graphed in a power triangle in Figure 7–62. Verification of these values may be accomplished as follows

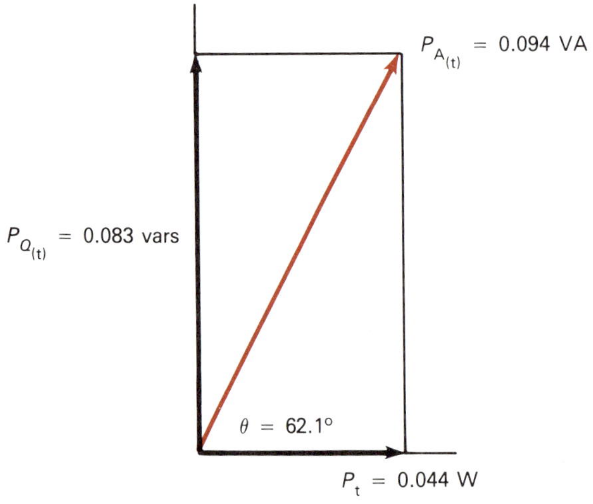

FIGURE 7–62

$$P_A^2 = P^2 + P_Q^2$$
$$P_A = \sqrt{P^2 + P_Q^2}$$
$$0.094 = \sqrt{(0.044)^2 + (0.083)^2}$$
$$0.094 = 0.094$$

Example: Compute the total power in watts, vars, and volt-amperes for the series *RCL* circuit in Figure 7–63.

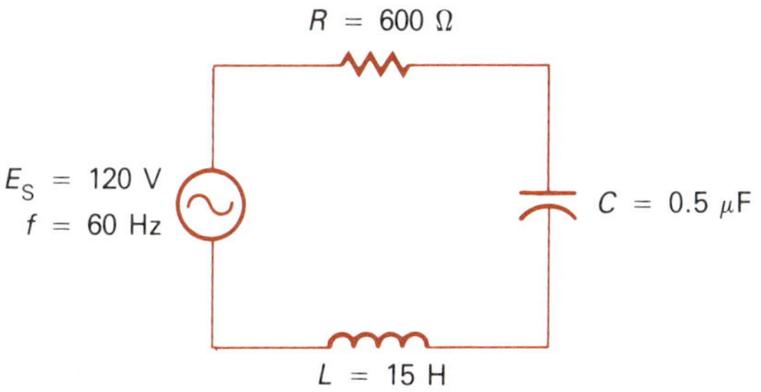

FIGURE 7–63

Solution:

$$X_L = 2\pi fL = 5\,654.87\ \Omega$$
$$X_C = \frac{1}{2\pi fC} = 5\,305.16\ \Omega$$
$$X = X_L - X_C = 5\,654.87\ \Omega - 5\,305.16\ \Omega = 349.7\ \Omega$$
$$Z = \sqrt{R^2 + X^2} = \sqrt{(600\ \Omega)^2 + (349.7\ \Omega)^2} = 694.47\ \Omega$$
$$I = \frac{E_S}{Z} = \frac{120\ \text{V}}{694.47\ \Omega} = 172.79\ \text{mA}$$
$$\tan\theta = \frac{X}{R} = \frac{349.7\ \Omega}{600\ \Omega} = 0.582\,8,\ \theta = 30.24°$$
$$P_t = EI\cos\theta = (120\ \text{V})(172.79\ \text{mA})(\cos 30.24°) = 17.91\ \text{W}$$
$$P_{Q(t)} = EI\sin\theta = (120\ \text{V})(172.79\ \text{mA})(\sin 30.24°) = 10.44\ \text{vars}$$
$$P_{A(t)} = EI = (120\ \text{V})(172.79\ \text{mA}) = 20.73\ \text{VA}$$

EXERCISE 7·3

Note that when performing multiple calculations, your answers may differ from the answer key due to rounding within the problem. Problems 1–10 are based upon the circuit in Figure 7–64. For each problem, calculate the missing value of voltage, current, or resistance.

1. $e = 120\ \text{V}\sin\omega t$, $R = 25\ \Omega$

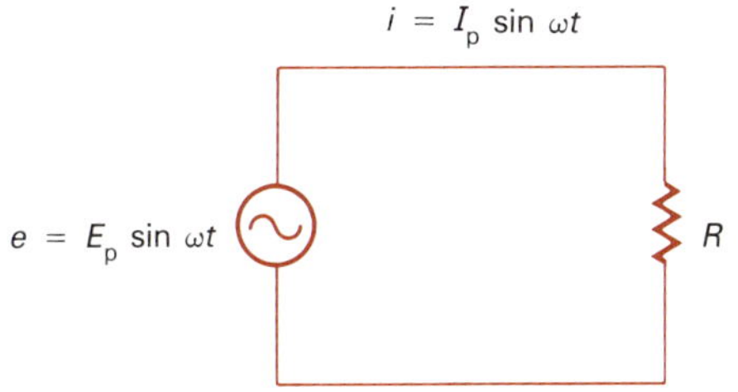

FIGURE 7–64

2. $i = 32\text{ mA} \sin \omega t,\ R = 5\text{ k}\Omega$
3. $e = 10\text{ V} \sin \omega t,\ i = 50\ \mu\text{A} \sin \omega t$
4. $i = 1.2\text{ A} \sin \omega t,\ R = 10\ \Omega$
5. $e = 45\text{ V} \sin (\omega t - 15°),\ i = 5\text{ mA} \sin (\omega t - 15°)$
6. $i = 65\ \mu\text{A} \sin (\omega t + 45°),\ R = 500\text{ k}\Omega$
7. $R = 2.7\text{ k}\Omega,\ e = 12.73\text{ V} \sin \omega t$
8. $R = 49\ \Omega,\ i = 0.8\text{ A} \sin (\omega t - 40°)$
9. $e = 15.8\text{ V} \sin \omega t,\ i = 1.2\text{ mA} \sin \omega t$
10. $i = 140\text{ mA} \sin (\omega t + 30°),\ e = 20\text{ V} \sin (\omega t + 30°)$

In problems 11–20, find the indicated circuit values using the configuration found in Figure 7–65.

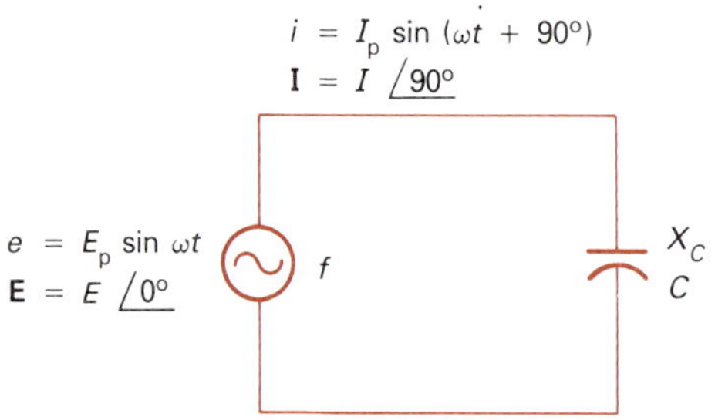

FIGURE 7–65

11. Given $f = 60$ Hz and $C = 10\ \mu$F. Find X_C.
12. Given $f = 120$ kHz and $C = 0.05\ \mu$F. Find X_C.
13. Given $f = 5$ MHz and $X_C = 63.67\ \Omega$. Find C.
14. Given $X_C = 35.37\ \Omega$ and $C = 0.1\ \mu$F. Find f.
15. Given $\mathbf{E} = 120\text{ V} \angle 0°$ and $i = 16.97\text{ A} \sin (\omega t + 90°)$. Find X_C.
16. Given $\mathbf{I} = 10\text{ mA} \angle 90°$, $e = 28.28\text{ V} \sin \omega t$, and $C = 80$ pF. Find X_C and f.
17. Given $C = 120$ pF, $\mathbf{I} = 0.2\text{ mA} \angle 90°$, and $f = 30$ kHz. Find X_C and $\mathbf{E_S}$.
18. Given $i = 79.88\text{ mA} \sin (188.5t + 90°)$ and $e = 21.21\text{ V} \sin 188.5t$. Find X_C, f, C, $\mathbf{E_S}$, and $\mathbf{I}$.
19. Given $\mathbf{E} = 10\text{ V} \angle 0°$, $X_C = 530.52\ \Omega$ and $f = 1.5$ MHz. Find $\mathbf{I}$ and C.
20. Given $e = 8.484\text{ V} \sin \omega t$, $f = 18$ kHz, and $C = 0.2\ \mu$F. Find X_C and i.

Problems 21–26 are based upon the circuit in Figure 7–66. For each problem, find the indicated values.

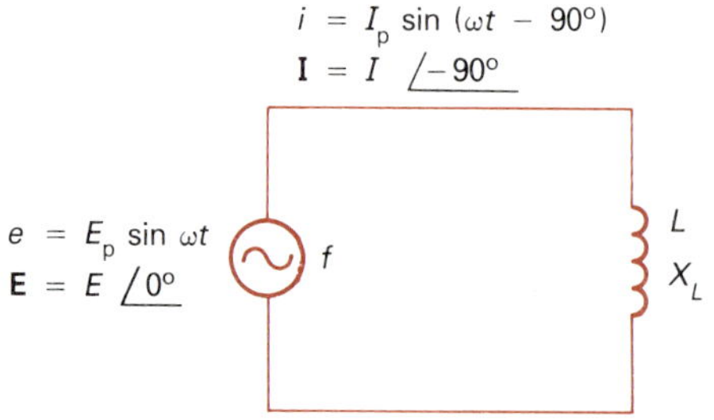

FIGURE 7–66

21. Given $\mathbf{E} = 120\text{ V} \angle 0°$, $L = 5$ H, and $i = 90.03\text{ mA}\sin(\omega t - 90°)$. Find X_L and f.
22. Given $f = 8$ MHz, $L = 20$ mH, and $\mathbf{E} = 5\text{ V} \angle 0°$. Find X_L and $\mathbf{I}$.
23. Given $X_L = 3.14\text{ k}\Omega$, $I = 3.82$ mA, and $f = 50$ kHz. Find E and L.
24. Given $E = 6$ V, $L = 0.5$ H, and $X_L = 4.71\text{ k}\Omega$. Find I and f.
25. Given $f = 400$ Hz, $L = 100$ mH, and $\mathbf{I} = 0.2\text{ A} \angle -90°$. Find $\mathbf{E}$ and X_L.
26. Given $L = 75\ \mu$H, $X_L = 117.81\ \Omega$, and $i = 0.29\text{ A}\sin(\omega t - 90°)$. Find e and f.

Problems 27–32 are based upon the series *RC* circuit in Figure 7–67. For each problem, find the indicated values.

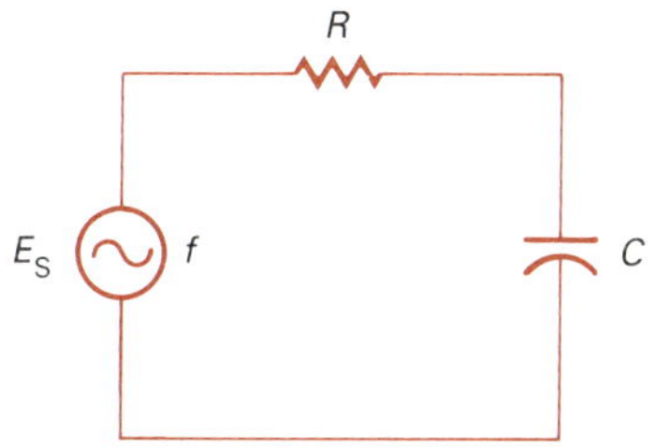

FIGURE 7–67

27. Given $E_S = 120$ V, $f = 60$ Hz, $R = 500\ \Omega$, and $C = 5\ \mu$F. Find X_C, $\mathbf{Z}$, $\mathbf{I}$, $\mathbf{E_R}$, and $\mathbf{E_C}$.
28. Given $E_S = 15$ V, $f = 20$ kHz, $\mathbf{Z} = 113.2\text{ k}\Omega \angle -27.95°$. Find R, X_C, C, $\mathbf{I}$, $\mathbf{E_R}$, and $\mathbf{E_C}$.
29. Given $\mathbf{I} = 84.32\text{ mA} \angle 32.46°$, $E_S = 0.5$ V, and $C = 0.005\ \mu$F. Find $\mathbf{Z}$, R, X_C, and f.
30. Given $\mathbf{Z} = 6.43\text{ k}\Omega \angle -45°$, $I = 1.55$ mA, and $C = 0.1\ \mu$F. Find R, X_C, E_S, and f.
31. Given $\mathbf{E_R} = 18.75\text{ V} \angle 38.66°$, $\mathbf{E_C} = 15\text{ V} \angle -51.34°$, and $\mathbf{I} = 18.75\ \mu\text{A} \angle 38.66°$. Find E_S, $\mathbf{Z}$, R, and X_C.
32. Given $Z = 165.94\ \Omega$, $X_C = 159.15\ \Omega$, $I = 54.24$ mA, and f = 50 kHz. Find R, E_S, and C.

Problems 33–38 are based upon the series *RL* circuit in Figure 7–68. For each problem, find the indicated values.

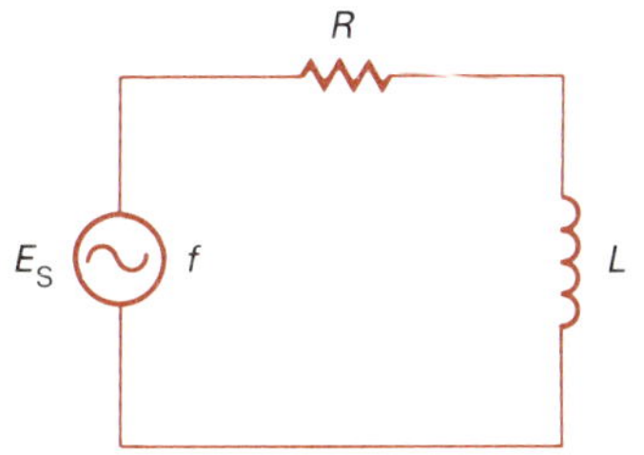

FIGURE 7–68

33. Given $E_S = 9$ V, $f = 75$ kHz, $R = 2.7\text{ k}\Omega$, and $L = 5$ mH. Find X_L, $\mathbf{Z}$, $\mathbf{I}$, $\mathbf{E_R}$, and $\mathbf{E_L}$.

34. Given $e_S = 84.84$ V sin 753.98t and $\mathbf{Z} = 250.48\ \Omega\ \angle 37.02^\circ$. Find E_S, X_L, f, R, **I**, $\mathbf{E_R}$, and $\mathbf{E_L}$.
35. Given $\mathbf{Z} = 9.47\ \text{k}\Omega\ \angle 83.94^\circ$, $\mathbf{I} = 0.53\ \text{mA}\ \angle -83.94^\circ$, and $f =$ 5 mHz. Find R, X_L, L, and E_S.
36. Given $Z = 9.05\ \text{k}\Omega$, $R = 5\ \text{k}\Omega$, $E_S = 18.1$ V, and $L = 120$ mH. Find X_L, I, and f.
37. Given $\mathbf{I} = 4.07\ \text{mA}\ \angle -42.91^\circ$, $E_S = 15$ V, and $f = 800$ Hz. Find **Z,** R, X_L, L, $\mathbf{E_R}$, and $\mathbf{E_L}$.
38. Given $\mathbf{E_R} = 2.81\ \text{V}\ \angle -64.89^\circ$, $\mathbf{E_L} = 5.3\ \text{V}\ \angle 25.11^\circ$, and $i =$ 39.76 μA sin $(\omega t - 64.89^\circ)$. Find E_S, **I, Z,** R, and X_L.

Problems 39–45 are based upon the series *RCL* circuit in Figure 7–69. For each problem, find the indicated values.

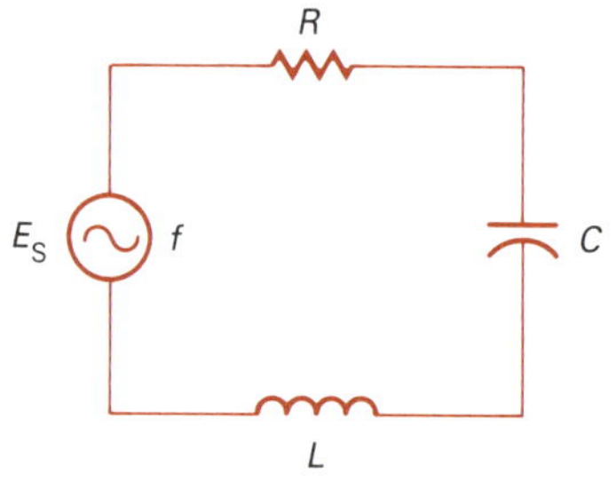

FIGURE 7–69

39. Given $E_S = 120$ V, $f = 60$ Hz, $R = 100\ \Omega$, $C = 2\ \mu\text{F}$, and $L =$ 4 H. Find X_C, X_L, X, **Z, I,** $\mathbf{E_R}$, $\mathbf{E_C}$, and $\mathbf{E_L}$.
40. Given $X_L = 23.56\ \text{k}\Omega$, $\mathbf{X} = 8.27\ \text{k}\Omega\ \angle -90^\circ$, $\mathbf{Z} = 12.98\ \text{k}\Omega\ \angle -39.59^\circ$, $E_S = 10$ V, and $L = 75$ mH. Find X_C, R, f, C, and **I.**
41. Given $\mathbf{I} = 10\ \text{mA}\ \angle -48.24^\circ$, $\mathbf{E_R} = 10\ \text{V}\ \angle -48.24^\circ$, $\mathbf{E_C} = 26.5\ \text{V}\ \angle -138.24^\circ$, $\mathbf{E_L} = 37.7\ \text{V}\ \angle 41.76^\circ$, and $C = 0.05\ \mu\text{F}$. Find R, X_C, X_L, f, L, **X, Z,** and E_S.
42. Given $f = 4.5$ MHz, $R = 25\ \Omega$, $\mathbf{E_C} = 3.24\ \text{V}\ \angle -130.27^\circ$, $\mathbf{E_R} =$ 2.29 V $\angle -40.27^\circ$ and $L = 2\ \mu\text{H}$. Find **I,** X_C, C, X_L, **X, Z,** $\mathbf{E_L}$, and E_S.
43. Given $R = 50\ \Omega$, $\mathbf{E_R} = 3.28\ \text{V}\ \angle 56.92^\circ$, $\mathbf{E_C} = 17.37\ \text{V}\ \angle -33.08^\circ$, $\mathbf{E_L} = 12.35\ \text{V}\ \angle 146.92^\circ$, and $f = 30$ kHz. Find E_S, **I,** X_C, X_L, C, L, **X,** and **Z.**
44. Given $E_S = 18$ V, $R = 2\ \text{k}\Omega$, C = 0.05 μF, L = 25 mH, and the circuit at resonance. Find f_r, X_C, X_L, X, Z, I, E_R, E_C, and E_L.
45. Given $f_r = 183.78$ kHz, $C = 150$ pF, $R = 3\ \text{k}\Omega$, and $E_L =$ 17.31 V. Find L, X_C, X_L, I, E_S, Z, E_C, and E_R.
46. For the series *RL* circuit in Figure 7–70, calculate total power in watts, vars, volt-amperes, and compute the power factor.
47. Compute total power in watts, vars, and volt-amperes for the circuit in Figure 7–70 if $R = 5\ \text{k}\Omega$ and $L = 10$ H.
48. For the series *RC* circuit in Figure 7–71, calculate total power in watts, vars, and volt-amperes. Also compute the power factor.

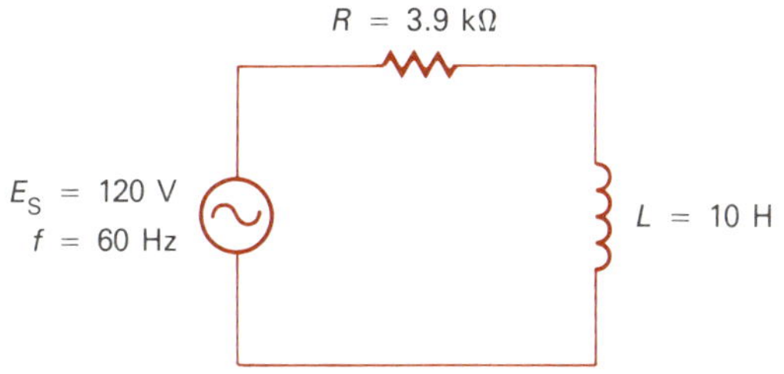

FIGURE 7–70

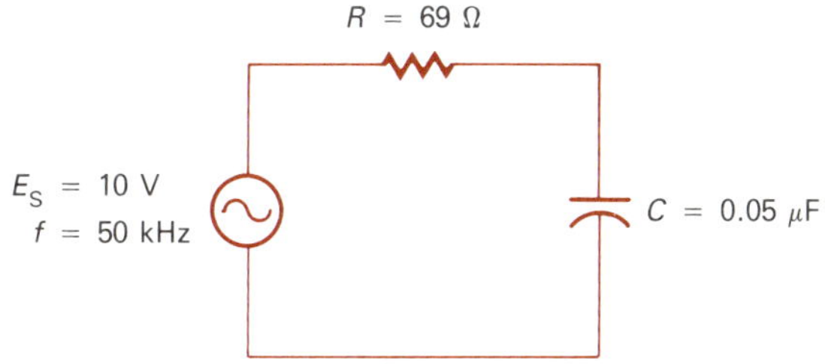

FIGURE 7–71

49. Find the total power in watts, vars, and volt-amperes for the series *RCL* circuit in Figure 7–72.

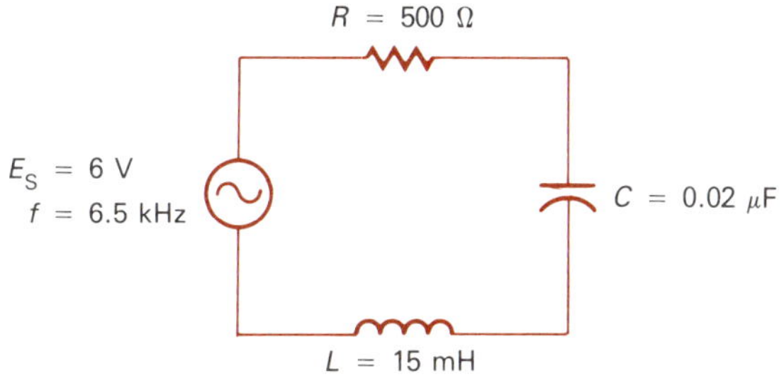

FIGURE 7–72

50. Find the total power in watts, vars, and volt-amperes for the circuit in Figure 7–72 if C is changed to 0.01 μF.

UNIT 7·4 PROBLEM REVIEW

In problems 1–5, find the wavelength of the given frequency.

1. 5.5 kHz
2. 120 kHz
3. 100 Hz
4. 2.5 MHz
5. 200 MHz

In problems 6–10, find the frequency for the given wavelength.

6. 3×10^3 m
7. 60 m
8. 20 cm
9. 5×10^6 m
10. 5×10^3 m

In problems 11–15, find the period of the given frequency.

11. 100 kHz
12. 25 MHz
13. 2.4 kHz
14. 500 MHz
15. 75 kHz

In problems 16–20, find the frequency for the given period.

16. 15 μs
17. 100 ps
18. 2 ms
19. 450 μs
20. 0.8 ms

In problems 21–25, convert the given value to the indicated value.

21. 20 $V_{p\text{-}p}$ to V
22. 60 mA_p to mA_{av}
23. 120 V to $V_{p\text{-}p}$
24. 15 V_{av} to $V_{p\text{-}p}$
25. 10 V to V_{av}

In problems 26–30, express the given voltage and frequency in angular velocity time notation.

26. 120 V, 60 Hz
27. 5 V, 400 Hz
28. 10 V_p, 5 kHz
29. 18 $V_{p\text{-}p}$, 750 Hz
30. 12 V, 50 kHz
31. What is the instantaneous voltage in problem 26 after 1 ms?
32. What is the instantaneous voltage in problem 27 after 250 μs?
33. What is the instantaneous voltage in problem 28 after 30 μs?
34. What is the instantaneous voltage in problem 29 after 0.1 ms?
35. What is the instantaneous voltage in problem 30 after 1.5 μs?
36. Find e after 10 μs for the waveform $e = E_p \sin(\omega t - 20°)$ if f = 10 kHz and E = 5 V.
37. Find i after 1 ms for the waveform $i = I_p \sin(\omega t + 50°)$ if f = 60 Hz and I = 1.2 A.
38. What is the instantaneous current in problem 37 after 150 μs?
39. What is the instantaneous current in problem 37 after 1.51 s?
40. Convert $\mathbf{E} = 20 \text{ V} \angle 65°$ to a time domain expression.
41. Convert $\mathbf{I} = 100 \text{ mA} \angle -20°$ to a time domain expression.
42. Convert $\mathbf{E} = 120 \text{ V} \angle 0°$ to a time domain expression.
43. Convert $e = 6 \text{ V} \sin \omega t$ to a phasor domain expression.

44. Convert $i = 5.5\ \mu A \sin(\omega t + 45°)$ to a phasor domain expression.
45. Convert $e = 15\ mV \sin(\omega t - 62°)$ to a phasor domain expression.

In problems 46–50, combine the two waveforms and express the composite waveform in phasor and time domain notation.

46. $e_1 = 10\ V \sin(\omega t + 45°)$
    $e_2 = 15\ V \sin \omega t$
47. $i_1 = 1.8\ mA \sin(\omega t + 10°)$
    $i_2 = 5\ mA \sin(\omega t - 15°)$
48. $e_1 = 40\ V \sin \omega t$
    $e_2 = 62\ V \sin(\omega t - 90°)$
49. $i_1 = 150\ \mu A \sin(\omega t + 90°)$
    $i_2 = 125\ \mu A \sin(\omega t - 90°)$
50. $e_1 = 9.8\ V \sin(\omega t + 30°)$
    $e_2 = 9.8\ V \sin(\omega t + 22°)$
51. What is the current expression for a circuit with a voltage of $e = 10\ V \sin(\omega t + 10°)$ and a resistance of 2 kΩ?
52. What is the voltage expression if the current through a 100-Ω resistor is $i = 2.5\ mA \sin(\omega t - 40°)$?
53. Given $f = 1.5$ MHz and $C = 150$ pF. Find X_C.
54. Given $f = 60$ Hz and $X_C = 500\ \Omega$. Find C.
55. Given $C = 0.5\ \mu F$ and $X_C = 1.2\ k\Omega$. Find f.
56. For the circuit in Figure 7–73, find **I** if $e = 14.14\ V \sin \omega t$, $f = 5$ kHz, and $C = 0.1\ \mu F$.

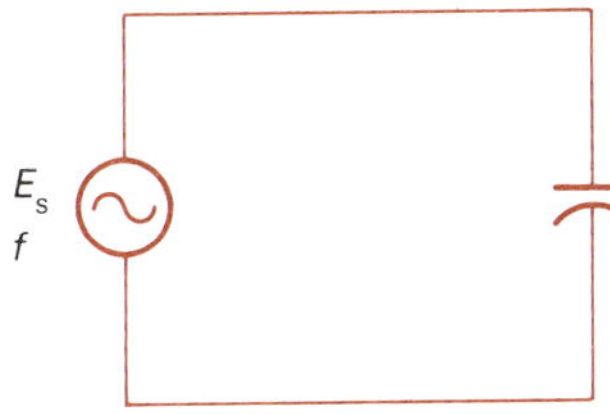

FIGURE 7–73

57. For the circuit in Figure 7–73, find i if $\mathbf{E} = 50\ V \angle 0°$, $f = 60$ Hz, and $C = 2\ \mu F$.
58. For the circuit in Figure 7–74, find e if $\mathbf{I} = 1.2\ mA \angle 90°$, $f = 12$ kHz, and $L = 10$ mH.

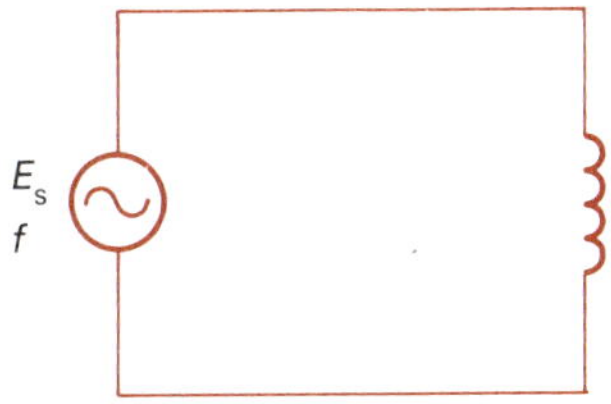

FIGURE 7–74

59. For the circuit in Figure 7–74, find i if $\mathbf{E} = 10\ V \angle 0°$, $f = 1$ MHz, and $L = 500\ \mu H$.

Note that when performing multiple calculations, your answers may differ from the answer key due to rounding within the problem. Problems 60–65 are based upon the circuit in Figure 7–75.

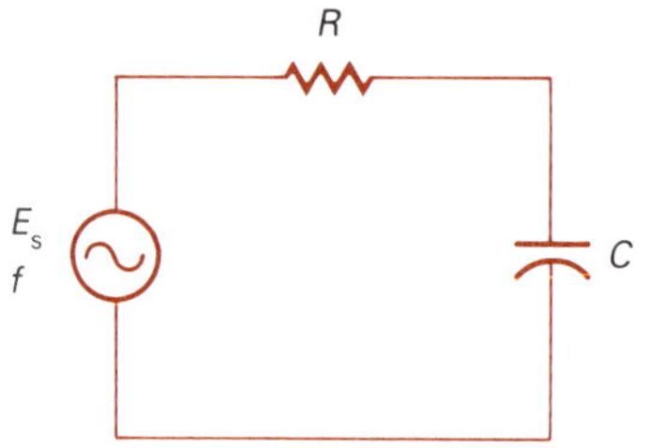

FIGURE 7–75

60. Given E_S = 16 V, f = 15 kHz, C = 0.03 μF, and R = 150 Ω. Find X_C, **Z, I,** E_R, and E_C.
61. Given E_R = 3.24 V, E_C = 3.82 V, f = 1 kHz and **I** = 1.2 mA $\angle 49.67°$. Find E_S, **Z,** R, X_C, and C.
62. Given $\mathbf{E_R}$ = 6.74 V $\angle 41.48°$, R = 50 kΩ, E_C = 5.96 V, and f = 24 kHz. Find **I,** E_S, **Z,** X_C, and C.
63. Given **I** = 545.95 μA $\angle 24.47°$, E_S = 6 V, and C = 0.01 μF. Find **Z,** X_C, R, E_R, E_C, and f.
64. Given $\mathbf{E_C}$ = 112.39 V $\angle -20.66°$, f = 60 Hz, C = 2 μF, and R = 500 Ω. Find X_C, **I,** $\mathbf{E_R}$, E_S, and **Z.**
65. Given Z = 10.13 kΩ, X_C = 1.59 kΩ, I = 1.48 mA, and f = 2 MHz. Find R, E_S, and C.

Problems 66–70 are based upon the circuit in Figure 7–76.

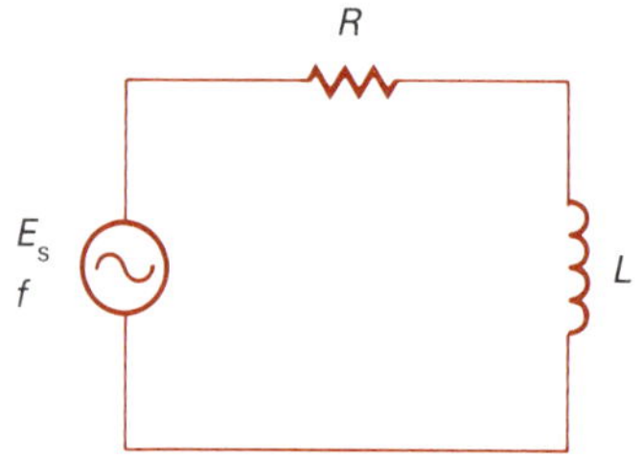

FIGURE 7–76

66. Given f = 150 kHz, L = 4 mH, R = 5 kΩ, and E_S = 10 V. Find X_L, **Z, I,** $\mathbf{E_R}$, and $\mathbf{E_L}$.
67. Given **Z** = 904.7 Ω $\angle 56.45°$, E_R = 66.32 V, and f = 60 Hz. Find X_L, R, **I,** E_L, E_S, and L.
68. Given $\mathbf{E_R}$ = 6.36 V $\angle -45.06°$, E_L = 6.37 V, R = 4.7 kΩ and L = 30 mH. Find E_S, **I, Z,** X_L, and f.
69. Given $\mathbf{E_L}$ = 9.67 V $\angle 14.86°$, f = 400 Hz, L = 1.5 H, and R = 1 kΩ. Find X_L, **I,** $\mathbf{E_R}$, E_S, and **Z.**
70. Given Z = 25.43 kΩ, X_L = 15.71 kΩ, I = 196.62 μA, and f = 2.5 MHz. Find R, E_S, and L.

Problems 71–77 are based upon the circuit in Figure 7–77.

71. Given E_S = 6 V, f = 100 kHz, R = 50 Ω, C = 0.02 μF and L = 200 μH. Find X_C, X_L, **X, Z, I,** $\mathbf{E_R}$, $\mathbf{E_C}$, and $\mathbf{E_L}$.

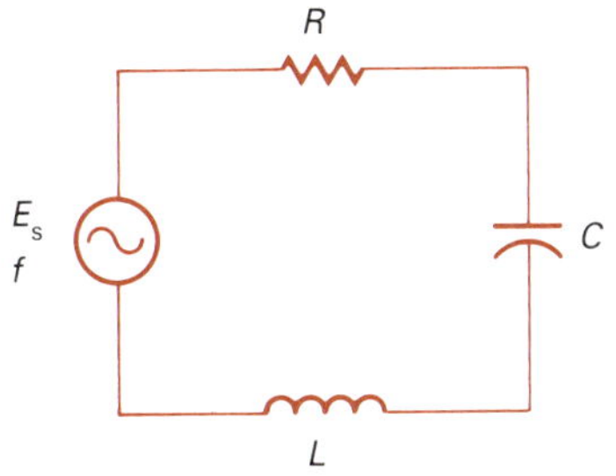

FIGURE 7–77

72. Given $\mathbf{Z}$ = 68.47 Ω $\angle -68.58°$, f = 18 kHz, L = 1 mH, and $\mathbf{I}$ = 131.44 mA $\angle 68.58°$. Find R, $\mathbf{X}$, X_L, X_C, E_S, C, $\mathbf{E_R}$, $\mathbf{E_C}$, and $\mathbf{E_L}$.
73. Given X_L = 23.56 kΩ, $\mathbf{X}$ = 2.34 kΩ $\angle 90°$, $\mathbf{Z}$ = 2.78 kΩ $\angle 57.33°$, E_S = 10 V, and L = 5 H. Find X_C, R, f, C, and $\mathbf{I}$.
74. Given $\mathbf{E_R}$ = 0.64 V $\angle -77.76°$, $\mathbf{E_C}$ = 5.06 V $\angle -167.76°$, $\mathbf{E_L}$ = 7.99 V $\angle 12.24°$, and R = 100 Ω. Find E_S, $\mathbf{I}$, $\mathbf{Z}$, and $\mathbf{X}$.
75. Given f = 1.4 kHz, R = 10 Ω, $\mathbf{E_C}$ = 646.53 mV $\angle -108.57°$, $\mathbf{E_R}$ = 11.37 mV $\angle -18.57°$ and L = 65 mH. Find $\mathbf{I}$, X_C, C, X_L, $\mathbf{X}$, $\mathbf{Z}$, $\mathbf{E_L}$, and E_S.
76. Given E_S = 5 V, C = 0.5 μF, L = 15 mH, R = 2 kΩ, and the circuit at resonance. Find f_r, X_C, X_L, X, Z, I, E_R, E_C, and E_L.
77. Given f_r = 318.31 kHz and L = 0.5 mH. Find C.
78. Compute total power in watts, vars, and volt-amperes for the circuit in Figure 7–78.

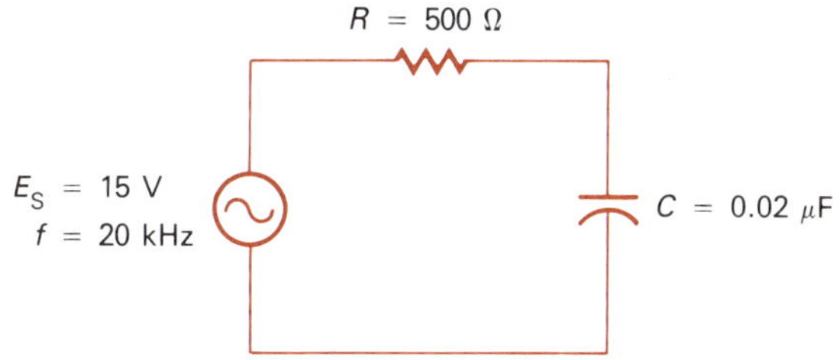

FIGURE 7–78

79. What is the power factor for the circuit in Figure 7–78?
80. Compute total power in watts, vars, and volt-amperes for the circuit in Figure 7–79.

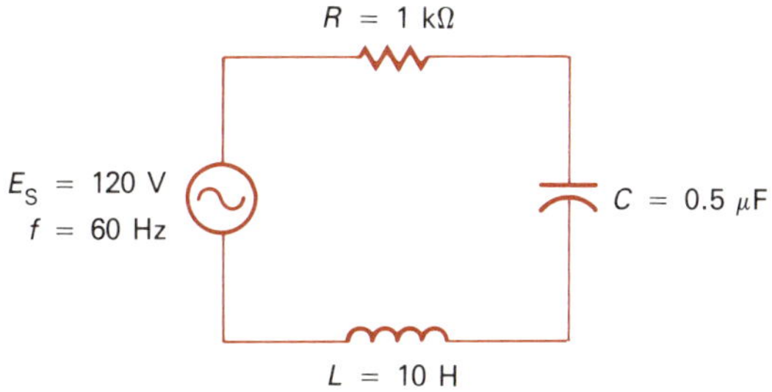

FIGURE 7–79

CHAPTER 8

TRIGONOMETRY FOR ALTERNATING CURRENT CIRCUITS IN PARALLEL

Applying the trigonometric fundamentals presented in previous chapters will help the modern electronics technician compute current and voltage values in complex circuits. Parallel alternating current circuits and series alternating circuits react differently. Through trigonometric calculations, it can be shown how these complex circuits function.

UNIT 8·1 ALTERNATING CURRENT CIRCUITS IN PARALLEL

Objectives:

After studying this unit, you should be able to

- calculate voltage, current, and impedance values in parallel ac circuits.
- calculate the resonant frequency for parallel ac circuits.
- calculate the Q factor of a parallel resonant circuit.
- calculate the line current in a parallel resonant circuit.

Parallel Resistive Circuits

The circuit in Figure 8–1 shows an ac power source connected to two resistors in parallel. As in series ac circuits, the voltages and currents in a purely resistive parallel circuit are in phase. The following discussions are presented in order to lay a foundation for reactive parallel circuits.

In series circuits, current is the same at any point in the circuit, and volt-

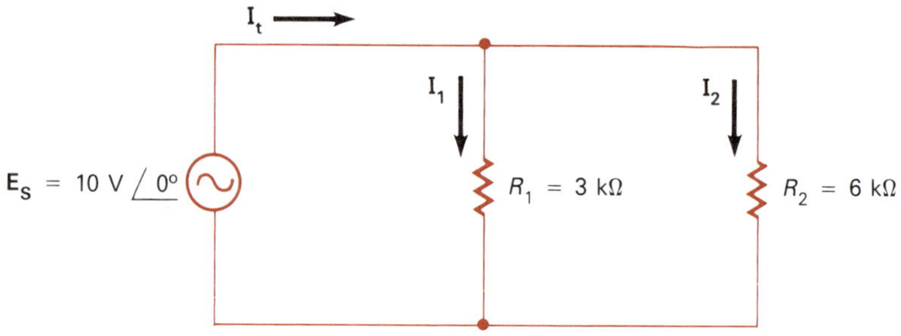

FIGURE 8–1

ages vary depending upon each component's resistance or reactance. In parallel circuits, the voltage across each branch is the same, and branch currents vary depending upon each component's resistance or reactance.

Example: For the circuit in Figure 8–1, find (a) the branch currents, (b) the total current, and (c) the total resistance.

Solution: a. $\mathbf{I_1} = \dfrac{\mathbf{E_S}}{\mathbf{R_1}} = \dfrac{10\text{ V}\angle 0°}{3\text{ k}\Omega\angle 0°} = 3.33\text{ mA}\angle 0°$

$\mathbf{I_2} = \dfrac{\mathbf{E_S}}{\mathbf{R_2}} = \dfrac{10\text{ V}\angle 0°}{6\text{ k}\Omega\angle 0°} = 1.67\text{ mA}\angle 0°$

b. $\mathbf{I_t} = \mathbf{I_1} + \mathbf{I_2} = 3.33\text{ mA}\angle 0° + 1.67\text{ mA}\angle 0° = 5\text{ mA}\angle 0°$

or, in terms of rectangular notation

$\mathbf{I_t} = (3.33 + j0)\text{ mA} + (1.67 + j0)\text{ mA} = (5 + j0)\text{ mA} = 5\text{ mA}\angle 0°$

c. $\mathbf{Z_t} = \mathbf{R_t} = \dfrac{\mathbf{E_S}}{\mathbf{I_t}} = \dfrac{10\text{ V}\angle 0°}{5\text{ mA}\angle 0°} = 2\text{ k}\Omega$

Parallel *RC* Circuits

The schematic in Figure 8–2 shows a parallel *RC* circuit. The voltages across *R* and *C* are the same and, therefore, are in phase. Current through the resistor is in phase with the voltage across the resistor. Since current through the capacitor leads the applied voltage by 90°, it will also lead current through the resistor by 90°.

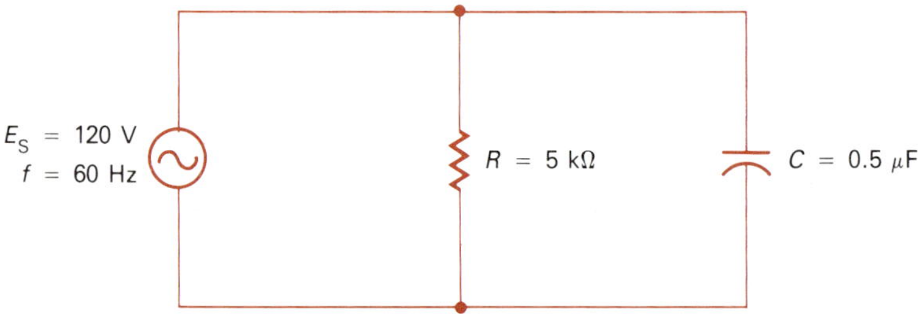

FIGURE 8–2

Total current delivered by the source is dependent upon total circuit impedance. The phase angle between applied voltage and total current is dependent upon the magnitude of the resistance and capacitive reactance. If X_C is much greater than R then the angle will be close to zero degrees. On the other hand, if R is much larger than X_C then the angle will approach ninety degrees. This will be easier to comprehend by examining Figure 8–2. When R is very large the majority of circuit current will flow through the capacitor. At this point the circuit appears capacitive, and the phase angle approaches ninety degrees. When X_C is greater than R, the majority of the total current flows through the resistor, the circuit appears resistive, and the phase angle is closer to zero degrees.

Example: For the circuit in Figure 8–2, calculate $\mathbf{X_C}$, $\mathbf{I_C}$, $\mathbf{I_R}$, $\mathbf{I_t}$, and **Z.** Also, graph the time domain expressions for the applied voltage and circuit currents.

Solution:

$$\mathbf{X_C} = \frac{1}{2\pi f C}\angle -90^\circ$$
$$= \frac{1}{(2)(\pi)(60\text{ Hz})(0.5 \times 10^{-6}\text{ F})}\angle -90^\circ$$
$$= 5.31\text{ k}\Omega \angle -90^\circ$$

$$\mathbf{I_C} = \frac{\mathbf{E_S}}{\mathbf{X_C}} = \frac{120\text{ V}\angle 0^\circ}{5.31\text{ k}\Omega\angle -90^\circ} = 22.6\text{ mA}\angle 90^\circ$$

$$\mathbf{I_R} = \frac{\mathbf{E_S}}{\mathbf{R}} = \frac{120\text{ V}\angle 0^\circ}{5\text{ k}\Omega\angle 0^\circ} = 24\text{ mA}\angle 0^\circ$$

$$\mathbf{I_t} = \mathbf{I_C} + \mathbf{I_R}$$
$$= 22.6\text{ mA}\angle 90^\circ + 24\text{ mA}\angle 0^\circ$$
$$= (0 + j22.6)\text{ mA} + (24 + j0)\text{ mA}$$
$$= (24 + j22.6)\text{ mA}$$
$$= 32.97\text{ mA}\angle 43.28^\circ$$

$$\mathbf{Z} = \frac{\mathbf{E_S}}{\mathbf{I_t}} = \frac{120\text{ V}\angle 0^\circ}{32.97\text{ mA}\angle 43.28^\circ} = 3.64\text{ k}\Omega\angle -43.28^\circ$$

The time domain expressions for the source voltage and circuit currents are

$e_S = 169.68\text{ V}\sin \omega t$

$i_R = 33.94\text{ mA}\sin \omega t$

$i_C = 31.96\text{ mA}\sin(\omega t + 90^\circ)$

$i_t = 46.62\text{ mA}\sin(\omega t + 43.28^\circ)$

These waveforms are graphed in Figure 8–3.

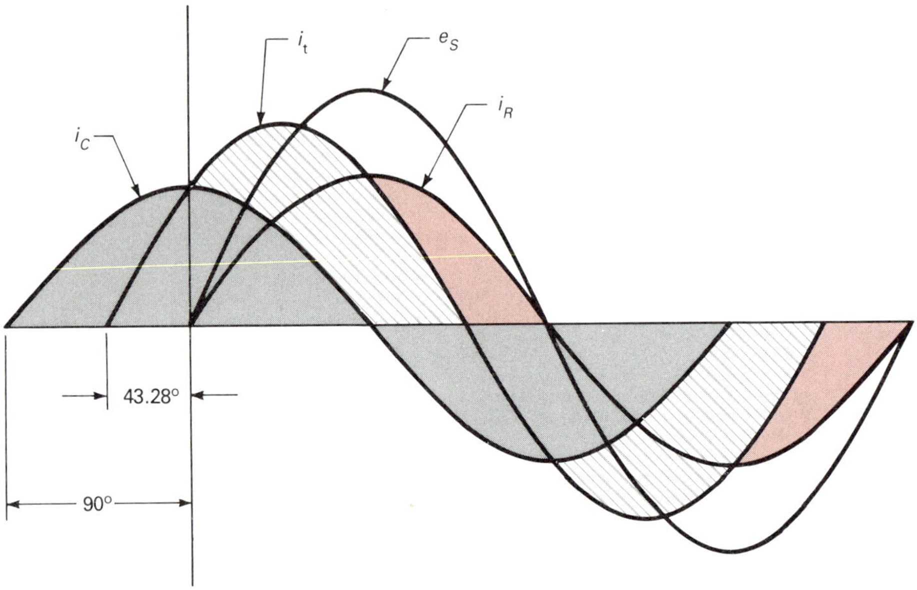

FIGURE 8–3

Example: For the circuit in Figure 8–4, calculate

a. $\mathbf{X_C}$ b. $\mathbf{I_C}$ c. $\mathbf{I_R}$ d. $\mathbf{I_t}$ e. $\mathbf{Z}$

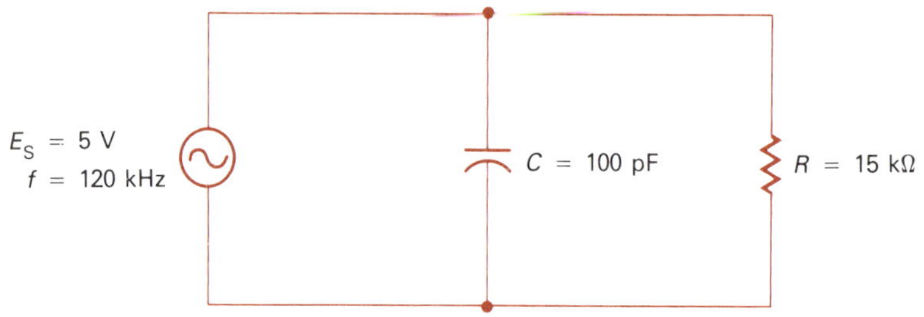

FIGURE 8–4

Solution:

a. $\mathbf{X_C} = \frac{1}{2\pi fC} \angle -90° = 13.26 \text{ k}\Omega \angle -90°$

b. $\mathbf{I_C} = \frac{\mathbf{E_S}}{\mathbf{X_C}} = \frac{5 \text{ V} \angle 0°}{13.26 \text{ k}\Omega \angle -90°} = 0.38 \text{ mA} \angle 90°$

c. $\mathbf{I_R} = \frac{\mathbf{E_S}}{\mathbf{R}} = \frac{5 \text{ V} \angle 0°}{15 \text{ k}\Omega \angle 0°} = 0.33 \text{ mA} \angle 0°$

d. $\mathbf{I_t} = \mathbf{I_C} + \mathbf{I_R}$

$= 0.38 \text{ mA} \angle 90° + 0.33 \text{ mA} \angle 0°$

$= (0 + j0.38) \text{ mA} + (0.33 + j0) \text{ mA}$

$= (0.33 + j0.38) \text{ mA}$

$= 0.5 \text{ mA} \angle 49.03°$

e. $\mathbf{Z} = \frac{\mathbf{E_S}}{\mathbf{I_t}} = \frac{5 \text{ V} \angle 0°}{0.5 \text{ mA} \angle 49.03°} = 10 \text{ k}\Omega \angle -49.03°$

In a previous chapter, an equation was developed to calculate total resistance in a parallel circuit containing two resistances. A similar expression for two parallel impedances is

$$\mathbf{Z}_t = \frac{\mathbf{Z}_1 \times \mathbf{Z}_2}{\mathbf{Z}_1 + \mathbf{Z}_2} \qquad (8\text{–}1)$$

where, $\mathbf{Z}_t$ = phasor expression for total impedance
$\mathbf{Z}_1$ = impedance for branch one
$\mathbf{Z}_2$ = impedance for branch two

Example: Find the total impedance for the circuit in Figure 8–4.

Solution:

$$\mathbf{Z}_1 = \mathbf{X}_C = \frac{1}{2\pi f C}\angle -90^\circ = 13.26\ \text{k}\Omega \angle -90^\circ$$

$$\mathbf{Z}_2 = \mathbf{R} = 15\ \text{k}\Omega \angle 0^\circ$$

$$\mathbf{Z}_t = \frac{\mathbf{Z}_1 \times \mathbf{Z}_2}{\mathbf{Z}_1 + \mathbf{Z}_2}$$

$$= \frac{(13.26\ \text{k}\Omega \angle -90^\circ)(15\ \text{k}\Omega \angle 0^\circ)}{(13.26\ \text{k}\Omega \angle -90^\circ) + (15\ \text{k}\Omega \angle 0^\circ)}$$

$$= \frac{198.9 \angle -90^\circ}{(0 - j13.26) + (15 + j0)}$$

$$= \frac{198.9 \angle -90^\circ}{15 - j13.26}$$

$$= \frac{198.9 \angle -90^\circ}{20.02 \angle -41.48^\circ}$$

$$= 9.94\ \text{k}\Omega \angle -48.52^\circ$$

Note in the previous example that the impedance value differed slightly from the value calculated previously. This is again due to the rounding of various values during the computation process.

Parallel *RL* Circuits

The schematic in Figure 8–5 shows a parallel *RL* circuit. Again, the voltage is the same across both components with the current through the resistor being in phase with the source voltage. In the coil current lags voltage by ninety

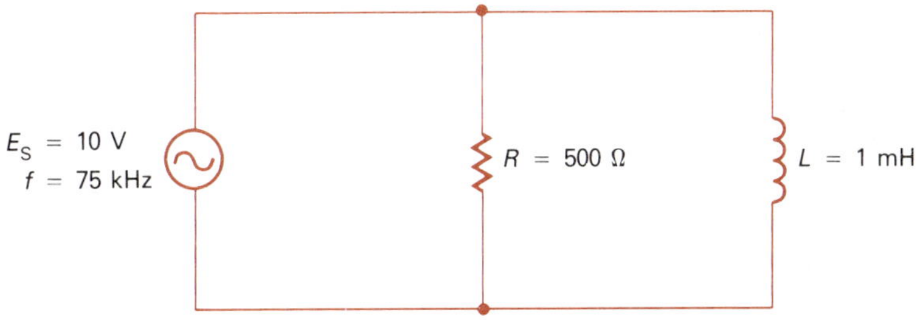

FIGURE 8–5

degrees. This results in coil current lagging resistor current by ninety degrees. Total circuit current is the vector sum of these currents as shown in Figure 8–6.

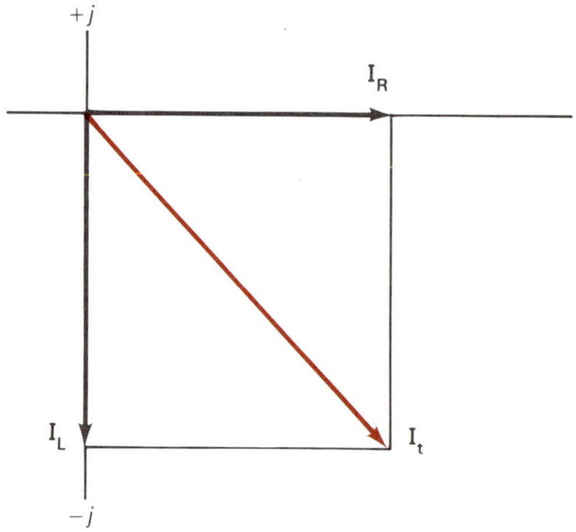

FIGURE 8–6

Example: For the circuit in Figure 8–5, compute
a. $\mathbf{X_L}$ b. $\mathbf{I_L}$ c. $\mathbf{I_R}$ d. $\mathbf{I_t}$ e. $\mathbf{Z}$

Solution:
a. $\mathbf{X_L} = 2\pi f L \angle 90°$
$= (2)(\pi)(75 \times 10^3 \text{ Hz})(1 \times 10^{-3} \text{ H}) \angle 90°$
$= 471.24\ \Omega \angle 90°$

b. $\mathbf{I_L} = \dfrac{\mathbf{E_S}}{\mathbf{X_L}} = \dfrac{10 \text{ V} \angle 0°}{471.24\ \Omega \angle 90°} = 21.22 \text{ mA} \angle -90°$

c. $\mathbf{I_R} = \dfrac{\mathbf{E_S}}{\mathbf{R}} = \dfrac{10 \text{ V} \angle 0°}{500\ \Omega \angle 0°} = 20 \text{ mA} \angle 0°$

d. $\mathbf{I_t} = \mathbf{I_L} + \mathbf{I_R}$
$= 21.22 \text{ mA} \angle -90° + 20 \text{ mA} \angle 0°$
$= (0 - j21.22) \text{ mA} + (20 + j0) \text{ mA}$
$= (20 - j21.22) \text{ mA}$
$= 29.16 \text{ mA} \angle -46.7°$

e. $\mathbf{Z} = \dfrac{\mathbf{E_S}}{\mathbf{I_t}} = \dfrac{10 \text{ V} \angle 0°}{29.16 \text{ mA} \angle -46.7°} = 342.94\ \Omega \angle 46.7°$

Example: For the circuit in Figure 8–7, find
a. $\mathbf{I_t}$ c. $\mathbf{I_L}$ e. L
b. $\mathbf{I_R}$ d. $\mathbf{X_L}$ f. P_t

Solution:
a. $\mathbf{I_t} = \dfrac{\mathbf{E_S}}{\mathbf{Z}} = \dfrac{9 \text{ V} \angle 0°}{2.44 \text{ k}\Omega \angle 51.17°} = 3.69 \text{ mA} \angle -51.17°$

b. $\mathbf{I_R} = \dfrac{\mathbf{E_S}}{\mathbf{R}} = \dfrac{9 \text{ V} \angle 0°}{3.9 \text{ k}\Omega \angle 0°} = 2.31 \text{ mA} \angle 0°$

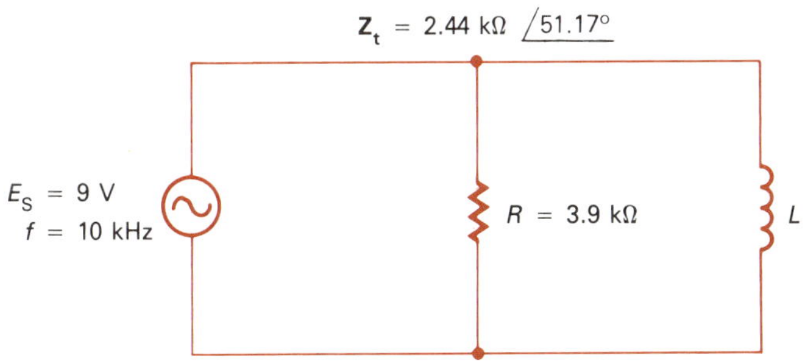

FIGURE 8–7

c. $\mathbf{I_L} = \mathbf{I_t} - \mathbf{I_R}$

$$= 3.69 \text{ mA } \angle -51.17° - 2.31 \text{ mA } \angle 0°$$
$$= (2.31 - j2.87) \text{ mA} - (2.31 + j0) \text{ mA}$$
$$= (0 - j2.87) \text{ mA}$$
$$= 2.87 \text{ mA } \angle -90°$$

d. $$\mathbf{X_L} = \frac{\mathbf{E_S}}{\mathbf{I_L}} = \frac{9 \text{ V } \angle 0°}{2.87 \text{ mA } \angle -90°} = 3.14 \text{ k}\Omega \angle 90°$$

e. $$L = \frac{X_L}{2\pi f} = \frac{3.14 \text{ k}\Omega}{(2)(\pi)(10 \times 10^3 \text{ Hz})} = 50 \text{ mH}$$

f. $P_t = EI \cos \theta = (9 \text{ V})(3.69 \text{ mA})(\cos -51.17°) = 20.82 \text{ mW}$

Parallel *RCL* Circuits

The schematic shown in Figure 8–8 is a parallel *RCL* circuit. The mathematical analysis of this configuration is similar to previous examples. As these three components are connected in parallel, the applied voltage will be the same across each branch at any point in time. Current through the resistor will be in phase with the applied voltage while the capacitor current will *lead* resistor current by ninety degrees. Current through the inductor will *lag* resistor current by ninety degrees.

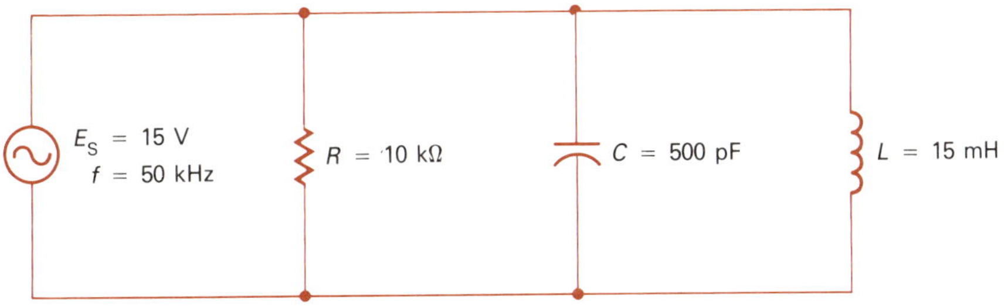

FIGURE 8–8

The phasor diagram in Figure 8–9 shows the relationship among the currents and the applied voltage in a parallel *RCL* circuit. Examining the diagram, it is apparent that X_L is less than X_C so that the coil conducts more current. The net current delivered to the parallel combination of L and C is I_X. This

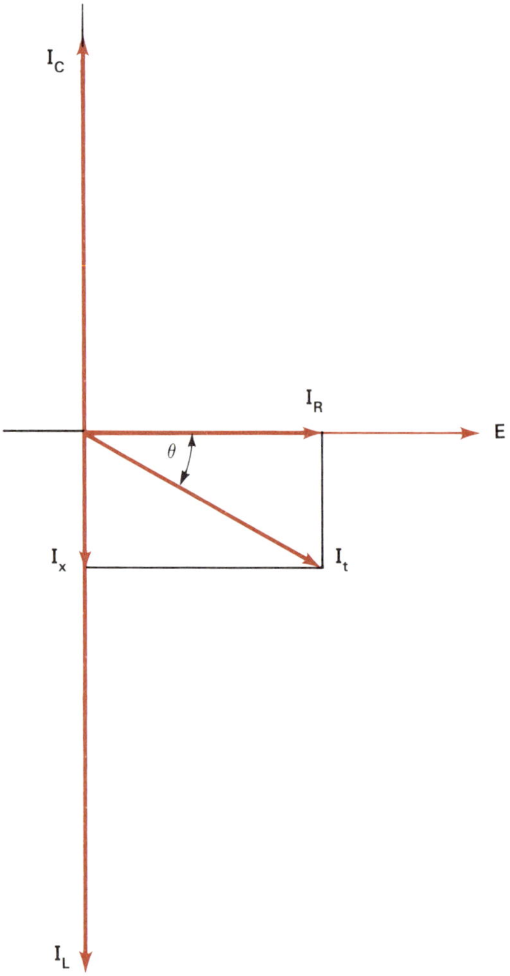

FIGURE 8–9

current drawn by the *net* reactance is the difference between I_C and I_L. For the example in Figure 8–9, this net current is inductive and is combined with I_R to form total circuit current.

At first this may seem confusing in that currents larger than I_X are obviously moving through the coil and capacitor. This is possible due to the exchange of current between the capacitor and inductor. As the capacitor discharges through the coil a magnetic field is produced. After the capacitor is discharged, the inductor's field begins to collapse and recharge the capacitor. As this current *oscillates* back-and-forth the only additional current required is I_X. The special case when $X_L = X_C$ will be discussed later in this unit.

Example: For the circuit in Figure 8–8, calculate

a. $\mathbf{X_C}$ c. $\mathbf{I_R}$ e. $\mathbf{I_L}$ g. $\mathbf{I_t}$
b. $\mathbf{X_L}$ d. $\mathbf{I_C}$ f. $\mathbf{I_X}$ h. $\mathbf{Z}$

Solution:

a. $\mathbf{X_C} = \dfrac{1}{2\pi f C} \angle -90^\circ$

$= \dfrac{1}{(2)(\pi)(50 \times 10^3 \text{ Hz})(500 \times 10^{-12} \text{ F})} \angle -90^\circ$

$= 6.37 \text{ k}\Omega \angle -90^\circ$

b. $\mathbf{X_L} = 2\pi f L \angle 90^\circ$

$= (2)(\pi)(50 \times 10^3 \text{ Hz})(15 \times 10^{-3} \text{ H}) \angle 90^\circ$

$= 4.71 \text{ k}\Omega \angle 90^\circ$

c. $\mathbf{I_R} = \dfrac{\mathbf{E_S}}{\mathbf{R}} = \dfrac{15 \text{ V} \angle 0^\circ}{10 \text{ k}\Omega \angle 0^\circ} = 1.5 \text{ mA} \angle 0^\circ$

d. $\mathbf{I_C} = \dfrac{\mathbf{E_S}}{\mathbf{X_C}} = \dfrac{15 \text{ V} \angle 0^\circ}{6.37 \text{ k}\Omega \angle -90^\circ} = 2.35 \text{ mA} \angle 90^\circ$

e. $\mathbf{I_L} = \dfrac{\mathbf{E_S}}{\mathbf{X_L}} = \dfrac{15 \text{ V} \angle 0^\circ}{4.71 \text{ k}\Omega \angle 90^\circ} = 3.18 \text{ mA} \angle -90^\circ$

f. $\mathbf{I_X} = \mathbf{I_L} - \mathbf{I_C}$

$= 3.18 \text{ mA} \angle -90^\circ - 2.35 \text{ mA} \angle 90^\circ$

$= 0.83 \text{ mA} \angle -90^\circ$

g. $\mathbf{I_t} = \mathbf{I_R} + \mathbf{I_X}$

$= 1.5 \text{ mA} \angle 0^\circ + 0.83 \text{ mA} \angle -90^\circ$

$= (1.5 + j0) \text{ mA} + (0 - j0.83) \text{ mA}$

$= (1.5 - j0.83) \text{ mA}$

$= 1.71 \text{ mA} \angle -28.96^\circ$

h. $\mathbf{Z} = \dfrac{\mathbf{E_S}}{\mathbf{I_t}} = \dfrac{15 \text{ V} \angle 0^\circ}{1.71 \text{ mA} \angle -28.96^\circ} = 8.77 \text{ k}\Omega \angle 28.96^\circ$

Example: For the circuit in Figure 8–10, calculate

a. $\mathbf{X_C}$ c. $\mathbf{I_R}$ e. $\mathbf{I_L}$ g. $\mathbf{I_t}$

b. $\mathbf{X_L}$ d. $\mathbf{I_C}$ f. $\mathbf{I_X}$ h. $\mathbf{Z}$

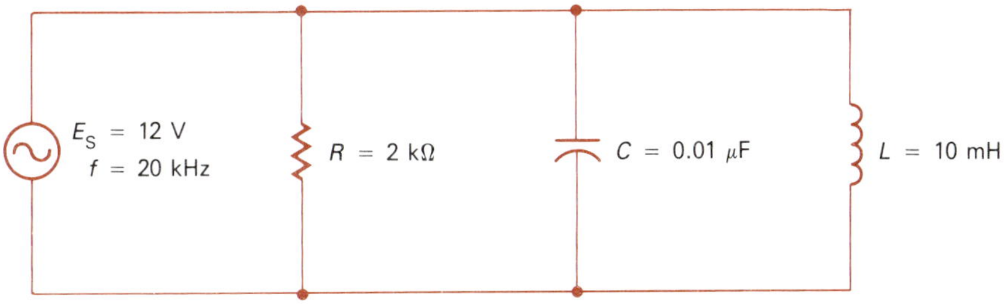

FIGURE 8–10

Solution:

a. $\mathbf{X_C} = \dfrac{1}{2\pi f C} \angle -90^\circ = 795.77\ \Omega \angle -90^\circ$

b. $\mathbf{X_L} = 2\pi f L \angle 90^\circ = 1\,256.64\ \Omega \angle 90^\circ$

c. $\mathbf{I_R} = \dfrac{\mathbf{E_S}}{\mathbf{R}} = \dfrac{12 \text{ V} \angle 0^\circ}{2 \text{ k}\Omega \angle 0^\circ} = 6 \text{ mA} \angle 0^\circ$

d. $\mathbf{I_C} = \dfrac{\mathbf{E_S}}{\mathbf{X_C}} = \dfrac{12\text{ V}\ \angle 0°}{795.77\ \Omega\ \angle -90°} = 15.08\text{ mA}\ \angle 90°$

e. $\mathbf{I_L} = \dfrac{\mathbf{E_S}}{\mathbf{X_L}} = \dfrac{12\text{ V}\ \angle 0°}{1\ 256.64\ \Omega\ \angle 90°} = 9.55\text{ mA}\ \angle -90°$

f. $\mathbf{I_X} = \mathbf{I_C} - \mathbf{I_L}$
$= 15.08\text{ mA}\ \angle 90° - 9.55\text{ mA}\ \angle -90°$
$= 5.53\text{ mA}\ \angle 90°$

g. $\mathbf{I_t} = \mathbf{I_R} + \mathbf{I_X}$
$= 6\text{ mA}\ \angle 0° + 5.53\text{ mA}\ \angle 90°$
$= (6 + j0)\text{ mA} + (0 + j5.53)\text{ mA}$
$= (6 + j5.53)\text{ mA}$
$= 8.16\text{ mA}\ \angle 42.67°$

h. $\mathbf{Z} = \dfrac{\mathbf{E_S}}{\mathbf{I_t}} = \dfrac{12\text{ V}\ \angle 0°}{8.16\text{ mA}\ \angle 42.67°} = 1.47\text{ k}\Omega\ \angle -42.67°$

Example: For the circuit in Figure 8–11, calculate

a. $\mathbf{I_R}$ c. E_S e. $\mathbf{I_L}$ g. $\mathbf{X_C}$ i. $\mathbf{Z}$
b. $\mathbf{I_X}$ d. $\mathbf{X_L}$ f. $\mathbf{I_C}$ h. C

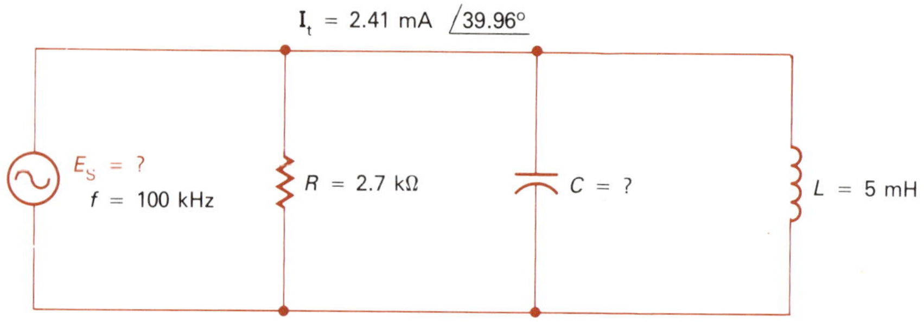

FIGURE 8–11

Solution: a. Since $\mathbf{I_t} = \mathbf{I_R} + \mathbf{I_X}$ then

$\mathbf{I_t} = 2.41\text{ mA}\ \angle 39.96°$ (Polar notation)
$= (1.85 + j1.55)\text{ mA}$ (Rectangular notation)
$= \mathbf{I_R} + \mathbf{I_X}$
$= 1.85\text{ mA}\ \angle 0° + 1.55\text{ mA}\ \angle 90°$ (Converting each term to polar)

$\mathbf{I_R} = 1.85\text{ mA}\ \angle 0°$

b. $\mathbf{I_X} = 1.55\text{ mA}\ \angle 90°$ (So that I_C is greater than I_L)

c. $E_S = I_R R = (1.85\text{ mA})(2.7\text{ k}\Omega) = 5\text{ V}$

d. $\mathbf{X_L} = 2\pi f L\ \angle 90°$
$= (2)(\pi)(100 \times 10^3\text{ Hz})(5 \times 10^{-3}\text{ H})\ \angle 90°$
$= 3.14\text{ k}\Omega\ \angle 90°$

e. $\mathbf{I_L} = \dfrac{\mathbf{E_S}}{\mathbf{X_L}} = \dfrac{5\text{ V}\ \angle 0°}{3.14\text{ k}\Omega\ \angle 90°} = 1.59\text{ mA}\ \angle -90°$

f. $I_X = I_C - I_L$

$I_C = I_X + I_L = 1.55 \text{ mA} + 1.59 \text{ mA} = 3.14 \text{ mA}$

$\mathbf{I_C} = 3.14 \text{ mA} \angle 90°$

g. $\mathbf{X_C} = \dfrac{\mathbf{E_S}}{\mathbf{I_C}} = \dfrac{5 \text{ V} \angle 0°}{3.14 \text{ mA} \angle 90°} = 1.59 \text{ k}\Omega \angle -90°$

h. $C = \dfrac{1}{2\pi f X_C} = \dfrac{1}{(2)(\pi)(100 \times 10^3 \text{ Hz})(1.59 \text{ k}\Omega)} = 0.001 \ \mu\text{F}$

i. $\mathbf{Z} = \dfrac{\mathbf{E_S}}{\mathbf{I_t}} = \dfrac{5 \text{ V} \angle 0°}{2.41 \text{ mA} \angle 39.96°} = 2.07 \text{ k}\Omega \angle -39.96°$

Ideal Parallel Resonance

The circuit shown in Figure 8–12 is an ideal parallel resonant circuit. This circuit, often called a *tank* circuit, contains no parallel resistance. At very low frequencies the inductor's reactance is very low, the source delivers a specific amount of current and the circuit appears inductive. At extremely high frequencies, the capacitive reactance is very low, the source delivers current, and the circuit appears capacitive.

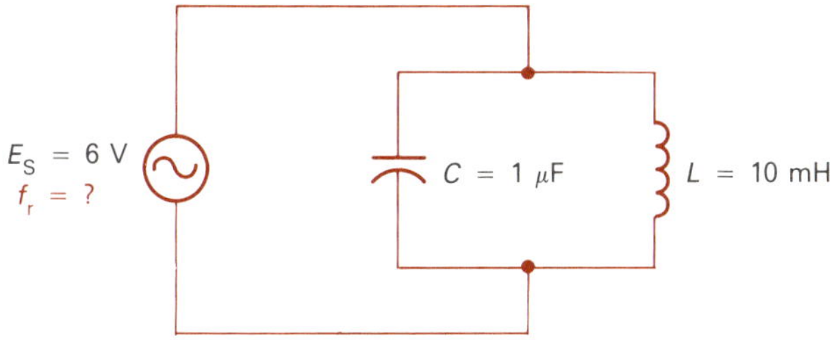

FIGURE 8–12

Resonance is defined as the frequency at which $X_L = X_C$. At this frequency (f_r) the current through the capacitor (I_C) is 180° out of phase with inductor current (I_L). Both of these currents are out of phase with the applied voltage.

The ideal resonant circuit will oscillate by converting the electrostatic charge stored in the capacitor to a current, which creates the electromagnetic field in the inductor. The field surrounding the inductor then collapses, developing a current to once again charge the capacitor. As this is an ideal circuit, there is no loss of current and the tank oscillates indefinitely.

Calculating the resonant frequency of a tank circuit is the same as in a series resonant circuit. That is

$$f_r = \frac{1}{2\pi\sqrt{LC}} \tag{8–2}$$

Example: For the circuit in Figure 8–12, calculate

a. f_r c. $\mathbf{X_C}$ e. $\mathbf{I_C}$ g. Z

b. $\mathbf{X_L}$ d. $\mathbf{I_L}$ f. I_X

Solution: a. $f_r = \dfrac{1}{2\pi\sqrt{LC}}$

$$= \frac{1}{(2)(\pi)\sqrt{(10 \times 10^{-3}\ \text{H})(1 \times 10^{-6}\ \text{F})}}$$

$$= 1\,591.55\ \text{Hz}$$

b. $\mathbf{X_L} = 2\pi fL\ \angle 90^\circ = 100\ \Omega\ \angle 90^\circ$

c. $\mathbf{X_C} = \dfrac{1}{2\pi fC}\angle -90^\circ = 100\ \Omega\ \angle -90^\circ$

d. $\mathbf{I_L} = \dfrac{\mathbf{E_S}}{\mathbf{X_L}} = \dfrac{6\ \text{V}\ \angle 0^\circ}{100\ \Omega\ \angle 90^\circ} = 60\ \text{mA}\ \angle -90^\circ$

e. $\mathbf{I_C} = \dfrac{\mathbf{E_S}}{\mathbf{X_C}} = \dfrac{6\ \text{V}\ \angle 0^\circ}{100\ \Omega\ \angle -90^\circ} = 60\ \text{mA}\ \angle 90^\circ$

f. $I_X = I_L - I_C = 60\ \text{mA} - 60\ \text{mA} = 0\ \text{mA}$

g. $Z = \dfrac{E_S}{I_X} = \dfrac{6\ \text{V}}{0\ \text{mA}} = \text{undefined}$

Note that, at this resonance, there are relatively large currents circulating in the tank, yet the source delivers no current.

Example: The frequency of an oscillator tank is to be 1 MHz. If a 100-pF capacitor is available, what value of inductance is needed?

Solution:

$$f_r = \frac{1}{2\pi\sqrt{LC}}$$

$$L = \frac{1}{(2\pi f_r)^2 C}$$

$$= \frac{1}{[(2)(\pi)(1 \times 10^6\ \text{Hz})]^2(100 \times 10^{-12}\ \text{F})}$$

$$= 253.3\ \mu\text{H}$$

Practical Parallel Resonance

In reality it is difficult to design an ideal parallel resonant circuit. This is due to the small resistance which is found in the wire used to construct the inductor. The circuit in Figure 8–13 shows the addition of this resistance to the inductive side of the tank. This added resistance will decrease the inductive current phase angle slightly, resulting in a small *line current* (I_{line}) from the source to the tank to sustain oscillations. The phasor diagram in Figure 8–14 shows this relationship. Note that the line current is in phase with the applied voltage.

The figure of merit (Q) of a parallel resonant circuit is an indication of the quality of the circuit. This quantity is the ratio of X_L to R. Hence,

$$Q = \frac{X_L}{R} \qquad (8\text{–}3)$$

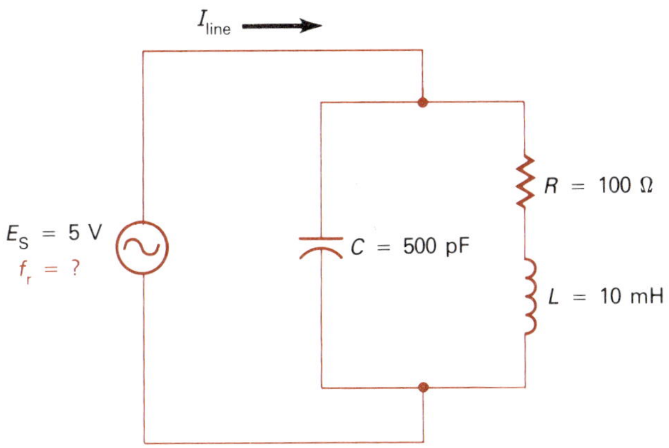

FIGURE 8–13

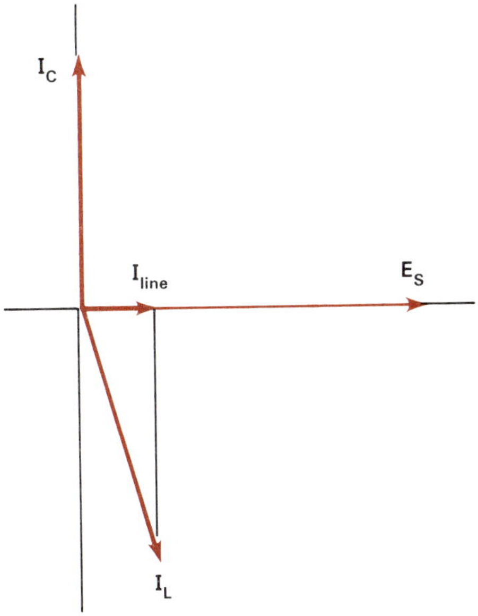

FIGURE 8–14

A high quality circuit would contain a very small amount of resistance when compared to the inductive reactance. This circuit would therefore conduct small amounts of line current. A low quality circuit would conduct larger amounts of line current. For circuits with Q values greater than ten, the effect of R is minimal.

There are a number of methods available to calculate circuit impedance. The following derivation will introduce them.

With a two-branch circuit the total impedance is

$$Z_t = \frac{Z_1 \times Z_2}{Z_1 + Z_2}$$

If Z_1 refers to the capacitive branch, then $Z_1 = X_C$. Assuming that Q is greater than ten, R can be ignored and $Z_2 = X_L$. This results in

$$Z_t = \frac{X_C \times X_L}{X_C + X_L}$$

At resonance $X_C = X_L$, with the phasors being 180° out of phase. When these are summed the result is zero, therefore leaving resistance. The expression for impedance then becomes

$$Z_t = \frac{X_C \times X_L}{R}$$

or,

$$Z_t = X_C \times \frac{X_L}{R}$$

Since $Q = \frac{X_L}{R}$, the impedance expression becomes

$$Z_t = QX_C$$

and since $X_C = X_L$

$$\mathbf{Z_t = QX_L} \qquad \textbf{(8–4)}$$

Substituting the expressions for X_L and X_C into this equation

$$Z_t = \frac{X_C \times X_L}{R} = \frac{2\pi fL}{2\pi fC \times R}$$

therefore,

$$\mathbf{Z_t = \frac{L}{CR}} \qquad \textbf{(8–5)}$$

A third expression for total circuit impedance is

$$\mathbf{Z_t = \frac{E_S}{I_{line}}} \qquad \textbf{(8–6)}$$

Example: For the circuit in Figure 8–13, calculate
a. f_r b. X_L c. Q d. Z_t e. I_{line}

Solution: a. $f_r = \dfrac{1}{2\pi\sqrt{LC}} = 71\ 176.25\ \text{Hz}$

b. $X_L = 2\pi fL = 4\ 472.14\ \Omega$

c. $Q = \dfrac{X_L}{R} = \dfrac{4\ 472.14\ \Omega}{100\ \Omega} = 44.72$

d. $Z_t = QX_L = (44.72)(4\ 472.14\ \Omega) = 200\ \text{k}\Omega$

Verifying this value of Z,

$$Z_t = \frac{L}{CR} = \frac{(10 \times 10^{-3}\ \text{H})}{(500 \times 10^{-12}\ \text{F})(100\ \Omega)} = 200\ \text{k}\Omega$$

e. $I_{\text{line}} = \dfrac{E_S}{Z_t} = \dfrac{5\ \text{V}}{200\ \text{k}\Omega} = 25\ \mu\text{A}$

Example: A 50-mH coil has a resistance of 75 Ω. It is in parallel with a 0.02-μF capacitor. Find

a. f_r b. X_L c. Q d. Z_t

Solution: a. $f_r = \dfrac{1}{2\pi\sqrt{LC}} = 5\ 032.92\ \text{Hz}$

b. $X_L = 2\pi fL = 1\ 581.14\ \Omega$

c. $Q = \dfrac{X_L}{R} = \dfrac{1\ 581.14\ \Omega}{75\ \Omega} = 21.08$

d. $Z_t = QX_L = (21.08)(1\ 581.14\ \Omega) = 33.33\ \text{k}\Omega$

EXERCISE 8·1

Note that when performing multiple calculations, your answers may differ from the answer key due to rounding within the problem. Problems 1–6 are based upon the circuit in Figure 8–15.

1. Given $E_S = 120$ V, $f = 60$ Hz, $C = 0.01$ μF, and $R = 200$ kΩ. Find $\mathbf{X_C}$, $\mathbf{I_C}$, $\mathbf{I_R}$, $\mathbf{I_t}$, and **Z.**
2. Given $\mathbf{Z} = 41.39\ \text{k}\Omega\ \angle -65.55°$, $R = 100$ kΩ, $I_R = 120$ μA, and $C = 100$ pF. Find E_S, $\mathbf{I_t}$, $\mathbf{I_C}$, X_C, and f.
3. Given $\mathbf{I_t} = 1.18\ \text{mA}\ \angle 32.21°$, $R = 5$ kΩ, and $f = 10$ kHz. Find I_C, I_R, E_S, X_C, and C.
4. Given $E_S = 9$ V, $I_R = 27.272$ mA, $X_C = 397.89$ Ω, and $C = 1$ μF. Find R, I_C, $\mathbf{I_t}$, **Z** and f.

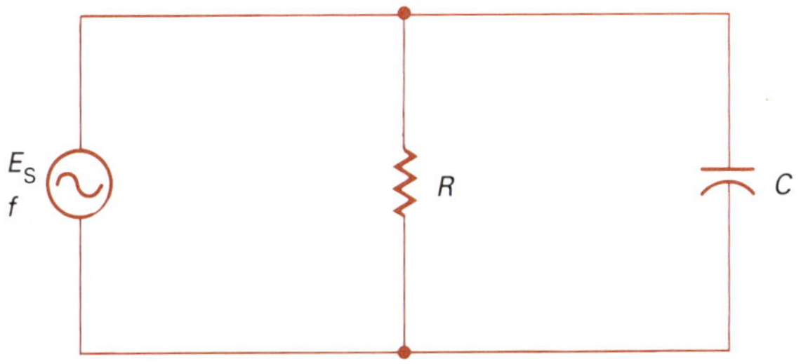

FIGURE 8–15

5. Given $\mathbf{Z} = 99.7\ \Omega \angle -4.5°$, $E_S = 10$ V, $R = 100\ \Omega$, and $f = 2.5$ kHz. Find $\mathbf{I_t}$, I_R, I_C, X_C, and C.
6. Given $\mathbf{Z} = 2.25\ \text{k}\Omega \angle -45°$ and $E_S = 15$ V. Find $\mathbf{I_t}$, I_C, I_R, R and X_C.

Problems 7–12 are based upon the circuit in Figure 8–16.

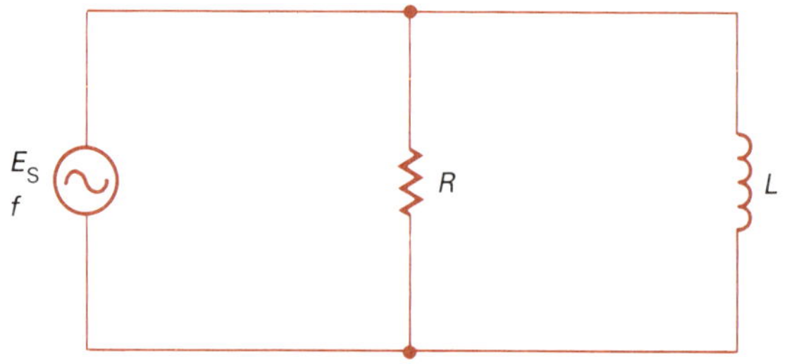

FIGURE 8–16

7. Given $E_S = 5$ V, $f = 350$ kHz, $L = 12\ \mu$H, and $R = 50\ \Omega$. Find $\mathbf{X_L}$, $\mathbf{I_L}$, $\mathbf{I_R}$, $\mathbf{I_t}$, and $\mathbf{Z}$.
8. Given $\mathbf{I_t} = 76.68\ \mu\text{A} \angle -38.51°$, $R = 100$ kΩ, and $f = 50$ kHz. Find I_L, I_R, E_S, X_L, and L.
9. Given $\mathbf{Z} = 374.58\ \Omega \angle 41.48°$, $R = 500\ \Omega$, $I_R = 240$ mA, and $L = 1.5$ H. Find E_S, $\mathbf{I_t}$, I_L, X_L, and f.
10. Given $Z = 4.72$ kΩ, $E_S = 10$ V, $R = 5$ kΩ, and $f = 455$ kHz. Find I_t, I_R, I_L, X_L, and L.
11. Given $E_S = 9$ V, $I_R = 9$ mA, $X_L = 942.48\ \Omega$, and $L = 20$ mH. Find R, I_L, $\mathbf{I_t}$, $\mathbf{Z}$, and f.
12. Given $\mathbf{Z} = 204.77\ \Omega \angle 40.68°$ and $E_S = 6$ V. Find $\mathbf{I_t}$, I_L, I_R, R, and X_L.

Problems 13–16 are based upon the circuit in Figure 8–17.

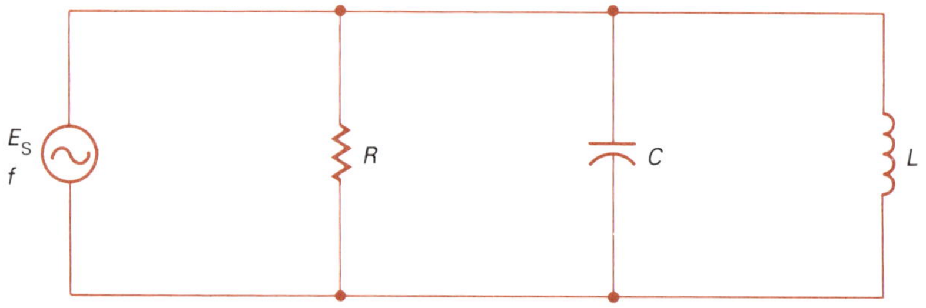

FIGURE 8–17

13. Given $E_S = 5$ V, $f = 20$ kHz, $R = 1$ kΩ, $C = 0.02\ \mu$F, and $L = 5$ mH. Find X_C, X_L, I_R, I_C, I_L, I_X, $\mathbf{I_t}$, and $\mathbf{Z}$.
14. Given $\mathbf{I_t} = 25.08\ \mu\text{A} \angle -53.27°$, $R = 400$ kΩ, $f = 100$ kHz, and $C = 20$ pF. Find I_R, $\mathbf{I_X}$, E_S, X_C, I_C, I_L, X_L, L, and $\mathbf{Z}$.

15. Given E_S = 10 V, R = 50 Ω, $\mathbf{I_X}$ = 107.84 mA $\angle -90°$, X_C = 167.53 Ω and f = 9.5 kHz. Find I_R, $\mathbf{I_t}$, $\mathbf{Z}$, X_L, I_C, I_L, C, and L.
16. Given $\mathbf{Z}$ = 3.7 kΩ $\angle -68.26°$ and E_S = 12 V. Find $\mathbf{I_t}$, I_R, I_X, and R.

Problems 17–20 are based upon the resonant tank circuit in Figure 8–18.

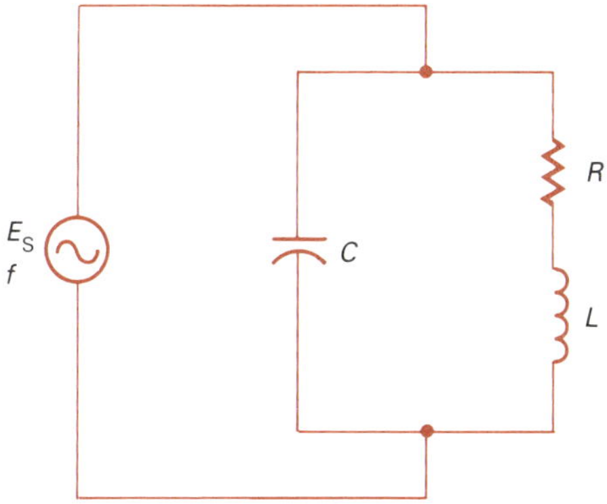

FIGURE 8–18

17. Given E_S = 9 V, R = 100 Ω, C = 0.02 μF, and L = 50 mH. Find f_r, X_L, Q, Z, and I_{line}.
18. Given E_S = 5 V, R = 1 000 Ω, C = 50 pF, and L = 15 mH. Find f_r, X_L, Q, Z, and I_{line}.
19. Given L = 0.1 H, f_r = 1 591.55 Hz, R = 10 Ω, and I_{line} = 100 μA. Find C, Z, E_S, X_L, and Q.
20. Given X_L = 5 kΩ, f_r = 31.83 kHz, R = 1 kΩ, and E_S = 10 V. Find L, C, Z, Q, and I_{line}.

UNIT 8·2 PROBLEM REVIEW

Note that when performing multiple calculations, your answers may differ from the answer key due to rounding within the problem. Problems 1–3 are based upon the circuit in Figure 8–19.

1. Given E_S = 12 V, R = 200 Ω, f = 10 kHz, and C = 0.08 μF. Find X_C, I_R, I_C, $\mathbf{I_t}$, and $\mathbf{Z}$.

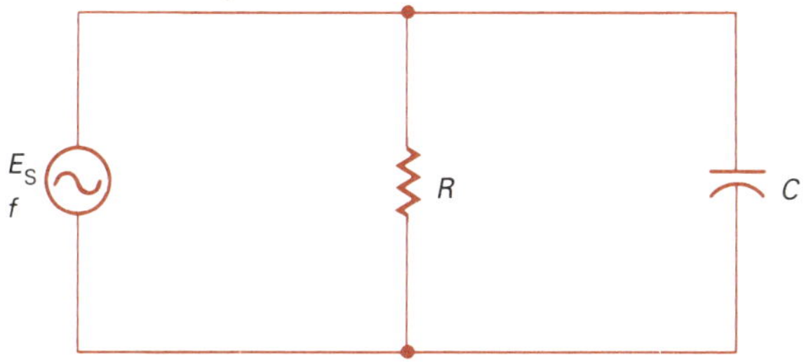

FIGURE 8–19

2. Given $\mathbf{I_t}$ = 164.03 mA $\angle 22.52°$, R = 33 Ω, and f = 500 Hz. Find I_R, I_C, E_S, **Z**, X_C, and C.
3. Given **Z** = 1.28 kΩ $\angle -75.13°$ and E_S = 3 V. Find $\mathbf{I_t}$, I_C, I_R, R, and X_C.

Problems 4–5 are based upon the circuit in Figure 8–20.

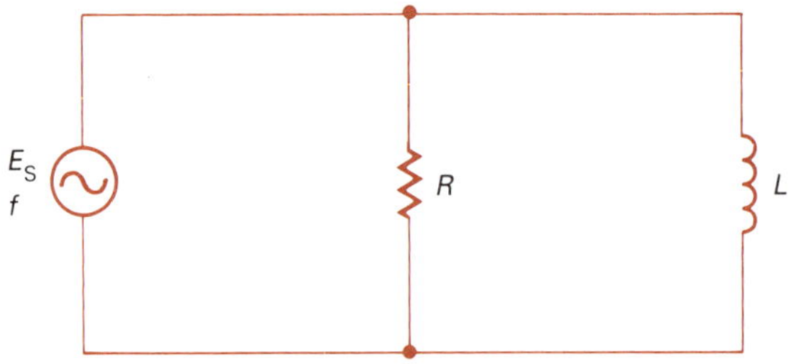

FIGURE 8–20

4. Given E_S = 120 V, f = 60 Hz, L = 2 H, and R = 500 Ω. Find X_L, I_R, I_L, $\mathbf{I_t}$, and **Z**.
5. Given **Z** = 9.67 kΩ $\angle 14.86°$ and E_S = 6 V. Find $\mathbf{I_t}$, I_L, I_R, R, and X_L.

Problems 6–10 are based upon the circuit in Figure 8–21.

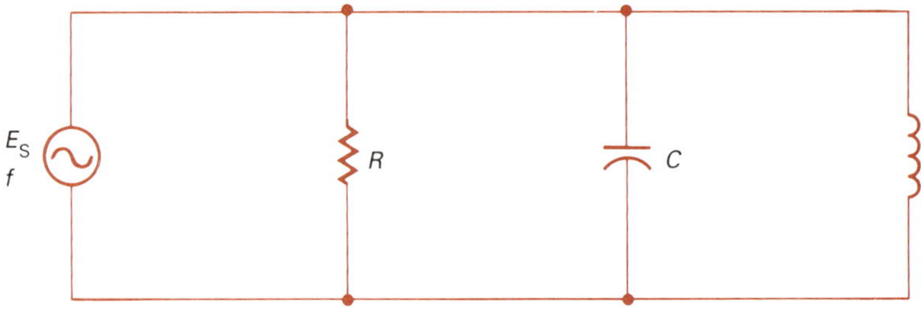

FIGURE 8–21

6. Given E_S = 9 V, f = 70 kHz, R = 1 kΩ, C = 0.005 μF, and L = 2 mH. Find I_R, I_C, I_L, $\mathbf{I_X}$, $\mathbf{I_t}$, and **Z**.
7. Given R = 5 kΩ, $\mathbf{I_t}$ = 2.57 mA $\angle -38.83°$, f = 500 kHz, and C = 50 pF. Find I_R, $\mathbf{I_X}$, E_S, **Z**, X_C, I_C, I_L, X_L, and L.
8. Given **Z** = 21.43 kΩ $\angle -43.53°$ and E_S = 6 V. Find $\mathbf{I_t}$, I_R, I_X, and R.
9. Given E_S = 12 V, I_C = 30.16 mA, f = 400 Hz, R = 500 Ω, and L = 0.1 H. Find I_R, X_C, X_L, I_L, $\mathbf{I_X}$, C, $\mathbf{I_t}$, and **Z**.
10. Given I_R = 3 mA, I_C = 3.39 mA, X_C = 884.19 Ω, and **Z** = 970.87 Ω $\angle -13.86°$. Find E_S, $\mathbf{I_t}$, I_X, I_L, and X_L.

Problems 11–15 are based upon the resonant tank circuit in Figure 8–22.

11. Given E_S = 6 V, R = 25 Ω, C = 0.5 μF, and L = 75 mH. Find f_r, X_L, Q, Z, and I_{line}.

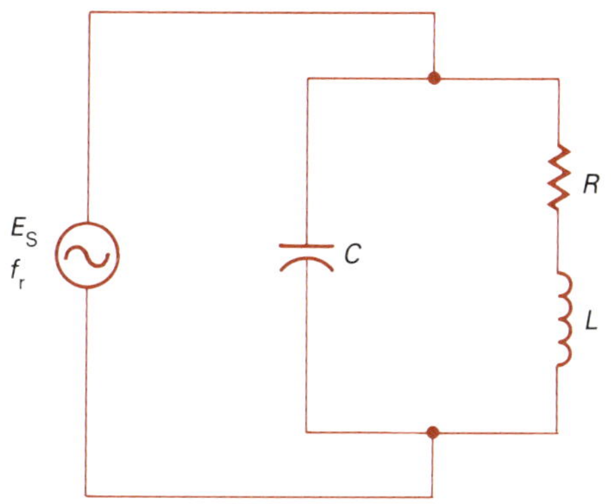

FIGURE 8–22

12. Given $E_S = 9$ V, $Q = 8.16$, $C = 150$ pF, and $L = 25$ μH. Find f_r, X_L, R, Z, and I_{line}.
13. Given $C = 0.002$ μF, $f_r = 35.59$ kHz, $R = 350$ Ω, and $I_{line} = 348.68$ μA. Find L, X_L, Q, E_S, and Z.
14. Given $E_S = 10$ V, $L = 10$ mH, $C = 0.01$ μF, and $R = 20$ Ω. Find Z and I_{line}.
15. Given $Z = 2$ kΩ, $L = 0.5$ mH, and $R = 5$ Ω. Find C.

CHAPTER 9

COMPUTER MATHEMATICS: BINARY NUMBER SYSTEM

As our world hurtles toward the twenty-first century, we are being inundated with computers, calculators, and a myriad of communications systems. Central to the advances in these areas is the world of digital electronics.

Thus far in our study of mathematics for electronics, we have utilized the decimal number system. The modern electronics technician, in order to study digital electronics, must also acquire an understanding of the mathematics used in computer circuits.

UNIT 9·1 THE BINARY NUMBER SYSTEM

Objectives:

After studying this unit, you should be able to

- identify binary numbers.
- convert binary numbers to decimal numbers.
- convert decimal numbers to binary numbers.

Binary Numbers

The decimal number system is a system based on 10, using symbols from 0 through 9 to represent each of the numbers in that system. A *binary* system, as the name implies, uses only two symbols, 0 and 1. The base or radix of the binary system is 2.

Every binary number, no matter what its value, consists only of symbols 0 and 1. Since the binary system uses two symbols that are also used in the decimal system, the subscript 2 is sometimes used to indicate a binary number and the subscript 10 for a decimal number. The binary number 1011 can be written as 1011_2. The function of the subscript 2 is to identify the number as binary. It can be omitted when the binary is otherwise specified or understood. Similarly, the decimal number 11 can be written as 11_{10}. The subscript 10 identifies the number as a decimal number. As in the case of the binary subscript, the subscript is omitted when it is quite clear that the number involved belongs to the decimal system.

The binary symbol 0 carries the same meaning as the symbol 0 used in the decimal system. In the decimal system, 0 times any number results in 0. This is also true in the binary system. The *binary point,* similar to a decimal point, is used to separate the whole number and the fraction.

Positive Powers of 2

An exponent indicates the number of times the base is used as a factor. For example, $2^3 = 2 \times 2 \times 2 = 8$. Since the number 2 is the base in the binary system, a list of the positive powers of 2 can be arranged in a manner similar to that done earlier for the decimal system.

$2^0 = 1$	$2^5 = 32$
$2^1 = 2$	$2^6 = 64$
$2^2 = 4$	$2^7 = 128$
$2^3 = 8$	$2^8 = 256$
$2^4 = 16$	$2^9 = 512$

The numbers to the left of the equal signs use 2 as a base, raised to various powers. The numbers to the right of the equal signs are the decimal equivalents. On the left side, note that each exponent is 1 larger than a preceding exponent: 2^6 follows 2^5; 2^4 follows 2^3. On the right side, each number is twice as great as the preceding number. For example, $2^2 = 4$, therefore the value of 2^3 is 4 times 2 or 8.

In the decimal system, the value of a symbol depends upon its position. A number such as 137 is equivalent to $(1 \times 100) + (3 \times 10) + (7 \times 1)$. The significance of this is that 1 is in the hundreds column, 3 is in the tens column and 7 is in the units or digits column. The columns are ordinarily omitted, although a number such as 137 could be written as

100's	10's	1's
1	3	7

The value of each symbol in the binary number also depends upon its position. Table 9–1 lists the column values for binary numbers. This table can be used to find the decimal equivalent of a binary number such as 1011_2. The first right-hand digit has a value of 1. Moving to the left, the next adjacent digit has a value of 2^1 or 1×2. Still moving to the left, the third digit is 0, and it has a value of 0×2^2 or $0 \times 4 = 0$. The final digit, at the extreme left

TABLE 9–1

Column Values of Binary Numbers

Column Number Left of Binary Point	11	10	9	8	7	6	5	4	3	2	1
Power of 2	2^{10}	2^9	2^8	2^7	2^6	2^5	2^4	2^3	2^2	2^1	2^0
Decimal Equivalent	1 024	512	256	128	64	32	16	8	4	2	1

is $1 \times 2^3 = 1 \times 8 = 8$. A number such as 1011_2, then, has an equivalent decimal value of $(1 \times 2^3) + (0 \times 2^2) + (1 \times 2^1) + (1 \times 2^0)$. This can also be written in decimal form as $(1 \times 8) + (0 \times 4) + (1 \times 2) + (1 \times 1)$. Performing the indicated arithmetic, we have $8 + 0 + 2 + 1 = 11$. Hence $1011_2 = 11_{10}$.

Example: What is the decimal equivalent of 0011101_2?

Solution: We can spread the number to show its relationship to the powers of 2.

$$\begin{matrix} 0 & 0 & 1 & 1 & 1 & 0 & 1 \\ 2^6 & 2^5 & 2^4 & 2^3 & 2^2 & 2^1 & 2^0 \end{matrix}$$

The decimal equivalent of 0011101_2 is

$$(0 \times 2^6) + (0 \times 2^5) + (1 \times 2^4) + (1 \times 2^3) + (1 \times 2^2) + (0 \times 2^1) + (1 \times 2^0) = 0 + 0 + 16 + 8 + 4 + 0 + 1 = 29$$

$$0011101_2 = 29_{10}$$

Table 9–2 shows decimal numbers ranging from 0 through 25 and their binary equivalents. Although the binary numbers are shown here in columnar form, they are only arranged in this way to emphasize their positional value. Ordinarily, a binary number is written in the same way as a decimal, that is, with the symbols immediately adjacent to each other. Using the table, the decimal number 5 has a binary equivalent of 00101_2. The zero at the extreme left and the one next to it indicate that these columns have a value of zero and can be disregarded since they do not contribute to the numerical value of the binary. Therefore $5_{10} = 101_2$.

Negative Powers of 2

The following list shows negative powers of 2, ranging from 2^{-1} to 2^{-10}, and their decimal equivalents.

$$2^{-1} = \frac{1}{2} = 0.5 \qquad 2^{-4} = \frac{1}{16} = 0.062\ 5$$

$$2^{-2} = \frac{1}{4} = 0.25 \qquad 2^{-5} = \frac{1}{32} = 0.031\ 25$$

$$2^{-3} = \frac{1}{8} = 0.125 \qquad 2^{-6} = \frac{1}{64} = 0.015\ 625$$

TABLE 9–2

Decimal and Binary Equivalents Table

Decimal Number	Binary Number 2^4 16	2^3 8	2^2 4	2^1 2	2^0 1
0	0	0	0	0	0
1	0	0	0	0	1
2	0	0	0	1	0
3	0	0	0	1	1
4	0	0	1	0	0
5	0	0	1	0	1
6	0	0	1	1	0
7	0	0	1	1	1
8	0	1	0	0	0
9	0	1	0	0	1
10	0	1	0	1	0
11	0	1	0	1	1
12	0	1	1	0	0
13	0	1	1	0	1
14	0	1	1	1	0
15	0	1	1	1	1
16	1	0	0	0	0
17	1	0	0	0	1
18	1	0	0	1	0
19	1	0	0	1	1
20	1	0	1	0	0
21	1	0	1	0	1
22	1	0	1	1	0
23	1	0	1	1	1
24	1	1	0	0	0
25	1	1	0	0	1

$$2^{-7} = \frac{1}{128} = 0.007\ 812 \qquad 2^{-9} = \frac{1}{512} = 0.001\ 953$$

$$2^{-8} = \frac{1}{256} = 0.003\ 906 \qquad 2^{-10} = \frac{1}{1\ 024} = 0.000\ 976$$

Examine the numbers and you will see that each negative exponent has a value that is one larger than its preceding number. On the right side of the equal signs, the equivalent values in the decimal system are given in the form of fractions and these, in turn, are followed by their decimal equivalents. Each succeeding fraction has a value that is one-half that of the preceding number.

Reciprocals

A reciprocal is the inverse of any number or simply the number inverted.

The reciprocal of 2 is $\frac{1}{2}$.

The reciprocal of $\frac{1}{2}$ is 2.

Positive and negative powers of 10 are reciprocals. The reciprocal of 10^{-4} is 10^4. The reciprocal of 10^9 is 10^{-9}. This concept applies to powers of 2 as well. The reciprocal of 2^8 is 2^{-8} and the reciprocal of 2^{-4} is 2^4.

Positive and negative powers of 2 can have their signs changed by transference from numerator to denominator or denominator to numerator. As an example,

$$2^4 = \frac{2^4}{1} = \frac{1}{2^{-4}}$$

Similarly:

$$2^{-5} = \frac{2^{-5}}{1} = \frac{1}{2^5}$$

The reciprocal of a number can be expressed in fractional or exponential form.

$$\text{The reciprocal of } 2^4 \text{ is } 2^{-4} \text{ or } \frac{1}{2^4}.$$

Column Values of Fractional Binary Numbers

Table 9–3 lists the column values for fractional binary numbers. This table can be used to find the decimal equivalent of a fractional binary number such as 0.1011_2. The decimal equivalent of 0.1011_2 is

$$\begin{aligned} & (1 \times 2^{-1}) + (0 \times 2^{-2}) + (1 \times 2^{-3}) + (1 \times 2^{-4}) \\ = & (1 \times 0.5) + (0 \times 0.25) + (1 \times 0.125) + (1 \times 0.062\ 5) \\ = & 0.5 + 0 + 0.125 + 0.062\ 5 \\ = & 0.687\ 5 \end{aligned}$$

Therefore $0.1011_2 = 0.687\ 5_{10}$.

A mixed binary number consists of a whole number portion plus a fraction. Mixed binary numbers always have a binary point, with the whole number

TABLE 9–3

Column Values of Fractional Binary Numbers

Column Number Right of Binary Point	1	2	3	4	5	6	7
Power of 2	2^{-1}	2^{-2}	2^{-3}	2^{-4}	2^{-5}	2^{-6}	2^{-7}
Fractional Equivalent	$\frac{1}{2}$	$\frac{1}{4}$	$\frac{1}{8}$	$\frac{1}{16}$	$\frac{1}{32}$	$\frac{1}{64}$	$\frac{1}{128}$
Decimal Equivalent	0.5	0.25	0.125	0.062 5	0.031 25	0.015 625	0.007 812

portion to the left of the point and the fractional part to the right. A binary such as 1101.101_2 is a mixed binary number. The part to the left of the binary point has a decimal equivalent value of 13 while the part to the right is equal to 0.625, and so the complete decimal number value is 13.625.

Converting Binary Numbers to Decimal Numbers

There are a number of conversion techniques for converting binary numbers to decimal numbers. One of the easiest, but also one of the most limited, is to use a binary-to-decimal conversion table. In a table of this kind, binary numbers are listed ranging from 0000 to some high value, with equivalent decimal values in an adjacent column. The disadvantage is that the number you want may not be listed. Further, such tables usually list whole numbers only, and not mixed binary numbers.

Another conversion method is illustrated in Figure 9–1. Write the binary number, and immediately above it write the decimal values of each place. The decimal values are added to find the equivalent decimal value of the binary number.

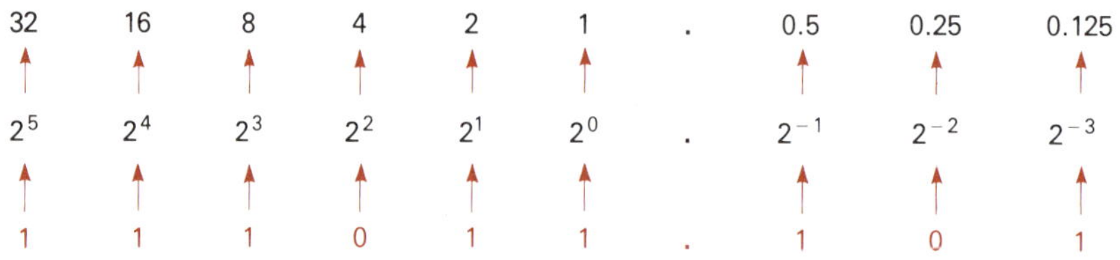

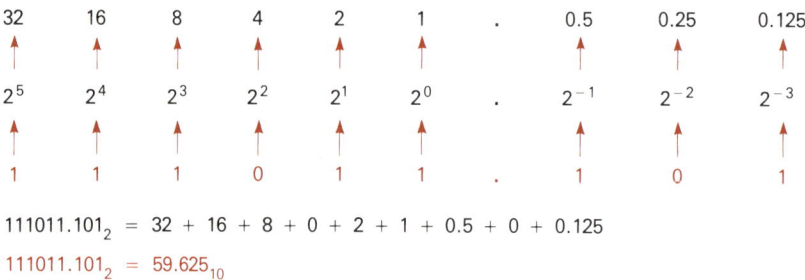

$111011.101_2 = 32 + 16 + 8 + 0 + 2 + 1 + 0.5 + 0 + 0.125$

$111011.101_2 = 59.625_{10}$

FIGURE 9–1 Method for converting binary numbers to decimal numbers

Example: Convert 1101001.1011_2 to a decimal number.

Solution: Constructing a conversion table

64	32	16	8	4	2	1		0.5	0.25	0.125	0.062 5
1	1	0	1	0	0	1	.	1	0	1	1

The decimal equivalent becomes

$64 + 32 + 0 + 8 + 0 + 0 + 1 + 0.5 + 0 + 0.125 + 0.062\,5$
$= 105.687\,5$

Example: Convert 10101.011_2 to a decimal number.

Solution:

16	8	4	2	1		0.5	0.25	0.125
1	0	1	0	1	.	0	1	1

Resulting in

$16 + 0 + 4 + 0 + 1 + 0 + 0.25 + 0.125 = 21.375$

Sometimes a binary number is written with more digits than is required. In the binary number 001010.0100, the zeros at the extreme left and the extreme right are unnecessary. The same binary number can be written as 1010.01 and will have the same equivalent decimal value as the binary with the extra zeros. Additional zeros are sometimes used when binary numbers are written in columnar form, such as

00101.0010
10001.0101
00010.1000
01011.0010

and left-right justification is wanted. Justification simply means that the left and right margins of the binary numbers will be straight.

Double-dabble is the name given to still another method of conversion. With this technique, each succeeding binary digit is multiplied by 2. The value of the next lower order digit is added in and the process is repeated. Figure 9–2 illustrates this technique.

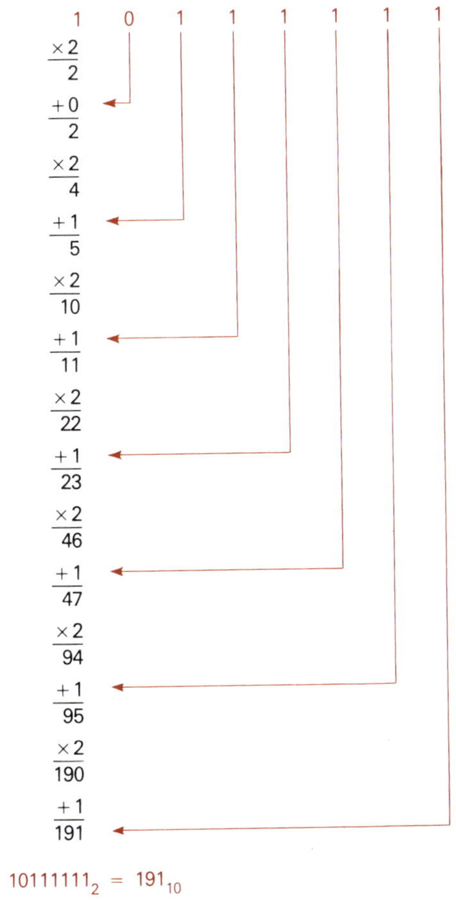

FIGURE 9–2 Double-dabble method of converting binary numbers to decimal numbers

Example: Convert 110101_2 to its decimal equivalent using the double-dabble method.

Solution: In order to apply the double-dabble method follow these steps.

1. Multiply the highest order value by 2: $1 \times 2 = 2$
2. Add the value of the next digit: $2 + 1 = 3$
3. Multiply by 2: $3 \times 2 = 6$
4. Add the next digit: $6 + 0 = 6$
5. Multiply by 2: $6 \times 2 = 12$
6. Add the next digit: $12 + 1 = 13$
7. Multiply by 2: $13 \times 2 = 26$
8. Add the next digit: $26 + 0 = 26$
9. Multiply by 2: $26 \times 2 = 52$
10. Add the next digit: $52 + 1 = 53$

So that $110101_2 = 53_{10}$

For practice, convert 101110101_2 using the double-dabble method. Your answer should be 373_{10}. As a check, using Table 9–1,

$$\begin{array}{ccccccccc} 1 & 0 & 1 & 1 & 1 & 0 & 1 & 0 & 1 \\ 256 + & 0 + & 64 + & 32 + & 16 + & 0 + & 4 + & 0 + & 1 = 373_{10} \end{array}$$

There is still another conversion technique for changing a binary number to a decimal number. This technique consists of writing the binary number in vertical form, with the lowest order value on top and the highest order value on the bottom. Figure 9–3 illustrates this method. For practice, use the vertical form to convert 1000101_2 to a decimal number. Your answer should be 69_{10}.

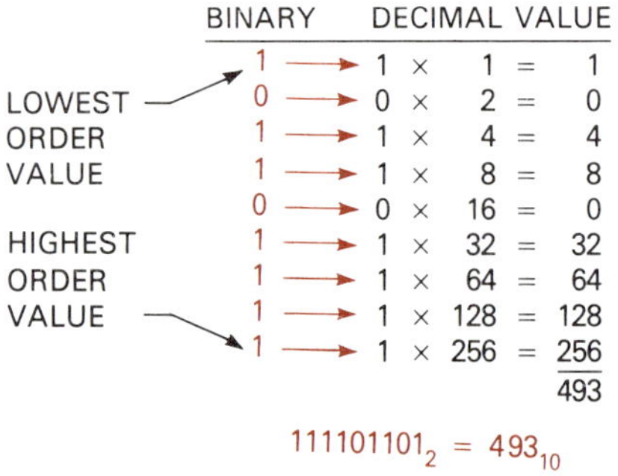

FIGURE 9–3 Vertical method of converting binary numbers to decimal numbers

Converting Decimal Numbers to Binary Numbers

As in the case of converting binary numbers to decimal numbers, there are also various ways in which decimal numbers can be changed into binary numbers. One method is a decimal-to-binary conversion table. Again, a disadvantage of these tables is that the number may not appear in the table. Another method of conversion involves working with powers of 2.

Example: What is the binary equivalent of 67_{10}?

Solution:
1. Find the nearest power of 2 that will be as close as possible to, but still less than 67.
   This is $2^6 = 64$.
2. Subtract 64 from the decimal number.
   $67 - 64 = 3$
3. Find the next power of 2 that will be as close as possible, but less than 3.
   This is $2^1 = 2$
4. Subtract 2 from 3.
   $3 - 2 = 1$
5. The power of 2 that is equal to 1 is 2^0.
6. Showing these columns with the symbol one and all others with a zero, the resulting binary number is

2^6	2^5	2^4	2^3	2^2	2^1	2^0
1	0	0	0	0	1	1

$67_{10} = 1000011_2$

For practice, express 133_{10} in binary form. Your answer should be 10000101_2.

Example: Convert 175_{10} to a binary number.

Solution:
1. The power of 2 closest to 175 is 2^7 or 128.
   $175 - 128 = 47$
2. The power of 2 closest to 47 is 2^5 or 32.
   $47 - 32 = 15$
3. The power of 2 closest to 15 is 2^3 or 8.
   $15 - 8 = 7$
4. The power of 2 closest to 7 is 2^2 or 4.
   $7 - 4 = 3$
5. The power of 2 closest to 3 is 2^1 or 2.
   $3 - 2 = 1$
6. The power of 2 equal to 1 is 2^0 or 1.
7. The binary equivalent is

2^7	2^6	2^5	2^4	2^3	2^2	2^1	2^0
1	0	1	0	1	1	1	1

$175_{10} = 10101111_2$

A second method of converting a decimal number to binary is to use the double-dabble process. This involves repeatedly dividing the number by 2, writing down the remainders, and then reading the remainders in *reverse* order as shown in Figure 9–4. Review these additional examples.

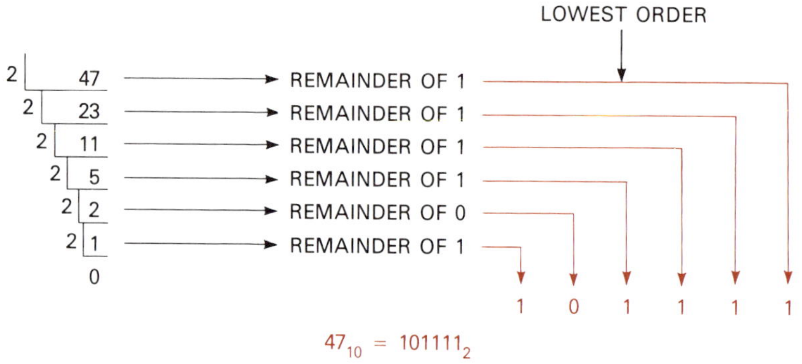

FIGURE 9–4 Double-dabble method of converting decimal numbers to binary numbers

Example: Convert 37_{10} to a binary number using the double-dabble method.

Solution: $37 \div 2 = 18$ with a remainder of 1

$18 \div 2 = 9$ with a remainder of 0

$9 \div 2 = 4$ with a remainder of 1

$4 \div 2 = 2$ with a remainder of 0

$2 \div 2 = 1$ with a remainder of 0

$1 \div 2 = 0$ with a remainder of 1

Reading these in *reverse* order, $37_{10} = 100101_2$

Example: Convert 84_{10} to binary using the double-dabble method.

Solution: $84 \div 2 = 42$, remainder $= 0$

$42 \div 2 = 21$, remainder $= 0$

$21 \div 2 = 10$, remainder $= 1$

$10 \div 2 = 5$, remainder $= 0$

$5 \div 2 = 2$, remainder $= 1$

$2 \div 2 = 1$, remainder $= 0$

$1 \div 2 = 0$, remainder $= 1$

Resulting in, $84_{10} = 1010100_2$

In order to convert a decimal fraction to its binary equivalent simply multiply the fraction by 2. If the product is greater than 1, list the 1, and move the *remaining* fraction forward. If the product is less than 1, record a 0 and move the fraction forward. This process is repeated until the product is 1.0, or sufficient precision is obtained. The binary number is read beginning with the first product. Follow these examples very closely.

Example: Convert 0.375_{10} to a binary number.

Solution:

$0.375 \times 2 = 0.75$, record a 0 (Product is less than 1)

$0.75 \times 2 = 1.5$, record a 1 (Product is greater than 1)

$0.5 \times 2 = 1.0$, record a 1 (Product is 1.0)

The binary equivalent is then 0.011_2.

Example: Convert 0.85_{10} to a binary number accurate to six places.

Solution:

$0.85 \times 2 = 1.7$, record a 1

$0.7 \times 2 = 1.4$, record a 1

$0.4 \times 2 = 0.8$, record a 0

$0.8 \times 2 = 1.6$, record a 1

$0.6 \times 2 = 1.2$, record a 1

$0.2 \times 2 = 0.4$, record a 0

Resulting in, $0.110110_2 \approx 0.85_{10}$

Example: Convert $19.312\ 5_{10}$ to a binary number.

Solution: Converting the whole number

$19 \div 2 = 9$, remainder $= 1$

$9 \div 2 = 4$, remainder $= 1$

$4 \div 2 = 2$, remainder $= 0$

$2 \div 2 = 1$, remainder $= 0$

$1 \div 2 = 0$, remainder $= 1$

Resulting in 10011_2

Converting the fraction

$0.312\,5 \times 2 = 0.625$, record a 0

$0.625 \times 2 = 1.25$, record a 1

$0.25 \times 2 = 0.5$, record a 0

$0.5 \times 2 = 1.0$, record a 1

Resulting in 0.0101_2

So that $19.312\,5_{10} = 10011.0101_2$

EXERCISE 9·1

In problems 1–15, convert each binary number to its decimal equivalent.

1. 1011_2
2. 1011011_2
3. 10001_2
4. 1001.1011_2
5. 0.011_2
6. 10101.01_2
7. 11111101_2
8. 100.101_2
9. 1_2
10. 0.101101_2
11. 111.1_2
12. 110101_2
13. 100111011_2
14. 101101.1101_2
15. 0.0011_2

In problems 16–30, convert each decimal number to its binary equivalent. If necessary, round binary fractions to five placcs.

16. 5.5
17. 17
18. 100
19. 67.8
20. 157
21. 101.1
22. 75
23. 36.6
24. 50
25. 284
26. 45.16
27. $92\frac{7}{8}$
28. $154\frac{1}{5}$
29. 84
30. 176

UNIT 9·2 BINARY ADDITION AND SUBTRACTION

Objectives:

After studying this unit, you should be able to

- add whole and mixed binary numbers.
- subtract whole and mixed binary numbers.
- use number complements to perform binary subtraction.

Adding Binary Nmbers

As in the decimal system, there are rules for addition. For binary numbers:

$$0_2 + 0_2 = 0_2 \quad (9\text{–}1)$$

$$0_2 + 1_2 = 1_2 \quad (9\text{–}2)$$

$$1_2 + 0_2 = 1_2 \quad (9\text{–}3)$$

$$1_2 + 1_2 = 10_2 \quad (9\text{–}4)$$

In the binary system, the number 10_2 is pronounced as *"one zero."* To call it ten is to designate it as a decimal number.

Example: Add 1011_2 and 1001_2. Check the result by using decimal numbers.

Solution:

$$\begin{array}{rcl} \underline{11\quad\;1} & & \underline{\text{CARRIES}} \\ 1011_2 & = & 11_{10} \\ +\;1001_2 & = & 9_{10} \\ \hline 10100_2 & = & 20_{10} \end{array}$$

Example: Add 10110_2 and 1111_2. Check the result by using decimal numbers.

Solution:

$$\begin{array}{rcl} 10110_2 & = & 22_{10} \\ +\;1111_2 & = & 15_{10} \\ \hline 100101_2 & = & 37_{10} \end{array}$$

As an alternative to remembering binary addition rules, it is possible to use a binary addition table, such as Table 9–4.The addends are the numbers to be added. Locate the binaries in the addend row and the addend column. Move down and across and where the two lines join you have the answer. As an example of how to use the table, assume you want to add $1_2 + 1_2$. Locate binary 1 in the addend row and also find binary 1 in the addend column. Move down from addend 1 and move across from addend 1. The point of intersection will be the sum, and as shown in the table, is 10_2.

TABLE 9–4

Binary Addition Table

+	0	1
0	0	1
1	1	10

Columnar binary addition is simplified if it is carried out in steps.

$$\begin{array}{r l}
1\;1 & \\
1\;1\;1\;1\;1 & \text{CARRIES} \\
\hline
1\;0\;1\;0_2 & \\
1\;0\;0\;1_2 & \\
1\;1\;0\;1\;0_2 & \\
+\quad 1\;1\;1\;0\;1_2 & \\
\hline
1\;0\;0\;1\;0\;1\;0_2 &
\end{array}$$

Start with the column at the far right. Begin at the bottom and add, moving up. $1_2 + 0_2 = 1_2$. Add this result to the next digit. $1_2 + 1_2 = 10_2$. We now have a two digit number, consisting of 1 and 0. Put 1 at the top of the adjacent column as a carry and add 0 to the last digit in the column. $0_2 + 0_2 = 0_2$. Put 0 in the sum or the answer.

Now proceed with the second column. Start at the bottom and add, moving up. $0_2 + 1_2 = 1_2$. Add this answer to the third digit. $1_2 + 0_2 = 1_2$. Add this result to the fourth digit. $1_2 + 1_2 = 10_2$. Put the 1 on the top of the adjacent column as a carry. Add the 0 and the carry number in the second column. $0_2 + 1_2 = 1_2$. Put 1 in the sum. The remaining columns are now added using the same technique. As a check, convert the binary numbers to decimals and add the decimals.

Mixed binaries are added in the same way as whole number binaries. The only rule is that the binary point of each number must be vertically aligned.

Example: Find the sum of 1001.1_2 and 10100.01_2.

Solution: The first step is to arrange the numbers in vertical form. Since it may be helpful to have the same number of digits in both numbers, zeros can be added to the first number without affecting the value. Note that the binary points must be aligned.

$$\begin{array}{r}
01001.10_2 \\
+\ 10100.01_2 \\
\hline
11101.11_2
\end{array}$$

Example: Find the sum of 110011.101_2 and 1101.101_2.

Solution:

$$\begin{array}{r l}
111111\;\;1 & \text{CARRIES} \\
\hline
110011.101_2 & \\
+\ 001101.101_2 & \\
\hline
1000001.010_2 &
\end{array}$$

Subtracting Binary Numbers

Binary subtraction is the inverse of binary addition and can be handled by using these rules.

$$\mathbf{0_2 - 0_2 = 0_2} \qquad \textbf{(9–5)}$$

$$1_2 - 1_2 = 0_2 \qquad (9\text{–}6)$$

$$1_2 - 0_2 = 1_2 \qquad (9\text{–}7)$$

$$10_2 - 1_2 = 1_2 \qquad (9\text{–}8)$$

In binary subtraction, subtract column by column borrowing from the column to the left when necessary. This is precisely the same technique used in decimal number subtraction. In this decimal number problem,

$$\begin{array}{r} 153 \\ -9 \\ \hline 144 \end{array}$$

change 3 in the first column to 13 by borrowing 10 (or 1 from the 10's column) from the adjacent column. The 5 in the ten's column becomes 4.

The following examples will show the process of binary subtraction. Note that when borrowing we actually borrow the decimal number 2, which is written as 10_2 in binary. The minus sign must always be used to indicate subtraction. Problems in subtraction can be proved by adding the difference to the subtrahend to obtain the minuend or by changing the binary numbers to their decimal equivalents.

Example: Subtract 101_2 from 1011_2. To check the answer, convert the binary numbers to their decimal equivalents.

Solution:

$$\begin{array}{rcr} 1011_2 & = & 11_{10} \\ -0101_2 & = & -\ 5_{10} \\ \hline 110_2 & = & 6_{10} \end{array}$$

In the first column we have 1 − 1 = 0.
In the second column we have 1 − 0 = 1.
In the third column we must borrow a 10 from the fourth column.

Example: Subtract 1010_2 from 11110_2. To check the answer, convert the binary numbers to their decimal equivalents.

Solution:

$$\begin{array}{rcr} 11110_2 & = & 30_{10} \\ -\ 1010_2 & = & -10_{10} \\ \hline 10100_2 & = & 20_{10} \end{array}$$

Example: Subtract 10011_2 from 110110_2. To check the answer, convert the binary numbers to their decimal equivalents.

Solution:

$$\begin{array}{rcr} 110110_2 & = & 54_{10} \\ -\ 10011_2 & = & -19_{10} \\ \hline 100011_2 & = & 35_{10} \end{array}$$

1st column: 10 − 1 = 1 (After borrow)
2nd column: 10 − 1 = 1 (After borrow)

3rd column: 0 − 0 = 0
4th column: 0 − 0 = 0
5th column: 1 − 1 = 0
6th column: 1 − 0 = 1

Example: Subtract 10110_2 from 10011_2. To check the answer, convert the binary numbers to their decimal equivalents.

Solution: In this example, the subtrahend is larger than the minuend. The difference will be a negative number.

$$\begin{array}{rcr} -10110_2 & = & -22_{10} \\ +10011_2 & = & +19_{10} \\ \hline -00011_2 & = & -\ 3_{10} \end{array}$$

1st column: 10 − 1 = 1
2nd column: 10 − 1 = 1
3rd column: 0 − 0 = 0
4th column: 0 − 0 = 0
5th column: 1 − 1 = 0

Subtraction With Number Complements

A second method of subtraction involves the use of number complements. This process is used in calculators and computers and utilizes both the 1's complement and 2's complement of a binary number. These complements are:

1's complement: Change each 0 to a 1, and each 1 to a 0
2's complement: Add a binary 1 to the 1's complement

Example: What is the 1's complement and the 2's complement of (a) 101101_2 and (b) 100_2?

Solution: a. Binary number: 101101_2
1's complement: 010010_2
2's complement: $010010_2 + 1_2 = 010011_2$

b. Binary number: 100_2
1's complement: 011_2
2's complement: $011_2 + 1_2 = 100_2$

The 2's complement of a binary number can be used in subtraction by following these steps:

1. Fill out the subtrahend with leading zeros until it has the same number of places as the minuend.
2. Change the subtrahend to its 2's complement.
3. Add the 2's complement to the minuend.
4. Omit the last carry.

Example: Subtract 11100_2 from 11110_2 using the 2's complement method.

Solution: The initial problem is

$$\begin{array}{rcr} 11110_2 & = & 30_{10} \\ \underline{-11100_2} & = & \underline{-28_{10}} \end{array}$$

Converting the subtrahend to its 1's complement

11100_2

00011_2

Adding 1 to form the 2's complement

$00011_2 + 1_2 = 00100_2$

Adding the 2's complement to the minuend and omitting the last carry

$$\begin{array}{r} 11110_2 \\ \underline{+00100_2} \\ \not{1}00010_2 \end{array} = 2_{10}$$

Example: Subtract 1101_2 from 11011_2 using the 2's complement method.

Solution:

$$\begin{array}{rcr} 11011_2 & = & 27_{10} \\ \underline{-01101_2} & = & \underline{-13_{10}} \end{array}$$

Converting the subtrahend to its 1's complement

01101_2

10010_2

The 2's complement is

$10010_2 + 1_2 = 10011_2$

Adding the 2's complement to the minuend and omitting the last carry

$$\begin{array}{r} 11011_2 \\ \underline{+10011_2} \\ \not{1}01110_2 \end{array} = 14_{10}$$

A third method of binary subtraction involves using only the 1's complement. The steps are as follows:

1. Fill out the subtrahend with leading zeros until it has the same number of places as the minuend.
2. Change the subtrahend to its 1's complement.
3. Add the 1's complement to the minuend.
4. Add the last carry to the sum.

Example: Subtract 1001_2 from 1011_2 using the 1's complement process.

Solution: The initial problem is

$$\begin{array}{rcr} 1011_2 & = & 11_{10} \\ \underline{-1001_2} & = & \underline{-\ 9_{10}} \end{array}$$

Converting the subtrahend to its 1's complement

1001_2

0110_2

Adding the 1's complement to the minuend

$$\begin{array}{r} 1011_2 \\ +0110_2 \\ \hline 10001_2 \end{array}$$

(1 ← FINAL CARRY; 0001 ← INTERIM SUM)

Adding the final carry to the interim sum results in

$$\begin{array}{r} 0001_2 \\ +\quad 1_2 \\ \hline 0010_2 \end{array} = 2_{10}$$

Example: Subtract 1111_2 from 11101_2 using the 1's complement process.

Solution:

$$\begin{array}{r} 11101_2 \\ -01111_2 \\ \hline \end{array} = \begin{array}{r} 29_{10} \\ -15_{10} \\ \hline \end{array}$$

Converting the subtrahend to its 1's complement

01111_2

10000_2

Adding the 1's complement to the minuend

$$\begin{array}{r} 11101_2 \\ +\ 10000_2 \\ \hline 101101_2 \end{array}$$

Adding the final carry to the interim sum results in

$$\begin{array}{r} 1101_2 \\ +\quad 1_2 \\ \hline 1110_2 \end{array} = 14_{10}$$

When the result of a subtraction problem is positive, there will be a final carry (often called the *end-around carry*). When there is no end-around carry the difference is negative and the number will be in the 1's complement form. In order to reach the actual difference a second 1's complement is taken, and a minus sign attached.

Example: Subtract 11001_2 from 10110_2.

Solution:

$$\begin{array}{r} 10110_2 \\ -11001_2 \\ \hline \end{array} = \begin{array}{r} 22_{10} \\ -25_{10} \\ \hline \end{array}$$

Converting the subtrahend to its 1's complement

11001_2

00110_2

Adding the 1's complement to the minuend

$$\begin{array}{r} 10110_2 \\ +00110_2 \\ \hline 11100_2 \end{array}$$

Since there is no end-around carry, the 1's complement is taken and the minus sign attached

11100_2

-00011_2, or $-11_2 = -3_{10}$

EXERCISE 9·2

In problems 1–10, perform the indicated additions and express the results in both binary and decimal notation.

1. $1011_2 + 1001_2$
2. $11011_2 + 10011_2$
3. $1101101_2 + 11101_2$
4. $101.11_2 + 1101.101_2$
5. $0.1011_2 + 0.011_2$
6. $111011_2 + 110_2 + 101110_2$
7. $1101_2 + 1001_2 + 1110_2 + 1111_2$
8. $10111010_2 + 10001110_2$
9. $1011101.1011_2 + 110011.11_2$
10. $11011_2 + 10001_2 + 11010_2 + 11111_2$

In problems 11–20, perform the indicated subtractions.

11. $1101_2 - 1001_2$
12. $10011_2 - 1101_2$
13. $11101_2 - 10111_2$
14. $100110_2 - 11101_2$
15. $1001_2 - 1101_2$
16. $1101_2 - 10110_2$
17. $1111.11_2 - 1000.01_2$
18. $110110_2 - 101110_2$
19. $1011101_2 - 1110111_2$
20. $1101.1011_2 - 1100.0111_2$

In problems 21–30, find the 1's and 2's complement for each binary value. Use zeros to retain the same number of places as in the original number.

21. 1001_2
22. 1101_2
23. 1111_2
24. 100110_2
25. 11011_2
26. 10001_2
27. 101101_2
28. 1110100_2
29. 101001_2
30. 101010_2

UNIT 9·3 BINARY MULTIPLICATION AND DIVISION

Objectives:

After studying this unit, you should be able to

- multiply whole and mixed binary numbers.
- divide whole and mixed binary numbers.

Multiplying Binary Numbers

The rules for binary multiplication are

$$0 \times 0 = 0 \qquad (9\text{–}9)$$

$$0 \times 1 = 0 \qquad (9\text{–}10)$$

$$1 \times 0 = 0 \qquad (9\text{–}11)$$

$$1 \times 1 = 1 \qquad (9\text{–}12)$$

The rules for binary multiplication are set up in table form as shown in Table 9–5. As in the addition table, the multiplication table lists the rules for binary multiplication in graphic form. Either the rules or the table can be used as an aid in multiplication. The result of the first step in multiplication is an answer called the *first partial product.* The second step produces a *second partial product.* There will be a partial product for each digit in the multiplier. If the multiplier has five digits, such as 10111_2, then there will be five partial products. The addition of all the partial products supplies the final product or answer.

TABLE 9–5

Binary Multiplication Table

×	0	1
0	0	0
1	0	1

Multiplication by zero, whether in the decimal or binary systems, always results in zero. The presence of zeros in binary multiplication helps simplify the work. As an example:

$$\begin{array}{rcr}
1001_2 & = & 9_{10} \\
\times\ \underline{101_2} & = & \underline{\times 5_{10}} \\
1001 & & \\
0000 & & \\
\underline{1001} & & \\
101101_2 & = & 45_{10}
\end{array}$$

Since the second partial product consists completely of zeros, there is no reason why this step cannot be eliminated. The only precaution to take is that the lowest-order digit of the partial product must line up exactly with the appropriate digit in the multiplier.

$$\begin{array}{rcr}
1001_2 & = & 9_{10} \\
\underline{\times\ 101_2} & = & \underline{\times 5_{10}} \\
1001 & & \\
\underline{1001} & & \\
101101_2 & = & 45_{10}
\end{array}$$

Example: Find the product of 10110_2 and 111_2.

Solution:

$$\begin{array}{rcr} 10110_2 & = & 22_{10} \\ \times \quad 110_2 & = & \times \ 6_{10} \\ \hline 101100 & & \\ 10110 & & \\ \hline 10000100_2 & = & 132_{10} \end{array}$$

In the multiplication of fractional binary numbers, the binary points of the multiplier and multiplicand do not need to be aligned vertically as they are for addition and subtraction of fractional binary numbers. The binary point position in the final product is equal to the sum of the number of binary places in the multiplier and multiplicand.

Example: Multiply 0.001_2 by 0.01_2 and check the product by using decimal equivalents.

Solution:

$$\begin{array}{rcr} 0.001_2 & = & 0.125_{10} \\ \times \ 0.01_2 & = & \times \ 0.25_{10} \\ \hline 0.00001_2 & = & 0.031\,25_{10} \end{array}$$

The previous example involves the multiplication of one fractional binary number by another. The same multiplication technique can be used for the multiplication of mixed binary numbers.

Example: Multiply 1101.01_2 by 101.1_2 and check the product by using decimal equivalents.

Solution:

$$\begin{array}{rcr} 1101.01_2 & = & 13.25_{10} \\ \times \quad 101.1_2 & = & \times \quad 5.5_{10} \\ \hline 110101 & & \\ 110101 & & \\ 000000 & & \\ 110101 & & \\ \hline 1001000.111_2 & = & 72.875_{10} \end{array}$$

Dividing Binary Numbers

Binary division is the inverse of binary multiplication. The number being divided is called the *dividend* while the number doing the dividing is the *divisor.* The answer is the *quotient.* To prove any problem in binary division, as in decimal division, multiply the divisor by the quotient. The answer will be the dividend.

To be able to do binary division involves the ability to do binary subtraction and multiplication. In a sense, binary division is simpler than decimal division, since only the numbers 1 and 0 are used.

Example: Divide 11001_2 by 101_2 and check the quotient by using decimal equivalents.

Solution: The first step, as in decimal division, is to determine the first number in the quotient. The divisor is smaller than 110, so that the first number in the quotient is 1. This 1 is multiplied by the divisor and the result is subtracted from 110. The next place (0) is brought down and the division process is repeated. As 10 is smaller than 101, a 0 is placed in the quotient and the next number is brought down.

$$
\begin{array}{rl}
 & 101_2 \\
101_2 & \overline{)\ 11001_2} \\
 & \underline{-101} \\
 & 101 \\
 & \underline{-101} \\
 & 000
\end{array}
\quad = \quad
\begin{array}{rl}
 & 5_{10} \\
5_{10} & \overline{)25_{10}}
\end{array}
$$

Example: Divide 1111110_2 by 1001_2 and check the quotient by using decimal equivalents.

Solution:

$$
\begin{array}{rl}
 & 1110_2 \\
1001_2 & \overline{)1111110_2} \\
 & \underline{1001} \\
 & 1101 \\
 & \underline{1001} \\
 & 1001 \\
 & \underline{1001} \\
 & 0000
\end{array}
\quad = \quad
\begin{array}{rl}
 & 14_{10} \\
9_{10} & \overline{)126_{10}}
\end{array}
$$

Example: Divide 11110.101 by 101 and check the quotient by using decimal equivalents.

Solution:

$$
\begin{array}{rl}
 & 110.001_2 \\
101_2 & \overline{)11110.101_2} \\
 & \underline{101} \\
 & 101 \\
 & \underline{101} \\
 & 0101 \\
 & \underline{101}
\end{array}
\quad = \quad
\begin{array}{rl}
 & 6.125_{10} \\
5_{10} & \overline{)30.625_{10}}
\end{array}
$$

EXERCISE 9·3

In problems 1–10, multiply the binary values.

1. $100_2 \times 111_2$
2. $1010_2 \times 110_2$
3. $1100_2 \times 1011_2$
4. $1111_2 \times 10_2$
5. $(101_2)^2$
6. $(1000_2)^2$
7. $(1110_2)(10_2)(101_2)$
8. $(10111_2)(1001_2)(11_2)$
9. $(10101_2)(1011_2)(111_2)$
10. $(100000_2)(1000_2)(100_2)$

In problems 11–20, divide the binary values.

11. $1001_2 \div 11_2$
12. $10010_2 \div 10_2$
13. $100111_2 \div 11_2$
14. $111111_2 \div 111_2$
15. $1010100_2 \div 1110_2$
16. $1000100_2 \div 100_2$
17. $1110011_2 \div 101_2$
18. $1011000_2 \div 1011_2$
19. $10010000_2 \div 1100_2$
20. $1110000_2 \div 111_2$

In problems 21–26, perform the indicated operations with mixed binary numbers. Note: Fractional answers should not exceed 5 binary places.

21. $(100.1_2)(1100_2)$
22. $(111.011_2)(101.11_2)$
23. $(1001.01_2)(10.1_2)(110.11_2)$
24. $10010100.000101_2 \div 1110.101_2$
25. $11001.101101_2 \div 11.1_2$
26. $100010011.1010111_2 \div 10001.11_2$

UNIT 9·4 PROBLEM REVIEW

In problems 1–10, convert each binary number to its decimal equivalent. Round decimal fractions to five places.

1. 10011_2
2. 101101101_2
3. 1001011.101_2
4. 100110011_2
5. 1101.111_2
6. 111011_2
7. 0.01011_2
8. 1001011.1011_2
9. 1101010111.00111_2
10. 11100101100.110111_2

In problems 11–20, convert each decimal number to its binary equivalent. Round binary fractions to five places.

11. 42
12. 10.5
13. 27.3
14. 357
15. 86.72
16. $116\frac{15}{32}$
17. 1 296.95
18. $18\frac{53}{64}$
19. $\frac{191}{256}$
20. 1 000

In problems 21–25, find the 1's and 2's complement for each binary value given. Use zeros to retain the same number of places as in the original number.

21. 1010_2
22. 1100101_2
23. 1001101_2
24. 100010100_2
25. 10111101_2

In problems 26–50, perform the indicated operations. Round binary fractions to five places.

26. $1101101_2 + 1000111_2$
27. $10110_2 - 10011_2$
28. $(101_2)(1101_2)$
29. $10000101_2 \div 111_2$
30. $11101_2 + 10001_2 + 11011_2$
31. $1101011.1011_2 - 11101.11_2$
32. $(1101.11_2)^2$
33. $111111.1111_2 \div 111.11_2$

34. $11.011_2 + 1011.11_2 + 1101.011_2$
35. $10110101_2 - 11101011_2$
36. $(1001101_2)(100_2)$
37. $10000111_2 \div 101_2$
38. $1110101_2 + 11011_2 + 10111_2$
39. $100000_2 - 11111_2$
40. $(101_2)^3$
41. $10101_2 \div 100_2$
42. $10011_2 - 11101_2 + 10000_2$
43. $(1011_2)(11_2) - 10111_2$
44. $(110011_2 + 1110_2) \div 110_2$
45. $(100_2)^4$
46. $[(1101_2)(111_2) + 100011_2] \div 10101_2$
47. $(111011_2 - 101110_2)^2$
48. $[(11000_2)(0.01_2)]^2$
49. $(1011011_2 + 1000110_2) \div 1001_2$
50. $(0.1_2)^4$

CHAPTER 10

COMPUTER MATHEMATICS: OCTAL AND HEXADECIMAL NUMBERING SYSTEMS

The decimal and binary systems are just two of a large group of possible number systems. The binary system is of value in electronics since simple devices, such as a lamp or a switch, can be represented by the two number symbols in the binary system. A lamp that is off could be regarded as 0; a lighted lamp could be considered 1. A transistor not carrying current could be 0; the same transistor conducting current could be 1. A ferrous material magnetized in one direction could be 1 and when magnetized in the other direction could be 0.

Two other systems having practical value, particularly in connection with computers, are the *octal* and *hexadecimal* numbering systems.

UNIT 10·1 THE OCTAL NUMBER SYSTEM

Objectives:

After studying this unit, you should be able to

- convert decimal numbers to octal numbers.
- convert octal numbers to decimal numbers.

Octal Numbers

In the decimal system the base is 10, in the binary the base is 2, and in the octal it is 8. Just as the binary system uses two symbols from the decimal system, the octal system uses eight symbols from the decimal system. These symbols are 0, 1, 2, 3, 4, 5, 6, and 7. Subscripts are used to identify octal numbers, just as they are for binary and decimal numbers. A number such as 7356_8 is an octal number, marked as such by the subscript 8.

No matter what number system is selected, the final result must ultimately be converted to decimal, since decimal numbers are used in business and in technology. If a problem in electronics is solved in binary or octal form, the person reading the result thinks in terms of decimal numbers and therefore will make a conversion, even if it is only a mental conversion.

Table 10–1 shows the relationship between decimal and octal numbers from zero through 16. This small listing is enough to indicate just how octal symbol values are formed.

TABLE 10–1

Decimal Number and Octal Number Relationships (0–16)

Decimal	Octal	Decimal	Octal
0	0	9	11
1	1	10	12
2	2	11	13
3	3	12	14
4	4	13	15
5	5	14	16
6	6	15	17
7	7	16	20
8	10		

Converting Octal Numbers to Decimal Numbers

The decimal equivalents of octal numbers can be calculated by using the powers of 8 shown in Figure 10–1. The radix point, or octal point, separates whole octal numbers from octal fractions. The octal number 743_8, for example, is equivalent to $(7 \times 8^2) + (4 \times 8^1) + (3 \times 8^0) = 483_{10}$.

8^5	8^4	8^3	8^2	8^1	8^0	8^{-1}	8^{-2}	8^{-3}	8^{-4}
32 768	4 096	512	64	8	1	$\frac{1}{8}$	$\frac{1}{64}$	$\frac{1}{512}$	$\frac{1}{4\,096}$

OCTAL POINT

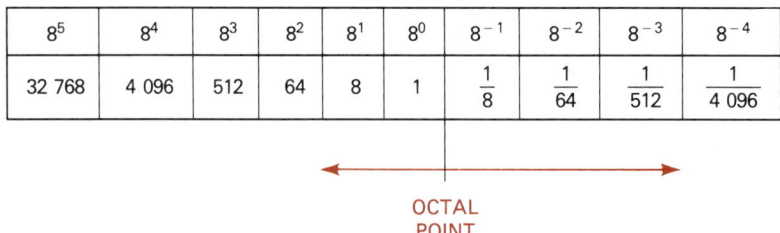

FIGURE 10–1 Powers of 8 and their decimal number equivalents

Example: Convert 5607_8 to its decimal equivalent.

Solution: Set up a conversion chart.

$8^3\ 8^2\ 8^1\ 8^0$

5 6 0 7

$$5607_8 = (5 \times 8^3) + (6 \times 8^2) + (0 \times 8^1) + (7 \times 8^0)$$
$$= 2\ 560 + 384 + 0 + 7$$
$$= 2\ 951_{10}$$

Example: Convert 364_8 to its decimal equivalent.

Solution:
$$364_8 = (3 \times 8^2) + (6 \times 8^1) + (4 \times 8^0)$$
$$= 192 + 48 + 4$$
$$= 244_{10}$$

Example: Convert 7063_8 to its decimal equivalent.

Solution:
$$7063_8 = (7 \times 8^3) + (0 \times 8^2) + (6 \times 8^1) + (3 \times 8^0)$$
$$= 3\ 584 + 0 + 48 + 3$$
$$= 3\ 635_{10}$$

Converting Decimal Numbers to Octal Numbers

The conversion of decimal numbers to their octal equivalents follows a process similar to the double-dabble method used in binary conversions. This involves repeatedly dividing the number by 8, writing down the remainders, and then reading the remainders in *reverse* order.

Example: Convert 270_{10} to its octal equivalent.

Solution: $270 \div 8 = 33$ with a remainder of 6

$33 \div 8 = 4$ with a remainder of 1

$4 \div 8 = 0$ with a remainder of 4

Reading the remainders in reverse order results in 416_8. To verify the answer convert 416_8 to a decimal.

$$416_8 = (4 \times 8^2) + (1 \times 8^1) + (6 \times 8^0)$$
$$= 256 + 8 + 6$$
$$= 270_{10}$$

Example: Convert $18\ 096_{10}$ to its octal equivalent.

Solution: $18\ 096 \div 8 = 2\ 262$ with a remainder of 0

$2\ 262 \div 8 = 282$ with a remainder of 6

$282 \div 8 = 35$ with a remainder of 2

$35 \div 8 = 4$ with a remainder of 3

$4 \div 8 = 0$ with a remainder of 4

Resulting in $43260_8 = 18\ 096_{10}$

Fractional Conversions

As in binary and decimal systems, fractional octal numbers exist to the right of the octal point. Figure 10–1 indicates that a symbol in the first place to the right of the octal point would be multiplied by 8^{-1} or $\frac{1}{8}$.

Given one or more symbols to the right of the octal point, the decimal equivalent is found through the same process as is used with whole octal numbers. Review these examples closely.

Example: Convert 0.372_8 to its decimal equivalent.

Solution: Set up a conversion chart.

8^{-1}	8^{-2}	8^{-3}
0.3	7	2

$$\begin{aligned} 0.372_8 &= (3 \times 8^{-1}) + (7 \times 8^{-2}) + (2 \times 8^{-3}) \\ &= \frac{3}{8} + \frac{7}{64} + \frac{2}{512} \\ &= 0.488\ 281\ 25_{10} \end{aligned}$$

Example: Convert 0.706_8 to its decimal equivalent.

Solution:
$$\begin{aligned} 0.706_8 &= (7 \times 8^{-1}) + (0 \times 8^{-2}) + (6 \times 8^{-3}) \\ &= \frac{7}{8} + \frac{6}{512} \\ &= 0.886\ 718\ 75_{10} \end{aligned}$$

In order to convert a decimal fraction to its octal equivalent, simply multiply the fraction by 8. If the product is greater than 1, list the whole number and move the *remaining fraction* forward. If the product is less than 1, record a 0 and move the fraction forward. This process is repeated until the product is a whole number, or until sufficient accuracy is obtained. The octal number is read beginning with the first product. Follow these examples carefully.

Example: Convert $0.437\ 5_{10}$ to its octal equivalent.

Solution: $0.437\ 5 \times 8 = 3.5$, record a 3

$0.5 \times 8 = 4.0$, record a 4

The octal value is 0.34_8. To verify our results convert the octal fraction to its decimal equivalent:

$$\begin{aligned} 0.34_8 &= (3 \times 8^{-1}) + (4 \times 8^{-2}) \\ &= \frac{3}{8} + \frac{4}{64} \\ &= 0.437\ 5_{10} \end{aligned}$$

Example: Convert the decimal fraction $\frac{37}{64}$ to its octal equivalent.

Solution: $\frac{37}{64} = 0.578\ 125$

$0.578\ 125 \times 8 = 4.625$, record a 4

$0.625 \times 8 = 5.0$, record a 5

Thus, $0.578\ 125_{10} = 0.45_8$

Example: Convert $0.351\ 562\ 5_{10}$ to its octal equivalent.

Solution: $0.351\ 562\ 5 \times 8 = 2.812\ 5$, record a 2

$0.812\ 5 \times 8 = 6.5$, record a 6

$0.5 \times 8 = 4.0$, record a 4

Resulting in 0.264_8

EXERCISE 10·1

In problems 1–15, convert each octal number to its decimal equivalent. Round fractional decimal values to four decimal places.

1. 12_8
2. 207_8
3. 1.3_8
4. 6.35_8
5. 7164_8
6. 101_8
7. 2.57_8
8. 13.4_8
9. 1000_8
10. 66275_8
11. 3004_8
12. 0.3017_8
13. 517243_8
14. 4462.73_8
15. 5336.56_8

In problems 16–30, convert each decimal number to its octal equivalent. Round fractional octal values to four places if necessary.

16. 31
17. 86
18. 281
19. $44\frac{15}{32}$
20. 0.916
21. 5 277
22. 1 755.375
23. 0.262 939 5
24. $370\frac{3}{16}$
25. 22 011
26. 388.953 125
27. 4 161
28. 503.843 75
29. 217 340
30. 3 451.523 193

UNIT 10·2 THE HEXADECIMAL NUMBER SYSTEM

Objectives:

After studying this unit, you should be able to

- convert decimal numbers to hexadecimal numbers.
- convert hexadecimal numbers to decimal numbers.

Hexadecimal Numbers

Computers use the binary system to register data. A memory circuit in the computer can be the equivalent of four successive binary symbols, such as 1011_2 or 1110_2, etc. The maximum value of four successive binary symbols, 1111_2, is $(1 \times 2^3) + (1 \times 2^2) + (1 \times 2^1) + (1 \times 2^0) = 8 + 4 + 2 + 1 = 15$. The minimum value of four successive binary symbols, 0000_2, is zero. With four successive binary symbols, then, we can represent the equivalent of any decimal number from 0 through 15, or a total of 16 different numbers. Since there are 16 possible combinations using four binary symbols, it is convenient to set up a number system which can represent the 16 combinations. This is the *hexadecimal* system, which uses 16 different symbols to furnish the equivalent of the decimal numbers 0 through 15. Table 10–2 lists decimal numbers from 0 through 15 and the corresponding hexadecimal symbols.

TABLE 10–2

Decimal Number and Hexadecimal Number Relationships (0–15)

Decimal	Hexadecimal	Decimal	Hexadecimal
0	0	8	8
1	1	9	9
2	2	10	A
3	3	11	B
4	4	12	C
5	5	13	D
6	6	14	E
7	7	15	F

Hexadecimal numbers are identified in the same way as binary, octal, and decimal numbers—through the use of a subscript. As in these other number systems, the subscript indicates the total number of symbols used in the particular system. In binary the subscript is 2, in octal it is 8, in decimal, 10, and in hexadecimal, 16. A hexadecimal number can be written as $135A_{16}$ or $1498CD_{16}$. Since hexadecimal uses all the symbols of the decimal system, subscript notation becomes important whenever there is a possibility of misunderstanding the number system being used.

The relationships of numbers in the decimal, binary, octal, and hexadecimal systems, ranging from zero through 32 are shown in Table 10–3. This table shows whole numbers only, not fractions. While an extended table of this kind would be useful for conversions between the various number systems, it would need to be very lengthy to cover every practical number combination.

For positive exponents, a list of powers of 16 can be set up in this way.

$$16^0 = 1 \qquad 16^4 = 65\ 536$$
$$16^1 = 16 \qquad 16^5 = 1\ 048\ 576$$
$$16^2 = 256 \qquad 16^6 = 16\ 777\ 216$$
$$16^3 = 4\ 096 \qquad 16^7 = 268\ 435\ 456$$

TABLE 10–3

Decimal	Binary	Octal	Hexadecimal
0	00000	0	0
1	00001	1	1
2	00010	2	2
3	00011	3	3
4	00100	4	4
5	00101	5	5
6	00110	6	6
7	00111	7	7
8	01000	10	8
9	01001	11	9
10	01010	12	A
11	01011	13	B
12	01100	14	C
13	01101	15	D
14	01110	16	E
15	01111	17	F
16	10000	20	10
17	10001	21	11
18	10010	22	12
19	10011	23	13
20	10100	24	14
21	10101	25	15
22	10110	26	16
23	10111	27	17
24	11000	30	18
25	11001	31	19
26	11010	32	1A
27	11011	33	1B
28	11100	34	1C
29	11101	35	1D
30	11110	36	1E
31	11111	37	1F
32	100000	40	20

Number Relationships in Decimal, Binary, Octal, and Hexadecimal (0–32)

A list can also be set up for negative powers of 16.

$$16^{-1} = \frac{1}{16} = 0.062\ 5$$

$$16^{-2} = \frac{1}{256} = 0.003\ 906\ 25$$

$$16^{-3} = \frac{1}{4\ 096} = 0.000\ 244\ 14$$

$$16^{-4} = \frac{1}{65\ 536} = 0.000\ 015\ 25$$

Converting Hexadecimal Numbers to Decimal Numbers

Hexadecimal numbers, just as in the binary, octal, and decimal systems, have values depending on their position with respect to the radix point. Starting at the radix point and moving to the left, the farther a number is away from the point, the greater its value. Again, starting at the hexadecimal point and moving away from it to the right, the farther a number is from the point, the smaller its value.

Powers of 16 can be arranged as shown in Figure 10–2 to supply a better indication of the positional value. The hexadecimal number 483_{16}, for example, has a decimal equivalent value of $(4 \times 16^2) + (8 \times 16^1) + (3 \times 16^0) =$ $1\ 155_{10}$

16^4	16^3	16^2	16^1	16^0	16^{-1}	16^{-2}	16^{-3}	16^{-4}
65 536	4 096	256	16	1	$\frac{1}{16}$	$\frac{1}{256}$	$\frac{1}{4\ 096}$	$\frac{1}{65\ 536}$

HEXADECIMAL POINT

FIGURE 10–2 Powers of 16 and their decimal number equivalents

Example: Convert $BFA35_{16}$ to its decimal equivalent.

Solution: Set up a conversion table.

16^4	16^3	16^2	16^1	16^0
B	F	A	3	5

$$\begin{aligned} BFA35_{16} &= (11 \times 16^4) + (15 \times 16^3) + (10 \times 16^2) + (3 \times 16^1) \\ &\quad + (5 \times 16^0) \\ &= 720\ 896 + 61\ 440 + 2\ 560 + 48 + 5 \\ &= 784\ 949_{10} \end{aligned}$$

Figure 10–3 shows the same technique for converting a hexadecimal number

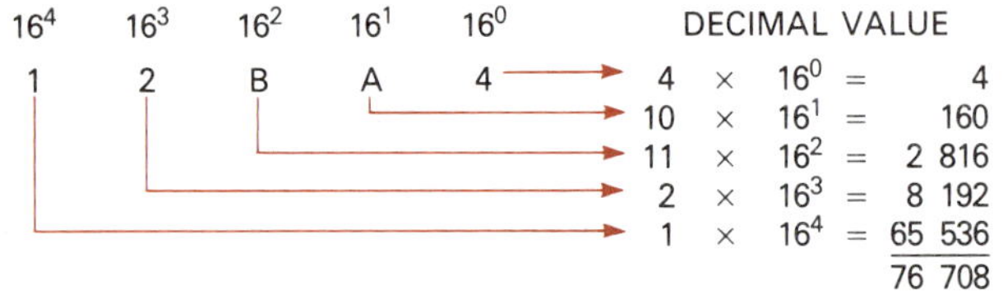

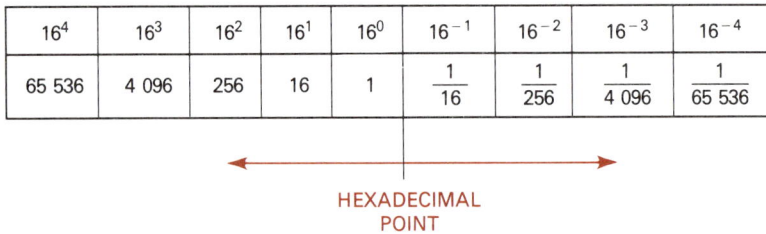

$12BA4_{16} = 76\ 708_{10}$

FIGURE 10–3 Method for converting hexadecimal numbers to decimal numbers

but somewhat more graphically arranged to emphasize the significance of the symbols and the positional values using the powers of 16.

Example: Convert $E8B.D5_{16}$ to its decimal equivalent.

Solution: Set up a conversion table.

16^2	16^1	16^0	.	16^{-1}	16^{-2}
E	8	B	.	D	5

$$E8B.D5_{16} = (14 \times 16^2) + (8 \times 16^1) + (11 \times 16^0) + (13 \times 16^{-1}) + (5 \times 16^{-2})$$
$$= 3\ 723.832\ 031\ 25_{10}$$

Converting Decimal Numbers to Hexadecimal Numbers

There are various ways of converting a decimal number to its hexadecimal equivalent. One method is to divide the given decimal repeatedly by 16, following the same technique previously described in connection with octal to decimal conversions. In the octal conversion process, the divisor was 8; in the hexadecimal, it is 16. The following problem represents one example, while still another is illustrated in Figure 10–4.

16 ⟌235 415	=	14 713	REMAINDER = 7
16 ⟌14 713	=	919	REMAINDER = 9
16 ⟌919	=	57	REMAINDER = 7
16 ⟌57	=	3	REMAINDER = 9
16 ⟌3	=	0	REMAINDER = 3

$235\ 415_{10} = 39797_{16}$

FIGURE 10–4 Method for converting decimal numbers to hexadecimal numbers

Example: Convert $2\ 879\ 106_{10}$ to its hexadecimal equivalent.

Solution:

2 879 106 ÷ 16 = 179 944, with a remainder of 2

179 944 ÷ 16 = 11 246, with a remainder of 8

11 246 ÷ 16 = 702, with a remainder of 14

702 ÷ 16 = 43, with a remainder of 14

43 ÷ 16 = 2, with a remainder of 11

2 ÷ 16 = 0, with a remainder of 2

Resulting in $2\ 879\ 106_{10} = 2BEE82_{16}$

Fractional Conversions

As in other numbering systems, fractional hexadecimal numbers exist to the right of the hexadecimal point. Figure 10–2 indicates that a symbol in the first place to the right of the point would be multiplied by 16^{-1} or $\frac{1}{16}$.

Given one or more symbols to the right of the hexadecimal point, the decimal equivalent is found through the same process as is used with whole hexadecimal numbers. One example was presented earlier in this unit. Review these additional examples.

Example: Convert $0.7BF_{16}$ to its decimal equivalent.

Solution:

$$\begin{aligned} 0.7BF_{16} &= (7 \times 16^{-1}) + (B \times 16^{-2}) + (F \times 16^{-3}) \\ &= (7 \times 16^{-1}) + (11 \times 16^{-2}) + (15 \times 16^{-3}) \\ &= 0.484\ 130\ 86_{10} \end{aligned}$$

Example: Convert $0.A92E_{16}$ to its decimal equivalent.

Solution:

$$\begin{aligned} 0.A92E_{16} &= (10 \times 16^{-1}) + (9 \times 16^{-2}) + (2 \times 16^{-3}) \\ &\quad + (14 \times 16^{-4}) \\ &= 0.660\ 858\ 15_{10} \end{aligned}$$

The purpose of discussing numbering systems in this text is to acquaint you with the systems often found in calculator and computer circuits. In the previous examples the decimal equivalents were carried out to eight decimal places. While this may be somewhat time consuming, computers and calculators will work at accuracies exceeding ±1%. In order to show the conversion process, the examples presented here will be rounded at eight decimal places.

The process to convert a decimal fraction to a hexadecimal fraction involves multiplying the fraction by 16. If the product is greater than 1, list the whole number and move the *remaining fraction* forward. If the product is less than 1, record a 0 and move the fraction forward. This process is repeated until the product is a whole number, or until sufficient accuracy is obtained. The hexadecimal number is read beginning with the first product. Follow these examples very closely.

Example: Convert $0.843\ 75_{10}$ to its hexadecimal equivalent.

Solution: $0.843\ 75 \times 16 = 13.5$, record 13

$0.5 \times 16 = 8.0$, record 8

Resulting in $0.843\ 75_{10} = 0.D8_{16}$

To verify our results convert this value back to a decimal.

$$\begin{aligned} 0.D8_{16} &= (13 \times 16^{-1}) + (8 \times 16^{-2}) \\ &= 0.843\ 75_{10} \end{aligned}$$

Example: Convert 0.37_{10} to its hexadecimal equivalent.

Solution: $0.37 \times 16 = 5.92$, record 5

$0.92 \times 16 = 14.72$, record 14

$0.72 \times 16 = 11.52$, record 11

$0.52 \times 16 = 8.32$, record 8

$0.32 \times 16 = 5.12$, record 5

$0.12 \times 16 = 1.92$, record 1

$0.92 \times 16 = 14.72$, record 14

As the fraction is repeating, the equivalent hexadecimal fraction will be rounded at the seventh place. The approximate equivalent fraction is:

$0.37_{10} = 0.5EB851E_{16}$

In order to check the accuracy of our conversion, simply reverse the process and convert the hexadecimal number to its decimal equivalent.

$$0.5EB851E_{16} = (5 \times 16^{-1}) + (14 \times 16^{-2}) + (11 \times 16^{-3}) + (8 \times 16^{-4}) + (5 \times 16^{-5}) + (1 \times 16^{-6}) + (14 \times 16^{-7}) = 0.37_{10}$$

EXERCISE 10·2

In problems 1–15, convert each hexadecimal number to its decimal equivalent. Round fractional values to eight decimal places.

1. 38_{16}
2. 127_{16}
3. $A12_{16}$
4. $A8B9_{16}$
5. $0.6B2_{16}$
6. $3E.54C_{16}$
7. $D604F_{16}$
8. 731.29_{16}
9. $61BA5_{16}$
10. $101.11E_{16}$
11. $2C6AB7_{16}$
12. $0.B75_{16}$
13. $45CF.B93_{16}$
14. $89BA.C4_{16}$
15. $A3DF18_{16}$

In problems 16–30, convert each decimal number to its hexadecimal equivalent. Round fractional values to three hexadecimal places if necessary.

16. 967
17. 159.695 312 5
18. 857 523
19. 0.457 031 25
20. 28 164
21. 569 234
22. 1 931.742 187 5
23. 14 091 237

24. 2 847.847 412 11
25. 1.062 5
26. 5 343 660
27. 176.125
28. 32 169
29. 0.556 884 77
30. 13 256 603.566 162 11

UNIT 10·3 BINARY-OCTAL-HEXADECIMAL CONVERSIONS

Objectives:

After studying this unit, you should be able to

- perform conversions between binary and octal numbers.
- perform conversions between octal and hexadecimal numbers.
- perform conversions between binary and hexadecimal numbers.

Decimal-Binary-Octal-Hexadecimal Relationships

Thus far in our study of number systems we have learned to count in binary, octal, and hexadecimal. In addition, we have demonstrated methods by which decimal numbers can be expressed in these three systems. The purpose of this unit is to demonstrate how conversions are performed among binary, octal, and hexadecimal number systems.

Conversions involving fairly low numbers are simple if a table, such as that shown in Table 10–4, is available. This table will be used extensively in this unit and displays the binary, octal, and hexadecimal equivalents of decimal numbers 0 through 15.

TABLE 10–4

Number Relationships in Decimal, Binary, Octal, and Hexadecimal (0–15)

Decimal	Binary	Octal	Hexadecimal
0	0000	0	0
1	0001	1	1
2	0010	2	2
3	0011	3	3
4	0100	4	4
5	0101	5	5
6	0110	6	6
7	0111	7	7
8	1000	10	8
9	1001	11	9
10	1010	12	A
11	1011	13	B
12	1100	14	C
13	1101	15	D
14	1110	16	E
15	1111	17	F

Converting Binary Numbers to Octal Numbers

Binary is easily converted to octal by grouping the binary numbers into *triads* (groups of three) and then converting directly to octal. Start the triad groups at the binary radix point, counting off in groups of three, moving outward from the radix point. If the highest order binary is either a single digit or two digits, add either one or two zeros to the left. This will not change the value of the binary but will supply a triad group.

The requirement for triad formation of the binary is based on the fact that the maximum value of a binary triad is 7, the largest number in the octal system. If the binaries were arranged in tetrads, groups of four, the largest number would be 15, a number that cannot be represented by a single symbol in the octal system.

Example: Convert 10110101_2 to octal form.

Solution: Counting off triads beginning at the binary point.

10 110 101

To complete the triad formation, add a zero to the left of the first triad.

010 110 101

This binary can now be converted directly to octal using Table 10–4.

010 110 101
↓ ↓ ↓
2 6 5

Resulting in $10110101_2 = 265_8$

To check the work, convert both binary and octal to decimal. Identical answers will show that the binary conversion to octal is correct. In this example:

$$
\begin{aligned}
10110101_2 &= (1 \times 2^7) + (0 \times 2^6) + (1 \times 2^5) + (1 \times 2^4) \\
&\quad + (0 \times 2^3) + (1 \times 2^2) + (0 \times 2^1) + (1 \times 2^0) \\
&= 128 + 0 + 32 + 16 + 0 + 4 + 0 + 1 \\
&= 181_{10}
\end{aligned}
$$

$$
\begin{aligned}
265_8 &= (2 \times 8^2) + (6 \times 8^1) + (5 \times 8^0) \\
&= 128 + 48 + 5 \\
&= 181_{10}
\end{aligned}
$$

Figure 10–5 shows another sample problem using this method.

Converting mixed binary numbers to mixed octal numbers follows the

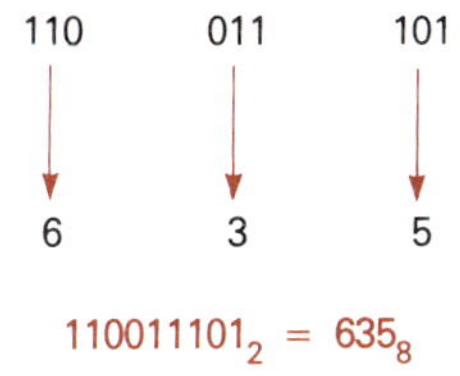

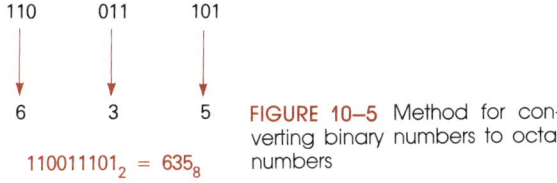

FIGURE 10–5 Method for converting binary numbers to octal numbers

same process. Beginning at the binary point and moving outward, group the binary values in groups of three. Add one or two zeros to the last value(s) if necessary. Convert each triad to its octal equivalent.

Example: Convert 1101100.01101_2 to its octal equivalent. Check the result by converting both values to their decimal equivalents.

Solution:

$$1101100.01101_2 = \begin{array}{ccccccc} 001 & 101 & 100 & . & 011 & 010 & \text{(Binary)} \\ \downarrow & \downarrow & \downarrow & & \downarrow & \downarrow & \\ 1 & 5 & 4 & . & 3 & 2 & \text{(Octal)} \end{array}$$

Resulting in $1101100.01101_2 = 154.32_8$

In order to verify our results:

$$1101100.01101_2 = (1 \times 2^6) + (1 \times 2^5) + (1 \times 2^3) + (1 \times 2^2) + (1 \times 2^{-2}) + (1 \times 2^{-3}) + (1 \times 2^{-5}) = 108.406\ 25_{10}$$

$$154.32_8 = (1 \times 8^2) + (5 \times 8^1) + (4 \times 8^0) + (3 \times 8^{-1}) + (2 \times 8^{-2}) = 108.406\ 25_{10}$$

Example: Convert 11110011001.1101_2 to its octal equivalent.

Solution:

$$\begin{array}{ccccccc} 011 & 110 & 011 & 001 & . & 110 & 100 \\ \downarrow & \downarrow & \downarrow & \downarrow & & \downarrow & \downarrow \\ 3 & 6 & 3 & 1 & . & 6 & 4 \end{array}$$

Resulting in $11110011001.1101_2 = 3631.64_8$

Converting Octal Numbers to Binary Numbers

The binary to octal process can be reversed to change numbers in the octal system to binary equivalents.

Example: Convert 6754_8 to binary.

Solution:

$$\begin{array}{cccc} 6 & 7 & 5 & 4 \\ \downarrow & \downarrow & \downarrow & \downarrow \\ 110 & 111 & 101 & 100 \end{array}$$

Resulting in $6754_8 = 110111101100_2$

Figure 10–6 illustrates another example of octal to binary conversion.

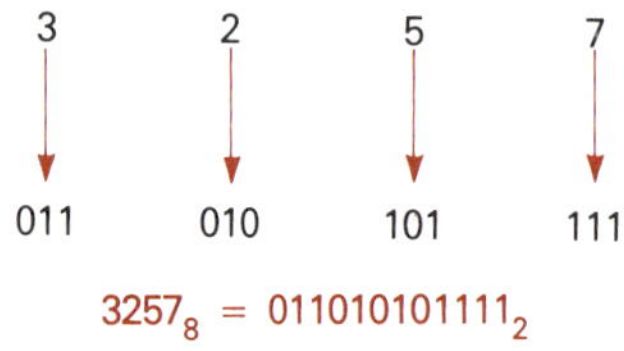

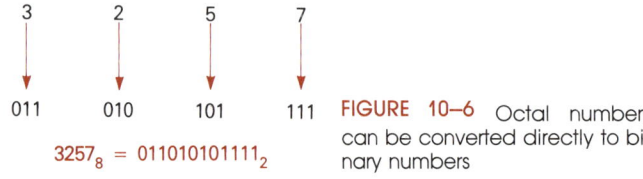

FIGURE 10–6 Octal numbers can be converted directly to binary numbers

Example: Convert 407.63_8 to its binary equivalent. Check the result by converting both values to their decimal equivalents.

Solution:

$$\begin{array}{ccccc} 4 & 0 & 7 & . & 6 & 3 \\ \downarrow & \downarrow & \downarrow & & \downarrow & \downarrow \\ 100 & 000 & 111 & . & 110 & 011 \end{array}$$

Resulting in $407.635_8 = 100000111.110011_2$

To verify our results:

$$\begin{aligned} 407.63_8 &= (4 \times 8^2) + (7 \times 8^0) + (6 \times 8^{-1}) + (3 \times 8^{-2}) \\ &= 263.796\ 875_{10} \end{aligned}$$

$$\begin{aligned} 100000111.110011_2 &= (1 \times 2^8) + (1 \times 2^2) + (1 \times 2^1) \\ &\quad + (1 \times 2^0) + (1 \times 2^{-1}) + (1 \times 2^{-2}) \\ &\quad + (1 \times 2^{-5}) + (1 \times 2^{-6}) \\ &= 263.796\ 875_{10} \end{aligned}$$

Example: Convert 7536_8 to its binary equivalent.

Solution: You should verify that the result is 111101011110_2.

Converting Hexadecimal Numbers to Binary Numbers

Hexadecimal numbers convert easily to binary form, actually so readily that it is possible to convert a mixed hexadecimal—that is, a hexadecimal consisting of a whole number and a fractional portion—directly. The only requirement is that the binary appear in *tetrad* form. A tetrad binary is one consisting of four binary symbols, such as 1101, 1001, etc. This is due to the fact that $2^4 = 16$, so that a tetrad binary number can be converted directly to a hexadecimal number.

Example: Convert $9B4.2B_{16}$ to binary form.

Solution:

$$\begin{array}{cccccccl} 9 & B & 4 & . & 2 & B & & \text{(Hexadecimal)} \\ \downarrow & \downarrow & \downarrow & & \downarrow & \downarrow & & \\ 1001 & 1011 & 0100 & . & 0010 & 1011 & & \text{(Binary)} \end{array}$$

Resulting in $9B4.2B_{16} = 100110110100.00101011_2$

Problems of this type can be checked by converting both binary and hexadecimal to decimal to see if the answers agree.

Example: Convert 25_{16} to binary form.

Solution:

$$\begin{array}{cc} 2 & 5 \\ \downarrow & \downarrow \\ 0010 & 0101 \end{array}$$

$25_{16} = 00100101_2$

Check:

$$\begin{aligned} 25_{16} &= (2 \times 16^1) + (5 \times 16^0) \\ &= 37_{10} \end{aligned}$$

$$\begin{aligned} 00100101_2 &= (1 \times 2^5) + (0 \times 2^4) + (0 \times 2^3) + (1 \times 2^2) \\ &\quad + (0 \times 2^1) + (1 \times 2^0) \\ &= 37_{10} \end{aligned}$$

Example: Convert $7EA1_{16}$ to its binary equivalent. Check the result by converting both numbers to their decimal equivalents.

Solution:

7	E	A	1
↓	↓	↓	↓
0111	1110	1010	0001

Resulting in $7EA1_{16} = 111111010100001_2$

You should verify that each of these is equivalent to $32\,417_{10}$.

Converting Binary Numbers to Decimal Numbers Using Hexadecimal Numbers

The hexadecimal number system provides a convenient way of converting from binary to decimal. The technique involves setting up the given binary number in tetrad form and adding zeros to the extreme left or extreme right of the binary, as required.

Example: Convert 1011011110.101_2 to its decimal equivalent, using the hexadecimal number system as an intermediary step.

Solution: Set up the binary in tetrad form, beginning at the radix point.

10 1101 1110.101

Add zeros left and right to complete the formation of tetrads.

0010 1101 1110.1010

Convert this tetrad binary group directly into hexadecimal.

0010	1101	1110	.	1010
↓	↓	↓		↓
2	D	E	.	A

Resulting in $1011011110.101_2 = 2DE.A_{16}$

Find the decimal equivalent.

$$\begin{aligned} 2DE.A_{16} &= (2 \times 16^2) + (13 \times 16^1) + (14 \times 16^0) + (10 \times 16^{-1}) \\ &= 512 + 208 + 14 + 0.625 \\ &= 734.625_{10} \end{aligned}$$

An alternative, of course, would be to convert directly from binary to decimal, using each binary symbol multiplied by the appropriate power of 2. While this would be a practical approach for a relatively small binary number, it becomes impractical for large binaries.

Converting Decimal Numbers to Binary Numbers Using Hexadecimal Numbers

The conversion of a decimal number to binary by using a hexadecimal number is the reverse process of the binary to decimal method just described. In this technique, the decimal number is repeatedly divided by 16, resulting in a hexadecimal number. The hexadecimal number is then converted to binary tetrads.

Example: Convert 98 555_{10} to binary form using the hexadecimal system as an intermediary step.

Solution:

16)98 555 = 6 159 (Remainder of 11 or B_{16})

16)6 159 = 384 (Remainder of 15 or F_{16})

16)384 = 24 (Remainder of 0)

16)24 = 1 (Remainder of 8)

16)1 = 0 (Remainder of 1)

Resulting in 98 555_{10} = $180FB_{16}$

Find the binary equivalent.

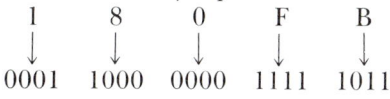

1	8	0	F	B
↓	↓	↓	↓	↓
0001	1000	0000	1111	1011

Resulting in 98 555_{10} = 11000000011111011_2

Converting Octal Numbers to Hexadecimal Numbers Using Binary Numbers

The process to convert an octal number to its hexadecimal equivalent can be accomplished through the use of binary numbers. Remember that octal is based upon 2^3 or 8, and that hexadecimal uses 2^4 or 16. With this in mind, it will be necessary to convert the octal number to binary triads, add additional zeros as required, and then change the triads to tetrads. The tetrads are then converted to the equivalent hexadecimal number. Follow these examples carefully.

Example: Convert 53_8 to its hexadecimal equivalent using binary numbers.

Solution:

5 3 (Octal)

↓ ↓

101 011 (Binary triads)

↘ ↙

101011 (Binary equivalent)

↙ ↘

0010 1011 (Binary tetrads)

↓ ↓

2 B (Hexadecimal)

Resulting in 53_8 = $2B_{16}$

Example: Convert 375_8 to its hexadecimal equivalent using binary numbers. Verify the result by expressing both values as decimal numbers.

Solution: $375_8 = 011\ 111\ 101$
$= 11111101$
$= 1111\ 1101$
$= FD_{16}$

You should verify that each of these is equivalent to 253_{10}.

Example: Convert 0.4207_8 to hexadecimal.

Solution: $0.4207_8 = 0.100\ 010\ 000\ 111$
$= 0.100010000111$
$= 0.1000\ 1000\ 0111$
$= 0.887_{16}$

Converting Hexadecimal Numbers to Octal Numbers Using Binary Numbers

In order to convert hexadecimal numbers to their octal equivalents simply reverse the process used for octal to hexadecimal conversions. The first step is to change the hexadecimal to binary tetrads. These tetrads are combined and, beginning at the binary point, seperated into binary triads. Additional zeros are added, if necessary, and the triads are converted back to their octal equivalents.

Example: Convert $38F_{16}$ to its octal equivalent using binary numbers.

Solution:

3 ↓ 0011, 8 ↓ 1000, F ↓ 1111	(Hexadecimal) → (Binary tetrads)
1110001111	(Binary equivalent)
001 110 001 111	(Binary triads)
1 6 1 7	(Octal)

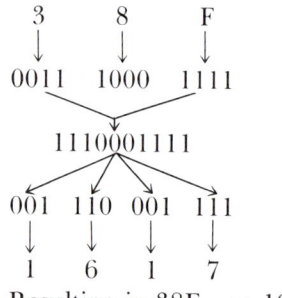

Resulting in $38F_{16} = 1617_8$

Example: Convert $8DF.5C_{16}$ to octal.

Solution: $8DF.5C_{16} = 1000\ 1101\ 1111\ .\ 0101\ 1100$
$= 100011011111.01011100$
$= 100\ 011\ 011\ 111\ .\ 010\ 111$
$= 4337.27_8$

EXERCISE 10·3

In problems 1–10, convert each binary number to its octal equivalent.

1. 110101_2
2. 11110110_2
3. 1011011_2
4. 1110001101_2
5. 11101100101_2
6. 10111011_2
7. 101100.11011_2
8. 110110111.01101_2
9. 0.01101110111_2
10. 10.01_2

In problems 11–20, convert each octal number to its binary equivalent.

11. 707_8
12. 265.034_8
13. 3166_8
14. 541_8
15. 0.627_8
16. 125.47_8
17. 4550_8
18. 1_8
19. 20.16_8
20. 1000_8

In problems 21–30, convert each hexadecimal number to its binary equivalent.

21. 10_{16}
22. $4C_{16}$
23. 36_{16}
24. $29B_{16}$
25. $10A_{16}$
26. $5B6.F3_{16}$
27. CAB_{16}
28. $82E9_{16}$
29. $0.78D_{16}$
30. $96F.D5_{16}$

In problems 31–40, convert each binary number to its hexadecimal equivalent.

31. 101101_2
32. 110111101.1101011_2
33. 111_2
34. 101111000111001_2
35. 10001110111101_2
36. 10100100.11101001_2
37. 11111111001011_2
38. 1_2
39. 10001101.000010111_2
40. 1010101010101.0101_2

In problems 41–50, convert each octal number to its hexadecimal equivalent.

41. 7_8
42. 13_8
43. 250_8
44. 116_8
45. 417_8
46. 6725_8
47. 0.532_8
48. 1430.6_8
49. 37165_8
50. 40042_8

In problems 51–60, convert each hexadecimal number to its octal equivalent.

51. $5B_{16}$
52. $9F.1C_{16}$
53. 4728_{16}
54. $1.D_{16}$
55. $0.3E9_{16}$
56. 100_{16}
57. $2A05_{16}$
58. $D707_{16}$
59. $4FF5_{16}$
60. $379AF_{16}$

UNIT 10·4 PROBLEM REVIEW

In problems 1–10, convert each decimal number to its octal equivalent.

1. 241
2. 454.375
3. 25 444
4. 3 990
5. $\frac{61}{64}$
6. 351.201 172
7. 15 633
8. 61.781 25
9. 240 515.593 75
10. 262 143

In problems 11–20, convert each octal number to its decimal equivalent. If necessary, round decimal fractions to five places.

11. 27_8
12. 561.34_8
13. 1064_8
14. 7134.267_8
15. 0.461_8
16. 0.7265_8
17. 44173_8
18. 65737.42_8
19. 621467_8
20. 773054_8

In problems 21–30, convert each decimal number to its hexadecimal equivalent.

21. 41
22. 2 214.5
23. $\frac{231}{256}$
24. 27 567
25. 137.875
26. 64 180
27. 403 273
28. 0.785 156 25
29. 36 117
30. 653 067

In problems 31–40, convert each hexadecimal number to its decimal equivalent. If necessary, round decimal fractions to five places.

31. 970_{16}
32. $B7F_{16}$
33. $0.8C_{16}$
34. $D76.A_{16}$
35. 4238_{16}
36. $6BFD_{16}$
37. $5F439_{16}$
38. $729.0F_{16}$
39. $906FA_{16}$
40. $8CB16_{16}$

In problems 41–50, convert each binary number to its octal equivalent.

41. 110111_2
42. 111011011_2
43. 1011011101_2
44. 0.11011001_2
45. 1111.101101_2
46. 1000100110001_2
47. 10011100011101_2
48. 111000111011.10011_2
49. 10.00000111011_2
50. 1111100011011001101_2

In problems 51–60, convert each octal number to its binary equivalent.

51. 736_8
52. 201.5_8
53. 1476_8
54. 3205_8
55. 0.37_8
56. 6.75_8
57. 64726_8
58. 1011.101_8
59. 542.172_8
60. 32760_8

In problems 61–70, convert each binary number to its hexadecimal equivalent.

61. 101_2
62. 111001110001_2
63. 11011.10001101_2
64. $111100110001.11011000011_2$
65. 1000001101111_2
66. 111111001000110101_2
67. 1000110001011001_2
68. 0.110010011000001101_2
69. $110001010110001110110 01_2$
70. $100000001110101101101111_2$

In problems 71–80, convert each hexadecimal number to its binary equivalent.

71. $7C_{16}$
72. 104_{16}
73. $62F_{16}$
74. $BA9_{16}$
75. 0.79_{16}
76. $16.2F8_{16}$
77. $327E_{16}$
78. 406_{16}
79. $1D.7A_{16}$
80. $5AF97_{16}$

In problems 81–90, convert each hexadecimal number to its octal equivalent.

81. $5F_{16}$
82. 297_{16}
83. $6BC.A_{16}$
84. $130A_{16}$
85. $BA9_{16}$
86. $9F7.3C_{16}$
87. $0.4D8_{16}$
88. $37CE_{16}$
89. $197F.8D_{16}$
90. $6247E_{16}$

In problems 91–100, convert each octal number to its hexadecimal equivalent.

91. 673_8
92. 1047.623_8
93. 25_8
94. 57703_8
95. 477.625_8
96. 3007.51_8
97. 1100.1001_8
98. 462573_8
99. 6621635_8
100. 77740167_8

CHAPTER 11

COMPUTER MATHEMATICS: BOOLEAN ALGEBRA

The mathematics of digital circuitry is based upon the logic work of, among others, De Morgan and Boole. The two-valued algebraic logic developed by Boole over one hundred years ago is known today as *Boolean Algebra.* This system makes it possible to construct circuits based upon mathematical expressions, as well as write the expressions for existing circuits. Through the use of Boolean algebra the technician is able to simplify and analyze complex electronic circuits.

UNIT 11·1 INTRODUCTION TO LOGIC GATES

Objectives:

After studying this unit, you should be able to

- diagram the basic logic gates.
- predict correctly the output of each gate given the inputs.
- use the NAND and NOR gates as universal logic elements.
- predict the output of EXCLUSIVE-OR and EXCLUSIVE-NOR circuits.

Digital Systems

As computers and their electronic spin-offs continue to expand into every facet of our personal and public lives, the importance of understanding the inner workings of these complex machines also grows. The modern electronics technician needs to acquire knowledge concerning both the electronic and mathematical principles at work within these circuits. Through the acquisition of this knowledge, the electronics technician will be better prepared to meet the challenges of the twenty-first century.

Modern electronics equipment, to a great extent, utilizes *digital* as opposed to *analog* signals. An analog signal is data represented in a continuous form while a digital signal is data in a discontinuous or discrete form. The mathematical equivalent of a digital circuit is expressed in the on-off, binary logic presented in previous chapters.

The term *logic* in electronics refers to the science dealing with the basic principles and applications of truth tables and switching circuits. A *gate* is defined as a circuit having two or more inputs and one output which is dependent upon the combination of signals present at the inputs. A *logic gate,* therefore, is a circuit that provides an input-output relationship corresponding to a Boolean-algebra logic function. These logic gates are critical building blocks in modern electronic circuits. In order to facilitate our study of Boolean algebra, it is appropriate that we study briefly some of the basic logic gates.

Logic algebra, or Boolean algebra, deals with circuits whose inputs and outputs can be only TRUE or FALSE, ON or OFF, HIGH or LOW, CLOSED or OPEN. All of these conditions can be represented by the binary digits 1 and 0.

Two voltage levels represent the two binary digits in digital systems. *Positive logic* results when the higher of the two voltages represents a 1 and the lower voltage represents a 0. In *negative logic,* the lower of the two voltages represents a 1 and the higher voltage represents a 0. If a digital system has logic level voltages of +5 V and 0 V, the +5 V may be represented by a 1 and the 0 V by a 0. The positive and negative logics are then defined as follows:

Positive Logic: HIGH = 1 (+5 V)
LOW = 0 (0 V)
Negative Logic: HIGH = 0 (0 V)
LOW = 1 (+5 V)

Both positive and negative logic are used in digital systems. However, positive logic is more common and will be used in this text.

Because logic circuits switch between the 0 and 1 states, they operate as switches that open and close. Therefore, they are known as logic gates, or more simply, gates. Actually, a logic gate is an electronic device that performs an operation in Boolean algebra on one or more inputs to produce an output. The output of the gate occurs when certain conditions are met. A *truth table* shows all possible inputs and the resultant outputs of the logic gate circuit. Such a table is used to analyze the operation of the gate.

Logic gate circuits are made using discrete combinations of diodes, transistors, and resistors. Digital integrated circuits (ICs) provide a complete circuit which is small in size. Users of digital logic circuits select ICs for both computer and industrial control applications. An integrated circuit is a single, monolithic chip of semiconductor material in which the electronic elements are fabricated.

An IC chip is about 0.01 inch thick and may vary in size from about 0.03 inch by 0.03 inch to 0.3 inch by 0.3 inch. A typical size is about 0.05 inch by 0.05 inch. The number of gates on one chip may vary from 1 to 2 to many thousands.

ICs are classified according to the number of gates they contain. *Small-scale integration* (SSI) refers to chips containing 12 or fewer gates. *Medium-scale integration* (MSI) refers to chips containing more than 12 gates, but fewer than 100 gates. *Large-scale integration* (LSI) refers to chips containing 100 or more gates.

Standard logic symbols are used to represent the various logic circuits. Three basic logic gates and their mathematical equivalents are analyzed in this unit; the AND gate, the OR gate, and the NOT gate.

The AND Gate

The basic operation of logic multiplication, commonly known as the AND operation or function, is performed by the AND gate. This gate is composed of two or more inputs and a single output. Figure 11–1A shows the standard logic symbol for an AND gate with two input variables *A* and *B*. The output of this gate is *AB* or *X*. The output is expressed in Boolean algebra notation and is read *A* and *B*, not *A* times *B*.

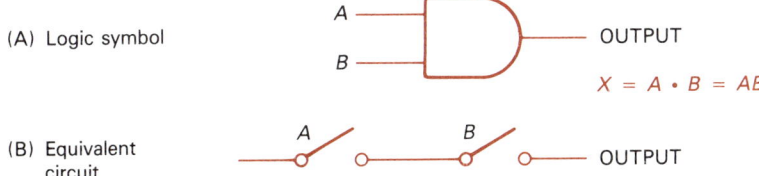

(A) Logic symbol

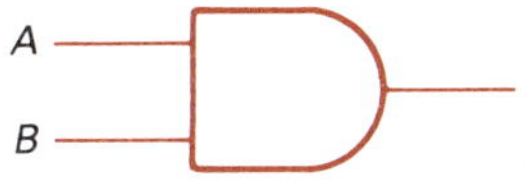

$X = A \cdot B = AB$

(B) Equivalent circuit

Input A	Input B	Output AB
0	0	0
0	1	0
1	0	0
1	1	1

(C) Positive logic truth table

Input A	Input B	Output AB
LOW	LOW	LOW
LOW	HIGH	LOW
HIGH	LOW	LOW
HIGH	HIGH	HIGH

(D) High-low truth table

FIGURE 11–1 Two-input AND gate

The operation of the AND gate is such that the output is HIGH (1) only when all of the inputs are HIGH (1). If any of the inputs are LOW (0), the output is LOW (0). An AND gate determines when certain conditions are simultaneously true. The AND gate is an all-or-nothing gate because there must be a 1 at all of the inputs to obtain a 1 at the output. The HIGH (1) level is the prime or active output level for the AND gate. An AND gate can have any number of inputs greater than one.

The operation of a gate, in terms of its logic, can be expressed by a truth table that lists all of the input combinations and the corresponding outputs. Figures 11–1C and 11–1D show the truth tables for a two-input AND gate. The truth tables are the same for an AND circuit composed of discrete components and for one consisting of an integrated circuit. Figure 11–1B illustrates an equivalent AND circuit using two switches.

Example: The following binary levels are applied to the three-input AND gate in Figure 11–2. What is the gate's output for each input?

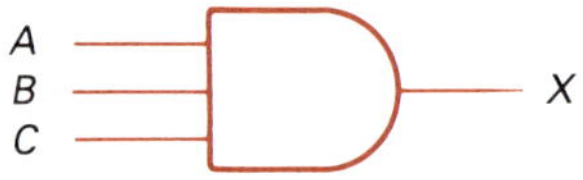

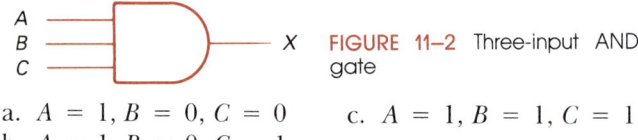

FIGURE 11–2 Three-input AND gate

a. $A = 1, B = 0, C = 0$
b. $A = 1, B = 0, C = 1$
c. $A = 1, B = 1, C = 1$

Solution: a. The output of the AND gate is 0 because at least one of the inputs is 0.
b. The output of the AND gate is 0 because at least one of the inputs is 0.
c. The output is 1 because all three inputs are equal to 1.

The OR Gate

The logic operation performed by the OR gate is logic addition, also known as the OR operation or function. The OR gate has two or more inputs and a single output. Figure 11–3 shows the standard logic symbol for an OR

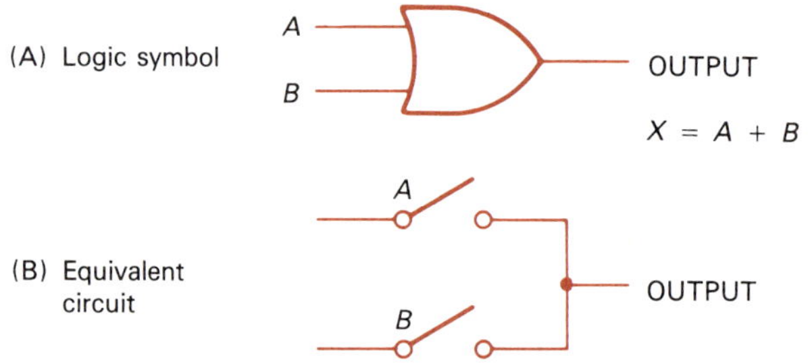

Input A	Input B	Output $A + B$
0	0	0
1	0	1
0	1	1
1	1	1

(C) Truth table

Input A	Input B	Output $A + B$
LOW	LOW	LOW
HIGH	LOW	HIGH
LOW	HIGH	HIGH
HIGH	HIGH	HIGH

(D) High-low truth table

FIGURE 11–3 Two-point OR gate

gate with two input variables A and B. The output of this gate is expressed in Boolean algebra notation as $A + B$, or X. This notation is read A or B, not as A plus B.

The operation of the OR gate is such that a HIGH (1) output is produced when any of the inputs is HIGH (1). The output is LOW (0) only when all of the inputs are LOW (0). The OR gate determines when certain conditions are met simultaneously. The OR gate is an any-or-all gate because a 1 at any input gives rise to a 1 at the output. The HIGH (1) level is the prime or active output level for the OR gate. An OR gate can have any number of inputs greater than one.

Figures 11–3C and 11–3D show the truth tables for a two-input OR gate. The truth tables are the same for an OR circuit composed of discrete components and one consisting of an integrated circuit. Figure 11–3B shows an equivalent OR circuit using two switches.

Example: What is the output of the four-input OR gate in Figure 11–4?

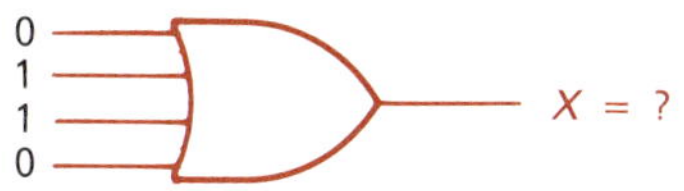

FIGURE 11–4 Four-input OR gate

Solution: With at least one of the inputs HIGH the output is HIGH.

The NOT Gate

Figure 11–5 gives the logic symbol and truth tables for a NOT or INVERTER gate. This gate has a single input, A. The output is the complement of the input and is expressed as $\overline{A}$. The small circle, or bubble, at the output of the NOT gate is the standard symbol representing inversion or complementation.

(A) Logic symbol

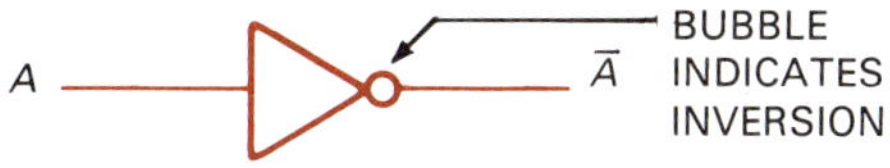

Input A	Output $\overline{A}$
0	1
1	0

(B) Positive logic truth table

Input A	Output $\overline{A}$
LOW	HIGH
HIGH	LOW

(C) High-low truth table

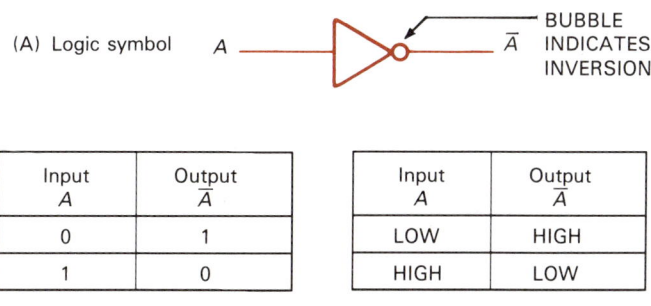

FIGURE 11–5 NOT (INVERTER) gate

An inverter gate changes or converts one logic level to the other logic level. If a HIGH input is applied, the output is LOW. If a LOW input is applied, the output is HIGH.

The NAND Gate

A very important Boolean logic function is the NAND operation. The NAND (a contraction of NOT AND) gate has two inputs (A and B) and an output $\overline{AB}$. As shown in Figure 11–6A, the NAND gate consists of an AND gate followed by a NOT or INVERTER gate. A single gate, called a NAND gate, is available to perform the important NAND operation. Figure 11–6B gives the standard logic symbol for the two-input NAND gate. The symbol consists of an AND gate followed by a small circle, or bubble, to indicate complementation. The truth table for the NAND gate is shown in Figure 11–6C.

(A) Combination of AND and NOT gates

(B) Logic symbol for NAND gates

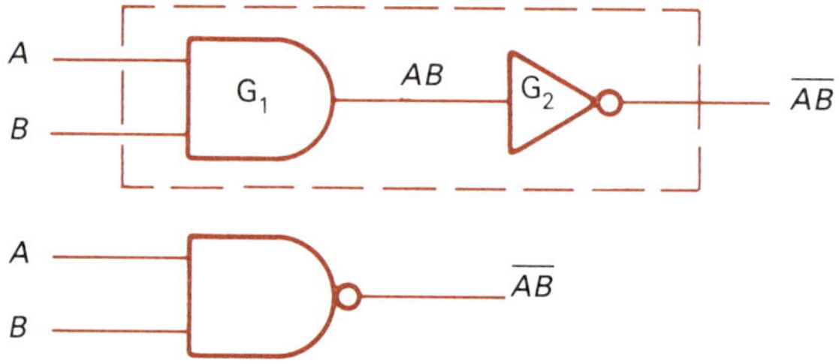

(C) Truth table

Input A	Input B	AB	Output $\overline{AB}$
0	0	0	1
0	1	0	1
1	0	0	1
1	1	1	0

FIGURE 11–6 NAND gate

In order to better understand the operation of the NAND gate consider the logic when $A = 0$ and $B = 1$. With these inputs the output of G_1 in Figure 11–6A will be $AB = 0$. With a LOW input to G_2 the NOT output will be HIGH, so that the output is not AB or $\overline{AB}$.

Example: What is the output of the three-input NAND gate in Figure 11–7?

Solution: The output is the NOT of the AND function. Since one of the inputs is LOW, the AND output is LOW, and the NAND output is HIGH. So $ABC = 0$ and $\overline{ABC} = 1$.

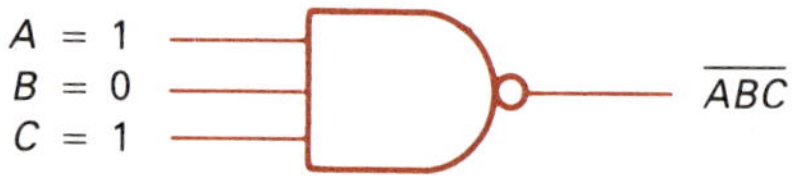

FIGURE 11–7 Three-input NAND gate

NAND gates can be used to produce any logic function. The NAND gate is known as a universal gate because it can be used to generate the NOT function, the AND function, and the OR function. Figure 11–8A shows how a NOT gate can be made from a NAND gate by connecting all of the inputs. In effect, a single common input is created. The AND function can be implemented using only NAND gates, as shown in Figure 11–8B. An OR function can be implemented using NAND gates as shown in Figure 11–8C.

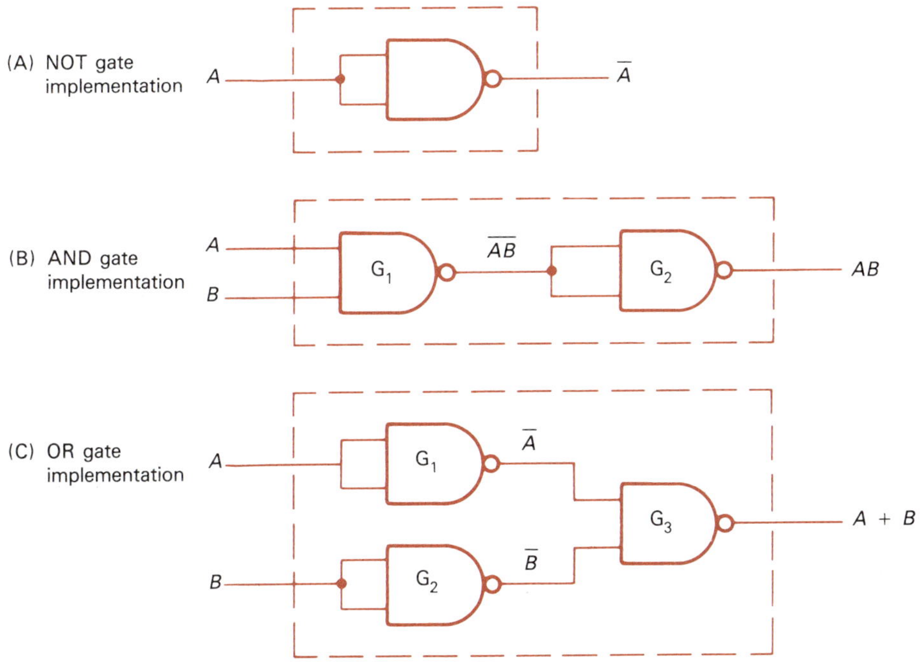

FIGURE 11–8 Universal application of NAND gates

The NOR Gate

Another very important Boolean logic operation is the NOR function. The NOR (a contraction of NOT OR) gate has two inputs (*A* and *B*) and an output $\overline{A + B}$. As shown in Figure 11–9A, the NOR gate consists of an OR gate followed by a NOT or INVERTER gate. However, as in the case of the NAND operation, a single NOR gate is available. Figure 11–9B shows the standard logic symbol for a two-input NOR gate. The symbol consists of an OR gate followed by a small circle, or bubble, to indicate complementation.

NOR gates may have more than two inputs. For this case, the output is the complement of the logic addition of the inputs. For example, the output of a three-input NOR gate is $\overline{A + B + C}$. The NOR gate has an output of 0 when any or all of the inputs are 1, and an output of 1 only when all of the inputs are 0.

NOR gates can be used to produce any logic function. The NOR gate is known as a universal gate because it can be used to generate the NOT function,

(A) Combination of OR and NOT gates

(B) Logic symbol for NOR gate

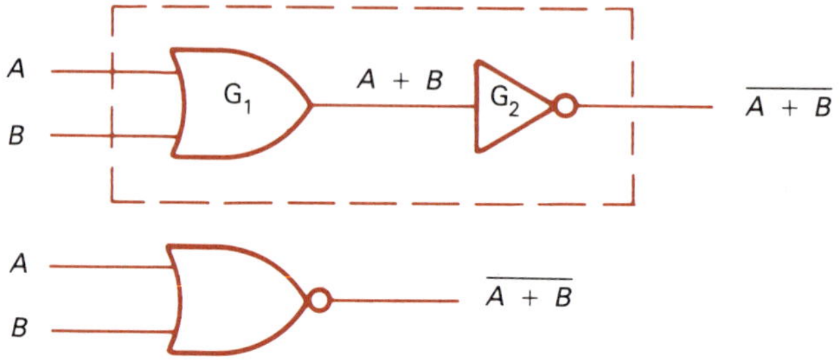

(C) Truth table

Input A	Input B	$A + B$	Output $\overline{A + B}$
0	0	0	1
0	1	1	0
1	0	1	0
1	1	1	0

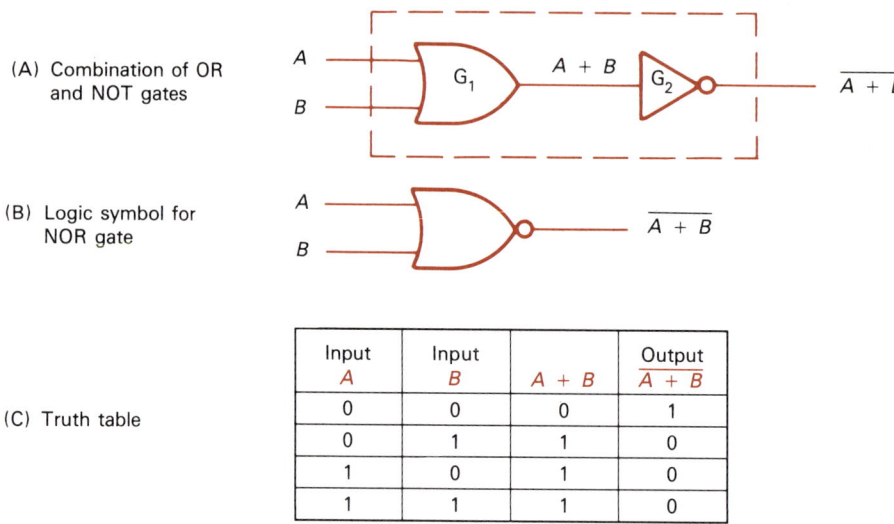

FIGURE 11–9 NOR gate

the AND function, and the OR function. Figure 11–10A shows how a NOT gate can be made from a NOR gate by connecting all of the inputs. In effect, a single common input is created. The AND function can be implemented using NOR gates, as shown in Figure 11–10B. An OR function can be implemented using NOR gates as shown in Figure 11–10C.

(A) NOT gate implementation

(B) AND gate implementation

(C) OR gate implementation

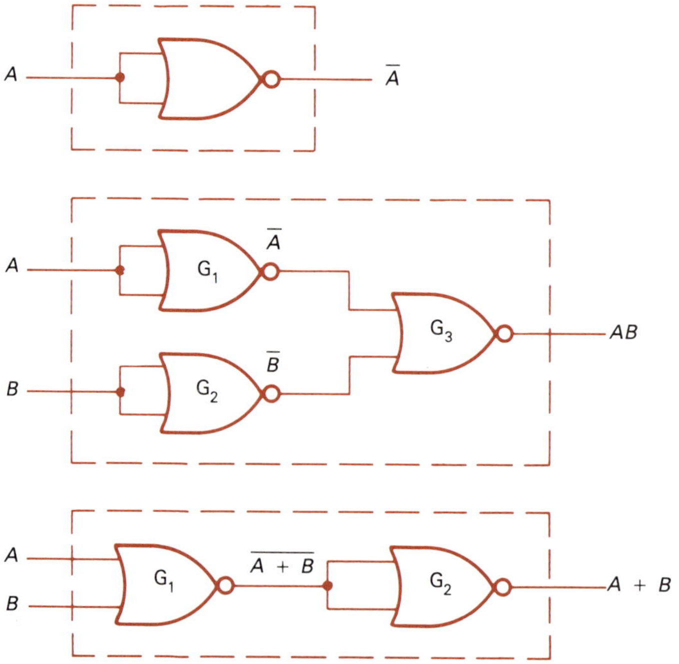

FIGURE 11–10 Universal application of NOR gates

Example: What is the output of a three-input NOR gate with the following input values: $A = 1$, $B = 0$, and $C = 0$?

Solution: As one of the inputs is HIGH, the NOR output is LOW.

The EXCLUSIVE-OR Gate

The normal OR operation applied to two input variables A and B results in an output equal to 1 when either $A = 1$ or $B = 1$, or when $A = 1$ and $B = 1$. A very useful Boolean operation, the EXCLUSIVE-OR function, excludes the case when $A = B = 1$. The output of the EXCLUSIVE-OR operation is 1 when $A = 1$ or $B = 1$, but not when both inputs are equal to 1. Figure 11–11 gives the truth table for the EXCLUSIVE-OR function, for the case of two variable inputs. The EXCLUSIVE-OR operation is used in digital logic systems to decide if two binary numbers are equal or not equal.

The symbol for the EXCLUSIVE-OR operation is $\oplus$. With two inputs the expression is $X = A \oplus B$

Input A	Input B	Output $A \oplus B$
0	0	0
0	1	1
1	0	1
1	1	0

FIGURE 11–11 Truth Table for EXCLUSIVE-OR function

One method of implementing the EXCLUSIVE-OR function is to use two AND gates, two NOT gates, and one OR gate as seen in Figure 11–12. Figure 11–13 shows the standard logic symbol for a single gate which can implement this function.

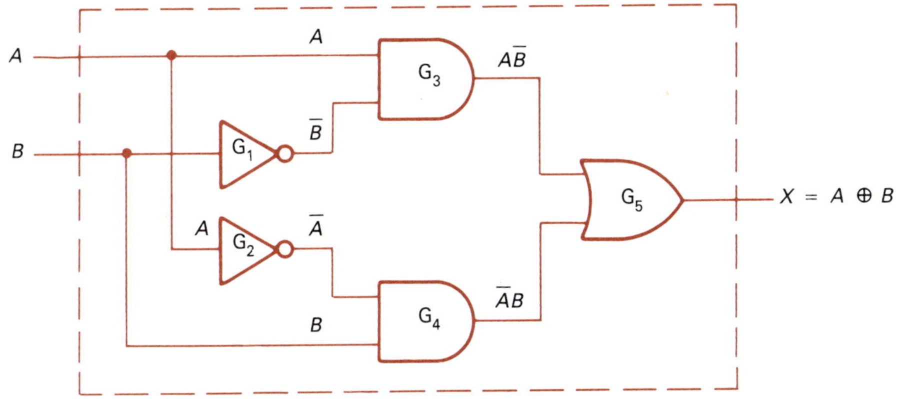

FIGURE 11–12 Implementation of the EXCLUSIVE-OR function

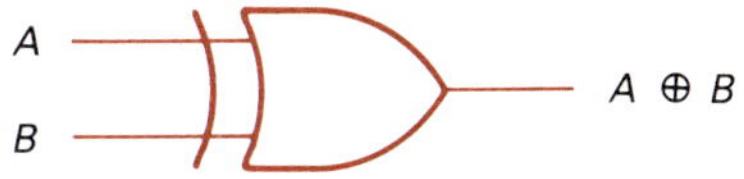

FIGURE 11–13 Logic symbol for a two-input EXCLUSIVE-OR gate

Example: Use the EXCLUSIVE-OR circuit in Figure 11–12 to prove
a. that when $A = 1$ and $B = 0$, $A \oplus B = 1$.
b. that when $A = B = 1$, $A \oplus B = 0$.

Solution: a. Given that $A = 1$ and $B = 0$ the following logic is presented:
1. Since $B = 0$, the output of G_1 is 1.
2. Since $A = 1$, the output of G_2 is 0.
3. Since the inputs to G_3 are 1 and 1, the output is 1.
4. Since the inputs to G_4 are both 0, the output is 0.
5. With one of the inputs to G_5 being 1, the output of the EXCLUSIVE-OR circuit is 1.

b. Given that $A = B = 1$, the following logic is presented:
1. Due to the NOT gates the inputs to G_3 and G_4 are both 0.
2. The outputs of G_3 and G_4 are both 0.
3. With both inputs to $G_5 = 0$, the circuit output is 0.

The EXCLUSIVE-NOR Gate

When an INVERTER or NOT gate is added to the EXCLUSIVE-OR circuit, the result is an EXCLUSIVE-NOR function. A method of implementing the EXCLUSIVE-NOR function is to use two NAND gates, two AND gates, and a NOT gate as in Figure 11–14A. The standard logic symbol for a single gate which can implement the EXCLUSIVE-NOR function is shown in Figure 11–14B.

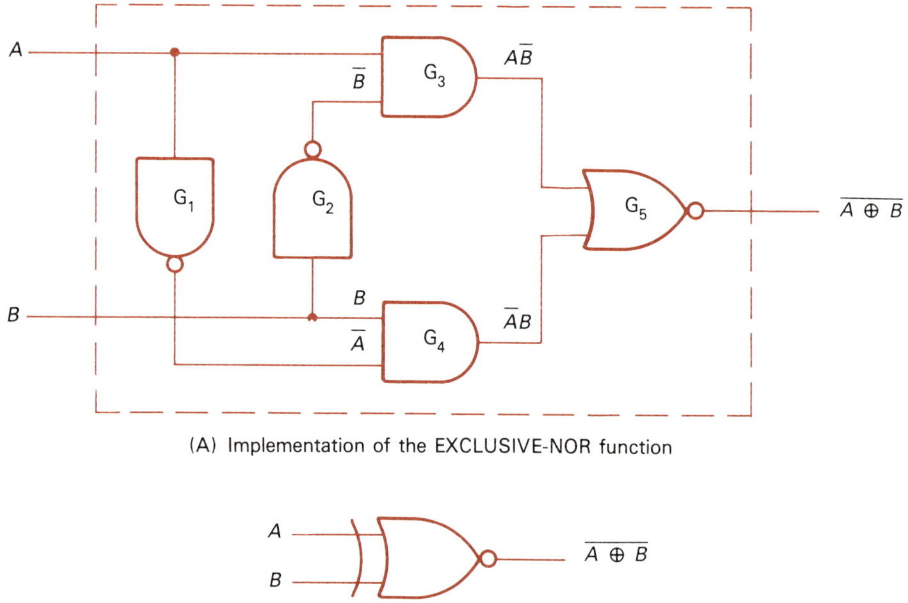

(A) Implementation of the EXCLUSIVE-NOR function

(B) Logic symbol for a two-input EXCLUSIVE-NOR gate

FIGURE 11–14 The EXCLUSIVE-NOR function

Example: Draw a truth table for the EXCLUSIVE-NOR gate in Figure 11–14A.

Solution:
1. Given that $A = B = 0$, the output of G_1 (a single-input NAND gate acting as an INVERTER) is 1 and the output of G_2 is also 1.
2. With the inputs to G_3 and G_4 both being 1 and 0, the outputs of both gates are 0.
3. With both inputs to G_5 being 0, the output of the EXCLUSIVE-NOR is 1. This value is entered in the truth table shown in Figure 11–15.

You should verify the remaining values given in the truth table.

Input A	Input B	Output $\overline{A \oplus B}$
0	0	1
0	1	0
1	0	0
1	1	1

FIGURE 11–15 EXCLUSIVE-NOR gate truth table

EXERCISE 11·1

1. A technician measures a binary 1 on both inputs of an AND gate. What is the gate output?
2. The output of a two-input AND gate is a 0. If one of the inputs is a 1, what must the other input be?
3. The output of a three-input OR gate is a 1. If one of the inputs is a 0, what must the other two inputs be?
4. In order for a three-input OR gate to have a HIGH output, what must the inputs be?
5. The input to a NOT gate is a 1. What is the output?
6. A technician measures a LOW and a HIGH on the input leads of a two-input NAND gate. What would be expected at the output?
7. A NOT gate is connected to the output of a two-input NAND gate. Two HIGHS are present on the NAND gate input leads. What will be measured at the output of the NOT gate?
8. Construct a truth table for the circuit described in problem 7.
9. A technician is checking a new, unmarked, two-input gate. With power applied and the inputs disconnected the output is 1. Considering the gates presented in this unit, which type of gate is the technician testing?
10. The technician in problem 9 continues to test the gate and finds that when the two inputs are HIGH the output is HIGH. Which type of gate is now being tested?
11. The output of a four-input NOR gate is 1. What must the inputs be?

12. The inputs to the EXCLUSIVE-OR gate in Figure 11–12 are both 0. What are the outputs of G_3 and G_4?
13. The inputs to a two-input EXCLUSIVE-OR gate are 1 and 0. What is the output of the gate?
14. The output of an EXCLUSIVE-NOR gate is a 1. What must the inputs be?
15. The inputs to the EXCLUSIVE-NOR gate in Figure 11–14 are $A = 1$ and $B = 0$. What are the outputs of G_3 and G_4?

UNIT 11·2 BOOLEAN EXPRESSIONS

Objectives:

After studying this unit, you should be able to

- write the Boolean expression for a basic logic circuit.
- draw the logic circuit given the Boolean expression.
- construct the truth table for a basic logic circuit.
- write the Boolean expression given a truth table.

Writing Boolean Expressions and Truth Tables for Simple Circuits

The previous unit pointed out the importance of Boolean algebra and presented the basic logic gates. Circuits utilizing these gates will be used in this unit to demonstrate how a logic circuit may be represented as a mathematical expression. The seven logic gates and their output expressions are summarized in Figure 11–16.

Practical electronic circuits will contain combinations of various logic gates in order to perform a specific function. Given a combination circuit, the technician may find it helpful to write the Boolean expression to gain an insight into what is occurring within the circuit. Once the mathematical equivalent of a circuit is derived, a truth table to assist in troubleshooting may be developed. In the next unit, we will examine methods by which these Boolean expressions may be simplified.

As a general rule, the process to write the Boolean expression for a simple circuit is as follows:

- Determine the point at which the circuit is to be evaluated.
- Move back into the circuit one gate at a time. Referring to Figures 11–16 and 11–17, write the expression for the inputs and outputs of each gate. Build the final expression as each gate is considered.
- Construct the circuit truth table by substituting 1's and 0's into the expression.

Logic Symbol	Function	Input/Output Expression	The Expression Indicates
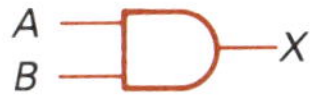	AND	$X = A \cdot B = AB$	X is a 1 when A AND B are both equal to 1
	OR	$X = A + B$	X is a 1 when A OR B is a 1
	NOT	$X = \overline{A}$	X is a 1 when A is a 0
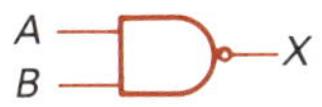	NAND	$X = \overline{AB}$	X is a 1 when A AND/OR B equals 0
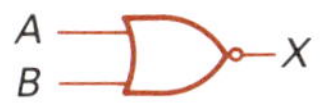	NOR	$X = \overline{A + B}$	X is a 1 when both A AND B are equal to 0
	EXCLUSIVE-OR	$X = A \oplus B$	X is a 1 when A OR B (but not both) equals 1
	EXCLUSIVE-NOR	$X = \overline{A \oplus B}$	X is a 1 when both A AND B equal 0 OR 1

FIGURE 11–16 Summary of the basic logic functions and their Boolean expressions

AND Rules	OR Rules	NOT Rules
$0 \cdot 0 = 0$	$0 + 0 = 0$	$\overline{1} = 0$
$1 \cdot 0 = 0$	$1 + 0 = 1$	$\overline{0} = 1$
$0 \cdot 1 = 0$	$0 + 1 = 1$	
$1 \cdot 1 = 1$	$1 + 1 = 1$	

FIGURE 11–17 Summary of the AND, OR, and NOT rules

Example: Write the Boolean expression and construct a truth table for the circuit in Figure 11–18.

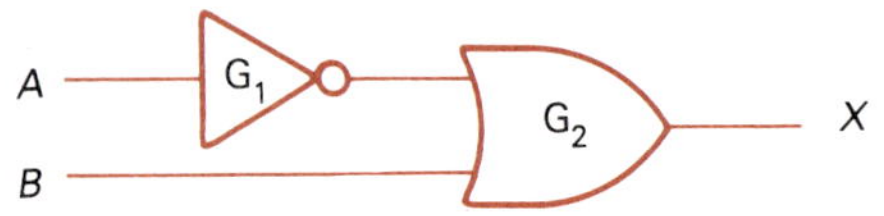

FIGURE 11–18

Solution: Referring to Figure 11–16, the output expression for an OR gate is $X = A + B$. Considering that input A passes through a NOT gate the inputs to the OR gate are $\overline{A}$ and B. The output expression becomes:

$X = \overline{A} + B$

The verbal equivalent is "X is a 1 when A is not a 1 or B is a 1." To evaluate this expression substitute values for A and B as follows:

When $A = 0$ and $B = 0$,

$$\begin{aligned} X &= \overline{A} + B \\ &= \overline{0} + 0 \\ &= 1 + 0 \\ &= 1 \end{aligned}$$

When $A = 1$ and $B = 0$,

$$\begin{aligned} X &= \overline{A} + B \\ &= \overline{1} + 0 \\ &= 0 + 0 \\ &= 0 \end{aligned}$$

You should verify that the two remaining combinations produce outputs equal to 1. The truth table is:

Input A	Input B	Output X
0	0	1
1	0	0
0	1	1
1	1	1

Example: Write the Boolean expression for the circuit in Figure 11–19.

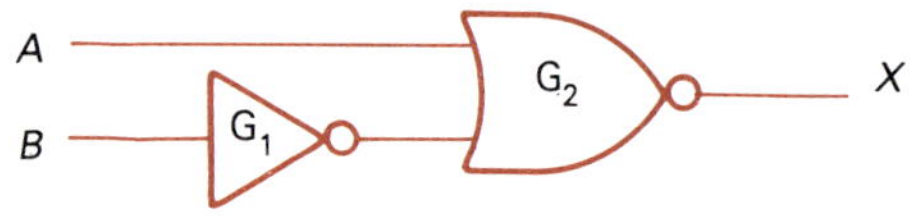

FIGURE 11–19

Solution: The output expression for a NOR gate is

$X = \overline{A + B}$

Considering that the inputs are A and $\overline{B}$ the expression becomes

$X = \overline{A + \overline{B}}$

When $A = B = 0$ the expression is

$$\begin{aligned} X &= \overline{A + \overline{B}} \\ &= \overline{0 + \overline{0}} \\ &= \overline{0 + 1} \\ &= \overline{1} \\ &= 0 \end{aligned}$$

When $A = 0$ and $B = 1$, the output is

$$\begin{aligned} X &= \overline{A + \overline{B}} \\ &= \overline{0 + \overline{1}} \\ &= \overline{0 + 0} \\ &= \overline{0} \\ &= 1 \end{aligned}$$

You should evaluate the expression for the two remaining input combinations.

Example: Write the Boolean expression for the logic circuit in Figure 11–20. Also construct a truth table.

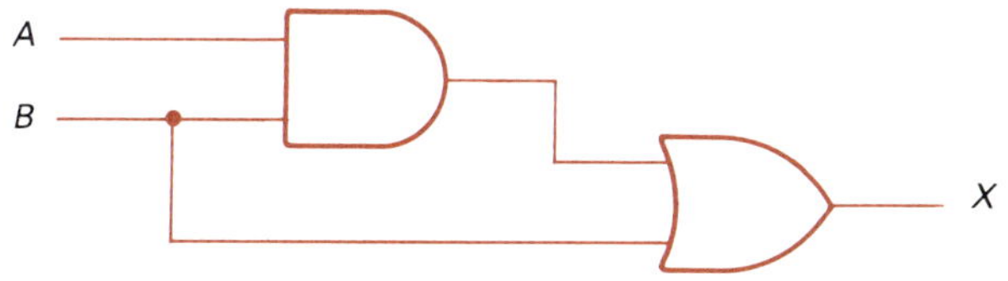

FIGURE 11–20

Solution: It should be obvious that one input of the OR gate is B. The other is the output of the AND gate. The circuit expression becomes:

$X = AB + B$

The verbal equivalent is "X is a 1 when A *and* B are both equal to 1, *or* B is equal to 1." In the next unit we will examine methods of simplifying this expression. To evaluate this expression, and develop a truth table, substitute values for A and B as follows:

When $A = B = 0$,

$$\begin{aligned} X &= AB + B \\ &= (0 \cdot 0) + 0 \\ &= 0 + 0 \\ &= 0 \end{aligned}$$

When $A = 0$ and $B = 1$,

$$\begin{aligned} X &= AB + B \\ &= (0 \cdot 1) + 1 \\ &= 0 + 1 \\ &= 1 \end{aligned}$$

You should evaluate the expression for the two remaining input combinations. The truth table is

Input A	Input B	Output X
0	0	0
1	0	0
0	1	1
1	1	1

Example: Write the Boolean expression and truth table for the circuit in Figure 11–21.

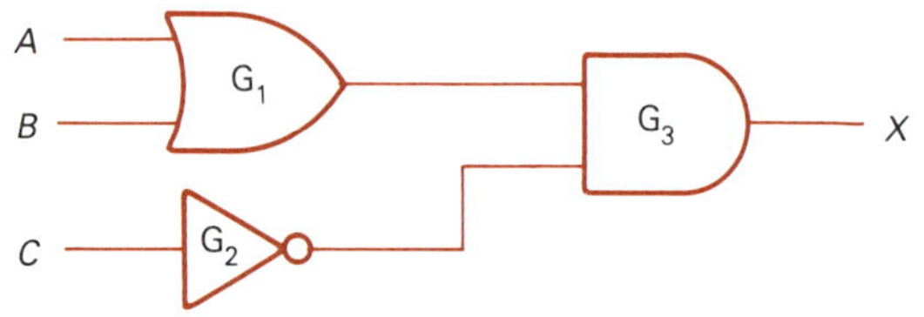

FIGURE 11–21

Solution: The output of G_1 is $A + B$, while the output of G_2 is $\overline{C}$. As these are the inputs to G_3, the output expression becomes:

$X = (A + B)\overline{C}$

The parentheses indicate the AND function between the outputs of G_1 and G_2. In order to evaluate this expression we could substitute all possible combinations of 1 and 0. Although, for X to be a 1, the input at C must be 0 and A and/or B also be equal to 1. Two examples will be presented:

When $A = 1$, $B = 0$, and $C = 1$,

$$\begin{aligned} X &= (A + B)\overline{C} \\ &= (1 + 0)\overline{1} \\ &= 1 \cdot \overline{1} \\ &= 1 \cdot 0 \\ &= 0 \end{aligned}$$

When $A = 0$, $B = 1$, and $C = 0$,

$$\begin{aligned} X &= (A + B)\overline{C} \\ &= (0 + 1)\overline{0} \\ &= 1 \cdot \overline{0} \\ &= 1 \cdot 1 \\ &= 1 \end{aligned}$$

You should verify each of the entries in this truth table:

Input A	Input B	Input C	Output X
0	0	0	0
1	0	0	1
0	1	0	1
0	0	1	0
1	1	0	1
1	0	1	0
0	1	1	0
1	1	1	0

Example: Write the Boolean expression and truth table for the circuit in Figure 11–22.

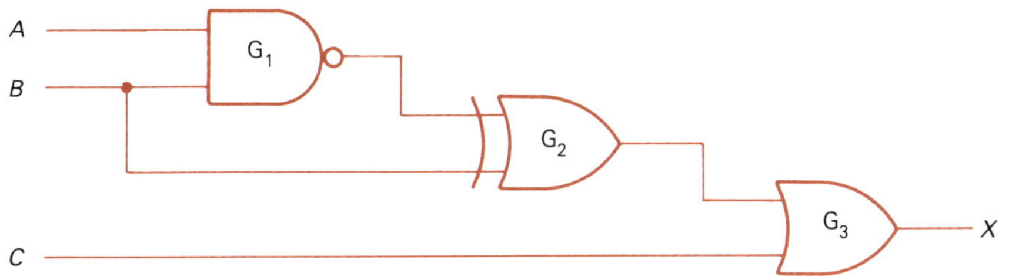

FIGURE 11–22

Solution: The output of G_1 is $\overline{AB}$, making the output of G_2 equal to $\overline{AB} \oplus B$. The output of G_3 then becomes:

$X = (\overline{AB} \oplus B) + C$

When $A = 0$, $B = 0$, and $C = 0$, the output is

$$
\begin{aligned}
X &= (\overline{AB} \oplus B) + C \\
&= (\overline{0 \cdot 0} \oplus 0) + 0 \\
&= (\overline{0} \oplus 0) + 0 \\
&= (1 \oplus 0) + 0 \\
&= 1 + 0 \\
&= 1
\end{aligned}
$$

When $A = 0$, $B = 1$, and $C = 0$, the output is

$$
\begin{aligned}
X &= (\overline{AB} \oplus B) + C \\
&= (\overline{0 \cdot 1} \oplus 1) + 0 \\
&= (\overline{0} \oplus 1) + 0
\end{aligned}
$$

$$= (1 \oplus 1) + 0$$
$$= 0 + 0$$
$$= 0$$

You should confirm the following truth table

Input A	Input B	Input C	Output X
0	0	0	1
1	0	0	1
0	1	0	0
0	0	1	1
1	1	0	1
1	0	1	1
0	1	1	1
1	1	1	1

Example: Write the Boolean expression for the circuit in Figure 11–23.

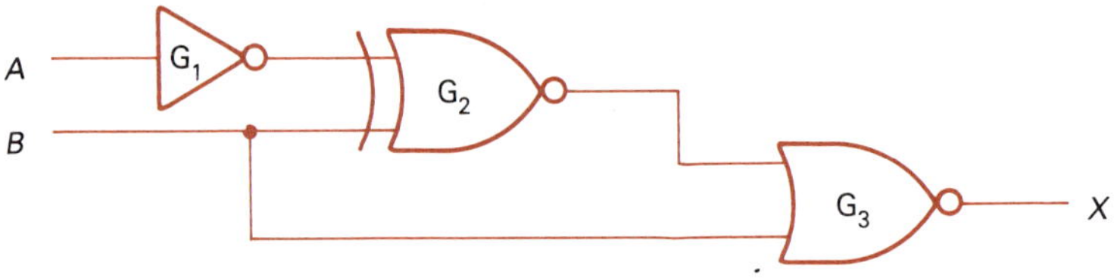

FIGURE 11–23

Solution: The input to G_2 is $\overline{A}$ and B, so that the output of G_2 becomes $\overline{\overline{A} \oplus B}$. The output of G_3 then becomes

$$X = \overline{\overline{\overline{A} \oplus B} + B}$$

When $A = B = 0$, the output is

$$X = \overline{\overline{\overline{A} + B} + B}$$
$$= \overline{\overline{\overline{0} \oplus 0} + 0}$$
$$= \overline{\overline{1 \oplus 0} + 0}$$
$$= \overline{0 + 0}$$
$$= 1$$

You should verify that any other combination of A and B will result in a 0 output.

Writing Boolean Expressions from Truth Tables

In practical design applications you will often use high-density integrated circuits to perform a specific function. In simpler circuits you may start with a truth table which describes the operation of a circuit. Given the truth table, it is possible to write the corresponding Boolean expression, simplify the expression if necessary, and then draw the logic circuit. Techniques for simplifying Boolean expressions will be addressed in the next unit of this chapter.

Example: Given the following truth table, write the corresponding Boolean expression.

Input A	Input B	Output X
0	0	1
1	0	1
0	1	0
1	1	1

Note that the circuit is to be LOW when input A is LOW and input B is HIGH. All other input combinations produce HIGH output.

Solution: There may be many circuit combinations that behave according to this truth table. To begin, we could list the AND and OR expressions from the table that will produce a HIGH output. These would be:

Input A	Input B	Output X	AND Expression	OR Expression
0	0	1	$X = \overline{A}\,\overline{B}$	$X = \overline{A} + \overline{B}$
1	0	1	$X = A\overline{B}$	$X = A + \overline{B}$
0	1	0	$X = \overline{A}B$	$X = \overline{A} + B$
1	1	1	$X = AB$	$X = A + B$

Examining the AND expressions shows that none will satisfy the four circuit states. Of the OR expressions, only $X = A + \overline{B}$ will meet the requirements of the circuit. Note that this Boolean expression is true for each statement in the truth table while the other seven expressions are true for only one or two of the input combinations.

Example: Draw the logic circuit for the Boolean expression developed in the previous example.

Solution: The expression is $X = A + \overline{B}$ and indicates a two-input OR circuit with an inverted B input. The circuit is

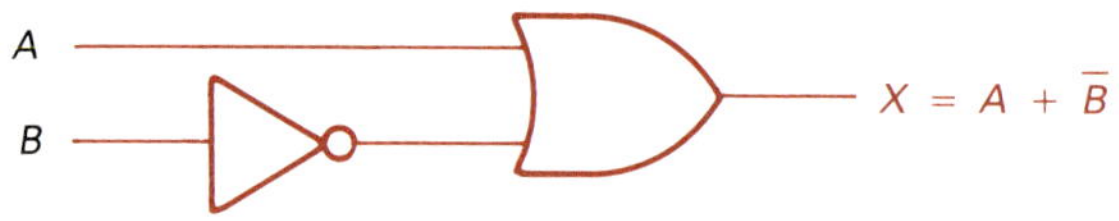

Example: Given the following truth table write a corresponding Boolean expression and draw a logic circuit capable of producing the required outputs.

Input A	Input B	Input C	Output X
0	0	0	1
1	0	0	0
0	1	0	1
0	0	1	1
1	1	0	0
1	0	1	0
0	1	1	1
1	1	1	1

Solution: Examining the truth table it appears that the output is HIGH when $A = 0$, regardless of the inputs at B and C. As a Boolean expression we have

$X = \overline{A}$

Also note that when B and C are HIGH the output is HIGH, regardless of the condition at A. This can be expressed as

$X = BC$

Our combined expression becomes

$X = \overline{A} + BC$

In drawing our circuit begin with the OR function.

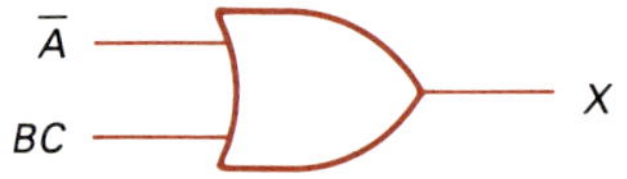

To complete the circuit we must add an inverter and an AND gate as shown.

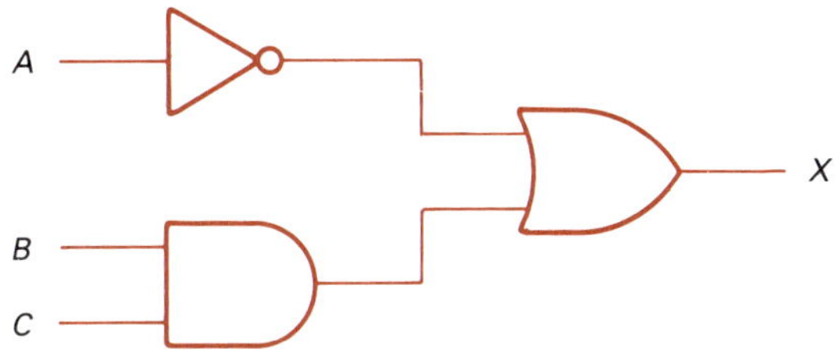

Example: Draw the logic circuit for the expression $X = \overline{A}BC + A\overline{B}$

Solution: Starting with the OR function we have

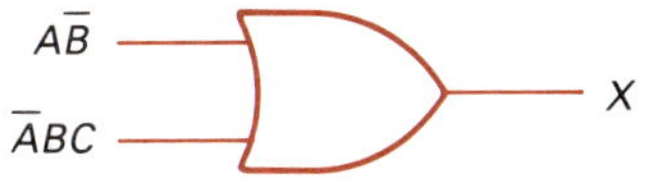

The circuit for $A\overline{B}$ is

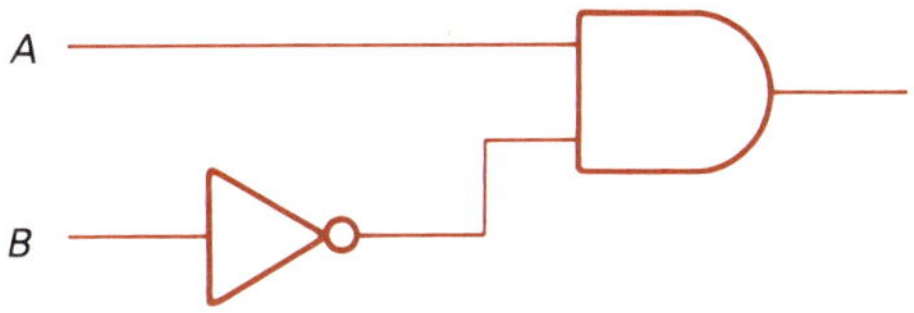

The circuit for $\overline{A}BC$ is

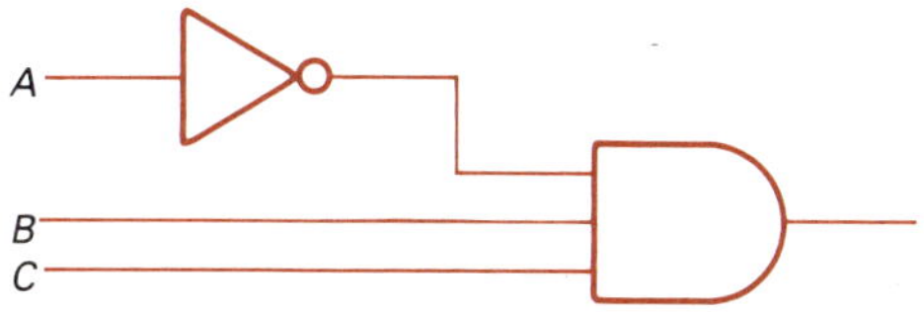

The combined circuit becomes

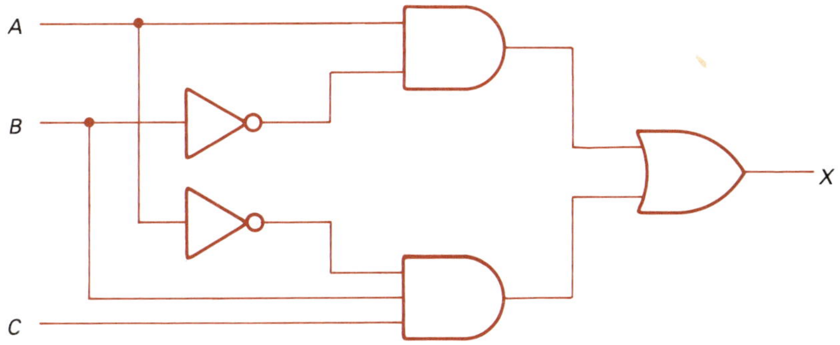

EXERCISE 11·2

Use Figure 11–24 for problems 1–3.

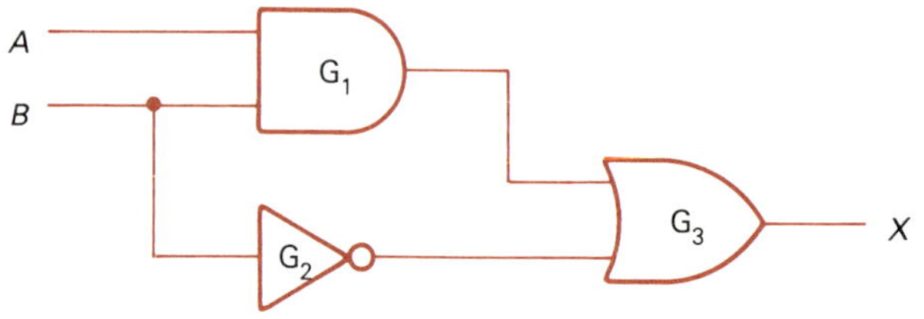

FIGURE 11–24

1. Write the Boolean expression for the circuit in Figure 11–24.
2. What are the inputs to G_3 in Figure 11–24 when $A = B = 1$?
3. What is the output of G_3 in Figure 11–24 when $A = 1$ and $B = 0$?

Use Figure 11–25 for problems 4–9.

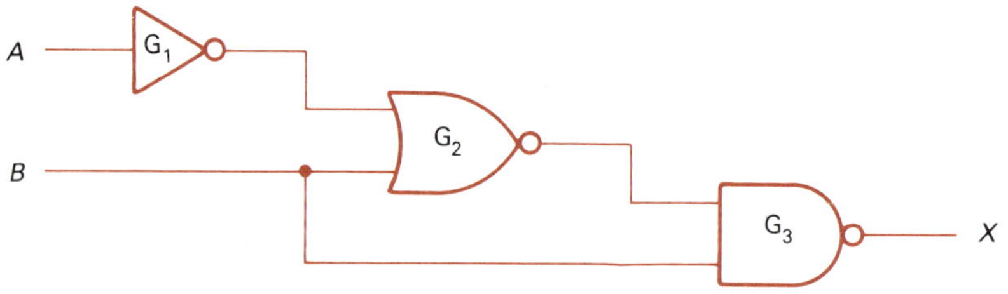

FIGURE 11–25

4. Write the Boolean expression for the output of G_2 in Figure 11–25.
5. What is the output expression for the circuit in Figure 11–25?
6. Given that $A = 0$ and $B = 0$ find the inputs to G_3 in Figure 11–25.
7. Find the output of the circuit in Figure 11–25 when $A = 0$ and $B = 1$.
8. What is the output in Figure 11–25 when $A = 1$ and $B = 0$?
9. Given the circuit in Figure 11–25, is it possible to have an output equal to 0?

Use Figure 11–26 for problems 10–13.

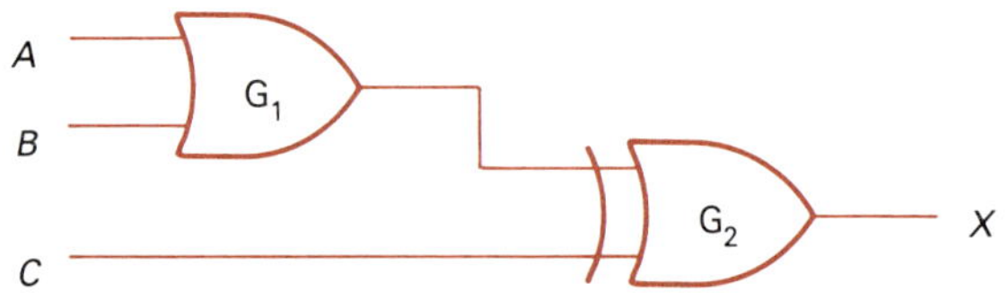

FIGURE 11–26

10. Write the output expression for the circuit in Figure 11–26.
11. What is the output of the circuit in Figure 11–26 when $A = 0$, $B = 1$, and $C = 1$?
12. What is the output in Figure 11–26 when $A = 1$, $B = 1$ and $C = 0$?
13. What combinations of A, B, and C will produce a 1 output in Figure 11–26?

Use Figure 11–27 for problems 14–20.

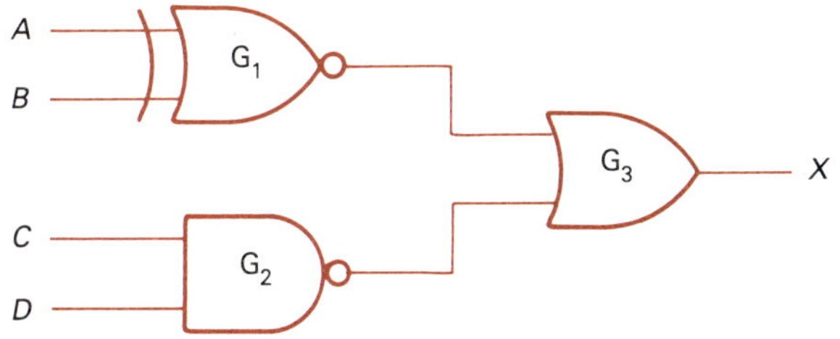

FIGURE 11–27

14. Write the Boolean expression for the output of G_1 in Figure 11–27.
15. Write the Boolean expression for the output of G_2 in Figure 11–27.
16. Write the output expression for G_3 in Figure 11–27.
17. What is the output of G_3 in Figure 11–27 when all four circuit inputs are equal to 0?
18. What is the output of G_3 in Figure 11–27 when $A = 1$, $B = 0$, $C = 1$, and $D = 1$?
19. Find the output of G_3 in Figure 11–27 when $A = 1$, $B = 1$, $C = 0$, and $D = 1$.
20. List the input combinations that will produce a 0 output of G_3 in Figure 11–27.
21. Given the following truth table, write the corresponding Boolean algebra expression.

Input A	Input B	Output X
0	0	0
1	0	1
0	1	0
1	1	0

22. Given the following truth table, write the corresponding Boolean algebra expression.

Input *A*	Input *B*	Input *C*	Output *X*
0	0	0	0
1	0	0	0
0	1	0	0
0	0	1	0
1	1	0	0
1	0	1	1
0	1	1	1
1	1	1	1

23. Draw the equivalent logic circuit for the truth table in problem 21.
24. Draw the equivalent logic circuit for the truth table in problem 22.

UNIT 11·3 BOOLEAN ALGEBRA FUNDAMENTALS

Objectives:

After studying this unit, you should be able to

- apply De Morgan's Theorems.
- apply the commutative, associative, and distributive laws.
- apply the Boolean identities.

Introduction

Boolean algebra is similar to traditional algebra in that symbols representing numbers are manipulated to simplify an expression. In conventional algebra, a system of identities, rules, and theorems are used to solve complex expressions. Once simplified, the unknowns in these expressions may be replaced by numerical values to evaluate the expression. The same process occurs in Boolean algebra, although the numbers 1 and 0 are the only values which may be substituted.

In order to use Boolean algebra in practical circuit applications, it is necessary to first gain an understanding of the algebraic rules and laws that govern this area of mathematics. Each of these will be presented along with examples.

Following the discussion of these laws and theorems we will apply our knowledge to some digital circuits.

AND, OR, and NOT Rules

The basic AND, OR and NOT rules developed in a previous unit are used extensively in Boolean algebra. These are presented again in Figure 11–28. Note the addition of the double NOT rules. These indicate that inverting a value twice is equivalent to the original value.

AND Rules	OR Rules	NOT Rules
$0 \cdot 0 = 0$	$0 + 0 = 0$	$\overline{1} = 0$
$1 \cdot 0 = 0$	$1 + 0 = 1$	$\overline{0} = 1$
$0 \cdot 1 = 0$	$0 + 1 = 1$	$\overline{\overline{1}} = 1$
$1 \cdot 1 = 1$	$1 + 1 = 1$	$\overline{\overline{0}} = 0$

FIGURE 11–28 The AND, OR, and NOT rules

De Morgan's Theorems

De Morgan, a friend of George Boole, developed two theorems which have had an impact on the mathematics of logic. The first of these is written

$$\overline{A \cdot B} = \overline{A} + \overline{B} \tag{11–1}$$

and indicates "the NOT of *A* AND *B* is equal to the NOT of *A* OR the NOT of *B*." Circuit equivalents for these expressions are shown in Figure 11–29. In order to prove that these two circuits are equivalent, substitute values of 1 and 0 for *A* and *B* and then develop a truth table.

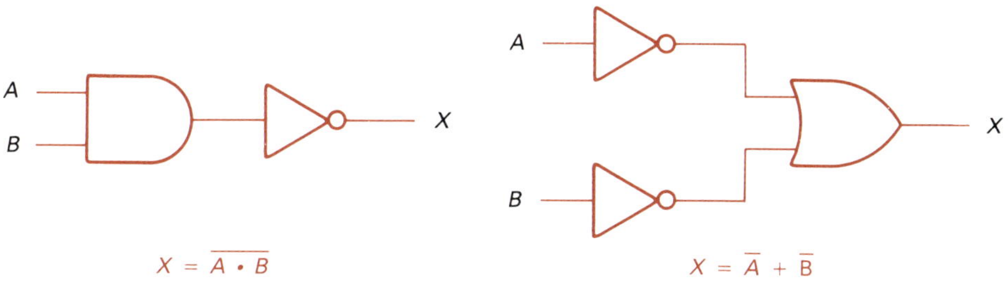

FIGURE 11–29

The outputs when $A = B = 0$ are

$$\overline{A \cdot B} = \overline{A} + \overline{B}$$
$$\overline{0 \cdot 0} = \overline{0} + \overline{0}$$
$$\overline{0} = 1 + 1$$
$$1 = 1$$

When $A = 1$ and $B = 0$,

$$\begin{aligned} \overline{A \cdot B} &= \overline{A} + \overline{B} \\ \overline{1 \cdot 0} &= \overline{1} + \overline{0} \\ \overline{0} &= 0 + 1 \\ 1 &= 1 \end{aligned}$$

You should verify that the two expressions are equivalent for the two remaining combinations. The truth table for these circuits is shown in Figure 11–30. Note that regardless of the input combination, the outputs of the two circuits are identical.

Input A	Input B	Output $\overline{A \cdot B}$	Output $\overline{A} + \overline{B}$
0	0	1	1
1	0	1	1
0	1	1	1
1	1	0	0

FIGURE 11–30

De Morgan's second theorem states

$$\overline{A + B} = \overline{A} \cdot \overline{B} \qquad (11\text{–}2)$$

and indicates that "the NOT of A OR B is equal to the NOT of A AND the NOT of B." The circuit equivalents of these expressions are shown in Figure 11–31. The truth table in Figure 11–31 indicates that for all possible combinations the outputs of the two circuits are identical.

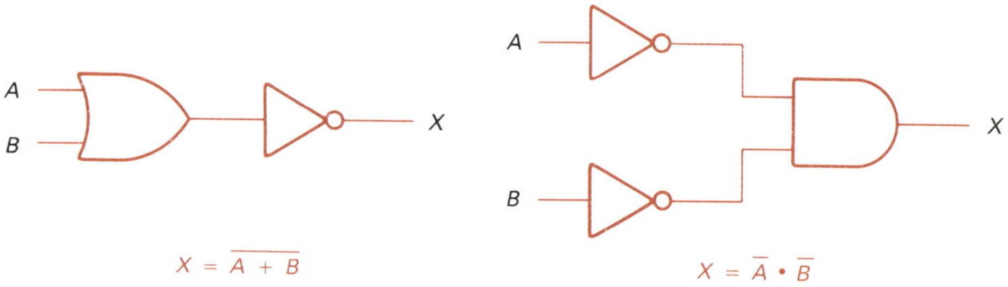

Input A	Input B	Output $\overline{A + B}$	Output $\overline{A} \cdot \overline{B}$
0	0	1	1
1	0	0	0
0	1	0	0
1	1	0	0

FIGURE 11–31

The verbal equivalents of these important theorems are

- The NOT of two or more values in an AND state is equal to the OR of the individual values in a NOT state.
- The NOT of two or more values in an OR state is equal to the AND of the individual values in a NOT state.

Example: Write an equivalent expression for $X = \overline{AB(C + D)}$.

Solution: This expression states that the output (X) is a 1 when the input is the NOT of the quantity A AND B, AND the quantity C OR D. Equivalent expressions include

$$X = \overline{AB(C + D)}$$

$$= \overline{A} + \overline{B} + \overline{(C + D)} \qquad \text{(Applying 11–1)}$$

$$= \overline{A} + \overline{B} + \overline{C} \cdot \overline{D} \qquad \text{(Applying 11–2)}$$

To check our results, let $A = 1$, $B = 0$, $C = 1$, and $D = 0$.

$$\overline{AB(C + D)} = \overline{A} + \overline{B} + \overline{C} \cdot \overline{D}$$

$$\overline{1 \cdot 0 \cdot (1 + 0)} = \overline{1} + \overline{0} + \overline{1} \cdot \overline{0}$$

$$\overline{1 \cdot 0 \cdot 1} = 0 + 1 + 0 \cdot 1$$

$$\overline{0} = 0 + 1 + 0$$

$$1 = 1$$

Example: Verify that $\overline{A + \overline{B} + \overline{C}} = \overline{A} \cdot B \cdot C$

Solution:

$$\overline{A + \overline{B} + \overline{C}} = \overline{A} \cdot \overline{\overline{B}} \cdot \overline{\overline{C}} \qquad \text{(Applying 11–1)}$$

$$= \overline{A} \cdot B \cdot C \qquad \text{(Removing double-NOTs)}$$

Commutative Laws

As in traditional algebra, the *commutative* laws state the order of addition or multiplication is unimportant. In Boolean algebra these laws indicate the AND and OR operation may occur in any order. The commutative laws are

$$A + B = B + A \qquad (11\text{–}3)$$

$$A \cdot B = B \cdot A \qquad (11\text{–}4)$$

The first of these states that A OR B is equal to B OR A. The second indicates that A AND B is equal to B AND A.

Associative Laws

The *associative* laws in algebra state that three or more values may be added or multiplied in any order, regardless of the manner in which they are grouped. These laws in Boolean algebra apply to the AND and OR functions as follows:

$$(A + B) + C = A + (B + C) \qquad (11\text{–}5)$$

$$(A \cdot B) \cdot C = A \cdot (B \cdot C) \qquad (11\text{–}6)$$

Equation 11–5 states the quantity *A* OR *B* and then OR *C* is equivalent to *A* OR the quantity *B* OR *C*. Equation 11–6 indicates that the quantity *A* AND *B* and then AND *C* is equal to *A* AND the quantity *B* AND *C*.

Example: Verify that $(A + B) + C = A + (B + C)$

Solution: Letting $A = 1$, $B = 1$, and $C = 0$,

$$(1 \cdot 1)0 = 1(1 \cdot 0)$$
$$1 \cdot 0 = 1 \cdot 0$$
$$0 = 0$$

Example: Show that $(AB)C = A(BC)$

Solution: Letting $A = 1$, $B = 1$, and $C = 0$,

$$(1 \cdot 1)0 = 1(1 \cdot 0)$$
$$1 \cdot 0 = 1 \cdot 0$$
$$0 = 0$$

You should verify that in the two previous examples the expressions are equivalent for all combinations of 0 and 1.

Distributive Law

The third law is extremely important and is known as the *distributive* law. This law states that the product of a value multiplied by the sum of two or more other values is equivalent to the sum of the individual products of the first value and each of the other values. In symbols,

$$A(B + C) = AB + AC \qquad (11\text{–}7)$$

The Boolean terminology for this law indicates that *A* AND the quantity *B* OR *C* is equal to *A* AND *B*, OR *A* AND *C*. The circuit equivalents for these expressions are shown in Figure 11–32.

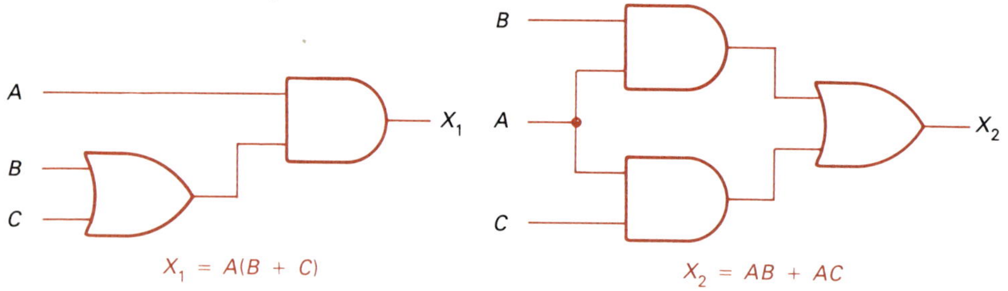

FIGURE 11–32

Example: Show that the two circuits in Figure 11–32 are equal.

Solution: Letting $A = 1$, $B = 0$, and $C = 1$,

$$X_1 = A(B + C) \qquad X_2 = AB + AC$$
$$= 1(0 + 1) \qquad = (1 \cdot 0) + (1 \cdot 1)$$
$$= 1 \cdot 1 \qquad = 0 + 1$$
$$= 1 \qquad = 1$$

You should verify that the circuits produce identical outputs for all combinations of A, B, and C.

Boolean Identities

Thus far we have examined logic rules, De Morgan's theorems, and the commutative, associative, and distributive laws. In addition to these, there are eight identities which are extremely important in the study of Boolean algebra. Four of these involve the OR operation and are

$$A + 0 = A \tag{11–8}$$

$$A + 1 = 1 \tag{11–9}$$

$$A + A = A \tag{11–10}$$

$$A + \overline{A} = 1 \tag{11–11}$$

The first of these identities indicates that $A + 0 = A$. Thinking of the OR gate with a 0 on one input, the output will be determined by the second input. When this input is a 1 the output will also be a 1. Similarly, if the second input is a 0, then the output will be a 0. The same logic can be followed for Equation 11–9. In this case one of the OR inputs is a 1, so that the output is a 1 regardless of the second input.

The third identity states that $A + A = A$. Given that both inputs are equal, there are only two possible combinations. When the inputs are 0, the output is also 0. When the inputs are both 1's, the output is a 1. Equation 11–11 states that $A + \overline{A} = 1$. As the two values are complements of each other the inputs will always be 0 and 1. Given that this is an OR function the output will always be a 1.

Just as there are four OR identities, there are also four AND identities. These are

$$A \cdot 0 = 0 \tag{11–12}$$

$$A \cdot 1 = A \tag{11–13}$$

$$A \cdot A = A \tag{11–14}$$

$$A \cdot \overline{A} = 0 \tag{11–15}$$

Equation 11–12 states that $A \cdot 0 = 0$. In the AND function all inputs must be HIGH to produce a HIGH output. With one input LOW, the output will also be LOW. The second identity indicates that when one input to an AND function is HIGH the output is determined by the second input.

Equation 11–14 states that when both inputs are the same the output is the same as the inputs. For example, when both inputs are LOW, the output is LOW. The opposite is true when both inputs are HIGH. The fourth AND identity indicates that when complements are presented to an AND function the output will always be LOW.

Boolean Algebra Applications

The laws and identities presented in this unit can assist the electronics technician in simplifying complex circuits. The following examples will attempt to demonstrate some applications of Boolean algebra.

Example: The circuit in Figure 11–33 has been designed to perform a specific function. The technician's job is to see if the circuit can be simplified.

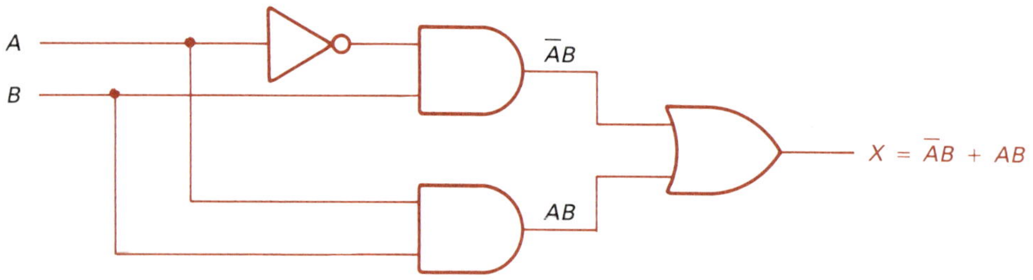

FIGURE 11–33

Solution: The output expression is

$$X = \overline{A}B + AB$$
$$= B(\overline{A} + A) \quad \text{(Factor out } B\text{)}$$
$$= B \cdot 1 \quad \text{(Applying 11–11)}$$
$$= B \quad \text{(Applying 11–13)}$$

So that the output is equal to the value of B.

Example: Verify that the circuit in Figure 11–34A is equivalent to that in Figure 11–34B.

Solution: The output expression of the first circuit is

$$X = \overline{A + B + \overline{C}} + A\overline{B}C$$
$$= \overline{A} \cdot \overline{B} \cdot \overline{\overline{C}} + A\overline{B}C \quad \text{(Applying 11–2)}$$
$$= \overline{AB}C + A\overline{B}C \quad \text{(Remove double-NOT)}$$
$$= \overline{B}C(\overline{A} + A) \quad \text{(Factor out } \overline{B}C\text{)}$$
$$= \overline{B}C(1) \quad \text{(Applying 11–11)}$$
$$= \overline{B}C \quad \text{(Applying 11–13)}$$

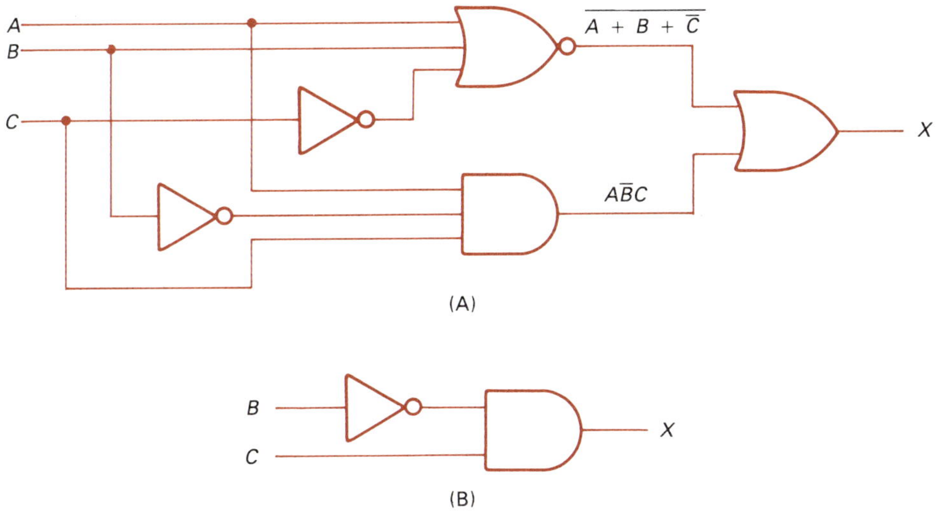

FIGURE 11–34

Letting $A = 1$, $B = 0$, and $C = 1$, the outputs become

$$
\begin{aligned}
X &= \overline{A + B + \overline{C}} + A\overline{B}C \\
&= \overline{1 + 0 + \overline{1}} + 1 \cdot \overline{0} \cdot 1 \\
&= \overline{1 + 0 + 0} + 1 \cdot 1 \cdot 1 \\
&= \overline{1} + 1 \\
&= 1 \\
X &= \overline{B}C \\
&= \overline{0} \cdot 1 \\
&= 1 \cdot 1 \\
&= 1
\end{aligned}
$$

You should verify that for all combinations of the inputs the two expressions and circuits are equal.

Example: Simplify the circuit in Figure 11–35A.

Solution:

$$
\begin{aligned}
X &= \overline{(\overline{\overline{A} + B})(AC)} \\
&= \overline{(\overline{\overline{A}} \cdot \overline{B})(AC)} && \text{(Applying 11–2)} \\
&= \overline{(A\overline{B})(AC)} && \text{(Remove the double-NOT)} \\
&= \overline{A\overline{B}C} && \text{(Remove the parentheses)} \\
&= \overline{A} + \overline{\overline{B}} + \overline{C} && \text{(Applying 11–1)} \\
&= \overline{A} + B + \overline{C} && \text{(Remove the double-NOT)}
\end{aligned}
$$

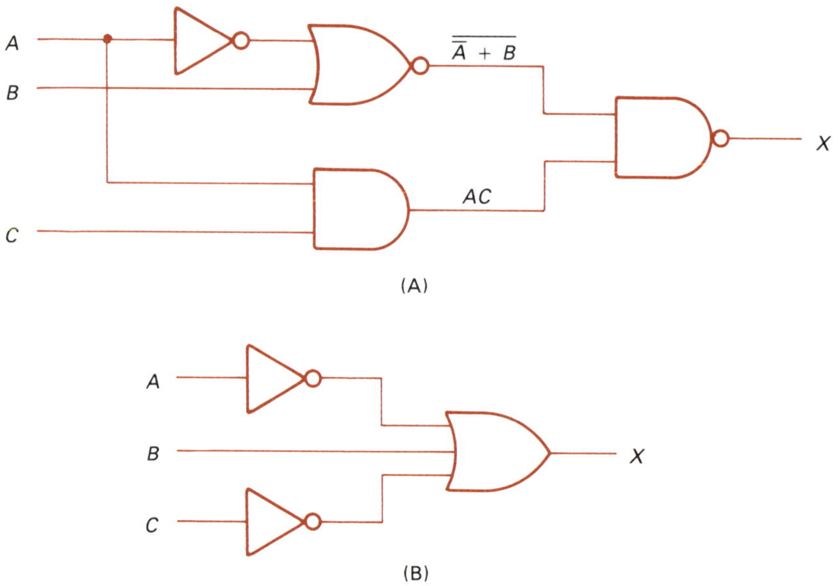

FIGURE 11–35

Let $A = 1$, $B = 0$, and $C = 0$ to check that the two circuits in Figure 11–35 are equivalent.

$$
\begin{aligned}
X &= \overline{(\overline{\overline{A} + B})(AC)} \\
&= \overline{(\overline{\overline{1} + 0})(1 \cdot 0)} \\
&= \overline{(\overline{0 + 0})(0)} \\
&= \overline{(\overline{0})(0)} \\
&= \overline{1 \cdot 0} \\
&= \overline{0} \\
&= 1 \\
X &= \overline{A} + B + \overline{C} \\
&= \overline{1} + 0 + \overline{0} \\
&= 0 + 0 + 1 \\
&= 1
\end{aligned}
$$

Example: Simplify the circuit in Figure 11–36.

Solution: $X = (\overline{A}B + A\overline{B} + AB)(A + \overline{B})$

Simplifying the first expression we have

$$
\begin{aligned}
\overline{A}B + A\overline{B} + AB &= \overline{A}B + A(\overline{B} + B) && \text{(Factor out } A\text{)} \\
&= \overline{A}B + A \cdot 1 && \text{(Applying 11–11)} \\
&= \overline{A}B + A && \text{(Applying 11–13)}
\end{aligned}
$$

So our expression now is

$X = (\overline{A}B + A)(A + \overline{B})$

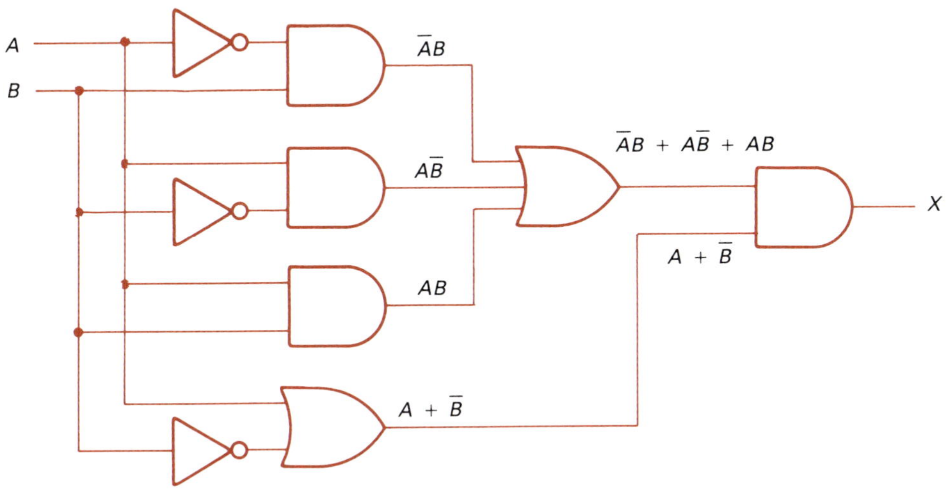

FIGURE 11–36

Multiplying (ANDing) each term in the first expression by each in the second is the reverse of the distributive law (11–7). Our expression now becomes

$$\begin{aligned} X &= A\overline{A}B + AA + \overline{BA}B + \overline{B}A \\ &= 0 + A + 0 + \overline{B}A \\ &= A + \overline{B}A \\ &= A(1 + \overline{B}) \\ &= A \cdot 1 \\ &= A \end{aligned}$$

This indicates that when A is HIGH the output is HIGH. You should verify this by tracing out the circuit in Figure 11–36.

EXERCISE 11·3

In problems 1–5, remove the long NOT bar and simplify the expression.

1. $\overline{A\overline{\overline{BC}}}$
2. $\overline{\overline{\overline{A}}B\overline{C}}$
3. $\overline{(A + B)(C + D)}$
4. $\overline{AB + C}$
5. $\overline{AB + A}$

In problems 6–10, simplify the given expressions.

6. $X = \overline{\overline{A} + B} + AB$
7. $X = A\overline{B}C + \overline{AB}C$
8. $X = \overline{\overline{AC} + \overline{BC}}$
9. $X = A\overline{B}C + \overline{AB}C + A\overline{B}C + A\overline{BC}$
10. $X = \overline{A(A + B)}$

11. Write the simplified output expression for the circuit in Figure 11–37.

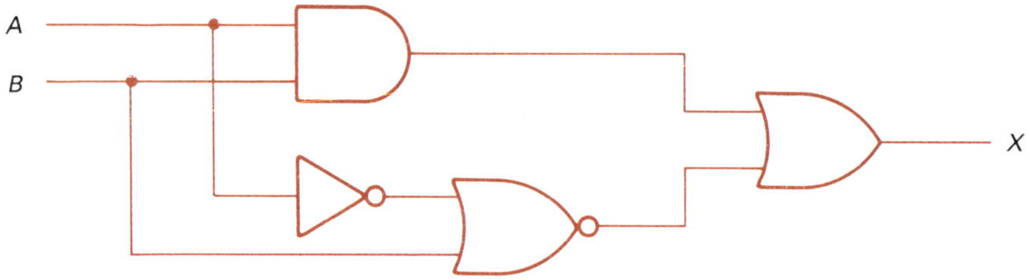

FIGURE 11–37

12. Write the simplified output expression for the circuit in Figure 11–38.

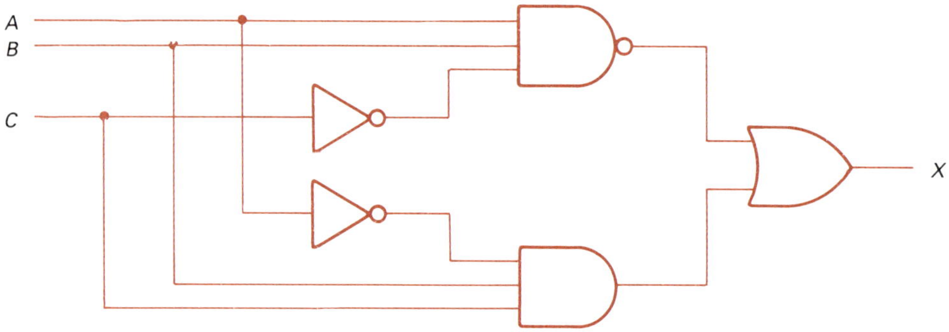

FIGURE 11–38

13. Write the simplified output expression for the circuit in Figure 11–39.

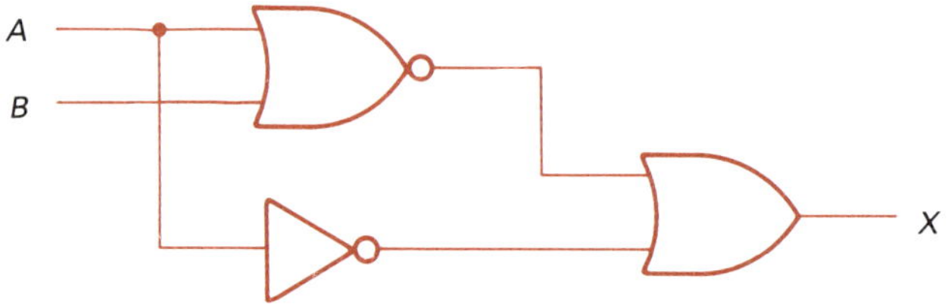

FIGURE 11–39

14. Write the simplified output expression for the circuit in Figure 11–40.

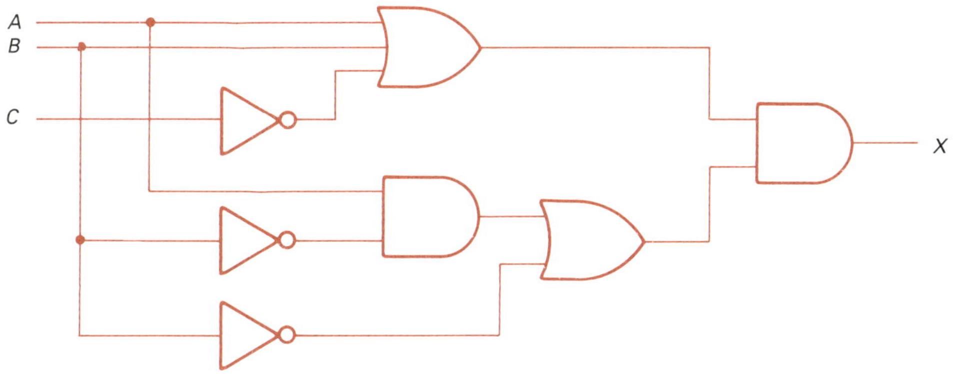

FIGURE 11–40

15. Write the simplified output expression for the circuit in Figure 11–41.

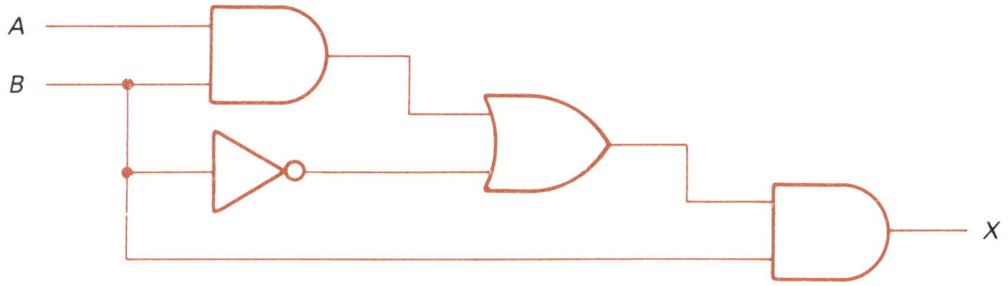

FIGURE 11–41

UNIT 11·4 PROBLEM REVIEW

1. The output of a NOT gate is HIGH. What is the input?
2. The output of a two-input NOR gate is LOW. If one of the inputs is HIGH, what will the other be?

Use Figure 11–42 for problems 3–4.

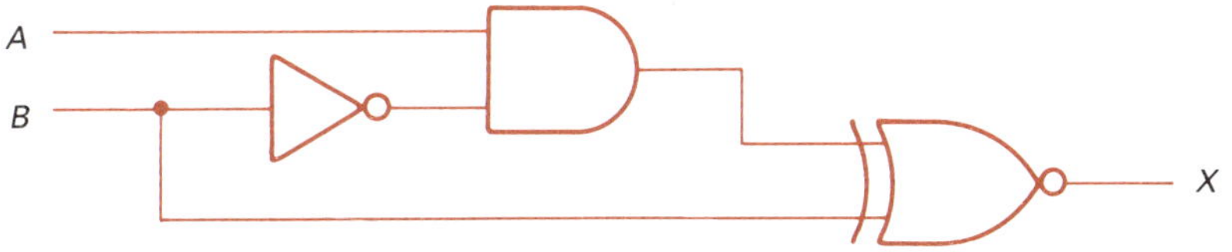

FIGURE 11–42

3. Given the circuit in Figure 11–42. Find the output when $A = 1$ and $B = 0$.
4. Write the output expression for the circuit in Figure 11–42.

5. A technician measures two LOWS on the input leads of a NOR gate. What is expected at the output?

Use Figure 11–43 for problems 6–8.

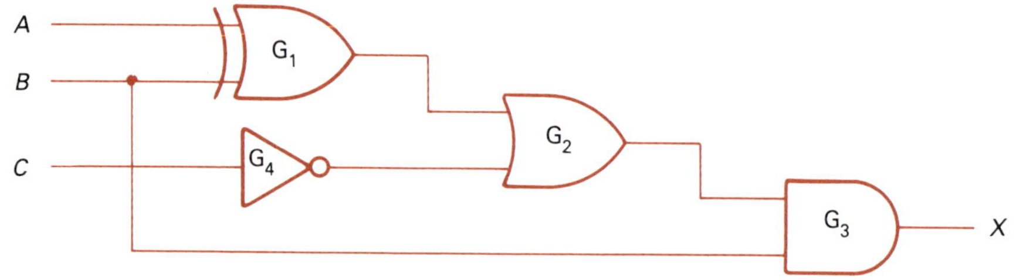

FIGURE 11–43

6. Given the circuit in Figure 11–43. Write the output expression for G_2.
7. Write the output expression for the circuit in Figure 11–43.
8. Given $A = 0$, $B = 1$, and $C = 1$. Find the output of the circuit in Figure 11–43.
9. The two input leads of a NOR gate are wired together. This configuration will function as what other type of gate?
10. The output of an EXCLUSIVE-NOR gate is LOW. What must the inputs be?

Use Figure 11–44 for problems 11–14.

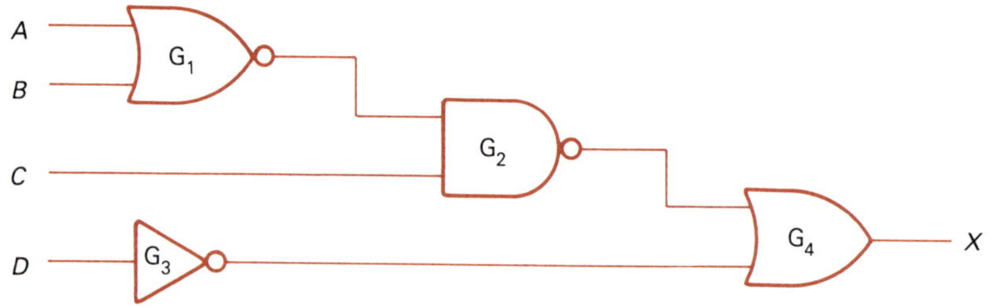

FIGURE 11–44

11. Write the simplified output expression for G_2 in Figure 11–44.
12. Write the output expression for G_4 in Figure 11–44.
13. Find the output of G_4 in Figure 11–44 when $A = 0$ and the remaining inputs are equal to 1.
14. What is the output of G_4 in Figure 11–44 when $A = 0$, $B = 0$, $C = 1$, and $D = 1$?

15. Write the simplified output expression for the circuit in Figure 11–45.

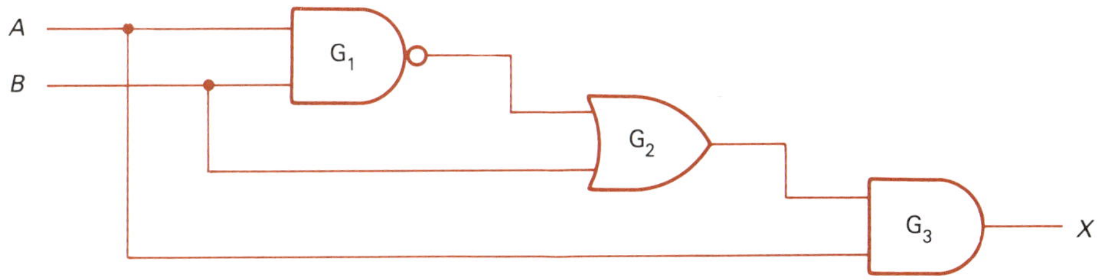

FIGURE 11–45

Use Figure 11–46 for problems 16–19.

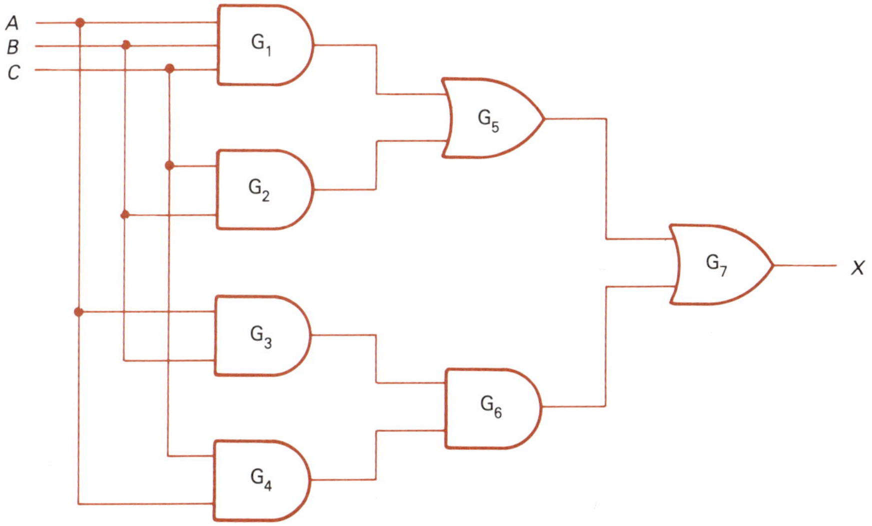

FIGURE 11–46

16. What is the simplified output expression for G_5 in Figure 11–46?
17. Find the output expression for G_6 in Figure 11–46.
18. Write the output expression for G_7 in Figure 11–46.
19. What gate could be used to replace the circuit in Figure 11–46?
20. Simplify this Boolean expression: $AB + \overline{\overline{A}B} + B$.

APPENDIX

TABLE 1

Electronics Abbreviations

Term	Abbreviation	Term	Abbreviation
Alpha	α	Inductive reactance	X_L
Alternating current	ac	Kilo	k
Ampere	A	Logarithm, common	log
Antilogarithm	antilog	Logarithm, natural	ln
Bel	B	Mega	M
Beta	β	Metre	m
Capacitance	C	Metres per second	m/s
Capacitive reactance	X_C	Micro	μ
Change of	Δ	Milli	m
Charge	Q	Nano	n
Conductance	G	Ohm	Ω
Cosecant	csc	Peak-to-peak	p-p
Cosine	cos	Period	T
Cotangent	cot	Pico	p
Coulomb	C	Power	P
Current	I, i	Reactance	X
Cycles per second	Hz	Resistance	R, r
Decibel	dB	Resonant frequency	f_r
Decibel (1 mW ref)	dBm	Root mean square	rms
Degrees Celsius	°C	Secant	sec
Degrees Fahrenheit	°F	Second	s
Direct current	dc	Siemens	S
Electromotive force	emf, E, e	Sine	sin
Farad	F	Tangent	tan
Feet per second	ft/s	Time	t
Frequency	f	Var	var
Giga	G	Volt	V, v
Henry	H	Voltampere	VA
Hertz	Hz	Wavelength	λ
Impedance	Z	Watt	W
Inductance	L		

TABLE 2

Conversion Factors

To Convert From	To	Multiply Column 1 by
Amperes	Microamperes	10^{6}
Amperes	Milliamperes	10^{3}
Centimetres	Feet	3.281×10^{-2}
Centimetres	Inches	0.393 7
Centimetres	Kilometres	10^{-5}
Centimetres	Metres	0.01
Centimetres	Millimetres	10
Circular mils	Square centimetres	5.067×10^{-6}
Circular mils	Square inches	7.854×10^{-7}
Circular mils	Square millimetres	5.067×10^{-4}
Circular mils	Square mils	0.785 4
Decimetres	Metres	0.1
Degrees	Minutes	60
Degrees	Radians	1.745×10^{-2}
Degrees	Seconds	3 600
Dekametres	Metres	10
Farads	Microfarads	10^{6}
Farads	Picofarads	10^{12}
Feet	Centimetres	30.48
Feet	Inches	12
Feet	Kilometres	3.048×10^{-4}
Feet	Metres	0.304 8
Feet	Millimetres	304.8
Feet	Yards	0.333 3
Henrys	Microhenrys	10^{6}
Henrys	Millihenrys	10^{3}
Hertz	Kilohertz	10^{-3}
Hertz	Megahertz	10^{-6}
Horsepower	Watts	746
Hours	Minutes	60
Hours	Seconds	3 600
Inches	Centimetres	2.540
Inches	Feet	8.333×10^{-2}
Inches	Kilometres	2.540×10^{-5}
Inches	Metres	2.540×10^{-2}
Inches	Miles	1.578×10^{-5}
Inches	Millimetres	25.40
Inches	Yards	2.778×10^{-2}
Kilohertz	Hertz	10^{3}
Kilohertz	Megahertz	10^{-3}
Kilohms	Megohms	10^{-3}
Kilohms	Ohms	10^{3}
Kilometres	Centimetres	10^{5}
Kilometres	Feet	3 281
Kilometres	Inches	3.937×10^{4}
Kilometres	Metres	10^{3}
Kilometres	Millimetres	10^{6}
Kilometres	Yards	1 094
Kilovolts	Megavolts	10^{-3}
Kilovolts	Volts	10^{3}
Kilowatt-hours	Watt-hours	10^{3}
Kilowatts	Megawatts	10^{-3}
Kilowatts	Watts	10^{3}
Megahertz	Hertz	10^{6}
Megawatts	Kilowatts	10^{3}
Megawatts	Milliwatts	10^{9}
Megawatts	Watts	10^{6}
Megohms	Kilohms	10^{3}
Megohms	Ohms	10^{6}
Metres	Centimetres	10^{2}

To Convert From	To	Multiply Column 1 by
Metres	Feet	3.281
Metres	Inches	39.37
Metres	Kilometres	10^{-3}
Metres	Millimetres	10^{3}
Metres	Yards	1.094
Microamperes	Amperes	10^{-6}
Microamperes	Milliamperes	10^{-3}
Microfarads	Farads	10^{-6}
Microfarads	Picofarads	10^{6}
Microhenrys	Henrys	10^{-6}
Microhenrys	Millihenrys	10^{-3}
Microsiemens	Siemens	10^{-6}
Microvolts	Millivolts	10^{-3}
Microvolts	Volts	10^{-6}
Miles	Centimetres	1.609×10^{5}
Miles	Feet	5 280
Miles	Inches	6.336×10^{4}
Miles	Kilometres	1.609
Miles	Metres	1 609
Miles	Yards	1 760
Milliamperes	Amperes	10^{-3}
Milliamperes	Microamperes	10^{3}
Millihenrys	Henrys	10^{-3}
Millihenrys	Microhenrys	10^{3}
Millimetres	Centimetres	0.1
Millimetres	Feet	3.281×10^{-3}
Millimetres	Inches	3.937×10^{-2}
Millimetres	Kilometres	10^{-6}
Millimetres	Metres	10^{-3}
Millimetres	Miles	6.214×10^{-7}
Millimetres	Yards	1.094×10^{-3}
Millivolts	Kilovolts	10^{-6}
Millivolts	Megavolts	10^{-9}
Millivolts	Microvolts	10^{3}
Millivolts	Volts	10^{-3}
Milliwatts	Megawatts	10^{-9}
Milliwatts	Watts	10^{-3}
Minutes (time)	Hours	0.016 6
Minutes (time)	Seconds	60
Ohms	Kilohms	10^{-3}
Ohms	Megohms	10^{-6}
Ohms	Microhms	10^{6}
Picofarads	Farads	10^{-12}
Picofarads	Microfarads	10^{-6}
Radians	Degrees	57.30
Radians	Degrees, minutes, seconds	57°,17′,44.8″
Radians	Minutes	3 438
Seconds (time)	Hours	0.000 2
Seconds (time)	Minutes	0.016 6
Siemens	Microsiemens	10^{6}
Volts	Kilovolts	10^{-3}
Volts	Megavolts	10^{-6}
Volts	Microvolts	10^{6}
Volts	Millivolts	10^{3}
Watts	Horsepower	0.001 341 0
Watts	Kilowatts	10^{-3}
Watts	Megawatts	10^{-6}
Watts	Microwatts	10^{6}
Watts	Milliwatts	10^{3}

TABLE 3

Resistor Color Code				
Color	1st Band 1st Digit	2nd Band 2nd Digit	3rd Band Number of Zeros	4th Band Tolerance
Black	—	0	0	—
Brown	1	1	1	—
Red	2	2	2	—
Orange	3	3	3	—
Yellow	4	4	4	—
Green	5	5	5	—
Blue	6	6	6	—
Violet	7	7	7	—
Gray	8	8	8	—
White	9	9	9	—
Gold	—	—	x.1	5%
Silver	—	—	x.01	10%
None	—	—	—	20%

TABLE 4

Powers of Ten			
Number	Power	Prefix	Letter
1 000 000 000 000	10^{12}	tera	T
1 000 000 000	10^{9}	giga	G
1 000 000	10^{6}	mega	M
1 000	10^{3}	kilo	k
1	10^{0}		
0.001	10^{-3}	milli	m
0.000 001	10^{-6}	micro	μ
0.000 000 001	10^{-9}	nano	n
0.000 000 000 001	10^{-12}	pico	p

TABLE 5

The Greek Alphabet	
Greek Letter	Greek Name
Α α	Alpha
Β β	Beta
Γ γ	Gamma
Δ δ	Delta
Ε ε	Epsilon
Ζ ζ	Zeta
Η η	Eta
Θ θ	Theta
Ι ι	Iota
Κ κ	Kappa
Λ λ	Lambda
Μ μ	Mu
Ν ν	Nu
Ξ ξ	Xi
Ο ο	Omicron
Π π	Pi
Ρ ρ	Rho
Σ σ	Sigma
Τ τ	Tau
Υ υ	Upsilon
Φ φ	Phi
Χ χ	Chi
Ψ ψ	Psi
Ω ω	Omega

TABLE 6

Logarithms*

N	0	1	2	3	4	5	6	7	8	9	N	0	1	2	3	4	5	6	7	8	9
10	0000	0043	0086	0128	0170	0212	0253	0294	0334	0374	55	7404	7412	7419	7427	7435	7443	7451	7459	7466	7474
11	0414	0453	0492	0531	0569	0607	0645	0682	0719	0755	56	7482	7490	7497	7505	7513	7520	7528	7536	7543	7551
12	0792	0828	0864	0899	0934	0969	1004	1038	1072	1106	57	7559	7566	7574	7582	7589	7597	7604	7612	7619	7627
13	1139	1173	1206	1239	1271	1303	1335	1367	1399	1430	58	7634	7642	7649	7657	7664	7672	7679	7686	7694	7701
14	1461	1492	1523	1553	1584	1614	1644	1673	1703	1732	59	7709	7716	7723	7731	7738	7745	7752	7760	7767	7774
15	1761	1790	1818	1847	1875	1903	1931	1959	1987	2014	60	7782	7789	7796	7803	7810	7818	7825	7832	7839	7846
16	2041	2068	2095	2122	2148	2175	2201	2227	2253	2279	61	7853	7860	7868	7875	7882	7889	7896	7903	7910	7917
17	2304	2330	2355	2380	2405	2430	2455	2480	2504	2529	62	7924	7931	7938	7945	7952	7959	7966	7973	7980	7987
18	2553	2577	2601	2625	2648	2672	2695	2718	2742	2765	63	7993	8000	8007	8014	8021	8028	8035	8041	8048	8055
19	2788	2810	2833	2856	2878	2900	2923	2945	2967	2989	64	8062	8069	8075	8082	8089	8096	8102	8109	8116	8122
20	3010	3032	3054	3075	3096	3118	3139	3160	3181	3201	65	8129	8136	8142	8149	8156	8162	8169	8176	8182	8189
21	3222	3243	3263	3284	3304	3324	3345	3365	3385	3404	66	8195	8202	8209	8215	8222	8228	8235	8241	8248	8254
22	3424	3444	3464	3483	3502	3522	3541	3560	3579	3598	67	8261	8267	8274	8280	8287	8293	8299	8306	8312	8319
23	3617	3636	3655	3674	3692	3711	3729	3747	3766	3784	68	8325	8331	8338	8344	8351	8357	8363	8370	8376	8382
24	3802	3820	3838	3856	3874	3892	3909	3927	3945	3962	69	8388	8395	8401	8407	8414	8420	8426	8432	8439	8445
25	3979	3997	4014	4031	4048	4065	4082	4099	4116	4133	70	8451	8457	8463	8470	8476	8482	8488	8494	8500	8506
26	4150	4166	4183	4200	4216	4232	4249	4265	4281	4298	71	8513	8519	8525	8531	8537	8543	8549	8555	8561	8567
27	4314	4330	4346	4362	4378	4393	4409	4425	4440	4456	72	8573	8579	8585	8591	8597	8603	8609	8615	8621	8627
28	4472	4487	4502	4518	4533	4548	4564	4579	4594	4609	73	8633	8639	8645	8651	8657	8663	8669	8675	8681	8686
29	4624	4639	4654	4669	4683	4698	4713	4728	4742	4757	74	8692	8698	8704	8710	8716	8722	8727	8733	8739	8745
30	4771	4786	4800	4814	4829	4843	4857	4871	4886	4900	75	8751	8756	8762	8768	8774	8779	8785	8791	8797	8802
31	4914	4928	4942	4955	4969	4983	4997	5011	5024	5038	76	8808	8814	8820	8825	8831	8837	8842	8848	8854	8859
32	5051	5065	5079	5092	5105	5119	5132	5145	5159	5172	77	8865	8871	8876	8882	8887	8893	8899	8904	8910	8915
33	5185	5198	5211	5224	5237	5250	5263	5276	5289	5302	78	8921	8927	8932	8938	8943	8949	8954	8960	8965	8971
34	5315	5328	5340	5353	5366	5378	5391	5403	5416	5428	79	8976	8982	8987	8993	8998	9004	9009	9015	9020	9025
35	5441	5453	5465	5478	5490	5502	5514	5527	5539	5551	80	9031	9036	9042	9047	9053	9058	9063	9069	9074	9079
36	5563	5575	5587	5599	5611	5623	5635	5647	5658	5670	81	9085	9090	9096	9101	9106	9112	9117	9122	9128	9133
37	5682	5694	5705	5717	5729	5740	5752	5763	5775	5786	82	9138	9143	9149	9154	9159	9165	9170	9175	9180	9186
38	5798	5809	5821	5832	5843	5855	5866	5877	5888	5899	83	9191	9196	9201	9206	9212	9217	9222	9227	9232	9238
39	5911	5922	5933	5944	5955	5966	5977	5988	5999	6010	84	9243	9248	9253	9258	9263	9269	9274	9279	9284	9289
40	6021	6031	6042	6053	6064	6075	6085	6096	6107	6117	85	9294	9299	9304	9309	9315	9320	9325	9330	9335	9340
41	6128	6138	6149	6160	6170	6180	6191	6201	6212	6222	86	9345	9350	9355	9360	9365	9370	9375	9380	9385	9390
42	6232	6243	6253	6263	6274	6284	6294	6304	6314	6325	87	9395	9400	9405	9410	9415	9420	9425	9430	9435	9440
43	6335	6345	6355	6365	6375	6385	6395	6405	6415	6425	88	9445	9450	9455	9460	9465	9469	9474	9479	9484	9489
44	6435	6444	6454	6464	6474	6484	6493	6503	6513	6522	89	9494	9499	9504	9509	9513	9518	9523	9528	9533	9538
45	6532	6542	6551	6561	6571	6580	6590	6599	6609	6618	90	9542	9547	9552	9557	9562	9566	9571	9576	9581	9586
46	6628	6637	6646	6656	6665	6675	6684	6693	6702	6712	91	9590	9595	9600	9605	9609	9614	9619	9624	9628	9633
47	6721	6730	6739	6749	6758	6767	6776	6785	6794	6803	92	9638	9643	9647	9652	9657	9661	9666	9671	9675	9680
48	6812	6821	6830	6839	6848	6857	6866	6875	6884	6893	93	9685	9689	9694	9699	9703	9708	9713	9717	9722	9727
49	6902	6911	6920	6928	6937	6946	6955	6964	6972	6981	94	9731	9736	9741	9745	9750	9754	9759	9763	9768	9773
50	6990	6998	7007	7016	7024	7033	7042	7050	7059	7067	95	9777	9782	9786	9791	9795	9800	9805	9809	9814	9818
51	7076	7084	7093	7101	7110	7118	7126	7135	7143	7152	96	9823	9827	9832	9836	9841	9845	9850	9854	9859	9863
52	7160	7168	7177	7185	7193	7202	7210	7218	7226	7235	97	9868	9872	9877	9881	9886	9890	9894	9899	9903	9908
53	7243	7251	7259	7267	7275	7284	7292	7300	7308	7316	98	9912	9917	9921	9926	9930	9934	9939	9943	9948	9952
54	7324	7332	7340	7348	7356	7364	7372	7380	7388	7396	99	9956	9961	9965	9969	9974	9978	9983	9987	9991	9996

*This table gives the mantissas of numbers with the decimal point omitted in each case. Characteristics are determined by inspection from the numbers.

TABLE 7

Trigonometric Functions

deg	*function*	0.0°	0.1°	0.2°	0.3°	0.4°	0.5°	0.6°	0.7°	0.8°	0.9°
0	sin	0.0000	0.0017	0.0035	0.0052	0.0070	0.0087	0.0105	0.0122	0.0140	0.0157
	cos	1.0000	1.0000	1.0000	1.0000	1.0000	1.0000	0.9999	0.9999	0.9999	0.9999
	tan	0.0000	0.0017	0.0035	0.0052	0.0070	0.0087	0.0105	0.0122	0.0140	0.0157
1	sin	0.0175	0.0192	0.0209	0.0227	0.0244	0.0262	0.0279	0.0297	0.0314	0.0332
	cos	0.9998	0.9998	0.9998	0.9997	0.9997	0.9997	0.9996	0.9996	0.9995	0.9995
	tan	0.0175	0.0192	0.0209	0.0227	0.0244	0.0262	0.0279	0.0297	0.0314	0.0332
2	sin	0.0349	0.0366	0.0384	0.0401	0.0419	0.0436	0.0454	0.0471	0.0488	0.0506
	cos	0.9994	0.9993	0.9993	0.9992	0.9991	0.9990	0.9990	0.9989	0.9988	0.9987
	tan	0.0349	0.0367	0.0384	0.0402	0.0419	0.0437	0.0454	0.0472	0.0489	0.0507
3	sin	0.0523	0.0541	0.0558	0.0576	0.0593	0.0610	0.0628	0.0645	0.0663	0.0680
	cos	0.9986	0.9985	0.9984	0.9983	0.9982	0.9981	0.9980	0.9979	0.9978	0.9977
	tan	0.0524	0.0542	0.0559	0.0577	0.0594	0.0612	0.0629	0.0647	0.0664	0.0682
4	sin	0.0698	0.0715	0.0732	0.0750	0.0767	0.0785	0.0802	0.0819	0.0837	0.0854
	cos	0.9976	0.9974	0.9973	0.9972	0.9971	0.9969	0.9968	0.9966	0.9965	0.9963
	tan	0.0699	0.0717	0.0734	0.0752	0.0769	0.0787	0.0805	0.0822	0.0840	0.0857
5	sin	0.0872	0.0889	0.0906	0.0924	0.0941	0.0958	0.0976	0.0993	0.1011	0.1028
	cos	0.9962	0.9960	0.9959	0.9957	0.9956	0.9954	0.9952	0.9951	0.9949	0.9947
	tan	0.0875	0.0892	0.0910	0.0928	0.0945	0.0963	0.0981	0.0998	0.1016	0.1033
6	sin	0.1045	0.1063	0.1080	0.1097	0.1115	0.1132	0.1149	0.1167	0.1184	0.1201
	cos	0.9945	0.9943	0.9942	0.9940	0.9938	0.9936	0.9934	0.9932	0.9930	0.9928
	tan	0.1051	0.1069	0.1086	0.1104	0.1122	0.1139	0.1157	0.1175	0.1192	0.1210
7	sin	0.1219	0.1236	0.1253	0.1271	0.1288	0.1305	0.1323	0.1340	0.1357	0.1374
	cos	0.9925	0.9923	0.9921	0.9919	0.9917	0.9914	0.9912	0.9910	0.9907	0.9905
	tan	0.1228	0.1246	0.1263	0.1281	0.1299	0.1317	0.1334	0.1352	0.1370	0.1388
8	sin	0.1392	0.1409	0.1426	0.1444	0.1461	0.1478	0.1495	0.1513	0.1530	0.1547
	cos	0.9903	0.9900	0.9898	0.9895	0.9893	0.9890	0.9888	0.9885	0.9882	0.9880
	tan	0.1405	0.1423	0.1441	0.1459	0.1477	0.1495	0.1512	0.1530	0.1548	0.1566
9	sin	0.1564	0.1582	0.1599	0.1616	0.1633	0.1650	0.1668	0.1685	0.1702	0.1719
	cos	0.9877	0.9874	0.9871	0.9869	0.9866	0.9863	0.9860	0.9857	0.9854	0.9851
	tan	0.1584	0.1602	0.1620	0.1638	0.1655	0.1673	0.1691	0.1709	0.1727	0.1745
10	sin	0.1736	0.1754	0.1771	0.1778	0.1805	0.1822	0.1840	0.1857	0.1874	0.1891
	cos	0.9848	0.9845	0.9842	0.9839	0.9836	0.9833	0.9829	0.9826	0.9823	0.9820
	tan	0.1763	0.1781	0.1799	0.1817	0.1835	0.1853	0.1871	0.1890	0.1908	0.1926
11	sin	0.1908	0.1925	0.1942	0.1959	0.1977	0.1994	0.2011	0.2028	0.2045	0.2062
	cos	0.9816	0.9813	0.9810	0.9806	0.9803	0.9799	0.9796	0.9792	0.9789	0.9785
	tan	0.1944	0.1962	0.1980	0.1998	0.2016	0.2035	0.2053	0.2071	0.2089	0.2107
12	sin	0.2079	0.2096	0.2113	0.2130	0.2147	0.2164	0.2181	0.2198	0.2215	0.2232
	cos	0.9781	0.9778	0.9774	0.9770	0.9767	0.9763	0.9759	0.9755	0.9751	0.9748
	tan	0.2126	0.2144	0.2162	0.2180	0.2199	0.2217	0.2235	0.2254	0.2272	0.2290
13	sin	0.2250	0.2267	0.2284	0.2300	0.2318	0.2334	0.2351	0.2368	0.2385	0.2402
	cos	0.9744	0.9740	0.9736	0.9732	0.9728	0.9724	0.9720	0.9715	0.9711	0.9707
	tan	0.2309	0.2327	0.2345	0.2364	0.2382	0.2401	0.2419	0.2438	0.2456	0.2475
14	sin	0.2419	0.2436	0.2453	0.2470	0.2487	0.2504	0.2521	0.2538	0.2554	0.2571
	cos	0.9703	0.9699	0.9694	0.9690	0.9686	0.9681	0.9677	0.9673	0.9668	0.9664
	tan	0.2493	0.2512	0.2530	0.2549	0.2568	0.2586	0.2605	0.2623	0.2642	0.2661
15	sin	0.2588	0.2605	0.2622	0.2639	0.2656	0.2672	0.2689	0.2706	0.2723	0.2740
	cos	0.9659	0.9655	0.9650	0.9646	0.9641	0.9636	0.9632	0.9627	0.9622	0.9617
	tan	0.2679	0.2698	0.2717	0.2736	0.2754	0.2773	0.2792	0.2811	0.2830	0.2849
16	sin	0.2756	0.2773	0.2790	0.2807	0.2823	0.2840	0.2857	0.2874	0.2890	0.2907
	cos	0.9613	0.9608	0.9603	0.9598	0.9593	0.9588	0.9583	0.9578	0.9573	0.9568
	tan	0.2867	0.2886	0.2905	0.2924	0.2943	0.2962	0.2981	0.3000	0.3019	0.3038
17	sin	0.2924	0.2940	0.2957	0.2974	0.2990	0.3007	0.3024	0.3040	0.3057	0.3074
	cos	0.9563	0.9558	0.9553	0.9548	0.9542	0.9537	0.9532	0.9527	0.9521	0.9516
	tan	0.3057	0.3076	0.3096	0.3115	0.3134	0.3153	0.3172	0.3191	0.3211	0.3230
18	sin	0.3090	0.3107	0.3123	0.3140	0.3156	0.3173	0.3190	0.3206	0.3223	0.3239
	cos	0.9511	0.9505	0.9500	0.9494	0.9489	0.9483	0.9478	0.9472	0.9466	0.9461
	tan	0.3249	0.3269	0.3288	0.3307	0.3327	0.3346	0.3365	0.3385	0.3404	0.3424
19	sin	0.3256	0.3272	0.3289	0.3305	0.3322	0.3338	0.3355	0.3371	0.3387	0.3404
	cos	0.9455	0.9449	0.9444	0.9438	0.9432	0.9426	0.9421	0.9415	0.9409	0.9403
	tan	0.3443	0.3463	0.3482	0.3502	0.3522	0.3541	0.3561	0.3581	0.3600	0.3620
20	sin	0.3420	0.3437	0.3453	0.3469	0.3486	0.3502	0.3518	0.3535	0.3551	0.3567
	cos	0.9397	0.9391	0.9385	0.9379	0.9373	0.9367	0.9361	0.9354	0.9348	0.9342
	tan	0.3640	0.3659	0.3679	0.3699	0.3719	0.3739	0.3759	0.3779	0.3799	0.3819
21	sin	0.3584	0.3600	0.3616	0.3633	0.3649	0.3665	0.3681	0.3697	0.3714	0.3730
	cos	0.9336	0.9330	0.9323	0.9317	0.9311	0.9304	0.9298	0.9291	0.9285	0.9278
	tan	0.3839	0.3859	0.3879	0.3899	0.3919	0.3939	0.3959	0.3979	0.4000	0.4020
22	sin	0.3746	0.3762	0.3778	0.3795	0.3811	0.3827	0.3843	0.3859	0.3875	0.3891
	cos	0.9272	0.9265	0.9259	0.9252	0.9245	0.9239	0.9232	0.9225	0.9219	0.9212
	tan	0.4040	0.4061	0.4081	0.4101	0.4122	0.4142	0.4163	0.4183	0.4204	0.4224
23	sin	0.3907	0.3923	0.3939	0.3955	0.3971	0.3987	0.4003	0.4019	0.4035	0.4051
	cos	0.9205	0.9198	0.9191	0.9184	0.9178	0.9171	0.9164	0.9157	0.9150	0.9143
	tan	0.4245	0.4265	0.4286	0.4307	0.4327	0.4348	0.4369	0.4390	0.4411	0.4431

Trigonometric Functions

deg	*function*	0.0°	0.1°	0.2°	0.3°	0.4°	0.5°	0.6°	0.7°	0.8°	0.9°
	sin	0.4067	0.4083	0.4099	0.4115	0.4131	0.4147	0.4163	0.4179	0.4195	0.4210
24	cos	0.9135	0.9128	0.9121	0.9114	0.9107	0.9100	0.9092	0.9085	0.9078	0.9070
	tan	0.4452	0.4473	0.4494	0.4515	0.4536	0.4557	0.4578	0.4599	0.4621	0.4642
	sin	0.4226	0.4242	0.4258	0.4274	0.4289	0.4305	0.4321	0.4337	0.4352	0.4368
25	cos	0.9063	0.9056	0.9048	0.9041	0.9033	0.9026	0.9018	0.9011	0.9003	0.8996
	tan	0.4663	0.4684	0.4706	0.4727	0.4748	0.4770	0.4791	0.4813	0.4834	0.4856
	sin	0.4384	0.4399	0.4415	0.4431	0.4446	0.4462	0.4478	0.4493	0.4509	0.4524
26	cos	0.8988	0.8980	0.8973	0.8965	0.8957	0.8949	0.8942	0.8934	0.8926	0.8918
	tan	0.4877	0.4899	0.4921	0.4942	0.4964	0.4986	0.5008	0.5029	0.5051	0.5073
	sin	0.4540	0.4555	0.4571	0.4586	0.4602	0.4617	0.4633	0.4648	0.4664	0.4679
27	cos	0.8910	0.8902	0.8894	0.8886	0.8878	0.8870	0.8862	0.8854	0.8846	0.8838
	tan	0.5095	0.5117	0.5139	0.5161	0.5184	0.5206	0.5228	0.5250	0.5272	0.5295
	sin	0.4695	0.4710	0.4726	0.4741	0.4756	0.4772	0.4787	0.4802	0.4818	0.4833
28	cos	0.8829	0.8821	0.8813	0.8805	0.8796	0.8788	0.8780	0.8771	0.8763	0.8755
	tan	0.5317	0.5340	0.5362	0.5384	0.5407	0.5430	0.5452	0.5475	0.5498	0.5520
	sin	0.4848	0.4863	0.4879	0.4894	0.4909	0.4924	0.4939	0.4955	0.4970	0.4985
29.	cos	0.8746	0.8738	0.8729	0.8721	0.8712	0.8704	0.8695	0.8686	0.8678	0.8669
	tan	0.5543	0.5566	0.5589	0.5612	0.5635	0.5658	0.5681	0.5704	0.5727	0.5750
	sin	0.5000	0.5015	0.5030	0.5045	0.5060	0.5075	0.5090	0.5105	0.5120	0.5135
30	cos	0.8660	0.8652	0.8643	0.8634	0.8625	0.8616	0.8607	0.8599	0.8590	0.8581
	tan	0.5774	0.5797	0.5820	0.5844	0.5967	0.5890	0.5914	0.5938	0.5961	0.5985
	sin	0.5150	0.5165	0.5180	0.5195	0.5210	0.5225	0.5240	0.5255	0.5270	0.5284
31	cos	0.8572	0.8563	0.8554	0.8545	0.8536	0.8526	0.8517	0.8508	0.8499	0.8490
	tan	0.6009	0.6032	0.6056	0.6080	0.6104	0.6128	0.6152	0.6176	0.6200	0.6224
	sin	0.5299	0.5314	0.5329	0.5344	0.5358	0.5373	0.5388	0.5402	0.5417	0.5432
32	cos	0.8480	0.8471	0.8462	0.8453	0.8443	0.8434	0.8425	0.8415	0.8406	0.8396
	tan	0.6249	0.6273	0.6297	0.6322	0.6346	0.6371	0.6395	0.6420	0.6445	0.6469
	sin	0.5446	0.5461	0.5476	0.5490	0.5505	0.5519	0.5534	0.5548	0.5563	0.5577
33	cos	0.8387	0.8377	0.8368	0.8358	0.8348	0.8339	0.8329	0.8320	0.8310	0.8300
	tan	0.6494	0.6519	0.6544	0.6569	0.6594	0.6619	0.6644	0.6669	0.6694	0.6720
	sin	0.5592	0.5606	0.5621	0.5635	0.5650	0.5664	0.5678	0.5693	0.5707	0.5721
34	cos	0.8290	0.8281	0.8271	0.8261	0.8251	0.8241	0.8231	0.8221	0.8211	0.8202
	tan	0.6745	0.6771	0.6796	0.6822	0.6847	0.6873	0.6899	0.6924	0.6950	0.6976
	sin	0.5736	0.5750	0.5764	0.5779	0.5793	0.5807	0.5821	0.5835	0.5850	0.5864
35	cos	0.8192	0.8181	0.8171	0.8161	0.8151	0.8141	0.8131	0.8121	0.8111	0.8100
	tan	0.7002	0.7028	0.7054	0.7080	0.7107	0.7133	0.7159	0.7186	0.7212	0.7239
	sin	0.5878	0.5892	0.5906	0.5920	0.5934	0.5948	0.5962	0.5976	0.5990	0.6004
36	cos	0.8090	0.8080	0.8070	0.8059	0.8049	0.8039	0.8028	0.8018	0.8007	0.7997
	tan	0.7265	0.7292	0.7319	0.7346	0.7373	0.7400	0.7427	0.7454	0.7481	0.7508
	sin	0.6018	0.6032	0.6046	0.6060	0.6074	0.6088	0.6101	0.6115	0.6129	0.6143
37	cos	0.7986	0.7976	0.7965	0.7955	0.7944	0.7934	0.7923	0.7912	0.7902	0.7891
	tan	0.7536	0.7563	0.7590	0.7618	0.7646	0.7673	0.7701	0.7729	0.7757	0.7785
	sin	0.6157	0.6170	0.6184	0.6198	0.6211	0.6225	0.6239	0.6252	0.6266	0.6280
38	cos	0.7880	0.7869	0.7859	0.7848	0.7837	0.7826	0.7815	0.7804	0.7793	0.7782
	tan	0.7813	0.7841	0.7869	0.7898	0.7926	0.7954	0.7983	0.8012	0.8040	0.8069
	sin	0.6293	0.6307	0.6320	0.6334	0.6347	0.6361	0.6374	0.6388	0.6401	0.6414
39	cos	0.7771	0.7760	0.7749	0.7738	0.7727	0.7716	0.7705	0.7694	0.7683	0.7672
	tan	0.8098	0.8127	0.8156	0.8185	0.8214	0.8243	0.8273	0.8302	0.8332	0.8361
	sin	0.6428	0.6441	0.6455	0.6468	0.6481	0.6494	0.6508	0.6521	0.6534	0.6547
40	cos	0.7660	0.7649	0.7638	0.7627	0.7615	0.7604	0.7593	0.7581	0.7570	0.7559
	tan	0.8391	0.8421	0.8451	0.8481	0.8511	0.8541	0.8571	0.8601	0.8632	0.8662
	sin	0.6561	0.6574	0.6587	0.6600	0.6613	0.6626	0.6639	0.6652	0.6665	0.6678
41	cos	0.7547	0.7536	0.7524	0.7513	0.7501	0.7490	0.7478	0.7466	0.7455	0.7443
	tan	0.8693	0.8724	0.8754	0.8785	0.8816	0.8847	0.8878	0.8910	0.8941	0.8972
	sin	0.6691	0.6704	0.6717	0.6730	0.6743	0.6756	0.6769	0.6782	0.6794	0.6807
42	cos	0.7431	0.7420	0.7408	0.7396	0.7385	0.7373	0.7361	0.7349	0.7337	0.7325
	tan	0.9004	0.9036	0.9067	0.9099	0.9131	0.9163	0.9195	0.9228	0.9260	0.9293
	sin	0.6820	0.6833	0.6845	0.6858	0.6871	0.6884	0.6896	0.6909	0.6921	0.6934
43	cos	0.7314	0.7302	0.7290	0.7278	0.7266	0.7254	0.7242	0.7230	0.7218	0.7206
	tan	0.9325	0.9358	0.9391	0.9424	0.9457	0.9490	0.9523	0.9556	0.9590	0.9623
	sin	0.6947	0.6959	0.6972	0.6984	0.6997	0.7009	0.7022	0.7034	0.7046	0.7059
44	cos	0.7193	0.7181	0.7169	0.7157	0.7145	0.7133	0.7120	0.7108	0.7096	0.7083
	tan	0.9657	0.9691	0.9725	0.9759	0.9793	0.9827	0.9861	0.9896	0.9930	0.9965
	sin	0.7071	0.7083	0.7096	0.7108	0.7120	0.7133	0.7145	0.7157	0.7169	0.7181
45	cos	0.7071	0.7059	0.7046	0.7034	0.7022	0.7009	0.6997	0.6984	0.6972	0.6959
	tan	1.0000	1.0035	1.0070	1.0105	1.0141	1.0176	1.0212	1.0247	1.0283	1.0319
	sin	0.7193	0.7206	0.7218	0.7230	0.7242	0.7254	0.7266	0.7278	0.7290	0.7302
46	cos	0.6947	0.6934	0.6921	0.6909	0.6896	0.6884	0.6871	0.6858	0.6845	0.6833
	tan	1.0355	1.0392	1.0428	1.0464	1.0501	1.0538	1.0575	1.0612	1.0649	1.0686

Trigonometric Functions

deg	*function*	0.0°	0.1°	0.2°	0.3°	0.4°	0.5°	0.6°	0.7°	0.8°	0.9°
	sin	0.7314	0.7325	0.7337	0.7349	0.7361	0.7373	0.7385	0.7396	0.7408	0.7420
47	cos	0.6820	0.6807	0.6794	0.6782	0.6769	0.6756	0.6743	0.6730	0.6717	0.6704
	tan	1.0724	1.0761	1.0799	1.0837	1.0875	1.0913	1.0951	1.0990	1.1028	1.1067
	sin	0.7431	0.7443	0.7455	0.7466	0.7478	0.7490	0.7501	0.7513	0.7524	0.7536
48	cos	0.6691	0.6678	0.6665	0.6652	0.6639	0.6626	0.6613	0.6600	0.6587	0.6574
	tan	1.1106	1.1145	1.1184	1.1224	1.1263	1.1303	1.1343	1.1383	1.1423	1.1463
	sin	0.7547	0.7559	0.7570	0.7581	0.7593	0.7604	0.7615	0.7627	0.7638	0.7649
49	cos	0.6561	0.6547	0.6534	0.6521	0.6508	0.6494	0.6481	0.6468	0.6455	0.6441
	tan	1.1504	1.1544	1.1585	1.1626	1.1667	1.1708	1.1750	1.1792	1.1833	1.1875
	sin	0.7660	0.7672	0.7683	0.7694	0.7705	0.7716	0.7727	0.7738	0.7749	0.7760
50	cos	0.6428	0.6414	0.6401	0.6388	0.6374	0.6361	0.6347	0.6334	0.6320	0.6307
	tan	1.1918	1.1960	1.2002	1.2045	1.2088	1.2131	1.2174	1.2218	1.2261	1.2305
	sin	0.7771	0.7782	0.7793	0.7804	0.7815	0.7826	0.7837	0.7848	0.7859	0.7869
51	cos	0.6293	0.6280	0.6266	0.6252	0.6239	0.6225	0.6211	0.6198	0.6184	0.6170
	tan	1.2349	1.2393	1.2437	1.2482	1.2527	1.2572	1.2617	1.2662	1.2708	1.2753
	sin	0.7880	0.7891	0.7902	0.7912	0.7923	0.7934	0.7944	0.7955	0.7965	0.7976
52	cos	0.6157	0.6143	0.6129	0.6115	0.6101	0.6088	0.6074	0.6060	0.6046	0.6032
	tan	1.2799	1.2846	1.2892	1.2938	1.2985	1.3032	1.3079	1.3127	1.3175	1.3222
	sin	0.7986	0.7997	0.8007	0.8018	0.8028	0.8039	0.8049	0.8059	0.8070	0.8080
53	cos	0.6018	0.6004	0.5990	0.5976	0.5962	0.5948	0.5934	0.5920	0.5906	0.5892
	tan	1.3270	1.3319	1.3367	1.3416	1.3465	1.3514	1.3564	1.3613	1.3663	1.3713
	sin	0.8090	0.8100	0.8111	0.8121	0.8131	0.8141	0.8151	0.8161	0.8171	0.8181
54	cos	0.5878	0.5864	0.5850	0.5835	0.5821	0.5807	0.5793	0.5779	0.5764	0.5750
	tan	1.3764	1.3814	1.3865	1.3916	1.3968	1.4019	1.4071	1.4124	1.4176	1.4229
	sin	0.8192	0.8202	0.8211	0.8221	0.8231	0.8241	0.8251	0.8261	0.8271	0.8281
55	cos	0.5736	0.5721	0.5707	0.5693	0.5678	0.5664	0.5650	0.5635	0.5621	0.5606
	tan	1.4281	1.4335	1.4388	1.4442	1.4496	1.4550	1.4605	1.4659	1.4715	1.4770
	sin	0.8290	0.8300	0.8310	0.8320	0.8329	0.8339	0.8348	0.8358	0.8368	0.8377
56	cos	0.5592	0.5577	0.5563	0.5548	0.5534	0.5519	0.5505	0.5490	0.5476	0.5461
	tan	1.4826	1.4882	1.4938	1.4994	1.5051	1.5108	1.5166	1.5224	1.5282	1.5340
	sin	0.8387	0.8396	0.8406	0.8415	0.8425	0.8434	0.8443	0.8453	0.8462	0.8471
57	cos	0.5446	0.5432	0.5417	0.5402	0.5388	0.5373	0.5358	0.5344	0.5329	0.5314
	tan	1.5399	1.5458	1.5517	1.5577	1.5637	1.5697	1.5757	1.5818	1.5880	1.5941
	sin	0.8480	0.8490	0.8499	0.8508	0.8517	0.8526	0.8536	0.8545	0.8554	0.8563
58	cos	0.5299	0.5284	0.5270	0.5255	0.5240	0.5225	0.5210	0.5195	0.5180	0.5165
	tan	1.6003	1.6066	1.6128	1.6191	1.6255	1.6319	1.6383	1.6447	1.6512	1.6577
	sin	0.8572	0.8581	0.8590	0.8599	0.8607	0.8616	0.8625	0.8634	0.8643	0.8652
59	cos	0.5150	0.5135	0.5120	0.5105	0.5090	0.5075	0.5060	0.5045	0.5030	0.5015
	tan	1.6643	1.6709	1.6775	1.6842	1.6909	1.6977	1.7045	1.7113	1.7182	1.7251
	sin	0.8660	0.8669	0.8678	0.8686	0.8695	0.8704	0.8712	0.8721	0.8729	0.8738
60	cos	0.5000	0.4985	0.4970	0.4955	0.4939	0.4924	0.4909	0.4894	0.4879	0.4863
	tan	1.7321	1.7391	1.7461	1.7532	1.7603	1.7675	1.7747	1.7820	1.7893	1.7966
	sin	0.8746	0.8755	0.8763	0.8771	0.8780	0.8788	0.8796	0.8805	0.8813	0.8821
61	cos	0.4848	0.4833	0.4818	0.4802	0.4787	0.4772	0.4756	0.4741	0.4726	0.4710
	tan	1.8040	1.8115	1.8190	1.8265	1.8341	1.8418	1.8495	1.8572	1.8650	1.8728
	sin	0.8829	0.8838	0.8846	0.8854	0.8862	0.8870	0.8878	0.8886	0.8894	0.8902
62	cos	0.4695	0.4679	0.4664	0.4648	0.4633	0.4617	0.4602	0.4586	0.4571	0.4555
	tan	1.8807	1.8887	1.8967	1.9047	1.9128	1.9210	1.9292	1.9375	1.9458	1.9542
	sin	0.8910	0.8918	0.8926	0.8934	0.8942	0.8949	0.8957	0.8965	0.8973	0.8980
63	cos	0.4540	0.4524	0.4509	0.4493	0.4478	0.4462	0.4446	0.4431	0.4415	0.4399
	tan	1.9626	1.9711	1.9797	1.9883	1.9970	2.0057	2.0145	2.0233	2.0323	2.0413
	sin	0.8988	0.8996	0.9003	0.9011	0.9018	0.9026	0.9033	0.9041	0.9048	0.9056
64	cos	0.4384	0.4368	0.4352	0.4337	0.4321	0.4305	0.4289	0.4274	0.4258	0.4242
	tan	2.0503	2.0594	2.0686	2.0778	2.0872	2.0965	2.1060	2.1155	2.1251	2.1348
	sin	0.9063	0.9070	0.9078	0.9085	0.9092	0.9100	0.9107	0.9114	0.9121	0.9128
65	cos	0.4226	0.4210	0.4195	0.4179	0.4163	0.4147	0.4131	0.4115	0.4099	0.4083
	tan	2.1445	2.1543	2.1642	2.1742	2.1842	2.1943	2.2045	2.2148	2.2251	2.2355
	sin	0.9135	0.9143	0.9150	0.9157	0.9164	0.9171	0.9178	0.9184	0.9191	0.9198
66	cos	0.4067	0.4051	0.4035	0.4019	0.4003	0.3987	0.3971	0.3955	0.3939	0.3923
	tan	2.2460	2.2566	2.2673	2.2781	2.2889	2.2998	2.3109	2.3220	2.3332	2.3445
	sin	0.9205	0.9212	0.9219	0.9225	0.9232	0.9239	0.9245	0.9252	0.9259	0.9265
67	cos	0.3907	0.3891	0.3875	0.3859	0.3843	0.3827	0.3811	0.3795	0.3778	0.3762
	tan	2.3559	2.3673	2.3789	2.3906	2.4023	2.4142	2.4262	2.4383	2.4504	2.4627
	sin	0.9272	0.9278	0.9285	0.9291	0.9298	0.9304	0.9311	0.9317	0.9323	0.9330
68	cos	0.3746	0.3730	0.3714	0.3697	0.3681	0.3665	0.3649	0.3633	0.3616	0.3600
	tan	2.4751	2.4876	2.5002	2.5129	2.5257	2.5386	2.5517	2.5649	2.5782	2.5916
	sin	0.9336	0.9342	0.9348	0.9354	0.9361	0.9367	0.9373	0.9379	0.9385	0.9391
69	cos	0.3584	0.3567	0.3551	0.3535	0.3518	0.3502	0.3486	0.3469	0.3453	0.3437
	tan	2.6051	2.6187	2.6325	2.6464	2.6605	2.6746	2.6889	2.7034	2.7179	2.7326

Trigonometric Functions

deg	*function*	0.0°	0.1°	0.2°	0.3°	0.4°	0.5°	0.6°	0.7°	0.8°	0.9°
	sin	0.9397	0.9403	0.9409	0.9415	0.9421	0.9426	0.9432	0.9438	0.9444	0.9449
70	cos	0.3420	0.3404	0.3387	0.3371	0.3355	0.3338	0.3322	0.3305	0.3289	0.3272
	tan	2.7475	2.7625	2.7776	2.7929	2.8083	2.8239	2.8397	2.8556	2.8716	2.8878
	sin	0.9455	0.9461	0.9466	0.9472	0.9478	0.9483	0.9489	0.9494	0.9500	0.9505
71	cos	0.3256	0.3239	0.3223	0.3206	0.3190	0.3173	0.3156	0.3140	0.3123	0.3107
	tan	2.9042	2.9208	2.9375	2.9544	2.9714	2.9887	3.0061	3.0237	3.0415	3.0595
	sin	0.9511	0.9516	0.9521	0.9527	0.9532	0.9537	0.9542	0.9548	0.9553	0.9558
72	cos	0.3090	0.3074	0.3057	0.3040	0.3024	0.3007	0.2990	0.2974	0.2957	0.2940
	tan	3.0777	3.0961	3.1146	3.1334	3.1524	3.1716	3.1910	3.2106	3.2305	3.2506
	sin	0.9563	0.9568	0.9573	0.9578	0.9583	0.9588	0.9593	0.9598	0.9603	0.9608
73	cos	0.2924	0.2907	0.2890	0.2874	0.2857	0.2840	0.2823	0.2807	0.2790	0.2773
	tan	3.2709	3.2914	3.3122	3.3332	3.3544	3.3759	3.3977	3.4197	3.4420	3.4646
	sin	0.9613	0.9617	0.9622	0.9627	0.9632	0.9636	0.9641	0.9646	0.9650	0.9655
74	cos	0.2756	0.2740	0.2723	0.2706	0.2689	0.2672	0.2656	0.2639	0.2622	0.2605
	tan	3.4874	3.5105	3.5339	3.5576	3.5816	3.6059	3.6305	3.6554	3.6806	3.7062
	sin	0.9659	0.9664	0.9668	0.9673	0.9677	0.9681	0.9686	0.9690	0.9694	0.9699
75	cos	0.2588	0.2571	0.2554	0.2538	0.2521	0.2504	0.2487	0.2470	0.2453	0.2436
	tan	3.7321	3.7583	3.7848	3.8118	3.8391	3.8667	3.8947	3.9232	3.9520	3.9812
	sin	0.9703	0.9707	0.9711	0.9715	0.9720	0.9724	0.9728	0.9732	0.9736	0.9740
76	cos	0.2419	0.2402	0.2385	0.2368	0.2351	0.2334	0.2317	0.2300	0.2284	0.2267
	tan	4.0108	4.0408	4.0713	4.1022	4.1335	4.1653	4.1976	4.2303	4.2635	4.2972
	sin	0.9744	0.9748	0.9751	0.9755	0.9759	0.9763	0.9767	0.9770	0.9774	0.9778
77	cos	0.2250	0.2232	0.2215	0.2198	0.2181	0.2164	0.2147	0.2130	0.2113	0.2096
	tan	4.3315	4.3662	4.4015	4.4374	4.4737	4.5107	4.5483	4.5864	4.6252	4.6646
	sin	0.9781	0.9785	0.9789	0.9792	0.9796	0.9799	0.9803	0.9806	0.9810	0.9813
78	cos	0.2079	0.2062	0.2045	0.2028	0.2011	0.1994	0.1977	0.1959	0.1942	0.1925
	tan	4.7046	4.7453	4.7867	4.8288	4.8716	4.9152	4.9594	5.0045	5.0504	5.0970
	sin	0.9816	0.9820	0.9823	0.9826	0.9829	0.9833	0.9836	0.9839	0.9842	0.9845
79	cos	0.1908	0.1891	0.1874	0.1857	0.1840	0.1822	0.1805	0.1788	0.1771	0.1754
	tan	5.1446	5.1929	5.2422	5.2924	5.3435	5.3955	5.4486	5.5026	5.5578	5.6140
	sin	0.9848	0.9851	0.9854	0.9857	0.9860	0.9863	0.9866	0.9869	0.9871	0.9874
80	cos	0.1736	0.1719	0.1702	0.1685	0.1668	0.1650	0.1633	0.1616	0.1599	0.1582
	tan	5.6713	5.7297	5.7894	5.8502	5.9124	5.9758	6.0405	6.1066	6.1742	6.2432
	sin	0.9877	0.9880	0.9882	0.9885	0.9888	0.9890	0.9893	0.9895	0.9898	0.9900
81	cos	0.1564	0.1547	0.1530	0.1513	0.1495	0.1478	0.1461	0.1444	0.1426	0.1409
	tan	6.3138	6.3859	6.4596	6.5350	6.6122	6.6912	6.7720	6.8548	6.9395	7.0264
	sin	0.9903	0.9905	0.9907	0.9910	0.9912	0.9914	0.9917	0.9919	0.9921	0.9923
82	cos	0.1392	0.1374	0.1357	0.1340	0.1323	0.1305	0.1288	0.1271	0.1253	0.1236
	tan	7.1154	7.2066	7.3002	7.3962	7.4947	7.5958	7.6996	7.8062	7.9158	8.0285
	sin	0.9925	0.9928	0.9930	0.9932	0.9934	0.9936	0.9938	0.9940	0.9942	0.9943
83	cos	0.1219	0.1201	0.1184	0.1167	0.1149	0.1132	0.1115	0.1097	0.1080	0.1063
	tan	8.1443	8.2636	8.3863	8.5126	8.6427	8.7769	8.9152	9.0579	9.2052	9.3572
	sin	0.9945	0.9947	0.9949	0.9951	0.9952	0.9954	0.9956	0.9957	0.9959	0.9960
84	cos	0.1045	0.1028	0.1011	0.0993	0.0976	0.0958	0.0941	0.0924	0.0906	0.0889
	tan	9.5144	9.6768	9.8448	10.02	10.20	10.39	10.58	10.78	10.99	11.20
	sin	0.9962	0.9963	0.9965	0.9966	0.9968	0.9969	0.9971	0.9972	0.9973	0.9974
85	cos	0.0872	0.0854	0.0837	0.0819	0.0802	0.0785	0.0767	0.0750	0.0732	0.0715
	tan	11.43	11.66	11.91	12.16	12.43	12.71	13.00	13.30	13.62	13.95
	sin	0.9976	0.9977	0.9978	0.9979	0.9980	0.9981	0.9982	0.9983	0.9984	0.9985
86	cos	0.0698	0.0680	0.0663	0.0645	0.0628	0.0610	0.0593	0.0576	0.0558	0.0541
	tan	14.30	14.67	15.06	15.46	15.89	16.35	16.83	17.34	17.89	18.46
	sin	0.9986	0.9987	0.9988	0.9989	0.9990	0.9990	0.9991	0.9992	0.9993	0.9993
87	cos	0.0523	0.0506	0.0488	0.0471	0.0454	0.0436	0.0419	0.0401	0.0384	0.0366
	tan	19.08	19.74	20.45	21.20	22.02	22.90	23.86	24.90	26.03	27.27
	sin	0.9994	0.9995	0.9995	0.9996	0.9996	0.9997	0.9997	0.9997	0.9998	0.9998
88	cos	0.0349	0.0332	0.0314	0.0297	0.0279	0.0262	0.0244	0.0227	0.0209	0.0192
	tan	28.64	30.14	31.82	33.69	35.80	38.19	40.92	44.07	47.74	52.08
	sin	0.9998	0.9999	0.9999	0.9999	0.9999	1.000	1.000	1.000	1.000	1.000
89	cos	0.0175	0.0157	0.0140	0.0122	0.0105	0.0087	0.0070	0.0052	0.0035	0.0017
	tan	57.29	63.66	71.62	81.85	95.49	114.6	143.2	191.0	286.5	573.0

GLOSSARY

Abscissa. The horizontal or x value of a quantity plotted in a rectangular coordinate system.

Addend. A number that is to be added.

Addition. The mathematical operation of combining numbers to produce a total number.

Algebra. The branch of mathematics in which signs are used to denote arithmetic operations and letters are used to represent numbers and quantities.

Alternating Current (ac). An electric current that reverses direction at regularly recurring intervals.

Altitude. A measure of height along a vertical leg of a triangle.

Ammeter. An instrument for measuring electric current.

Ampere (A). A unit of electrical current.

Amplifier. A device used to increase the strength of an input current or voltage signal so that the output is stronger, but has the same features as the input.

Amplitude. The maximum value of a sine or other periodic wave.

Analog. A device or circuit in which the output varies as a continuous function of the input.

AND Gate. A binary circuit with two or more inputs and a single output. The output is logic 1 when *all* inputs are logic 1.

Angle. A figure formed by two lines extended from the same point or vertex.

Angle, Acute. An angle less than 90°.

Angle, Obtuse. An angle greater than 90° and less than 180°.

Angle, Right. An angle of 90°.

Angles, Complementary. Two angles whose sum is 90°.

Angles, Supplementary. Two angles whose sum is 180°.

Angular Velocity. The rate at which an angle changes. The angular velocity is the product of frequency and 2π. It is denoted by the lowercase Greek letter omega (ω), and is measured in radians per second.

Antilogarithm. A number whose logarithm is the given number.

Arc. Any part of a circle or other curved line.

Armature. The rotating portion of a generator or motor.

Associative Law for Addition. The sum of three or more numbers is the same, regardless of the order in which they are summed.

Asymptotic Curve. A line which continually approaches a curve but, though infinitely extended, never meets it.

Attenuator. An electrical or electronic device that reduces the power of a signal.

Average Power. The average (real) power in an ac circuit is the product of the voltage, current, and the cosine of the phase angle between the voltage and current.

Average Value. In a sine wave, the average value is 0.637 times the peak value.

Axis. One of the reference lines in a coordinate system.

Bandwidth. A group or band of frequencies which surround a center frequency.

Base. (1) A measure of length along a horizontal leg of a triangle. (2) When a number is raised to a power, that number is the base. (3) One region or element of a transistor.

Base Current (I_b). The current through the base of a transistor.

Base Voltage (E_b). The voltage present at the base of a transistor.

Battery. Two or more cells in one case producing voltage for a dc circuit.

Bel. The basic unit in a logarithmic scale for expressing the ratio of two amounts of power.

Beta (β). The current gain of a common emitter transistor circuit.

Binary Number System. A numbering system using the number 2 as a base.

Binomial. An algebraic expression containing two terms.

Boolean Algebra. An algebraic system dealing with on-off circuit elements. The system is named for George Boole who introduced it in 1847.

Boolean Identities. A series of mathematical identities used in Boolean algebra.

Borrowing. Pertaining to moving one from one column as an equivalent value in the next lower place column during the subtraction process.

Branch Current. The current flowing through one branch of a parallel circuit.

Capacitance (C). The property of a capacitor that permits the storage of electrostatic energy. The unit of capacitance is the farad.

Capacitive Reactance (X_C). The opposition to an alternating current due to capacitance. The capacitive reactance is expressed in ohms.

Capacitor. An electrical device consisting of two conducting surfaces which are oppositely charged and are separated by thin layers of dielectric material.

Carrying. When a column of numbers is added (or multiplied), there is a transfer of any higher place symbols to the next higher column.

Cartesian Coordinate System. This rectangular coordinate system consists of two axes and is used to display mathematical relationships.

Cell. A single unit that produces voltage for a dc circuit.

Characteristic. The integral part of a logarithm.

Charge. The electrical energy stored in a capacitor, battery, etc. Also, the quantity of charge carried by a single electron.

Chord. The part of a straight line between two of its intersections with a curve.

Circulating Decimal. A decimal in which groups of digits are used over and over again. For example, $15/37 = 0.405\ 405\ 405\ldots$

Circumference. The line that bounds a circle.

Coefficient. Any factor of an algebraic expression.

Coefficient, Literal. A constant factor of an algebraic expression which is a letter (*a, b,* etc.).

Coefficient, Numerical. A constant factor of an algebraic expression which is a number.

Coil. See Inductor.

Collector Current (I_c). The current through the collector of a transistor.

Collector Voltage (E_c). The voltage present at the collector of a transistor.

Common Log. Logarithms to the base 10.

Commutative Law for Addition. The sum of two numbers is the same, regardless of the order in which they are added.

Complement. Pertaining to a subtraction process used with binary numbers.

Complex Number. A number consisting of a real number and an imaginary number (known as a *j*-number).

Conditional Equation. An equation where some values of the variable will satisfy the equation while other values will not satisfy the equation.

Conductance (G). The ability of a material to carry electrical current. Conductance is the reciprocal of resistance and is measured in siemens.

Conductor. A material, such as copper, that conducts electricity due to a large number of free electrons.

Conjugate. The conjugate of the complex number $a + jb$ is $a - jb$. The sign of the imaginary part of the number is changed to obtain the conjugate.

Constant. A number or letter in an equation that does not change in value.

Coordinates. A pair of numbers used to locate a point in a plane.

Cosecant. A trigonometric function equal to the ratio of the hypotenuse and the side opposite the angle (inverse of sine).

Cosine. A trigonometric function equal to the ratio of the side adjacent to an angle and the hypotenuse.

Cotangent. A trigonometric function equal to the ratio of the side adjacent to an angle and the side opposite that angle (inverse of tangent).

Coulomb (C). The quantity of electric charge carried by 6.24×10^{18} electrons.

Cross-multiplication. Pertaining to a proportion in which the product of the means equals the product of the extremes.

Cube. The product of a number used as a factor three times. The cube of $3 = 3^3 = 3 \cdot 3 \cdot 3 = 27$.

Current (I). The transfer of electrical charge through a material. Current is measured in amperes.

Current Divider Rule. Given two parallel current paths, the current in the first branch (I_1) is equal to the ratio of the resistance in the second branch (R_2) and the total resistance in both branches ($R_1 + R_2$) times the total current (I_t). So that

$$I_1 = \frac{R_2}{R_1 + R_2} \times I_t$$

Current Gain. The ratio of output current to input current.

Cycle. A complete positive and a complete negative alternation of voltage or current.

Cycles Per Second. See Hertz.

Decibel (dB). One-tenth of a bel.

Decimal. (1) Pertaining to a number system with ten symbols. (2) Pertaining to a fraction with a denominator of 10, 100, etc.

Decimal Logarithms. A system of logarithms in which 10 is the base. Also referred to as common logarithms.

Decimal Point. A point separating the whole number from the decimal fraction of a mixed decimal number.

Degree (°). A unit of angular measurement.

Degrees Celsius (°C). Pertaining to a temperature scale in which the melting point of ice is 0°, and the boiling point of water is 100°.

Degrees Fahrenheit (°F). Pertaining to a temperature scale in which the melting point of ice is 32° above zero, and the boiling point of water is 212° above zero.

Delta (Δ). In electronics this symbol is used to denote a change in a quantity.

De Morgan's Theorems. Pertaining to those theorems developed by De Morgan and which are used in Boolean algebra.

Denominator. The lower portion of a fraction by which the upper portion (the numerator) is divided.

Dependent Variable. The value of the dependent variable is determined by the value of the independent variable in an equation.

Determinant. A special expression used to solve systems of linear equations.

Diameter. The length of a straight line when it passes through the center of a circle and intersects the circumference.

Difference. The result of subtracting two numbers.

Digital. A device or circuit in which the output varies in discrete steps (on/off).

Direct Current (dc). An electrical current in which there is a continuous transfer of charge in one direction only.

Direct Proportion. A statement of equality such that an increase in one quantity results in an increase in the other quantity.

Dissimilar Terms. Algebraic terms that do not contain the same letters and/or exponents.

Distributive Law. The product of one number and the sum of two or more other numbers is equal to the sum of the products of the first number and each of the other numbers. For example, $a(b + c) = ab + ac$.

Dividend. A number that is divided by another.

Division. The mathematical operation for finding the quotient of two numbers.

Divisor. A number by which another number (the dividend) is divided.

Double-dabble Process. A method of converting a decimal number to a binary number.

Effective Value. Also called the root mean square (rms) value. In a sine wave the effective value is 0.707 times the peak value.

Efficiency. The ratio of output to the input of a device expressed in percent.

Electromagnetic Field. The field produced around a conductor (or inductor) by the current passing through it.

Emitter Current (I_e). The current flowing through the emitter of a transistor.

Emitter Voltage (E_e). The voltage present at the emitter of a transistor.

Engineering Notation. A number expressed as the product of a number and a power of 10 with an exponent that is a multiple of three.

Equation. Two equivalent expressions separated by an equal sign.

EXCLUSIVE-NOR Gate. A binary circuit consisting of an EXCLUSIVE-OR gate followed by a NOT gate.

EXCLUSIVE-OR Gate. A binary circuit used in digital logic systems to decide if two binary numbers are equal.

Exponent. The number of times a term is to be used as a factor. In 3^4, 4 is the exponent and indicates that 3 is used as a factor 4 times, or $3 \times 3 \times 3 \times 3 = 81$.

Exponential Notation. Expressing a number as a base raised to a specific value denoted by an exponent.

Extraneous Root. The result of manipulating algebraic equations to obtain a solution (root) that is not a solution of the original equation.

Factor(s). When an expression (or number) is made up of the product of a number of symbols, each symbol is known as a factor.

Factoring. The process of removing common factors from numbers and algebraic expressions.

Farad (F). The unit of capacitance.

Filter. A circuit or device which blocks some frequencies and passes others.

Flux. Pertaining to the magnetic lines of force around an inductor.

Formula. An algebraic statement that two expressions are equivalent. These statements usually pertain to specific laws or rules.

Fraction. A two-part expression containing an upper portion (numerator) and a lower portion (denominator). A fraction directs that the numerator be divided by the denominator.

Frequency (f). The number of complete cycles in a unit of time. It is expressed in Hertz.

Generator. An electronic or electrical device or machine to produce electronic or electrical voltages. The most common generator output is the sine wave, although generators often produce other waveforms.

Graduation (Grad). A unit of angular measurement where 1 Grad = 1/400 of a circle.

Graph. A diagram representing a relationship between two variables.

Henry (H). The unit of inductance.

Hertz (Hz). The unit for frequency.

Hexadecimal Number System. A number system having 16 as its base.

Hypotenuse. The longest side of a right triangle.

Identity. An equation that is true for all values of an unknown.

Imaginary Number. Pertaining to the square roots of negative numbers.

Impedance (Z). The total opposition to an alternating current. Included may be inductive reactance, capacitive reactance, and resistance. It is expressed in ohms.

Independent Variable. That variable in an equation that may be chosen arbitrarily.

Indeterminate Equation. An equation having a limitless number of values that can satisfy the equation.

Inductance (L). The property of an alternating current circuit to induce an electromotive force by varying the current. It is expressed in henrys.

Inductive Reactance (X_L). The opposition of an inductor to an alternating current. It is expressed in ohms.

Inductor. An electrical device that acts upon another or is itself acted upon by induction. As the conductor is wound into a spiral or coil the inductive intensity increases.

Instantaneous Value. A voltage or current value along an alternating waveform at a specific point in time.

Integer. The positive and negative whole numbers and zero.

Integrated Circuit (IC). A combination of devices and elements in a miniature self-contained package. Usually these ICs are designed to perform a particular function.

Interpolation. A process to locate values in mathematical tables that fall between table values.

Inverse Proportion. A statement of equality such that an increase in one quantity results in a decrease in the other quantity.

Inverter Gate. See NOT Gate.

***j*-Operator.** The $\sqrt{-1}$ is defined as an imaginary number and is denoted by the symbol j.

Kirchhoff's Current Law. The algebraic sum of the currents entering a point is equal to the sum of the currents leaving the point.

Kirchhoff's Voltage Law. The sum of the voltage sources around a closed loop equals the sum of the voltage drops around that loop.

Lag. Given two waveforms of the same frequency, one is said to lag if it crosses the x axis after the other waveform. This lag is specified as a phase angle.

Large-scale integration (LSI). Integrated circuits containing large-area circuit chips of optimum density.

Lead. Given two waveforms of the same frequency, one is said to lead if it crosses the *x* axis before the other waveform. This lead is specified as a phase angle.

Least Significant Digit (LSD). The value occupying the column immediately to the left of the radix point. This may be the units column in the decimal number system.

Linear. Pertaining to a straight line.

Linear Equation. An equation where each term contains only one variable to the first power, or is a constant.

Line Current. That current supplied by a source to sustain oscillations in a tank circuit.

Load. The resistance, reactance, or impedance connected across the output of a circuit.

Log. See logarithm.

Logarithm. The logarithm of a number is the exponent to which a base number is raised to obtain the number.

Logic Gate. A device that produces an input-output relationship based upon Boolean algebra.

Loop Analysis. A process used to determine voltage and current values in electronic circuits containing closed loops.

Lowest Common Denominator (LCD). The product of all the prime factors that appear in the denominators. In the LCD each factor is raised to the highest power to which it appears in any one of the denominators.

Magnitude. The value of a quantity expressed as a number and a unit of measure.

Mantissa. The fractional part of a logarithm.

Medium-scale Integration (MSI). Integrated circuits that function as simple, self-contained systems.

Metre (m). The metric basic unit of distance.

Metric System. A decimal system of weights and measures based upon the metre and the gram.

Minuend. A number from which the subtrahend is subtracted.

Minute. A unit of circular measure equivalent to 1/60 of a degree.

Monomial. A one-term mathematical expression.

Most Significant Digit (MSD). The value occupying the left-most column of a number.

Multiplicand. The number being multiplied by the multiplier.

Multiplication. The mathematical operation to perform addition a given number of times.

Multiplier. The number by which the multiplicand is multiplied.

Mutual Inductance (M or L_M). The condition that exists in a circuit when the positions of the two inductors cause magnetic lines of force from one inductor to link the turns of the other.

NAND Gate. A binary circuit with two or more inputs and a single output. The NAND Gate consists of an AND gate followed by a NOT gate.

Naperian Logarithms. See Natural Logarithms.

Natural Logarithms. A logarithmic system using *e* as a base. The value of *e* is approximately 2.718.

Negative Logic. When the lower of two voltages represents a binary "1" and the higher a "0" in digital systems.

Net Reactance. The difference between inductive and capacitive reactance in a series ac circuit.

Network. A combination of electrical elements and components.

Node. A point or junction in a circuit.

Node Analysis. A process used to determine voltage and current values in electronic circuits containing more than one node or junction point.

NOR Gate. A binary circuit with two or more inputs and a single output. The NOR gate consists of an OR gate followed by a NOT gate.

Norton Current. The value of the current source in a Norton equivalent circuit.

NOT Gate. A binary circuit with a single input and single output. The output is a logic 1 when the input is a logic 0.

Numerator. The upper portion of a fraction which is divided by the lower portion (the denominator).

Octal Number System. A number system having 8 as its base.

Ohm (Ω). The unit of resistance.

Ohmmeter. A meter used to measure resistance in ohms.

Ohm's Law. Current is directly proportional to the voltage and inversely proportional to the resistance.

Open. Pertaining to a circuit or component that is "open" and will not allow current flow. An open has infinite resistance.

Open Circuit Voltage. The voltage at the output terminals of a Thévenin equivalent circuit.

Operational Amplifier (Op-Amp). An amplifier that can be used for many applications. Some of these are mathematical, such as addition, subtraction, and multiplication.

Ordered Pair. The values of the *x* and *y* coordinates of a point in a rectangular coordinate system.

Ordinate. The vertical or *y* value of a quantity plotted in a rectangular coordinate system.

OR Gate. A binary circuit with two or more inputs and a single output. The output is a logic 1 when *any* of the inputs equal a logic 1.

Origin. The intersection of the *x* and *y* axes in a rectangular coordinate system.

Oscillate. Periodic fluctuations in an electronic circuit due to the flow of current in opposite directions.

Oscilloscope. A test instrument using a cathode ray tube (CRT) to display signal patterns.

Parabola. A graph of the quadratic equation in the form $y = ax^2 + bx + c$.

Parallel. (1) Two straight lines in the same plane, the same distance apart, never meeting. (2) Pertaining to an electronic circuit with more than one path for current flow.

Peak. The point of maximum voltage, current, or power during a cycle.

Peak to Peak. The voltage, current, or power measured from positive peak to negative peak.

Percent. A fraction with the denominator equal to 100.

Period (T). The length of time for one cycle to occur.

Perpendicular. An intersection of two lines at an angle of 90°.

Phase Angle. The constant angle between waveforms that have the same frequency.

Phasor. An entity (voltage, current, and impedance) expressed as a vector with magnitude and direction.

Phasor Domain. Pertaining to situations where voltage, current, and impedance values are expressed as phasors.

Pi (π). A letter of the Greek alphabet used as the symbol for the ratio of the circumference of a circle to its diameter. The approximate value of $\pi = 3.141\ 59$.

Polar Notation. Expressing phasors as a magnitude (rms value) and an angle (phase angle) such as $\mathbf{I} = 5\ \text{A} \angle 18°$.

Polynomial. An expression containing more than one term.

Positive Logic. When the higher of two voltages represents a binary "1," and the lower a "0" in digital systems.

Potentiometer. A three-terminal variable resistor.

Power, Apparent (P_A). The power delivered to the circuit elements by the source. It is indicated by the product of effective voltage and effective current. Apparent power is expressed in volt-amperes (VA).

Power, Electrical (P). The rate at which energy is dissipated or generated. Power is measured in watts.

Power Factor. The cosine of the phase angle between voltage and current. Real power equals the product of the apparent power and the power factor.

Power Gain. The ratio of output power to input power.

Power, Reactive (P_{Q_L} and P_{Q_C}). That part of the apparent power that is stored in reactive elements of the circuit and returned to the source during a later part of the cycle. Reactive power is expressed in reactive volt-amperes (vars).

Power, Real (P). That part of the apparent power actually consumed by the resistive elements of the circuit. The rate at which energy is converted to heat, light, work, etc. Real power is measured in watts.

Prefix. An affix placed before a word or element to add to or qualify the meaning.

Primary. The input winding to a transformer which usually sets up voltages in one or more secondaries.

Prime. A superscript often used to differentiate among similar quantities. For example, I prime is written I'.

Principal Diagonal. Pertaining to the solution of a determinant.

Product. The result of multiplying two or more quantities.

Proportion. A relation of four quantities (two ratios) such that the product of the first and fourth quantities equals the product of the second and third quantities.

Pythagorean Theorem. This theorem refers to right triangles and states that the square of the hypotenuse is equal to the sum of the squares of thc two other sides.

Q Factor. The quality factor, in a parallel resonant circuit, has a value of X_L divided by R.

Quadrant. One-fourth of the rectangular coordinate system. One quadrant contains 90°.

Quadratic Equation. The general quadratic equation is written $ax^2 + bx + c$.

Quotient. A number obtained through the process of dividing.

Radian. A unit of circular measure such that the angle intercepts an arc on the circumference equal to the radius of the circle. One radian equals approximately 57.3°.

Radical Sign. The symbol used to denote a square root ($\sqrt{\ }$).

Radius. The distance along a straight line from the center of a circle to a point on the circumference of the circle.

Radix. A number taken as the base of a system of numbers.

Radix Point. The point which separates the whole and fractional portion of a number within a number system. Examples include the decimal point, octal point, etc.

Ratio. The quotient obtained by dividing one number by another.

RC Circuit. An electronic circuit consisting of resistance and capacitance.

RCL Circuit. An electronic circuit consisting of resistance, capacitance, and inductance.

Reactance (*X*). Opposition to ac due to inductance and/or capacitance. Reactance is measured in ohms.

Reciprocal. The reciprocal of a number is found by dividing one by that number. The reciprocal of x is $1/x$.

Rectangular Notation. Expressing phasors as a complex number with a real part and an imaginary part. For example: $9 + j12$.

Remainder. The quantity that remains after subtraction or division.

Repeating Decimal. A decimal in which a single number is repeated. For example: 1/3 = 0.333 . . . .

Resistance (*R*). The opposition a material has to current flow. Resistance is measured in ohms.

Resistor. A device that opposes the flow of an electric current. It is used for protection, operation, or current control.

Resonance. An ac circuit condition where inductive reactance and capacitive reactance are equal.

Resonant Frequency. The frequency at which the inductive and capacitive reactances in an ac circuit are equal.

Rheostat. A two-terminal variable resistor.

RL Circuit. An electronic circuit consisting of resistance and inductance.

Root. The root of an equation is any number that will satisfy the equation.

Root-mean-square (rms). The square root of the average of the squares of a waveform taken throughout one complete period.

Scalar. A quantity which may be described by specifying a magnitude. Examples include time, money, lengths, temperatures, etc.

Schematic. The diagram of an electrical or electronic circuit.

Scientific Notation. A number expressed as the product of a number between 1 and 10 and a power of 10. For example: $450 = 4.5 \times 10^2$.

Secant. (1) A trigonometric function equal to the ratio of the hypotenuse and the side adjacent to that angle (inverse of cosine). (2) Pertaining to a straight line passing through a circle and touching the circumference at two points.

Second. (1) The metric and English unit of time. (2) In terms of angular measure, 1 second is 1/60 of 1 minute, or 1/3 600 of 1 degree.

Secondary. Pertaining to the winding(s) of a transformer used as an output.

Secondary Diagonal. Pertaining to the solution of a determinant.

Sector. The area bounded by two radii and the arc which they intercept.

Segment. The area between a chord and its intercepted arc.

Series-aiding. Inductors connected in series such that the magnetic lines of force add to the total inductance. This additional inductance is termed mutual inductance.

Series Circuit. A method of connecting a circuit so that current has one path to follow.

Series-opposing. Inductors connected in series

such that the magnetic lines of force subtract from the total inductance. This is termed mutual inductance.

Series-parallel. An electronic circuit connected such that current flows through one path at some points, and through more than one path at other points.

Short. Pertaining to a circuit or component which has no opposition to current flow. A "short" circuit has no resistance.

Shunt. Any device or component placed in parallel with another device or component.

Significant Digits. Those digits in a number that are known to be exact.

Similar Terms. Algebraic terms containing the same letters and exponents.

Simultaneous Equations. Two or more equations which contain the same variables.

Sine. A trigonometric function equal to the ratio of the side opposite an angle and the hypotenuse.

Sine Wave. A wave which can be expressed as the sine of a linear function such as time.

Slope. The slope of a straight line is the vertical distance between two points on the line divided by the horizontal distance between the points.

Square. (1) The product of a number multiplied by itself. For example, 8 squared $= 8^2 = 8 \times 8 = 64$. (2) A rectangle with four equal sides.

Square Root. The square root of a number is the value, which when multiplied by itself, will produce the original number.

Subscript. A number or letter written below and to the right of a symbol and used to identify similar quantities. Resistor number two, for example, is written R_2.

Subtraction. The mathematical operation for finding the difference between the subtrahend and the minuend.

Subtrahend. The number subtracted from the minuend.

Sum. The total resulting from addition.

Superscript. A number or letter written above and to the right of a symbol which provides instructions pertaining to that symbol. To cube the a in $8a$, the expression would be written $8a^3$.

Tangent. (1) A straight line that touches the circumference at only one point. (2) A trigonometric function equal to the ratio of the side opposite an angle and the side adjacent to that angle.

Tank Circuit. A parallel-resonant circuit that oscillates at a frequency determined by the components in the circuit.

Terminals. A point of connection in an electronic circuit. The output terminals of a circuit indicate where a load or additional circuitry is to be connected.

Tetrad. A group of four.

Thévenin Resistance. A series resistance in a Thévenin equivalent circuit.

Thévenin Voltage. The value of the source voltage in a Thévenin equivalent circuit.

Time Constant. The time required for an exponential quantity to change by 63.2%.

Time Domain. Expressing instantaneous voltages and currents at a specific point in time. A time domain expression, such as $e = E_p \sin \omega t$, uses the peak voltage (E_p), the sine of the angular velocity (ω) and the time (t) of rotation.

Tolerance. Pertaining to the manufacturer's specifications which indicates how much a product (resistor, capacitor, etc.) may differ from the specified value.

Transformer. A device which utilizes electromagnetic induction to transform electric energy from one level to another. Transformers, for example, may step-up or step-down voltages at the same frequency.

Transistor. An electronic device consisting of a small block of semiconductors with a minimum of three electrodes.

Transpose. Pertaining to the movement of a term from one side of an equation to the other.

Triad. A group of three.

Triangle. A three-sided plane figure.

Trigonometry. The branch of mathematics that deals with the relationships between the sides and angles of triangles.

Trinomial. An algebraic expression containing three terms.

Truth Table. A table displaying the input and output combinations for electronic circuits. Truth tables are typically used with digital logic circuits.

Var. The unit of reactive power as opposed to real power which is measured in watts. The var is the abbreviation for volt ampere reactive.

Variable. Letters in algebraic expressions which may vary in a given problem.

Vector. A quantity described by both magnitude and direction. A vector is typically drawn as an arrow.

Vertex. Pertaining to the fixed point where the two sides of an angle intersect.

Vinculum. A bar used as a grouping symbol.

Volt (V). A unit of electrical potential or pressure.

Voltage (*E*). The electromotive force or electrical pressure. It is expressed in volts.

Voltage Divider Rule. The ratio between any two voltage drops in a series circuit is the same as the ratio of the two resistances across which these voltage drops occur.

Voltage Gain. The ratio of output voltage to input voltage.

Voltmeter. A meter used to measure voltage.

Watt (W). A unit of power.

Wavelength. The distance between two points measured at the same place on two consecutive cycles.

Wheatstone Bridge. A null-type precision instrument used to measure resistance.

***x* Intercept.** Pertaining to that point at which a graphed function crosses the x axis.

***y* Intercept.** Pertaining to that point at which a graphed function crosses the y axis.

ANSWERS TO ODD-NUMBERED PROBLEMS

CHAPTER 1

Unit 1·1

1. 84 V
3. 16 307 Ω
5. 1 865 mH
7. 1 880 Ω
9. 7 345 Ω
11. 468
13. 83 kilometres
15. 340 V_{p-p}
17. a. 1 144
    b. 1 394 V
    c. −4 998
    d. 6 600 Hz
    e. 7 590
    f. 1 221 mH
    g. −39 375
    h. 44 042
19. 5 400 μA
21. 66 V
23. 126 W
25. 420 W
27. 5 460 000 Hz
29. 1 230 V
31. a. 60 V
    b. −84
    c. 14 A
    d. 82
    e. 569 V
    f. 61
    g. −58 mA
    h. 524 mV
33. 116 A
35. 455 000 Hz
37. 10 570 Ω
39. 13 metres
41. 85 083 feet
43. 18 V
45. a. 29 W
    b. −80
    c. 101 g
    d. 276 A
    e. −357
    f. 12 V
    g. 173 Ω
    h. 128
    i. 412
    j. 159
47. 51
49. 13
51. 14 days
53. 12 A
55. 20 V
57. 16 A
59. a. 220
    b. 35
    c. −21
    d. 272
    e. 17
    f. 262 440
    g. −890
    h. 15
    i. 100
    j. −512
61. 24 Ω
63. a. 9 Ω
    b. 18 V
65. a. 5 A
    b. 4 Ω

Unit 1·2

1. a. 1:12
   b. 4:1
   c. 2:33
   d. 100:Þ
   e. 7:11
   f. 1:9
   g. 1:5
   h. 2:7
   i. 22:1
   j. 40:1
3. a. 16
   b. 200 V
   c. 12
   d. 28 Ω
   e. 8 mA
   f. 1
   g. 63
   h. 15 Ω
   i. 16
   j. 768
5. 10:1
7. 15 V
9. 480
11. 33 kΩ

Unit 1·3

1. a. $\frac{3}{8}$
   b. $\frac{7}{15}$
   c. $1\frac{7}{20}$
   d. $3\frac{5}{9}$
   e. $\frac{7}{16}$
3. a. 40
   b. 36
   c. 240
5. a. $6\frac{15}{16}$
   b. $7\frac{7}{8}$
   c. $12\frac{15}{32}$
   d. $4\frac{5}{6}$
   e. $17\frac{55}{256}$
   f. $4\frac{3}{8}$
   g. $42\frac{29}{40}$
   h. $14\frac{317}{360}$
7. $1\frac{1}{2}$ Ω
9. 95°F
11. 80°C
13. Increase
15. 86.01
17. −86.72
19. −327.25

Unit 1·4

1. 495 Ω to 605 Ω
3. 2 090 Ω to 2 310 Ω
5. 42.5 μF to 57.5 μF
7. 445 900 Hz to 464 100 Hz
9. 26 730 Ω to 27 270 Ω
11. 1
13. 2
15. 6
17. 5
19. 2
21. 1.4
23. 700.60
25. 1 990.

Unit 1·5

1. 9 870 Ω
3. 1 800 Ω
5. 99 500 Ω
7. 27 V
9. 1 326 240
11. 6 375 μA
13. 651
15. −19
17. 455 000 Hz
19. 8 H
21. a. 27
    b. 19 V
    c. −54
23. 7 A
25. 273 W
27. 7 A
29. a. − 4 780
    b. −47 712
    c. 1
31. 20 V
33. 30 A
35. a. $13\frac{41}{64}$
    b. $10\frac{9}{16}$
    c. $15\frac{3}{32}$
37. $\frac{1}{100}$ or 0.01 siemens
39. 792 W
41. a. 733
    b. 3.82
    c. 200
43. $\frac{3}{5}$
45. 170
47. 142.5 μF to 165 μF
49. 150.

CHAPTER 2

Unit 2·1

1. 221 760 inches
3. 14 inches
5. 19.05 cm/s
7. 5.833 yards
9. 533.4 mm
11. 475 488 mm
13. 0.304 miles per hour
15. 1 738 667.5 feet per hour
17. 4.556 metres
19. 3.776 feet

Unit 2·2

1. 0.093 75
3. 0.937 5
5. 0.8
7. $0.066\ 6\overline{6}$
9. $0.818\ \overline{181}$
11. $\frac{1}{2}$
13. $\frac{17}{25}$
15. $\frac{1}{125}$
17 $1\frac{11}{80}$
19. $\frac{63}{250}$

Unit 2·3

1. 5.91 Hz
3. 290.14 mA
5. 228.522 Ω
7. 1 367.303 kHz
9. 50 A
11. 2.916 kV
13. 336.08 Ω
15. 28.788 s
17. 32.8 μF
19. 1.96 A
21. 36.244
23. 1.44 A
25. 48.56
27. 49.742 4
29. 212.576 4
31. 400 Ω
33. 22.5
35. 16
37. 800 Ω
39. 14.8
41. 448.5
43. 3.25
45. 39.32
47. 2.5
49. 8.2
51. 3 409.82 Ω
53. 118.069 V
55. 90

Unit 2·4

1. 2.55×10^{1}
3. 2.053×10^{4}
5. 4.659×10^{4}
7. 5.31×10^{29}
9. 4.276
11. 3.551×10^{-7}
13. 2.5×10^{-9}
15. 6
17. 1.79×10^{5}
19. 0
21. 3.506×10^{1}
23. 1.83×10^{-5}
25. -1.201×10^{-4}

Unit 2·5

1. 0.38 A
3. 0.28 mA
5. 0.009 8 kV
7. 0.049 H
9. 62 500 mA
11. 75 000 W
13. 4.5 s
15. 590 000 kHz
17. 2 000 pF
19. 470 000 Ω
21. 13.32 V
23. 463 kW
25. 1 500 kHz
27. 0.078 75 A
29. 32 000 Hz
31. 1.875 mA
33. 25 kHz
35. 8.94 kΩ

Unit 2·6

1. 1 000
3. 16
5. $\frac{3}{8} = 0.375$
7. 11.391
9. 1.214
11. 1.331
13. 1.020
15. 1 296
17. 161 051
19. 1.357
21. 1.526×10^{11}
23. 1.297
25. 191.501
27. 2.513
29. 11.334
31. 12 566.371 Ω

Unit 2·7

1. 4.115 m
3. 79.539 miles
5. 1 800 mm
7. 4.6
9. 0.32
11. $\frac{3}{20}$
13. $\frac{7}{20}$
15. $\frac{5}{64}$
17. 35.71 W
19. 46.467 Ω
21. 25 V
23. 229.6
25. 500 Hz
27. −61.568
29. 64 056
31. 27 900 miles
33. 7.69×10^4
35. 6.444×10^1
37. $1.640\ 5 \times 10^3$
39. 3.2×10^5
41. 1.047×10^{-5}
43. 1.8×10^4
45. 1
47. 18 000 Hz
49. 4.12 kΩ
51. 2 500 μV
53. 35 A
55. 1 082 kHz
57. 36.3
59. 3.451 9
61. 274.625
63. 1.498
65. 694.694

CHAPTER 3

Unit 3·1

1. 12.8 mA
3. 10.88 V
5. 416.67 mA
7. 160 kΩ
9. 24 kΩ
11. 0.734 mW
13. 200 Ω
15. 0.5 A
17. 0.12 W
19. 100 Ω
21. $L = \dfrac{RA}{k}$
23. $\beta = \dfrac{I_{ceo}}{I_{cbo}} - 1$
25. $BW = 2(f_r - f_L)$

Unit 3·2

1. 30 kΩ
3. 6.5 V
5. 25 V
7. 1.62 mA
9. 15.45 kΩ
11. 1.21 kΩ
13. 668.57 Ω
15. 270 Ω
17. 112.67 mA
19. 332.18 Ω
21. 3 V
23. 40 mA
25. 0.9 W
27. 120 V
29. 98.18 V
31. 120 kΩ
33. 360 V
35. 43.99 mA
37. 25 mA
39. 9 kΩ
41. 764.71 Ω
43. 110 mA
45. 17.51 mA
47. 6.8 kΩ
49. 9.6 kΩ

Unit 3·3

1. 39.8 V
3. 25.30 V
5. 10 V, 5 V
7. 368.42 mA
9. 14.2 mA
11. 681.82 μA
13. 15 V

Unit 3·4

1. a. $P = 6.75$ W
   b. $E_L = 4.5$ V
   c. $I_L = 1.5$ A
3. a. $E_{TH} = 6.12$ V
   b. $R_{TH} = 1.632$ kΩ
5. a. $E_{TH} = 11.55$ V (*A* negative with respect to *B*)
   b. $R_{TH} = 1\ 658.31$ Ω
7. a. $I_{SL} = 6.44$ mA (from *B* to *A*)
   b. $R_{TH} = 10.29$ kΩ
9. 66.27 V

Unit 3•5

1. 0.01 μF
3. 0.001 7 μF
5. 14.58 μF
7. 0.007 7 μF
9. 0.002 7 μF
11. 50 μs
13. a. 5 s
    b. 50 ms
15. 0.38 H
17. 0.53
19. 40 mH
21. 57.69 mH
23. 3.7 μs

Unit 3•6

1. Positive x axis
3. IV
5. I
7. negative y axis
9. IV
11. 1 V
13. −12 V
15. 1 mA
17. 10 mA
19. 1 000 Ω
21. 200

23.

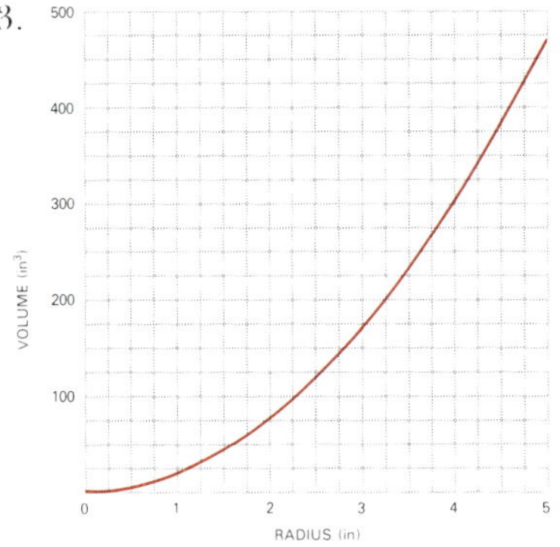

25.

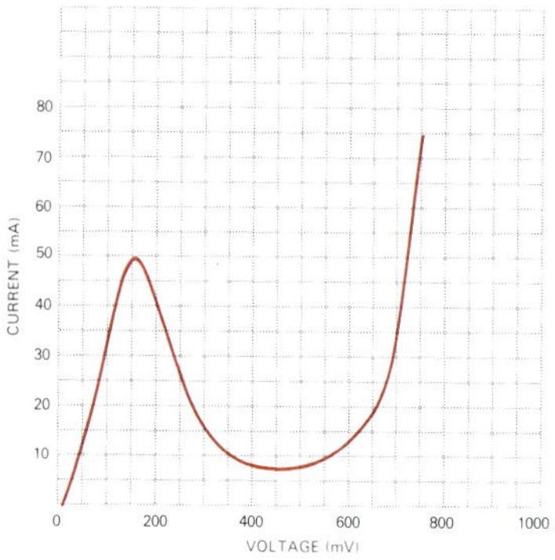

Unit 3•7

1. 83.33 mA
3. 144 Ω
5. 1.5 kΩ
7. 37.68 V
9. 27.79 kΩ
11. 40 kΩ
13. 66.09 V
15. R_2 = 1 kΩ
    R_5 = 6 kΩ
    R_6 = 4 kΩ
17. R_t = 5.43 kΩ
    E_s = 38 V
19. 0.75 μC
21. 0.035 μF
23. 3.69 μF
25. 0.6 ms
27. t = 1 s
    E_R = 1.22 V
29. 79.04 mH
31. 0.42
33. 25 Ω
35. 76.92 mA
37. 3 717.33 Ω
39. a. E_s = 65.52 V
    b. R_t = 568.49 Ω
41. 65 mA
43. 6 250 Ω
45. 21.06 kΩ
47. 1.04 mA
49. 3.2 Ω
51. 36 V
53. a. E_{TH} = 3.44 V
    b. R_{TH} = 2 643.5 Ω
55. a. E_{TH} = 39.75 V
    b. R_{TH} = 3.75 kΩ
    c. I_{SL} = 10.6 mA
57. a. I_{SL} = 1.77 mA
    b. R_{TH} = 464.59 Ω
59. a. IV
    b. I
    c. III
    d. II
61. −40 V
63. 0 V

CHAPTER 4

Unit 4•1

1. −56
3. −36
5. −6
7. Polynomial
9. $\dfrac{3xy^2}{x - y}$
11. 121
13. −251
15. −15.206

Unit 4•2

1. $16a^2b^4$
3. $\dfrac{a^7b^6}{c^{10}}$
5. 1
7. 19.725
9. $\dfrac{1}{4x^5y^3z^4}$
11. $5^6a^{12}x^6$
13. $\dfrac{1}{a^4b^8}$
15. $\dfrac{12^3y^9}{x^6}$
17. $\dfrac{-25x^8b^2}{a^2}$
19. $62.5a^2y^2$
21. $\dfrac{1}{b^{m+n}c^{m-1}}$ or $\dfrac{c^{1-m}}{b^{m+n}}$
23. $\dfrac{784b^{14}}{x^6}$
25. 3.929×10^{-5}
27. 1.537×10^{-23}
29. $6^3a^{27}x^{18}$

Unit 4·3

1. $13x$
3. $-x^2y - 9x$
5. $6y$
7. $-3x^2y^2 + 10x^2y + 5xy^2$
9. $-18xy^2z - 180x^2yz^2 + 19xyz^2$
11. 1 083
13. 74 964
15. $-5ab^2 - 11a^2b$
17. $10gh^2 - 24gh + 51$
19. $3y^2z^3 - 12y^2z^2 + 17yz + x^2z^2$

Unit 4·4

1. $36x^6y^3$
3. $\dfrac{y^8}{-81x^{16}}$
5. $12b^3 - 36b^6$
7. $a^3 + a^2b - ab^2 - b^3$
9. $2a^2b^2 + \dfrac{3}{2ab^6}$
11. $25a^2 - 4b^2$
13. $\dfrac{256a}{x^9b^4}$
15. $36x^{13}y^{12}z^6$
17. $11x^4y - \dfrac{4x^2}{3y^2} + \dfrac{2y}{3} + \dfrac{y^4}{12x^2}$
19. $\dfrac{a^2}{b^4} + \dfrac{5a}{12b^3} - \dfrac{1}{6b^2}$
21. $3x + 2$
23. $4a^4 - 11a^3 + 2a^2 - 5a + 1$

Unit 4·5

1. $3ax - 12a$
3. $-3y^5 + 2y^3 - y^2 + y$
5. $4x^3 - 8x^2 + 4x$
7. $a^3 + 2a^2b + ab^2$
9. $y^2 - 49$
11. $25x^4 - 81$
13. $56y + 224 - 8xy - 32x$
15. $18x^3 - 60x^2 + 3xy - 10y$
17. $a^2 + 5a - 14$
19. $-m^2 + 9$
21. $x^3 - 12x^2 + 48x - 64$
23. $324a^3 + 1\,296a^2 + 1\,728a + 768$
25. $x^9 - 3x^8 + 3x^7 - x^6$
27. $3a^3b^2c^2(9a^9b^4c + 3 - 5a^2b)$
29. $3x^4(x + 1)(x - 1)$
31. $(x + 1)(x - 1)$
33. $(7 - x)(7 - x)$
35. $b(8b + 3)(8b + 3)$
37. $(x + 3)(x - 1)$
39. $(5a^2 - 6)(5a^2 + 3)$
41. $(a - 1)(b - 1)$
43. $(b + 1)^3$
45. $x^3(3x + 1)^3$

Unit 4·6

1. $\dfrac{5}{24}$
3. $\dfrac{3xb - 5xa}{ab}$
5. $\dfrac{y^2 + 6y + 1}{y^2 - 1}$
7. $\dfrac{(3a - b)(a - b)^2 + (2a - 4b)^2(a + b)^2}{(a + b)^2(a - b)^2}$
9. $\dfrac{12x^4 + 5x^3 - 63x^2 - 20}{10x^3 - 20x^2 - 40x + 80}$
11. $\dfrac{a^3 + 23a^2 + a}{3a^3 + 9a^2 - 12a}$
13. $\dfrac{m^2 - 20m + 20}{m^2 - 4m - 12}$
15. $\dfrac{2x^3 + 5x}{x^3 + x^2 - x - 1}$
17. $\dfrac{x^2 + 3y^2}{x^2 - y^2}$
19. $\dfrac{4a^2 - 7}{16a^2 - 20a + 4}$

Unit 4·7

1. $\dfrac{4xb^6}{c^5}$
3. $\dfrac{36a^2}{xc}$
5. $2ac + 2ab$
7. $4x - 4$
9. $\dfrac{a^2 + 5a - 14}{a^2 - 6a + 9}$
11. $y^2 - 2y + 1$
13. $a^2 + 3a$
15. $\dfrac{3b - 21}{2b - 7}$
17. $\dfrac{b^2 + ab}{a^3}$
19. $\dfrac{y^4 + y^2}{2y - 2}$

Unit 4·8

1. a. $\log_7 2\,401 = 4$
   b. $\log_{1/8} 4\,096 = -4$
   c. $\log_4 2.828 = 3/4$
   d. $\log_5 2.627 = 3/5$
3. a. $9^4 = 6\,561$
   b. $(0.2)^{-3} = 125$
   c. $6^{3/2} = 14.697$
   d. $15^{1.2} = 25.782$
5. a. 2.883 7
   b. 1.568 2
   c. 2.340 4
   d. 5
   e. −0.0339
   f. 2.857 4
   g. −3.204 1
   h. 0.643 3

7. a. 23.599 3
   b. 11 614.486
   c. $6.717\ 4 \times 10^{-4}$
   d. 1×10^{8}
   e. 1.281 2
   f. 250.668 6
   g. 1×10^{-5}
   h. 0.018 6
9. 2.699 0 B
11. 23.979 4 dB
13. 15.105 5 dB
15. 62.498 8 dB
17. −3.010 3 dB
19. 1.2 W
21. 73.784 2 dB
23. 2.444 mW
25. 29.294 2 dBm
27. a. 11.968 0
    b. −11.250 6
    c. 5.740 3
    d. 1.975 9
    e. −5.116 0
    f. 22.863 3
    g. −2.302 6
    h. −9.675 6
29. a. 0.374 2
    b. 25 848.297
    c. 433.950 3
    d. 0.018 3
    e. 5.754 6
    f. 1 265.218 4
    g. 1.001 3
    h. 0.046 8
31. 10.376 V
33. 2.175 V
35. 98.168 mA

Unit 4•9

1. 1
3. 354.126
5. $8x^2y + 3x^3 - xy^2 - x^2y^2$
7. $3mn^2 - m^2n$
9. −165.2
11. −808 704
13. $\dfrac{9a^8}{b^4}$
15. 5.914
17. 1.415
19. $\dfrac{16a^6c}{b^4}$
21. $\dfrac{5c^{18}}{25^5a^{24}b^{12}}$
23. $9^{1/n}a^{2/n}b^3$
25. $\dfrac{432a^3}{b^{11}}$
27. $-b^2$
29. $15x^2y^2 - 7x^2y - 3y^2x$
31. 364
33. −4 208
35. $\dfrac{-12b^5c^2}{a}$
37. $-12a^2b + 12ab^2$
39. $x^2 - 2xy + y^2$
41. $a^3 + 7a^2b + 5ab^2 - b^3$
43. $25y^6z^6 + 50y^5z^7 + 25y^4z^8$
45. $\dfrac{4m}{n} - 4 + \dfrac{n}{m}$
47. $a - 9$
49. $x^2 + x - 6,\ r = -10$
51. $a^2 - b^2$
53. $25 - 10x + x^2$
55. $108w^3z - 72w^2z^2 + 12wz^3$
57. $xy - 2x + 3y - 6$
59. $y^3 - 6y^2 + 32$
61. $3a^3bc^2(2a - 5b + 3a^2c)$
63. $3(a + 2)(a - 2)$
65. $6(x^2 + 1)(x + 1)(x - 1)$
67. $-3(a + 4)(a + 5)$
69. $(x^2 + 2)(x^2 - 2)$
71. $-2(y - 9)(y + 2)$
73. $3y(y + 1)^3$
75. $\dfrac{1}{x - 3}$
77. $\dfrac{-8a + 44}{6a^2 - 30a + 24}$
79. $\dfrac{z^2 + z - 7}{2z^2 + 10z + 12}$
81. $\dfrac{x - 4}{2x^2 - 8}$
83. $\dfrac{185}{y^3 - y^2 - 16y + 16}$
85. $\dfrac{3a^7c^5}{b^7}$
87. $\dfrac{1}{x^2 - 2x}$
89. $\dfrac{m + 4}{m + 5}$
91. $\dfrac{y^2 - 1}{y^2 - 6y + 9}$
93. 1
95. $y^4 + y^2$
97. a. −1.295 4
    b. 3.000 4
    c. 5.942 6
    d. 15.822 8
99. a. 1.181 4
    b. 0.660 7
    c. $1.389\ 6 \times 10^5$
    d. 630.957 3
101. 0.126 W
103. 6.020 6 dB
105. 10.718 dB
107. 28.451 dBm
109. a. 64.715 5
     b. 0.345 6
     c. $6.461\ 8 \times 10^7$
     d. 0.535 4
111. 4.636 V
113. 0.218 V
115. 1.483 V, 8.9 V

CHAPTER 5

Unit 5•1

1. 3.5
3. 12
5. 5, −5
7. 5, 3
9. 2.4
11. 6
13. $w = -0.33\overline{3}$
15. −2, 7
17. −3
19. −2
21. 4, −2
23. −5.5
25. 1
27. $I = \sqrt{\dfrac{P}{R}}$
29. $L = \dfrac{RA}{K}$
31. $R = \dfrac{1}{G}$
33. $R_2 = \dfrac{E_bR_1}{E_{cc} - E_b}$
35. $C = \dfrac{1}{2\pi fX_C}$
37. $R^2 = Z^2 - (X_c - X_L)^2$
39. $\beta = \dfrac{I_{CEO}}{I_{CBO}} - 1$

41. $M = \sqrt{\frac{2P_t}{P_c} - 2}$

43. $L_m = \frac{L_t - L_1 - L_2}{2}$

45. $C_t = \frac{1}{\frac{1}{C_1} + \frac{1}{C_2}}$ or

$C_t = \frac{C_1 C_2}{C_1 + C_2}$

47. $I_s = \frac{I_B - I_{BQ}}{\sqrt{2}}$

49. $\beta = \frac{\alpha}{1 - \alpha}$

Unit 5•2

1. $x = -2$
   $y = 20$
3. $x = -0.6$
   $y = -5.6$
5. $x = 2$
   $y = 2$
7. $x = 8$
   $y = 6$
9. $x = 4$
   $y = -7$
11. $x = 4.16\overline{6}$,
    $y = -6$ and
    $x = 4.5$,
    $y = -2$
13. $x = 3$
    $y = -5$
15. $x = 0.875$,
    $y = -0.75$ and
    $x = -1.25$,
    $y = -5$
17. $x = -3$
    $y = -9$
19. $x = 4$
    $y = -3$
21. $x = -2$
    $y = 1$
    $z = 3$
23. $x = 5$
    $y = 0$
    $z = -6$
25. $x = 2$
    $y = -6$
    $z = 4$
27. $x = -4$
    $y = 1$
    $z = -4$
29. $x = 3$
    $y = 6$
    $z = 3$

Unit 5•3

1. $6I_1 - I_2 + 0I_3 = 12$
   $-I_1 + 13I_2 - 10I_3 = 0$
   $0I_1 - 10I_2 + 25I_3 = 0$
3. $E_{R_4} = 1.358$ V
5. Increase to 10.213 V
7. $I_1 = 0.532$ A
   $I_2 = 0.009$ A
   $I_3 = -0.397$ A
9. $I_{R_2} = 0.522$ A
11. $I_T = 3.011$ mA
13. $I_{R_3} = 0.187$ mA
15. $E_A = 9$ V
    $E_B = 2.3\overline{3}$ V
    $E_C = 2.1\overline{3}$ V
    $E_D = 6$ V
17. $I_{R_5} = 1.935$ mA
19. $I_{R_4} = 0.89$ A

Unit 5•4

1. $x = \pm 13$
3. $x = \pm 8$
5. $x = \pm 2.309$
7. $x = \pm 3$
9. $x = -7$
11. $x = -1, x = 5$
13. $x = -2.5$
15. $x = -1.75, x = 3$
17. $x = 6.236, x = 1.764$
19. $x = 3.414, x = 0.586$
21. $x = 4.162, x = -2.162$
23. $x = 2.791, x = -1.791$
25. $x = 1, x = 9$
27. $x = 1.391, x = -1.558$
29. $x = 2.479, x = -1.479$
31. $x = 4.639, x = -3.305$
33. $x = 15.307, x = -1.307$
35. $x = 2, x = 1.25$

Unit 5•5

1. x
3. a. II
   b. I
   c. III
   d. II
5. $y = -\frac{5}{3}x + \frac{14}{3}$

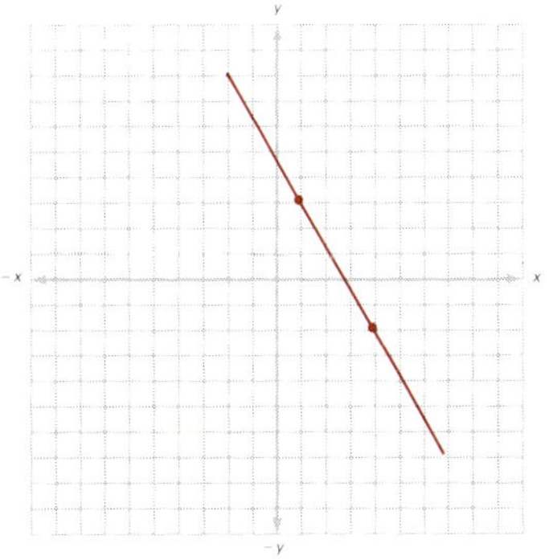

7. $y = -1.5x$

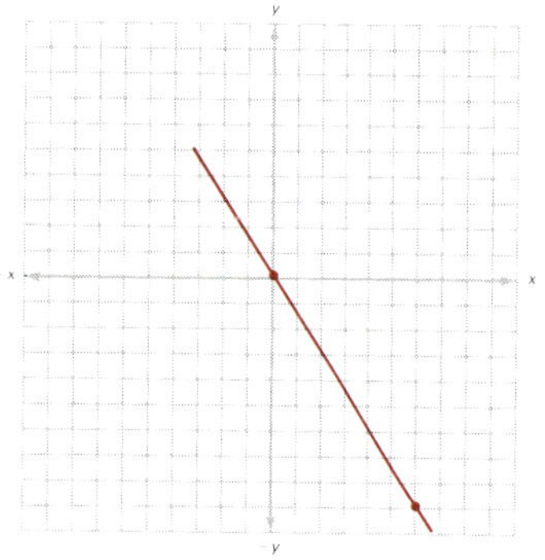

9. $y = 0.5x - 9$

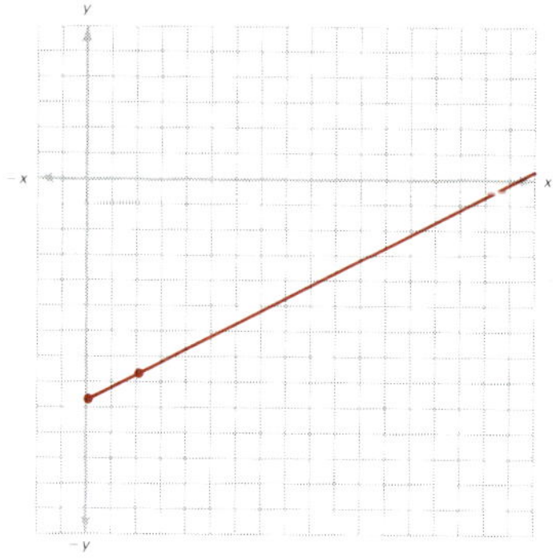

11. $y = -3$

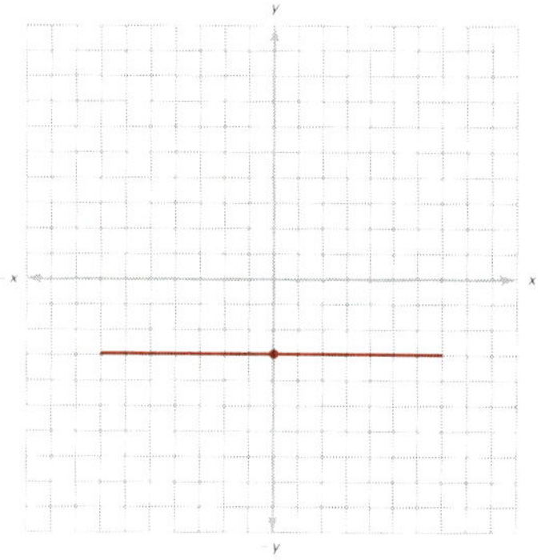

13. $x = 1$

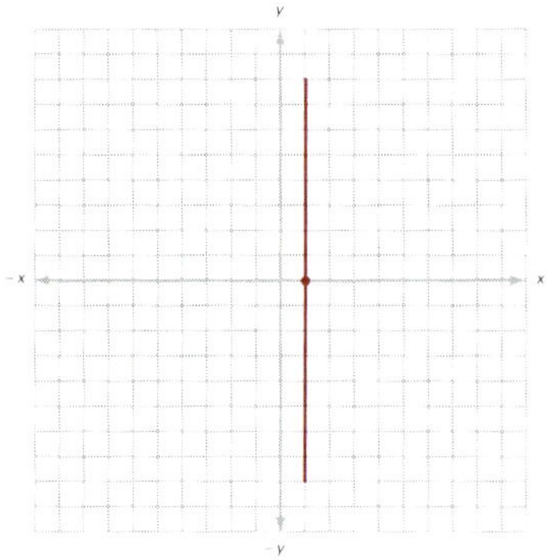

15. $m = 6$
    $b = 10$
17. $m = -7$
    $b = -7$
19. $m = -1\frac{1}{8}$
    $b = 1\frac{7}{8}$
21. $m = 0.6$
23. $y = -x - 7$
25. $y = 7$
27. $(\frac{31}{7}, 43)$
29. $b = -10$ (line 1)
    $b = 15$ (line 2)
31. $m = 0.5$
33. (1.14, 2.57)
35. $y = -\frac{4}{7}x + 2$

Unit 5 · 6

1.

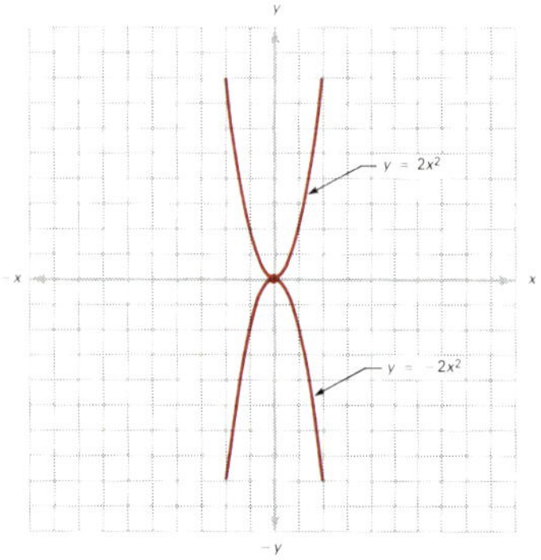

3.

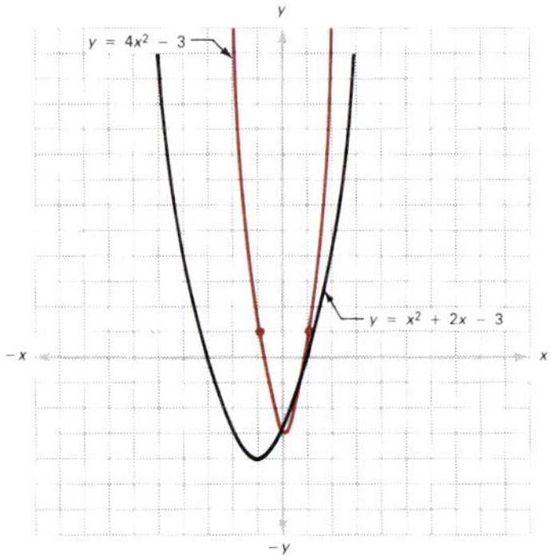

5.

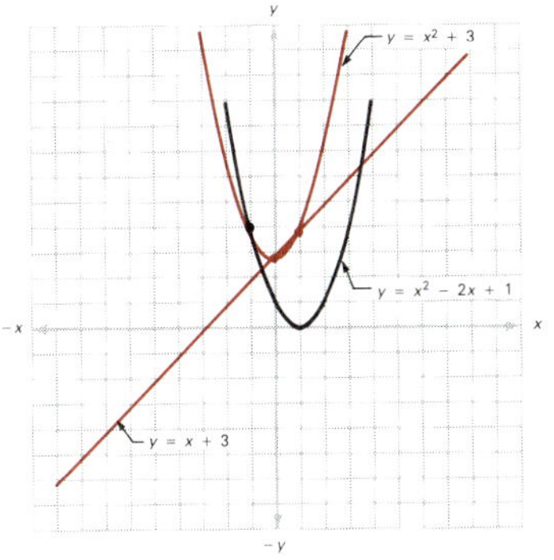

7. (−1, −4)

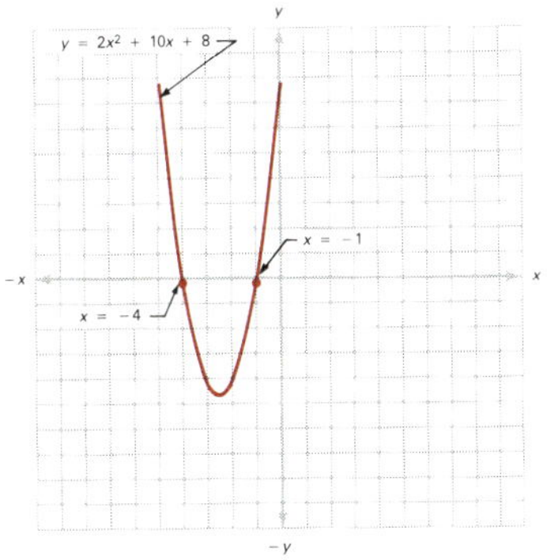

9. No roots

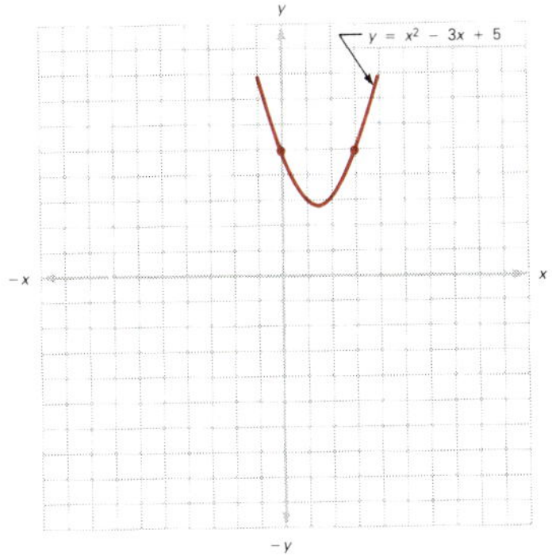

Unit 5 • 7

1. $V_c = \dfrac{p_c V_b}{p_b \beta_{dc}}$
3. $n = \sqrt{\dfrac{A}{V_n \pi} + \dfrac{1}{4}}$
5. $r_c = \dfrac{(0.707\ V_{CEQ})^2}{P_L} - r_e$
7. $R = \sqrt{\dfrac{X_C^2}{A^2} - X_C^2}$ or $R = X_C \sqrt{\dfrac{1}{A^2} - 1}$
9. $N_2 = N_1 \sqrt{\dfrac{R_2}{R_1}}$
11. $x = -7$
13. $x = 0.83\overline{3}$
15. $a = 6, -3$
17. $x = 5$
19. $x = 2$
21. $x = -4, y = 6$
23. $x = 0, y = -2$
25. $x = -3, y = 8$
27. $x = -y$
29. $x = -5, y = 9$
31. $x = -4, y = -3, z = 7$
33. $x = 2, y = 4, z = -5$
35. $x = 15, y = -12, z = -6$
37. 2.402 mA
39. 3.244 mA
41. $12.6I_1 - 3.6I_2 + 0I_3 = 48$
    $-3.6I_1 + 16I_2 - 2.7I_3 = 0$
    $0I_1 - 2.7I_2 + 15.2I_3 = 0$
43. 4.47 V
45. 4.08 mA
47. 2.628 mA
49. 79.5 μA, direction of I_3
51. $E_A = 12$ V, $E_B = 5.59$ V, $E_C = 4.91$ V, $E_D = 5.79$ V, $E_t = 15$ V
53. 0.19 mA from D to C

55. 2.37 mA
57. $y = 5x - 8$

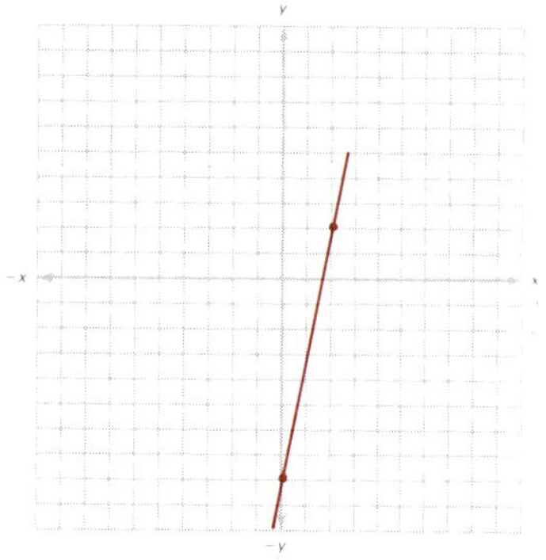

59. $y = x + 4$

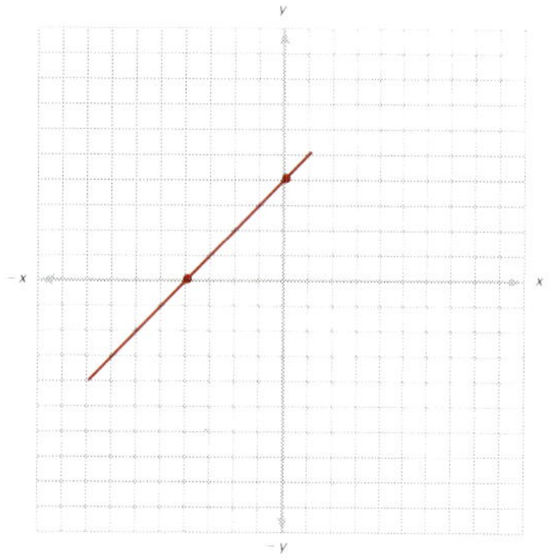

61. $x = 9$
63. $y = 3x + 15$
65. $m = -\frac{2}{3}$
67. $y = 0.8x - 0.4$
69. $m = 0.3$
    $b = -12$
71. $m - 1$
    $b = 4$
73. $m = -10$
    $b = -10$
75. $y = -0.3x$
77. $y = 0.5x + 3$
79. $m = 1$
81. $m = 2.5$
83. $y = 2.5x + 15$
85.

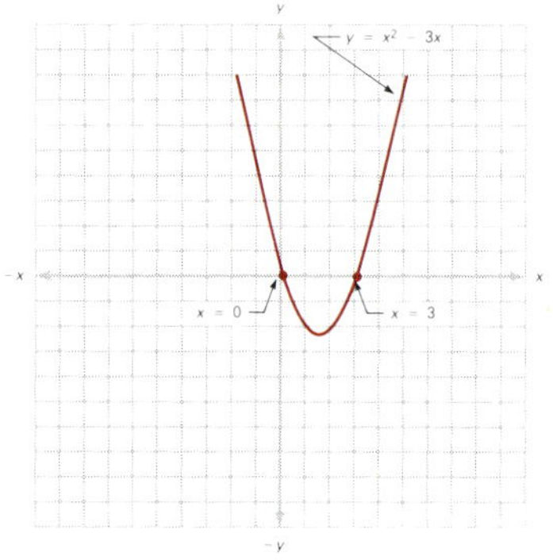

87.

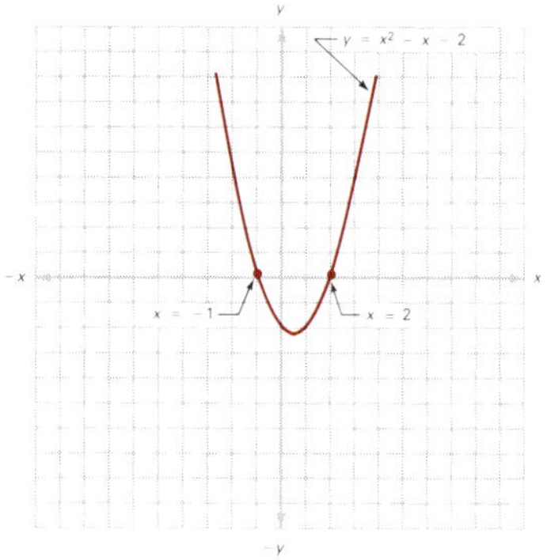

89. $x = 0, x = 0.25$
91. $x = \pm 4$
93. $x = 1, x = 7$
95. $x = 5.646, x = 0.354$
97. $x = 7.5, x = -0.5$
99. $x = 0.333, x = -1$
101. No roots
103. $x = -0.284, x = -0.348$

CHAPTER 6

Unit 6•1

1. a. 55°
   b. 17°22′46″
   c. 88°37′14″
3. a. 176°52′
   b. 4°59′20″
   c. 32.72°
5. 40 cm
7. 14.137 inches
9. a. 66.791°
   b. 142.338°
   c. 296.441°
   d. 166.836°
11. a. 1.08 radians
    b. 2.39 radians
    c. 4.26 radians
    d. 3.01 radians
    e. 6.98 radians
    f. 0.47 radian
    g. 0.63 radian
    h. 2 radians
13. a. 299.21 grads
    b. 334.8°
    c. 5.15 radians
    d. 3 580.99°
    e. 1.53 radians
    f. 203.69 grads

Unit 6•2

1. 90°, 76°
3. 90°, 67.75°, 22.25°
5. 22.46°, 67.54°
7. 83.24°, 6.76°
9. 59.14°, 30.86°
11. 69.09°, 20.91°
13. 45°, 45°
15. 60.07°, 29.93°
17. a. 1.154 7
    b. 0.965 9
    c. 572.957 1
    d. 1.019 4
    e. 0.857 2
    f. 57.29
    g. 1.219 8
    h. 0.113 9
19. a. 38.71°
    b. 75.79°
    c. 2.14°
    d. 33.65°
    e. 90°
    f. 6.81°
    g. 66°
    h. 20.19°
21. a. 0.267 9
    b. 3.599 5
    c. 0.707 1
    d. 0.707 1
23. a. 135°, 225°
    b. 294.85°, 114.85°
    c. 40°, 140°
    d. 88.85°, 91.15°

Unit 6•3

1. $c = 21.95''$
   $A = 30.07^\circ$
   $B = 59.93^\circ$
3. $a = 24.82''$
   $A = 83.11^\circ$
   $B = 6.89^\circ$
5. $b = 20.85$ cm
   $A = 33.5^\circ$
   $B = 56.5^\circ$
7. $c = 5.1''$
   $A = 11.31^\circ$
   $B = 78.69^\circ$
9. $a = 29.43$ cm
   $A = 74.79^\circ$
   $B = 15.21^\circ$
11. $b = 19.56''$
    $c = 25.91''$
    $B = 49^\circ$
13. $a = 21.81$ cm
    $b = 30.01$
    $B = 54^\circ$
15. $a = 4.05$ cm
    $c = 12.57$ cm
    $A = 18.8^\circ$
17. $b = 108.51''$
    $c = 114.97''$
    $B = 70.7^\circ$
19. $a = 35.55$ cm
    $b = 19.2$ cm
    $B = 28.38^\circ$

Unit 6•4

1. $j3$
3. $-j\frac{3}{4}$
5. $-j2x\sqrt{2}$
7. $-j0.1x^2$
9. $-j7\sqrt{3}$
11. 1
13. $-j$
15. -1
17. $j14$
19. -36
21. $16 + j12$
23. $-9 + j20$
25. $\dfrac{-j65}{13}$
27. $21 + j6$
29. $-1 - j3$
31. II
33. I
35. II
37. II
39. IV
41. $24\angle 80^\circ$
43. $40\angle 40^\circ$
45. $18\angle 90^\circ$
47. $3\angle -15^\circ$
49. $2.78\angle -51^\circ$
51. $17.89\angle 116.57^\circ$
53. $11.66\angle 239.04^\circ$
55. $14.87\angle 19.65^\circ$
57. $6.58 - j6.14$
59. $18.47 + j0.97$

Unit 6•5

1. a. 11.8°
   b. 76°
   c. 45.01°
   d. 89.92°
3. a. 101.8°
   b. 78.1°
   c. 138.98°
   d. 6°
5. 23.56 inches
7. 1.57
9. a. 102.3°
   b. 158.47°
   c. 12.57°
   d. 307.06°
11. a. 222.82 grads
    b. 385.2°
    c. 3.8 radians
    d. $572.96^\circ = 212.96^\circ$
13. 90°
15. $26.57^\circ, 63.43^\circ$
17. $66.08^\circ, 23.92^\circ$
19. $27.52^\circ, 62.48^\circ$
21. a. 0.922 5
    b. 0.199 9
    c. 0.469 5
    d. 14.101 4
    e. $-1.111\ 3$
    f. $-2.062\ 7$
23. a. $128.18^\circ, 308.18^\circ$
    b. $60.76^\circ, 119.24^\circ$
    c. $112.19^\circ, 247.81^\circ$
    d. $42.89^\circ, 222.89^\circ$
25. $c = 19.85''$
    $A = 49.09^\circ$
    $B = 40.91^\circ$
27. $a = 55.74$ cm
    $A = 53.01^\circ$
    $B = 36.99^\circ$
29. $a = 14.41''$
    $b = 20.43''$
    $A = 35.2^\circ$
31. $b = 4.1$ cm
    $c = 20.22$ cm
    $A = 78.3^\circ$
33. $a = 1.92$ cm
    $b = 8.01$ cm
    $B = 76.5^\circ$
35. $a = 12.36$ cm
    $c = 17.48$ cm
    $B = 45^\circ$
37. $10 - j4$
39. $-7 + j2$
41. $48 - j4$
43. -128
45. $\dfrac{-60 + j66}{34}$
47. $5\angle 0^\circ$
49. $100\angle 90^\circ$
51. $3\angle 225^\circ$
53. $4.63\angle -108$
55. $1.8\angle -90^\circ$
57. $8.54\angle -159.44^\circ$
59. $4\angle 90^\circ$
61. $16.55 + j12.93$
63. $14.14 + j14.14$
65. $2.25 - j9.74$

CHAPTER 7

Unit 7•1

1. 5×10^6 m
3. 1 000 m
5. 6.67 cm
7. 1 m
9. 8.57 cm
11. 400 Hz
13. 200 kHz
15. 500 MHz
17. 60 MHz
19. 1.5 MHz
21. 0.6 μs
23. 9.26 ns
25. 8.3 μs
27. 200 MHz
29. 2.5 MHz
31. 111 V
33. 30.576 mA_{av}
35. 17.82 V_{pp}
37. 108 V_{av}
39. 112.1 V_{pp}
41. 106.07 V
43. 0 V
45. −89.66 V
47. 195.41 V
49. 0 V
51. $e = 141.4$ V sin $377t$
53. $e = 23.55$ V sin $6.28 \times 10^3 t$
55. $e = 5$ V sin $9.42 \times 10^8 t$
57. 14.73 V
59. 0.87 V

Unit 7•2

1. i_2 leads i_1 by 30°
3. i leads e by 24°
5. i leads **E** by 35°
7. $\mathbf{E_1}$ leads $\mathbf{E_2}$ by 15°
9. **I** leads by 67.05°
11. 12.21 V
13. 33.14 V
15. $e = 162.61$ V sin $(\omega t - 18°)$
17. $e = 8.48$ V sin $(\omega t + 48°)$
19. $e = 70.7$ V sin $(\omega t - 27°)$
21. 7.07 A $\angle 15°$
23. 14.14 mA $\angle -72°$
25. 6.36 mA $\angle 0°$
27. $\mathbf{E_t} = 76.3$ V $\angle 0.21°$
    $e_t = 107.89$ V sin $(\omega t + 0.21°)$
29. $\mathbf{I_t} = 1.41$ A $\angle 14°$
    $i_t = 1.99$ A sin $(\omega t + 14°)$
31. $\mathbf{I_t} = 155.17$ μA $\angle -3.62°$
    $i_t = 219.41$ μA sin $(\omega t - 3.62°)$
33. $\mathbf{E_t} = 197.78$ V $\angle 47.86°$
    $e_t = 279.65$ V sin $(\omega t + \angle 47.86°)$

Unit 7•3

1. $i = 4.8$ A sin ωt
3. $R = 200$ kΩ
5. $R = 9$ kΩ
7. $i = 4.71$ mA sin ωt
9. $R = 13.17$ kΩ
11. $X_C = 265.26$ Ω
13. $C = 500$ pF
15. $X_C = 10$ Ω
17. $X_C = 44.21$ kΩ
    $\mathbf{E_S} = 8.84$ V $\angle 0°$
19. $I = 18.85$ mA $\angle 90°$
    $C = 200$ pF
21. $X_L = 1\,885.27$ Ω
    $f = 60$ Hz
23. $E = 11.99$ V
    $L = 10$ mH
25. $\mathbf{E} = 50.27$ V $\angle 0°$
    $X_L = 251.33$ Ω
27. $X_C = 530.52$ Ω
    $\mathbf{Z} = 729.01$ Ω $\angle -46.7°$
    $\mathbf{I} = 164.61$ mA $\angle 46.7°$
    $\mathbf{E_R} = 82.31$ V $\angle 46.7°$
    $\mathbf{E_C} = 87.33$ V $\angle -43.3°$
29. $\mathbf{Z} = 5.93$ Ω $\angle -32.46°$
    $R = 5$ Ω
    $X_C = 3.18$ Ω
    $f = 10$ MHz
31. $E_S = 24$ V
    $\mathbf{Z} = 1.28$ MΩ $\angle -38.66°$
    $R = 1$ MΩ
    $X_C = 0.8$ MΩ
33. $X_L = 2.36$ kΩ
    $\mathbf{Z} = 3.59$ kΩ $\angle 41.16°$
    $\mathbf{I} = 2.51$ mA $\angle -41.16°$
    $\mathbf{E_R} = 6.78$ V $\angle -41.16°$
    $\mathbf{E_L} = 5.92$ V $\angle 48.84°$
35. $R = 1$ kΩ
    $X_L = 9.42$ kΩ
    $L = 0.3$ mH
    $E_S = 5$ V
37. $\mathbf{Z} = 3.69$ kΩ $\angle 42.91°$
    $R = 2.7$ kΩ
    $X_L = 2.51$ kΩ
    $L = 0.5$ H
    $\mathbf{E_R} = 10.99$ V $\angle -42.91°$
    $\mathbf{E_L} = 10.22$ V $\angle 47.09°$
39. $X_C = 1\,326$ Ω
    $X_L = 1\,508$ Ω
    $X = 182$ Ω
    $\mathbf{Z} = 207.66$ Ω $\angle 61.21°$
    $\mathbf{I} = 0.58$ A $\angle -61.21°$
    $\mathbf{E_R} = 58$ V $\angle -61.21°$
    $\mathbf{E_C} = 769.08$ V $\angle -151.21°$
    $\mathbf{E}_L = 874.64$ V $\angle 28.79°$
41. $R = 1$ kΩ
    $X_C = 2.65$ kΩ
    $X_L = 3.77$ kΩ
    $f = 1.2$ kHz
    $L = 0.5$ H
    $\mathbf{X} = 1.12$ kΩ $\angle 90°$
    $\mathbf{Z} = 1.5$ kΩ $\angle 48.24°$
    $\mathbf{E_S} = 15$ V
43. $E_S = 6$ V
    $\mathbf{I} = 65.6$ mA $\angle 56.92°$
    $X_C = 264.78$ Ω
    $X_L = 188.26$ Ω
    $C = 0.02$ μF

$L = 1$ mH
$\mathbf{X} = 76.52\ \Omega\ \angle -90°$
$\mathbf{Z} = 91.4\ \Omega\ \angle -56.84°$

45. $L = 5$ mH
$X_C = 5.77$ kΩ
$X_L = 5.77$ kΩ

$I = 3$ mA
$E_S = 9$ V
$Z = 3$ kΩ
$E_C = 17.31$ V
$E_R = 9$ V

47. $P_t = 1.84$ W
$P_{Q(t)} = 1.38$ vars
$P_{A(t)} = 2.3$ VA

49. $P_t = 0.029$ W
$P_{Q(t)} = 0.035$ vars
$P_{A(t)} = 0.046$ VA

Unit 7•4

1. 5.45×10^4 m
3. 3×10^6 m
5. 1.5 m
7. 5 MHz
9. 60 Hz
11. 10 μs
13. 416.67 μs
15. 13.33 μs
17. 10 GHz
19. 2.22 kHz
21. 7.07 V
23. 339.36 $V_{p\text{-}p}$
25. 9 V_{av}
27. $e = 7.07\text{ V} \sin 2.51 \times 10^3 t$
29. $e = 9\text{ V} \sin 4.71 \times 10^3 t$
31. 62.46 V
33. 8.09 V
35. 7.7 V
37. 1.61 A
39. −1.69 A
41. $i = 141.4\text{ mA} \sin(\omega t - 20°)$
43. $\mathbf{E} = 4.24\text{ V}\ \angle 0°$
45. $\mathbf{E} = 10.61\text{ mV}\ \angle -62°$
47. $\mathbf{I_t} = 4.72\text{ mA}\ \angle -8.44°$
$i_t = 6.68\text{ mA} \sin(\omega t - 8.44°)$
49. $\mathbf{I_t} = 17.675\ \mu\text{A}\ \angle 90°$
$i_t = 25\ \mu\text{A} \sin(\omega t + 90°)$
51. $i = 5\text{ mA} \sin(\omega t + 10°)$
53. 707.36 Ω
55. 265.26 Hz
57. $i = 53.31\text{ mA} \sin(\omega t - 90°)$
59. $i = 4.5\text{ mA} \sin(\omega t + 90°)$
61. $E_S = 5$ V
$\mathbf{Z} = 4.17\text{ k}\Omega\ \angle -49.67°$
$R = 2.7$ kΩ
$X_C = 3.18$ kΩ
$C = 0.05$ μF
63. $\mathbf{Z} = 10.99\text{ k}\Omega\ \angle -24.47°$
$X_C = 4.55$ kΩ
$R = 10$ kΩ
$E_R = 5.46$ V
$E_C = 2.48$ V
$f = 3.5$ kHz
65. $R = 10$ kΩ
$E_S = 15$ V
$C = 50$ pF
67. $X_L = 753.98$ Ω
$R = 500$ Ω
$\mathbf{I} = 132.64\text{ mA}\ \angle -56.45°$
$E_L = 100.01$ V
$E_S = 120$ V
$L = 2$ H
69. $X_L = 3.77$ kΩ
$\mathbf{I} = 2.56\text{ mA}\ \angle -75.14°$
$\mathbf{E_R} = 2.56\text{ V}\ \angle -75.14°$
$E_S = 10$ V
$\mathbf{Z} = 3.9\text{ k}\Omega\ \angle 75.14°$
71. $X_C = 79.58$ Ω
$X_L = 125.66$ Ω
$\mathbf{X} = 46.08\ \Omega\ \angle 90°$
$\mathbf{Z} = 68\ \Omega\ \angle 42.66°$
$\mathbf{I} = 88.24\text{ mA}\ \angle -42.66°$
$\mathbf{E_R} = 4.41\text{ V}\ \angle -42.66°$
$\mathbf{E_C} = 7.02\text{ V}\ \angle -132.66°$
$\mathbf{E_L} = 11.09\text{ V}\ \angle 47.34°$
73. $X_C = 21.22$ kΩ
$R = 1.5$ kΩ
$f = 750$ Hz
$C = 0.01$ μF
$\mathbf{I} = 3.6\text{ mA}\ \angle -57.33°$
75. $\mathbf{I} = 1.137\text{ mA}\ \angle -18.57°$
$X_C = 568.63$ Ω
$C = 0.2$ μF
$X_L = 571.77$ Ω
$\mathbf{X} = 3.14\ \Omega\ \angle 90°$
$\mathbf{Z} = 10.48\ \Omega\ \angle 17.43°$
$\mathbf{E_L} = 650.10\text{ mV}\ \angle 71.43°$
$E_S = 12$ mV
77. $C = 500$ pF
79. 0.78

CHAPTER 8

Unit 8•1

1. $\mathbf{X_C} = 265.26\ \text{k}\Omega\ \angle -90^\circ$
   $\mathbf{I_C} = 452.39\ \mu\text{A}\ \angle 90^\circ$
   $\mathbf{I_R} = 600\ \mu\text{A}\ \angle 0^\circ$
   $\mathbf{I_t} = 751.44\ \mu\text{A}\ \angle 37.02^\circ$
   $\mathbf{Z} = 159.69\ \text{k}\Omega\ \angle -37.02^\circ$
3. $I_C = 0.63\ \text{mA}$
   $I_R = 1\ \text{mA}$
   $E_S = 5\ \text{V}$
   $X_C = 7.94\ \text{k}\Omega$
   $C = 0.002\ \mu\text{F}$
5. $\mathbf{I_t} = 100.3\ \text{mA}\ \angle 4.5^\circ$
   $I_R = 100\ \text{mA}$
   $I_C = 7.87\ \text{mA}$
   $X_C = 1.27\ \text{k}\Omega$
   $C = 0.05\ \mu\text{F}$
7. $\mathbf{X_L} = 26.39\ \Omega\ \angle 90^\circ$
   $\mathbf{I_L} = 189.47\ \text{mA}\ \angle -90^\circ$
   $\mathbf{I_R} = 100\ \text{mA}\ \angle 0^\circ$
   $\mathbf{I_t} = 214.24\ \text{mA}\ \angle -62.18^\circ$
   $\mathbf{Z} = 23.34\ \Omega\ \angle 62.18^\circ$
9. $E_S = 120\ \text{V}$
   $\mathbf{I_t} = 320.36\ \text{mA}\ \angle -41.48^\circ$
   $I_L = 212.19\ \text{mA}$
   $X_L = 565.53\ \Omega$
   $f = 60\ \text{Hz}$
11. $R = 1\ \text{k}\Omega$
    $I_L = 9.55\ \text{mA}$
    $\mathbf{I_t} = 13.12\ \text{mA}\ \angle -46.7^\circ$
    $\mathbf{Z} = 685.98\ \Omega\ \angle 46.7^\circ$
    $f = 7.5\ \text{kHz}$
13. $X_C = 397.88\ \Omega$
    $X_L = 628.32\ \Omega$
    $I_R = 5\ \text{mA}$
    $I_C = 12.57\ \text{mA}$
    $I_L = 7.96\ \text{mA}$
    $I_X = 4.61\ \text{mA}$
    $\mathbf{I_t} = 6.8\ \text{mA}\ \angle 42.68^\circ$
    $\mathbf{Z} = 735.29\ \Omega\ \angle -42.68^\circ$
15. $I_R = 200\ \text{mA}$
    $\mathbf{I_t} = 227.22\ \text{mA}\ \angle -28.33^\circ$
    $\mathbf{Z} = 44.01\ \Omega\ \angle 28.33^\circ$
    $X_L = 59.69\ \Omega$
    $I_C = 59.69\ \text{mA}$
    $I_L = 167.53\ \text{mA}$
    $C = 0.1\ \mu\text{F}$
    $L = 1\ \text{mH}$
17. $f_r = 5\,032.92\ \text{Hz}$
    $X_L = 1.58\ \text{k}\Omega$
    $Q = 15.81$
    $Z = 24.98\ \text{k}\Omega$
    $I_{line} = 360.26\ \mu\text{A}$
19. $C = 0.1\ \mu\text{F}$
    $Z = 100\ \text{k}\Omega$
    $E_S = 10\ \text{V}$
    $X_L = 1\,000\ \Omega$
    $Q = 100$

Unit 8•2

1. $X_C = 198.94$
   $I_R = 60\ \text{mA}$
   $I_C = 60.32\ \text{mA}$
   $\mathbf{I_t} = 85.08\ \text{mA}\ \angle 45.15^\circ$
   $\mathbf{Z} = 141.04\ \Omega\ \angle -45.15^\circ$
3. $\mathbf{I_t} = 2.34\ \text{mA}\ \angle 75.13^\circ$
   $I_C = 2.26\ \text{mA}$
   $I_R = 0.6\ \text{mA}$
   $R = 5\ \text{k}\Omega$
   $X_C = 1.33\ \text{k}\Omega$
5. $\mathbf{I_t} = 620.48\ \mu\text{A}\ \angle -14.86^\circ$
   $I_L = 159.15\ \mu\text{A}$
   $I_R = 600\ \mu\text{A}$
   $R = 10\ \text{k}\Omega$
   $X_L = 37.7\ \text{k}\Omega$
7. $I_R = 2\ \text{mA}$
   $\mathbf{I_X} = 1.61\ \text{mA}\ \angle -90^\circ$
   $E_S = 10\ \text{V}$
   $\mathbf{Z} = 3.89\ \text{k}\Omega\ \angle 38.83^\circ$
   $X_C = 6.37\ \text{k}\Omega$
   $I_C = 1.57\ \text{mA}$
   $I_L = 3.18\ \text{mA}$
   $X_L = 3.14\ \text{k}\Omega$
   $L = 1\ \text{mH}$
9. $I_R = 24\ \text{mA}$
   $X_C = 397.88\ \Omega$
   $X_L = 251.33\ \Omega$
   $I_L = 47.75\ \text{mA}$
   $\mathbf{I_X} = 17.59\ \text{mA}\ \angle -90^\circ$
   $C = 1\ \mu\text{F}$
   $\mathbf{I_t} = 29.76\ \text{mA}\ \angle -36.24^\circ$
   $\mathbf{Z} = 403.23\ \Omega\ \angle 36.24^\circ$
11. $f_r = 821.87\ \text{Hz}$
    $X_L = 387.3\ \Omega$
    $Q = 15.49$
    $Z = 6\ \text{k}\Omega$
    $I_{line} = 1\ \text{mA}$
13. $L = 10\ \text{mH}$
    $X_L = 2.24\ \text{k}\Omega$
    $Q = 6.4$
    $E_S = 5\ \text{V}$
    $Z = 14.34\ \text{k}\Omega$
15. $C = 0.05\ \mu\text{F}$

CHAPTER 9

Unit 9•1

1. 11
3. 17
5. 0.375
7. 253
9. 1
11. 7.5
13. 315
15. 0.187 5
17. 10001_2
19. 1000011.11001_2
21. 1100101.00011_2
23. 100100.10011_2
25. 100011100_2
27. 1011100.111_2
29. 1010100_2

Unit 9•2

1. 10100_2, 20_{10}
3. 10001010_2, 138_{10}
5. 1.0001_2, $1.062\ 5_{10}$
7. 110011_2, 51_{10}
9. 10010001.0111_2, $145.437\ 5_{10}$
11. 100_2
13. 110_2
15. -100_2
17. 111.1_2
19. 11010_2
21. 0110_2, 0111_2
23. 0000_2, 0001_2
25. 00100_2, 00101_2
27. 010010_2, 010011_2
29. 010110_2, 010111_2

Unit 9•3

1. 11100_2
3. 10000100_2
5. 11001_2
7. 10001100_2
9. 11001010001_2
11. 11_2
13. 1101_2
15. 110_2
17. 10111_2
19. 1100_2
21. 110110_2
23. 10011100.00011_2
25. 111.01011_2

Unit 9•4

1. 19
3. 75.625
5. 13.875
7. 0.343 75
9. 855.218 75
11. 101010_2
13. 11011.01001_2
15. 1010110.10111_2
17. 10100010000.1111_2
19. 0.10111_2
21. 0101_2, 0110_2
23. 0110010_2, 0110011_2
25. 01000010_2, 01000011_2
27. 11_2
29. 10011_2
31. 1001101.1111_2
33. 1000.01_2
35. -110110_2
37. 11011_2
39. 1_2
41. 101.01_2
43. 1010_2
45. 100000000_2
47. 10101001_2
49. 10001.111_2

CHAPTER 10

Unit 10•1

1. 10
3. 1.375
5. 3 700
7. 2.734 4
9. 512
11. 1 540
13. 171 683
15. 2 782.718 8
17. 126_8
19. 54.36_8
21. 12235_8
23. 0.2065_8
25. 52773_8
27. 10101_8
29. 650374_8

Unit 10·2

1. 56
3. 2 578
5. 0.418 457 03
7. 876 623
9. 400 293
11. 2 910 903
13. 17 871.723 388 67
15. 10 739 480
17. $9F.B2_{16}$
19. 0.75_{16}
21. $8AF92_{16}$
23. $D703E5_{16}$
25. 1.1_{16}
27. $B0.2_{16}$
29. 0.8E9

Unit 10·3

1. 65_8
3. 133_8
5. 3545_8
7. 54.66_8
9. 0.3356_8
11. 111000111_2
13. 11001110110_2
15. 0.110010111_2
17. 100101101000_2
19. 10000.00111_2
21. 10000_2
23. 110110_2
25. 100001010_2
27. 110010101011_2
29. 0.011110001101_2
31. $2D_{16}$
33. 7_{16}
35. $23BD_{16}$
37. $3FCB_{16}$
39. $8D.0B8_{16}$
41. 7_{16}
43. $A8_{16}$
45. $10F_{16}$
47. $0.AD_{16}$
49. $3E75_{16}$
51. 133_8
53. 43450_8
55. 0.1751_8
57. 25005_8
59. 47765_8

Unit 10·4

1. 361_8
3. 61544_8
5. 0.75_8
7. $36\ 421_8$
9. 725603.46_8
11. 23
13. 564
15. 0.595 70
17. 18 555
19. 205 623
21. 29_{16}
23. $0.E7_{16}$
25. $89.E_{16}$
27. 62749_{16}
29. $8D15_{16}$
31. 2 416
33. 0.546 88
35. 16 952
37. 390 201
39. 591 610
41. 67_8
43. 1335_8
45. 17.55_8
47. 23435_8
49. 2.0166_8
51. 111011110_2
53. 1100111110_2
55. 0.011111_2
57. 110100111010110_2
59. 101100010.001111010_2
61. 5_{16}
63. $1B.8D_{16}$
65. $106F_{16}$
67. $8C59_{16}$
69. $62B1D9_{16}$
71. 1111100_2
73. 11000101111_2
75. 0.01111001_2
77. 11001001111110_2
79. 11101.01111010_2
81. 137_8
83. 3274.5_8
85. 5651_8
87. 0.233_8
89. 14577.432_8
91. $1BB_{16}$
93. 15_{16}
95. $13F.CA8_{16}$
97. 240.201_{16}
99. $1B239D_{16}$

CHAPTER 11

Unit 11·1

1. 1
3. 1 on one or both inputs
5. 0
7. HIGH
9. NAND, NOR, or EXCLUSIVE-NOR
11. All 0's
13. 1
15. $G_3 = 1$, and $G_4 = 0$

Unit 11·2

1. $X = AB + \overline{B}$
3. 1
5. $X = \overline{(\overline{\overline{A} + B})\,B}$
7. 1
9. No
11. 0

13. $A = B = 0$ AND $C = 1$
A AND/OR $B = 1$ AND $C = 0$
15. $G_2 = \overline{CD}$
17. 1
19. 1
21. $X = A\overline{B}$
23.

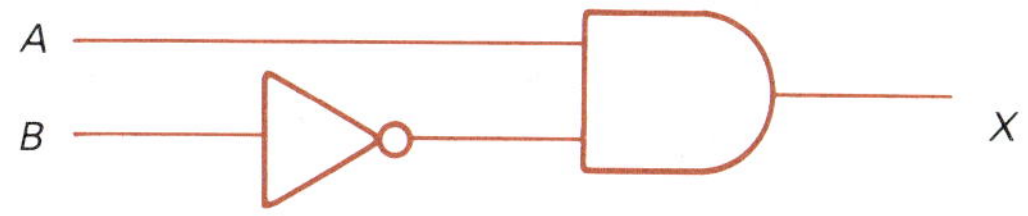

Unit 11•3

1. $\overline{A} + B + C$
3. $(\overline{A} \cdot \overline{B}) + (\overline{C} \cdot \overline{D})$
5. $\overline{A}$
7. $X = \overline{B}C$
9. $X = \overline{B}(A + C)$
11. $X = A$
13. $X = \overline{A}$
15. $X = AB$

Unit 11•4

1. LOW
3. $X = 0$
5. 1
7. $X = B[(A \oplus B) + \overline{C}]$
9. NOT gate
11. $G_2 = \overline{C} + A + B$
13. $X = 1$
15. $X = A$
17. $G_6 = ABC$
19. a two-input AND gate

INDEX

Abscissa, 164
Absolute error, 41–42
Acute angle, 308
Addition:
  associative law of, 181
  commutative law of, 181
Adjacent side, of triangle, 318
Algebra:
  basic laws of, 181
  Boolean, 495–531
  coefficients, 180–181
  for electronics, 179–233
  factoring, 202–207
  grouping signs, 181
  language of, 179–184
  signed numbers, 183
  special products, 199–202
Algebraic expressions, 180
  addition of, 191–194
  division of, 194–199
  evaluating, 183
  multiplication of, 194–199
  subtraction of, 191–194
Algebraic fractions:
  addition of, 207–210
  division of, 211–212
  multiplication of, 210–212
  subtraction of, 207–210
Algebraic symbols, translating English into, 182–183
Alternating current circuits, 383–423
  capacitive, 385–388
  inductive, 388–392
  in parallel, 429–445
  power in, 413–417
  resistive, 383–385
  in series, 383–423
Altitude, of triangle, 318
Analog, 496
Analysis:
  loop, 259–264
  node, 264–268
AND gate, 497–498
Angle(s), 305–306, 307–309
  acute, 308
  base, 306
  complementary, 308
  cosecant of, 326
  cosine of, 320–321
  cotangent of, 326
  forming an, 306
  negative, 306
  obtuse, 308
  phase, 370–371
  positive, 306
  right, 307
  secant of, 326
  sides, 305
  sine of, 319–320
  size, 306
  supplementary, 309
  symbols, 306
  tangent of, 321
  types of, 307–309
  vertex, 305
Antecedent, 16
Antilogarithms, 217–218
Apparent power, 414, 417
Armature, 353
Associative law of addition, 181
Associative law of multiplication, 182
Associative laws, in Boolean algebra, 521
Asymptotic curve, 153
Average power, 413

Base, 60
  of angle, 306
  of triangle, 318
Base number, finding, 35
Bel (B), 219
Bell Laboratories, 1

Binary number system, 47, 48, 449–471
  computer mathematics and, 449–471
  reciprocals, 452–453
Binary numbers:
  addition of, 461–462
  converting decimal numbers to, 457–460
    using hexadecimal numbers, 488–489
  converting hexadecimal numbers to, 487–488
  converting octal numbers to, 486–487
  converting to decimal numbers, 454
    using hexadecimal numbers, 488
  division of, 469–470
  fractional, common values of, 453–454
  multiplication of, 467–469
  subtraction of, 462–467
Binary-octal-hexadecimal conversions, 484–491
Binary point, 450
Boolean algebra:
  AND, OR and NOT rules, 519
  applications, 524–527
  associative laws, 521–522
  commutative laws, 521
  computer mathematics and, 495–531
  De Morgan's theorems, 519–521
  distributive law, 522–523
  fundamentals, 518–529
Boolean expressions, 506–518
  writing from truth tables, 513–515
Branch currents, 109
Bridge circuits, 119–122

Calculator(s):
  electric, 1–3
  programmable, 2
  selecting, 1–2
    guidelines for, 2
Capacitance, conversions of, 77
Capacitive reactance (X_C), 386
Capacitors;
  in direct current circuits, 149–150
  in parallel circuits, 151–152
  in series circuits, 150–151
Cartesian system, 162
Celsius scale, 32–33
Characteristic, 215
Chord, of circle, 310
Circle, 307–308, 309–317
  chord, 310
  circumference, 309
  degrees, 307
  diameter, 309
  parts of, 309–311
  quadrants, 307
  radius, 309
  secant, 311
  sector, 311
  segment, 311
  tangent, 311
Circuit(s):
  alternating current, 383–423
  bridge, 119–122
  direct current, 148–159
  multiple source, 144–146
  open, 145
  parallel, 104–109
  RC, 152–154, 393–399
  RCL, 403–410, 435–439
  resistive, parallel, 429–430
  RL, 158–159, 399–403
  series, 99–102, 150–157
  series-parallel, 109–115
  short, 145
Circuit analysis theorems, 136–146
Circular measure, 311
Circulating decimal, 51
Circumference, of circle, 310
Coefficients, 62
  in algebra, 180–181
Cofunctions, in right triangles, 327
Columnar position, decimal, 49
Common logs, 214
Commutative law of addition, 181
Commutative law of multiplication, 181
Commutative laws, in Boolean algebra, 521–522
Complementary angles, 308
Completing the square, 274–275
Complex numbers, 340–342
  addition of, 340
  division of, 341–342
  multiplication of, 341
  subtraction of, 340–341
Computer mathematics:
  binary number system, 449–471
  Boolean algebra and, 495–531
  hexadecimal number system, 477–484
  octal number system and, 473–477
Conditional equations, 240
Conductance *(C)*, 29–30, 115
Consequent, 16
Constant memory, 2
Constants, 239
Conversion(s):
  binary-octal-hexadecimal, 484–491
  of capacitance, 77
  of current, 74–75
  decimal point, 77–78
  double-dabble, 455

fractional hexadecimal numbers, 482–483
fractional octal numbers, 476
of frequency, 77
of inductance, 77
radian-degree, 315
rectangular-polar, 347–350
of resistance, 76
sine wave:
frequency to wavelength, 357
wavelength to frequency, 356
of voltage, 75–76
Cosecant, of angle, 326
Cosine, of angle, 320–321
Cotangent, of angle, 326
Cross multiplying, 18
Current:
conversion of, 74–75
sine wave, 358–360
Current divider rule, 132–133
Current law, Kirchhoff's, 130–133
Current ratios, voltage and, 223–225

Decibel (dB), 219
Decimal-binary-octal hexadecimal relationships, 484
Decimal equivalents, table of, 55–56
Decimal fraction, 24
Decimal log, 214
Decimal number system, 47–87
Decimal numbers, converting binary numbers to, using hexadecimal numbers, 488
Decimal point, 48–49
conversions, 77–78
Decimal system, fundamentals, 47–54
Decimals, 34–48
addition of, 57
circulating, 51
converting fractions to, 55
division of, 58
multiplication of, 57–58
repeating, 51
subtraction of, 57
Degree(s):
of circle, 307
circular measure and, 311
De Morgan's theorems, 519–521
Denominator, 24
lowest common (LCD), 26
Dependent variable, 280
Determinants, 252–259
second order, 253–254
third-order, 255–258
Diagonal:
principal, 253
secondary, 253
Diameter, of circle, 309
Digital systems, 496–497
Direct current circuits, 148–159
capacitors in, 149–150
inductors in, 154–156
mathematics for, 91–171
Direct relationships, 28, 30
Distributive law, 182
in Boolean algebra, 522–523
Double-dabble conversion, 455

Efficiency, electronic device and, 36
Electronic calculators, 1–3
Electronic device, efficiency and, 36
Electronic study, formulas used in, 244–245
Electronic symbols, 72–73
Electronic units:
addition of, 74
subtraction of, 74
Electronics:
algebra for, 179–233
basic mathematics for, 1–43
powers in, 80–83
roots in, 83–85
trigonometry for, 305–350
English system, 51–54
Equation(s), 180
conditional, 240
exponential, 212–213
indeterminate, 279
linear, 239–298
logarithmic, 212–213
quadratic, 239–298
simultaneous linear, 247–259, 287–291
solving, 241–244
with three unknowns, 251–252
with two unknowns, 247–251
types of, 239–241
Equation axioms, 94–96
Error:
absolute, 41–42
relative, 42
EXCLUSIVE-OR gate, 503–504
EXCLUSIVE-NOR gate, 504–505
Exponent(s), 60
fractional, 187–190
Exponential equation, 212–213
Exponential notation, 60–71
Exponential numbers:
addition of, 68–69
division of, 66–67, 186–187
multiplication of, 62, 184–185
raising to powers, 185–186
subtraction of, 68–69
Extremes, 17

Factoring, in algebra, 202–207
 common terms, 202
 special products, 203–207
Fahrenheit scale, 32–33
Formula relationships, 30–31
Formulas:
 in algebra, 179–180
 used in electronic study, 244–245
Fraction(s), 24–26, 27–28, 55, 207–210
 addition of, 25–26
 algebraic, 207–210
 converting decimals to, 55
 decimal, 24
 division of, 28
 improper, 24
  converting mixed numbers to, 27–28
 multiplication of, 27–28
 proper, 24
 rules for, 25–27
 subtraction of, 26–27
Frequency:
 conversions of, 77
 of sine wave, 355

Gate, 496
Grad measure, 316
Graph(s), 161–168, 279–298
 coordinates, 279–281
 intercepts, 284–287
 linear, 165
 of linear equations, 279–294
 locating a point, 164–165
 nonlinear, 166
 plotting, 167–171
 quadratic equations and, 294–298
 reading, 165–167
 slope of, 281–284
 types of, 161–164
Grouping signs, in algebra, 181

Hexadecimal number system, 477–484
 computer mathematics and, 477–484
Hexadecimal numbers, 478
 converting decimal numbers to, 481–483
 converting octal numbers to, using binary numbers, 489–490
 converting to decimal numbers, 480–481
 fractional conversions, 482–483
Hypotenuse, of triangle, 318

Ideal current source, 142
Imaginary numbers, 272, 337–340
 addition of, 339
 division of, 340
 multiplication of, 339–340
 subtraction of, 339
Impedance *(Z)*, 392–393
Improper fraction, 24
Independent variable, 280
Indeterminate equation, 279
Inductance, conversions of, 77
Inductive reactance (X_L), 389
Inductors:
 in direct current circuits, 154–156
 in parallel circuits, 157–158
 in series circuits, 156–157
Instantaneous power, 413
Intercepts, graphs and, 284–287
Interpolation, 218, 324–326
Inverse relationships, 28, 30

j-operator, 338
 powers of, 338–339

Kilo, 73
Kirchhoff's law(s):
 applications of, 133–134
 current law, 130–133
 and simultaneous equations, 259–270
 voltage law, 127–130

Lagging current, 371, 435
Large-scale integration (LSI), 497
Leading current, 370, 435
Least significant digit (LSD), 49
Light emitting diodes (LED), 2
Line current, 440
Linear equations, 239–298, 247–259
 dependent, 289
 graphing, 279–294
 inconsistent, 289
 independent, 288
Linear graph, 165
Liquid crystal displays (LCD), 2
Logarithmic equation, 212–213
Logarithms, common, 214–216
Logic, 496
Logic gate(s), 496
 AND gate, 497–498
 EXCLUSIVE-OR Gate, 503–504
 EXCLUSIVE-NOR gate, 504–505
 introduction to, 495–506
 NAND gate, 500–501
 NOR gate, 501–503
 NOT gate, 499–500
 OR gate, 498–499
Loop, 259
Loop analysis, 259–264
Loudness indicator, 219–220
Lowest common denominator (LCD), 26

Mantissa, 215
Mathematical operations:
  basic, 3–5
  combined, 5–7, 42
  four basic functions, 5–7
  with polar numbers, 344–347
Maximum power transfer theorem, 136–138
Means, 17
Measure:
  circular, 311–313
  grad, 316
  radian, 313–314
Measurement:
  conversions and, 72–80
  electronic units of, 72–80
Medium-scale integration (MSI), 497
Mega, 73
Metric system, 51–54
Micro, 73
Microprocessors, 1
Milli, 73
Minute, circular measure and, 311
Mixed number(s), 25
  addition of, 26–27
  converting decimals to, 55
  division of, 28
  multiplication of, 27–28
  subtraction of, 26–27
Monomials:
  addition of, 191–192
  division of, 196
  multiplication of, 194–195
  subtraction of, 192–193
Most significant digit (MSD), 9
Multiple source circuits, 144–146
Multiplication:
  associative law of, 181
  commutative law of, 181

NAND gate, 500–501
Naperian logs. *See* Natural logs
Natural logs, 226–233
Negative angles, 306
Negative logic, 496
Net reactance, 436
Node, 259
Node analysis, 264–268
Nonlinear graph, 166
NOR gate, 501–502
Norton equivalent circuit, 144
Norton's theorem, 142–144
NOT gate, 499–500
Notation:
  exponential, 60–71
  polar, 342–344
  rectangular, 342–344
  scientific, 65–66
Number systems, binary, 449–471
Numbers:
  power of, 80–83
  roots of, 83–85
  rounding off, 41
Numerator, 24

Obtuse angle, 308
Octal number system, 47, 48, 473–477
  computer mathematics and, 473–477
Octal numbers, 474
  converting binary numbers to, 485–486
  converting decimal numbers to, 475–477
  converting hexadecimal numbers to, 490
  converting to decimal numbers, 474–475
  fractional conversions, 476
Ohm, Georg Simon, 91
Ohm's law, 38, 41, 91–93, 127, 181
  in series circuits, 91–93
Open circuit, 145
Opposite side, of triangle, 318
OR gate, 498–499
Ordinate, 164
Origin, 164

Parabola, 295
Parallel circuits, 104–109
  capacitors in, 151–152
  inductors in, 157–158
  total resistance in, 106
Percent, 34–35
  conversion of fraction to, 34
  conversion of, to decimal or fraction, 34
  finding, 35
Percentage, finding, 35
Period, of sine wave, 357
Phase angle, 370–371
Phase relations, 374–377
Phasors, 377–383
  addition of, 379–383
  subtraction of, 379–383
Pico, 73
Polar notation, 342–344
Polar numbers, mathematical operations with, 344–347
Polynomials:
  addition of, 192–193
  division of, 196–199
  multiplication of, 195–196
  subtraction of, 192–193
Positive angles, 306
Positive logic, 496

Power:
apparent, 414, 417
average, 413
instantaneous, 413
reactive (P_Q), 415, 417
real, 413, 417
Power factor, 414
Power gain, 220–222
Power law, 96–98
in series circuits, 96–98
Power loss, 222
Power reference level, 222–223
Powers:
in electronics, 80–83
in numbers, 80–83
Powers of 2:
positive, 450–451
negative, 451–452
Powers of 10, 61–62
multiplication of positive and negative, 70–71
negative, 69–70
Primes, 61
Principal diagonal, 253
Proper fraction 24
Proportion, 17–18
direct, 28
rules of, 17–18
Proportionality, 31–32
Pythagorean theorem, 332–337

Quadrants, 163
of circle, 307
Quadratic equations, 239–298
form of, 271
graphs of, 294–298
roots of, 295–298
solving by completing the square, 274–275
solving by factoring, 273–274
solving by square root, 271–272
solving with quadratic formula, 275–278

Radian, 313
Radian-degree conversions, 315
Radian measure, 313–314
Radians, current and voltage in terms of, 365–369
Radicals, 187–190
Radius, of circle, 309
Radix, 47
Radix point, 48–49
Radix subscript, 48
Ratio, 16–17, 42
Ratio and proportion, 16–22
RC circuits, 152–154, 393–399
parallel, 430–433
RCL circuits, 403–410, 435–439
parallel, 435–439
series, power in, 417
Reactive power (P_Q), 415, 417
Real power, 413, 417
Reciprocal functions, in right triangles, 326
Reciprocal relationships, 28, 29
Reciprocals, in binary number system, 452–453
Rectangular coordinate system, 162
Rectangular notation, 342–344
Rectangular-polar conversions, 347–350
Relationships:
decimal-binary-octal hexadecimal, 484
direct, 28, 30
formula, 30–31
inverse, 28, 30
reciprocal, 28, 29
sine wave current, 362–363
sine wave voltage, 362–363
Relative error, 42
Repeating decimal, 51
Resistance (R), 29–30
conversions of, 76
Resonance, 410–413
ideal parallel, 439–440
practical parallel, 440–443
Resonant frequency (f_r), 410
Right angle, 307
Right triangles, 332–337
angles of, 318–319
cofunctions in, 327
Pythagorean theorem, 332–337
reciprocal functions in, 326
sides of, 318–319
RL circuits, 158–159, 399–403
parallel, 433–435
Roots:
in electronics, 83–85
of numbers, 83–85
square, 83, 85–87

Scientific notation, 65–66
Secant:
of angle, 326
of circle, 311
Second, circular measure and, 311
Secondary diagonal, 253
Sector, of circle, 311
Segment, of circle, 311
Series circuits, 99–102, 150–157
capacitors in, 150–151
inductors in, 156–157
Ohm's law in, 91–93
power law in, 96–98
Series-parallel circuits, 109–115

Series resonance, 410–413
Short circuit, 145
Shorted-load current, 142
Shunt law, 19–20
Sides, of angles, 305
Signed numbers, in algebra, 183
Significant figures, meaning of, 40
Simple circuits, writing Boolean expressions and truth tables for, 506–512
Simultaneous equations, 247–276
  elimination:
    through addition and subtraction, 247–248
    through comparison, 250–251
    through substitution, 249–250
  Kirchhoff's laws and, 259–270
  linear, 287–291
Sine, of angle, 319–320
Sine wave, 353–369
  conversions, 256–357
  current reference points, 358–360
  forming, 353–355
  frequency, 355
  period, 357
  periodic, 355
  voltage, 358–360
  wavelength, 355
Sine wave current:
  effective value, 361
  instantaneous values of, 365
  peak-to-peak value, 360
  peak value of, 358
  root-mean-square value, 361
  in terms of radians, 365–369
Sine wave current relationships, 362–363
Sine wave voltage:
  effective value, 361
  instantaneous values of, 363–365
  peak-to-peak values, 360
  peak value of, 358
  root-mean-square values, 361
  in terms of radians, 365–369
Sine wave voltage relationships, 362–363
Slide rules, 1
Slope, of graph, 281–284
Small-scale integration (SSI), 497
Special products, in algebra, 199–202
Square root(s), 83
  signs of, 85–87
Subscript(s), 61
  radix, 48
Superposition theorem, 139–142
Supplementary angles, 309
Symbols, electronic, 72–73

Tangent:
  of angle, 321
  of circle, 311
Tank circuit, 439
Temperature, effects of, on behavior of components, 32–33
Theorems:
  circuit analysis, 136–146
    maximum power transfer, 136–138
    Norton's, 142–144
  De Morgan's, 519–521
  Pythagorean, 332–337
  superposition, 139–142
  Thévenin's, 138–139
Theta, 370
Thévenin equivalent circuit, 144
Thévenin's theorem, 138–139
Tolerances, 38–39
Transformer, turns ratio of, 20–22
Transistor, 1
Trigonometric functions, 318–332
  ranges:
    first quadrant, 327–328
    other quadrants, 328–330
  values of, 321–324
Trigonometry:
  for alternating current circuits in parallel, 429–447
  for alternating current circuits in series, 353–423
  for electronics, 305–350
  language of, 305–317
Trinary system, 47, 48
Truth tables, 496, 506–512, 513–515

Unknown value, solving for, 18–19

Variables, 239–240
  dependent, 280
  independent, 280
Vector(s), 342, 372–374
  addition of, 372–374
    parallelogram method, 373
  of right-angle components, 373–374
Vertex, of angles, 305
Voltage:
  conversions of, 75–76
  and current ratios, 223–225
  sine wave, 358–360
Voltage divider rule, 129–130
Voltage dividers, 116–119
Voltage law, Kirchhoff's, 127–130
Volume units, 225–226

Wavelength, of sine wave, 355

Zero, use of, 50–51

Key Equation		Page
5–6	$m = \dfrac{y - y_1}{x - x_1}$	286
5–7	$y - y_1 = m(x - x_1)$	286
6–1	$r = \dfrac{D}{2}$	309
6–2	$C = 2\pi r$	310
6–3	$C = \pi D$	310
6–4	$\text{degrees} = \text{radians} \times \dfrac{180°}{\pi}$	315
6–5	$\text{radians} = \text{degrees} \times \dfrac{\pi}{180°}$	315
6–6	$\text{grads} = \text{degrees} \times 1.11\overline{1}\ \text{grads/degree}$	316
6–7	$\text{grads} = \text{radians} \times 63.662$	316
6–8	$\text{degrees} = \text{grads} \times 0.9$	316
6–9	$\text{radians} = \text{grads} \times 0.015\ 708\ \text{radians/grad}$	316
6–10	$c = \sqrt{a^2 + b^2}$	333
6–11	$a = \sqrt{c^2 - b^2}$	333
6–12	$b = \sqrt{c^2 - a^2}$	333
6–13	$\sqrt{-x} = \sqrt{x}\,j$	338
6–14	$(a + jb) + (c + jd) = (a + c) + j(b + d)$	341
6–15	$(a + jb)(c + jd) = (ac - bd) + j(ad + cb)$	342
6–16	$\dfrac{a + jb}{c + jd} = \dfrac{(ac + bd) + j(cb - ad)}{c^2 + d^2}$	342
6–17	$(r_1 \angle\theta_1)(r_2 \angle\theta_2) = (r_1)(r_2) \angle\theta_1 + \theta_2$	345
6–18	$\dfrac{r_1 \angle\theta_1}{r_2 \angle\theta_2} = \dfrac{r_1}{r_2} \angle\theta_1 - \theta_2$	347
6–19	$r = \sqrt{a^2 + b^2}$	347
6–20	$\tan\theta = \dfrac{b}{a}$	347
6–21	$a = r(\cos\theta)$	348
6–22	$b = r(\sin\theta)$	348

Key Equation		Page
7–1	$f = \dfrac{v}{\lambda}$	355
7–2	$f = \dfrac{300\ 000\ 000\ \text{m/s}}{\lambda} = \dfrac{3 \times 10^8\ \text{m/s}}{\lambda}$	356
7–3	$\lambda = \dfrac{300\ 000\ 000\ \text{m/s}}{f} = \dfrac{3 \times 10^8\ \text{m/s}}{f}$	357
7–4	$T = \dfrac{t}{N}$	357
7–5	$f = \dfrac{1}{T}$	358
7–6	$T = \dfrac{1}{f}$	358
7–7	$E_{av} = 0.637E_p$	359
7–8	$E_p = \dfrac{E_{av}}{0.637} = 1.57E_{av}$	359
7–9	$I_{av} = 0.637I_p$	359
7–10	$I_p = 1.57I_{av}$	359
7–11	$E_{p\text{-}p} = 2E_p$	360
7–12	$I_{p\text{-}p} = 2I_p$	360
7–13	$E_p = \dfrac{E_{p\text{-}p}}{2} = 0.5E_{p\text{-}p}$	360
7–14	$I_p = \dfrac{I_{p\text{-}p}}{2} = 0.5I_{p\text{-}p}$	360
7–15	$E_{av} = 0.637 \times \dfrac{E_{p\text{-}p}}{2} = 0.318\ 5E_{p\text{-}p}$	360
7–16	$I_{av} = 0.637 \times \dfrac{I_{p\text{-}p}}{2} = 0.318\ 5I_{p\text{-}p}$	360
7–17	$E_{rms} = 0.707E_p$	361
7–18	$I_{rms} = 0.707I_p$	361
7–19	$E_p = \dfrac{E_{rms}}{0.707} = 1.414E_{rms}$	362
7–20	$I_p = \dfrac{I_{rms}}{0.707} = 1.414I_{rms}$	362